Student Book

D1382281

EDEXCEL INTERNATIONAL GCSE MATHS

Chris Pearce

William Collins' dream of knowledge for all began with the publication of his first book in 1819. A self-educated mill worker, he not only enriched millions of lives, but also founded a flourishing publishing house. Today, staying true to this spirit, Collins books are packed with inspiration, innovation and practical expertise. They place you at the centre of a world of possibility and give you exactly what you need to explore it.

Collins. Freedom to teach.

Published by Collins
An imprint of HarperCollins*Publishers*
The News Building
1 London Bridge Street
London
SE1 9GF

Browse the complete Collins catalogue at
www.collins.co.uk

© HarperCollins*Publishers* Limited 2016

10 9 8 7 6 5

ISBN 978-0-00-820587-4

Chris Pearce, Brian Speed, Keith Gordon, Kevin Evans and Trevor Senior assert their moral rights to be identified as the authors of this work.

British Library Cataloguing in Publication Data
A Catalogue record for this publication is available from the British Library.

Commissioned by Lindsey Charles and Jennifer Hall

Edited by Lindsey Charles, Amanda Dickson and Helen Davey

Project editor Amanda Redstone

Answer check by Steven Matchett

Cover concept design by Angela English

Cover image PhotonCatcher/Shutterstock

Design and typesetting by
Jordan Publishing Design Limited and 2Hoots Publishing Services

Production by Lauren Crisp

Printed and bound by Grafica Veneta S.p.A.

nowledgements

publishers wish to thank Edexcel Limited for permission to reproduce questions from past national GCSE Mathematics papers.

publishers wish to thank the following for permission to reproduce photographs. Every t has been made to trace copyright holders and to obtain their permission for the use of right material. The publishers will gladly receive any information enabling them to rectify error or omission at the first opportunity.

er photograph PhotonCatcher/Shutterstock: p. 6 iStockphoto.com/©Jitalia17, iStockphoto. /©Robert Churchill, iStockphoto.com/©Izabela Habur, iStockphoto.com/ ©Elena Talberg; iStockphoto.com/© weareadventurers, iStockphoto.com/© eugeph, iStockphoto.com/© alskillet, iStockphoto.com/© Lisa F. Young; p. 38 iStockphoto.com/ ndy_lim, iStockphoto.com/© Irtati Hasan Wibisono, iStockphoto.com/ jerophotography, iStockphoto.com/© bojan fatur, iStockphoto.com/© Sean Locke, ckphoto.com/© Viktor Kitaykin, iStockphoto.com/© nullplus, iStockphoto.com/ as Kotzsch; p.52 iStockphoto.com/© Uyen Le, iStockphoto.com/© Peter van Wagner, ckphoto.com/© Andrew Howe, iStockphoto.com/© craftvision, iStockphoto.com/ artin McCarthy, iStockphoto.com/© Tan Kian Khoon; p. 64 iStockphoto.com/ tali Khamitsevich, iStockphoto.com/© rzdeb, © Billcasselman; p. 70 iStockphoto.com/ arren Hendley, iStockphoto.com/© gerenme, © Schutz; p.80 iStockphoto.com/ te Saloutos, iStockphoto.com/© Darren Pearson, iStockphoto.com/© Graeme Purdy, ckphoto.com/© Andrew Howe; p. 98 iStockphoto.com/© Lya_Cattel, © HarperCollins/ Fowler; p. 110 iStockphoto.com/ © Christian Anthony; p. 118 common.wikimedia.org/, ckphoto.com/© Patricia Burch, iStockphoto.com/© Terry Wilson, iStockphoto.com/ roslaw Wojcik; p. 136 from common.wikimedia.org/; p. 148 Dreamstime/© Panagiotis as, Dreamstime/© Konstantin Sutyagin; p. 174 iStockphoto.com/© P J Morley, iStockphoto. /© Dane Wirtzfield; p. 220 from common.wikimedia.org/; 32 iStockphoto.com/© Andreas Weber; iStockphoto.com/© Philip Beasley; p. 246 ckphoto.com/© Rick Szczechowski, iStockphoto.com/© Derek Dammann; p. 260 ckphoto.com/© Manfred Konrad, iStockphoto.com/© Rubén Hidalgo; p. 274 iStockphoto. /© Ivan Kmit, iStockphoto.com/© Scott Leigh, iStockphoto.com/ visual, iStockphoto.com/© Dan Barnes, iStockphoto.com/© SilentWolf, iStockphoto.com/ undsnaps, iStockphoto.com/© Gary Martin, iStockphoto.com/© Ivan Stevanovic, ckphoto.com/© porcorex; p. 284 iStockphoto.com/© narvikk, iStockphoto.com/ oktug Gurellier; p. 294 iStockphoto.com/© Tyler Olson, iStockphoto.com/© Eric Hood, ckphoto.com/© Eric Hood, iStockphoto.com/© Ugurhan Betin; p. 312 iStockphoto.com/ nplett, Shutterstock/© LilKar; p 322 iStockphoto.com/© Steve Cole, iStockphoto.com/ geny Terentev; p. 340 iStockphoto.com/© Karim Hesham, iStockphoto.com/© Luis Carlos es, iStockphoto.com/© magaliB; p. 362 iStockphoto.com/© Gene Chutka, iStockphoto.com/ ranko Miokovic; p. 378 iStockphoto.com/© fotoVoyager, iStockphoto.com/© Matthias nrich; p. 410 iStockphoto.com/© Robert Churchill, iStockphoto.com/© Wendell Franks; 40 iStockphoto.com/© Ludmila Yilmaz, iStockphoto.com/© Carmen Martínez Banús, ckphoto.com/© agoxa, Image Courtesy of Jill Britton, iStockphoto.com/© Dirk Freder, ckphoto.com/© Giorgio Fochesato; p. 450 iStockphoto.com/© Arthur Kwiatkowski, ckphoto.com/© Mark Evans, iStockphoto.com/© David Joyner, iStockphoto.com/ efan Weichelt; p. 462 iStockphoto.com/© Roberto Gennaro, iStockphoto.com/ ndrey Prokhorov; p. 510 iStockphoto.com/© Kevin Russ; p. 538 iStockphoto.com/ nDen2005, iStockphoto.com/© Domenico Pellegriti; p. 566 iStockphoto.com/ eorge Clerk, iStockphoto.com/© Owen Price.

CONTENTS

TRODUCTION

come to Collins International GCSE Maths for Edexcel. This page will introduce you
e key features of the book which will help you to succeed in your examinations and
joy your maths course.

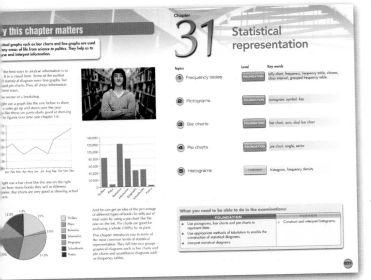

Why this chapter matters

This page is at the start of each chapter. It tells you why the mathematics in the chapter is important and how it is useful.

Chapter overviews

The overview at the start of each chapter shows what you will be studying, the key words you need to know and what you will be expected to know and do in the examination.

Worked examples

Worked examples take you through questions step by step and help you understand the topic before you start the practice questions.

Practice questions and answers

Every chapter has extensive questions to help you practise the skills you need for the examination. Many of the questions require you to solve problems which is an important part of mathematics.

Colour-coded levels

The colour coded panels at the side of the question pages show whether the questions are at Foundation (blue) or Higher level (yellow). The (H) on some topic headings shows that the content in that topic is at Higher level only.

Exam practice

Each of the four main sections in the book ends with sample exam questions from past examinations. These will show you the types of questions you will meet in the exams. Mark schemes are available in the teacher pack.

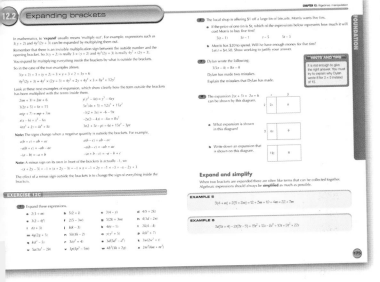

Why this chapter matters

A pattern is an arrangement of repeated parts. You see patterns every day in clothes, art and home furnishings. Patterns also occur in numbers.

There are many mathematical problems that can be solved using patterns in numbers. Some numbers have fascinating features.

Here is a pattern.

$3 + 5 = 8$ (5 miles ≈ 8 km)

$5 + 8 = 13$ (8 miles ≈ 13 km)

$8 + 13 = 21$ (13 miles ≈ 21 km)

Approximately how many kilometres are there in 21 miles?

Note: ≈ means 'approximately equal to'.

In the boxes are some more patterns. Can you work out the next line of each pattern?

Now look at these numbers and see why they are special.

$4096 = (4 + 0^9)^6$

$81 = (8 + 1)^2$

Some number patterns have special names.
Can you pair up these patterns and their names?

4, 8, 12, 16, …	Prime numbers
1, 4, 9, 16, …	Multiples (of 4)
2, 3, 5, 7, …	Cube numbers
1, 8, 27, 64, …	Square numbers

You will look at these in more detail in this chapter.

Below are four sets of numbers. Think about which number links together all the other numbers in each set.
(The mathematics that you cover in 1.2 'Factors of whole numbers' will help you to work this out!)

10, 5, 2, 1

18, 9, 6, 3, 2, 1

25, 5, 1

32, 16, 8, 4, 2, 1

$5^2 = 5 \times 5 = 25$
$50^2 = 50 \times 50 = 2500$
$500^2 = 500 \times 500 = 250\,000$

$10 \times 10 = 100$
$10 \times 10 \times 10 = 1000$
$10 \times 10 \times 10 \times 10 = 10\,000$

$1 \times 1 = 1$
$11 \times 11 = 121$
$111 \times 111 = 12\,321$

$1 \times 1 = 1$
$2 \times 2 = 1 + 3$
$3 \times 3 = 1 + 3 + 5$
$4 \times 4 = 1 + 3 + 5 + 7$

$1089 \times 9 = 9801$
$10\,989 \times 9 = 98\,901$
$109\,989 \times 9 = 989\,901$

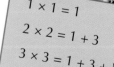

Number

cs	Level	Key words
Multiples of whole numbers	**FOUNDATION**	multiple, common multiple, even, odd
Factors of whole numbers	**FOUNDATION**	factor, factor pair, common factor
Prime numbers	**FOUNDATION**	prime number
Square numbers and cube numbers	**FOUNDATION**	square, square number, cube, cube number
Products of prime numbers	**FOUNDATION**	product
HCF and LCM	**FOUNDATION**	highest common factor, lowest common multiple

hat you need to be able to do in the examinations:

FOUNDATION

Use the terms odd, even and prime numbers, factors and multiples.

Identify prime factors, common factors and common multiples.

Express integers as the product of powers of prime factors.

Find Highest Common Factors (HCF) and Lowest Common Multiples (LCM).

Multiples of whole numbers

When you multiply any whole number by another whole number, the answer is called a **multiple** of either of those numbers. Multiples of 2 are **even** numbers.

For example, 5 × 7 = 35, which means that 35 is a multiple of 5 and it is also a multiple of 7
Here are some other multiples of 5 and 7:

multiples of 5 are   5   10   15   20   25   30   35   …

multiples of 7 are   7   14   21   28   35   42   …

35 is a **common multiple** of 5 and 7. Other common multiples of 5 and 7 are 70, 105, 140 a
so on.

EXERCISE 1A

1   Write out the first five multiples of:

    a   3   **b**   7   **c**   9   **d**   11   **e**   16

    Remember: the first multiple is the number itself.

2   Use your calculator to see which of the numbers below are:

    a   multiples of 4   **b**   multiples of 7   **c**   multiples of 6.

72	135	102	161	197
132	78	91	216	514

> **HINTS AND TIPS**
>
> **Odd** numbers cannot be multiples of even numbers. Whole number are either even or odd.

3   Find the biggest number that is smaller than 100 and that is:

    a   a multiple of 2   **b**   a multiple of 3

    c   a multiple of 4   **d**   a multiple of 5

    e   a multiple of 7   **f**   a multiple of 6.

4   A party of 20 people are getting into taxis. Each taxi holds the same number of passeng
If all the taxis fill up, how many people could be in each taxi? Give two possible answ

5   Here is a list of numbers.

      6   8   12   15   18   28

    a   From the list, write down a multiple of 9.

    b   From the list, write down a multiple of 7.

    c   From the list, write down a multiple of both 3 and 5.

6   How many numbers between 1 and 100 inclusive are multiples of both 6 and 9?
List the numbers.

Factors of whole numbers

A **factor** of a whole number is any whole number that divides into it exactly. So:

the factors of 20 are   1   2   4   5   10   20

the factors of 12 are   1   2   3   4   6   12

The **common factors** of 12 and 20 are 1, 2 and 4. They are factors of both numbers.

Factor facts

Remember these facts.

● 1 is always a factor and so is the number itself.

● When you have found one factor, there is always another factor that goes with it – unless the factor is multiplied by itself to give the number. For example, look at the number 20:

$1 \times 20 = 20$ so 1 and 20 are both factors of 20

$2 \times 10 = 20$ so 2 and 10 are both factors of 20

$4 \times 5 = 20$ so 4 and 5 are both factors of 20.

These are called **factor pairs**.

You may need to use your calculator to find the factors of large numbers.

EXAMPLE 1

Find the factors of 36.

Look for the factor pairs of 36. These are:

$1 \times 36 = 36$ $2 \times 18 = 36$ $3 \times 12 = 36$ $4 \times 9 = 36$ $6 \times 6 = 36$

6 is a repeated factor so it is counted only once.

So, the factors of 36 are 1, 2, 3, 4, 6, 9, 12, 18, 36.

EXERCISE 1B

1 What are the factors of each of these numbers?

a 10	**b** 28	**c** 18	**d** 17	**e** 25
f 40	**g** 30	**h** 45	**i** 24	**j** 16

2 What is the biggest factor that is less than 100 for each of these numbers?

a 110	**b** 201	**c** 145	**d** 117
e 130	**f** 240		

3 Find the common factors of each of the following pairs of numbers.

a 2 and 4   **b** 6 and 10   **c** 9 and 12

d 15 and 25   **e** 9 and 15   **f** 12 and 21

g 14 and 21   **h** 25 and 30   **i** 30 and 50

j 55 and 77

> **HINTS AND TIPS**
>
> Look for the largest number that has both numbers in its multiplication table.

4 Find the highest odd number that is a factor of 40 and a factor of 60.

1.3 Prime numbers

What are the factors of 2, 3, 5, 7, 11 and 13?

Notice that each of these numbers has only two factors: itself and 1. They are all examples of **prime numbers**.

So, a prime number is a whole number that has only two factors: itself and 1.

Note: 1 is *not* a prime number, since it has only one factor – itself.

The prime numbers up to 50 are:

2, 3, 5, 7, 11, 13, 17, 19, 23, 29, 31, 37, 41, 43, 47

EXERCISE 1C

1 Write down the prime numbers between 20 and 30.

2 Write down the only prime number between 90 and 100.

3 Decide which of these numbers are **not** prime numbers.

462   108   848   365   711

4 When three different prime numbers are multiplied together the answer is 105.

What are the three prime numbers?

5 A shopkeeper has 31 identical soap bars.

He is trying to arrange the bars on a shelf in rows, each with the same number of bars.

Is it possible?

Explain your answer.

Square numbers and cube numbers

What is the next number in this sequence?

1, 4, 9, 16, 25, …

Write each number as:

1 × 1, 2 × 2, 3 × 3, 4 × 4, 5 × 5, …

These factors can be represented by **square** patterns of dots:

From these patterns, you can see that the next pair of factors must be 6 × 6 = 36, therefore 36 is the next number in the sequence.

Because they form square patterns, the numbers 1, 4, 9, 16, 25, 36, … are called **square numbers**.

When you multiply any number by itself, the answer is called the *square of the number* or the *number squared*. This is because the answer is a square number. For example:

the square of 5 (or 5 squared) is 5 × 5 = 25

the square of 6 (or 6 squared) is 6 × 6 = 36

There is a short way to write the square of any number. For example:

5 squared (5 × 5) can be written as 5^2

13 squared (13 × 13) can be written as 13^2

So, the sequence of square numbers, 1, 4, 9, 16, 25, 36, …, can be written as:

$1^2, 2^2, 3^2, 4^2, 5^2, 6^2, …$

If dots are arranged in three dimensional **cubes** we get **cube numbers**.

We can write these as $1^3, 2^3, 3^3, 4^3, …$ and we read them as 'one cubed', 'two cubed' and so on.

1 × 1 × 1 = 1   2 × 2 × 2 = 8   3 × 3 × 3 = 27   4 × 4 × 4 = 64

 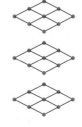 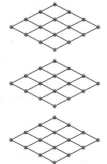

EXERCISE 1D

1 The square number pattern starts:

1   4   9   16   25   ...

Copy and continue the pattern above until you have written down the first 20 square numbers. You may use your calculator for this.

2 Work out the answer to each of these number sentences.

$1 + 3 =$

$1 + 3 + 5 =$

$1 + 3 + 5 + 7 =$

Look carefully at the pattern of the three number sentences. Then write down the next three number sentences in the pattern and work them out.

3 Find the next three numbers in each of these number patterns. (They are all based on square numbers.) You may use your calculator.

	1	4	9	16	25	36	49	64	81
a	2	5	10	17	26	37	...	...	...
b	2	8	18	32	50	72	...	...	...
c	3	6	11	18	27	38	...	...	...
d	0	3	8	15	24	35	...	...	...

> **HINTS AND TIPS**
>
> Look for the connection with the square numbers on the top line.

4 **a** Work out each of the following. You may use your calculator.

$3^2 + 4^2$ and 5^2 $\qquad$ $5^2 + 12^2$ and 13^2

$7^2 + 24^2$ and 25^2 $\qquad$ $9^2 + 40^2$ and 41^2

b Describe what you notice about your answers to part **a**.

5 Find:

a 5^3 $\qquad\qquad$ **b** 6^3 $\qquad\qquad$ **c** 10^3

6 Show that 1331 is a cube number.

7 Which is larger, 10^3 or 30^2? Find the difference between them.

8 **a** Show that $(1 + 2 + 3)^2 = 1^3 + 2^3 + 3^3$.

b Is it true that $(1 + 2 + 3 + 4)^2 = 1^3 + 2^3 + 3^3 + 4^3$?

9 How many cube numbers are there between 2000 and 4000?

10 4 and 81 are square numbers with a sum of 85.

Find two different square numbers with a sum of 85.

The following exercise will give you some practice on multiples, factors, square numbers, cube numbers and prime numbers.

EXERCISE 1E

1 Write out the first three numbers that are multiples of both of the numbers shown.

    a 3 and 4   **b** 4 and 5   **c** 3 and 5   **d** 6 and 9   **e** 5 and 7

2 Here are four numbers.

    14   16   35   49

Copy and complete the table by putting each of the numbers in the correct box.

	Square number	Factor of 70
Even number		
Multiple of 7		

3 Arrange these four number cards to make a square number.

4 One dog barks every 8 seconds and another dog barks every 12 seconds. If both dogs bark together, how many seconds will it be before they both bark together again?

5 A bell rings every 6 seconds. Another bell rings every 5 seconds. If they both ring together, how many seconds will it be before they both ring together again?

6 From this box, choose one number that fits each of these descriptions.

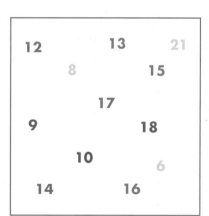

    a a multiple of 3 and a multiple of 4

    b a square number and an odd number

    c a factor of 24 and a factor of 18

    d a prime number and a factor of 39

    e an odd factor of 30 and a multiple of 3

    f a number with 5 factors exactly

    g a multiple of 5 and a factor of 20

    h a prime number that is one more than a square number

    i a cube number

    j a number which is a quarter of a cube number

7 Arrange these four cards to make a cube number.

FOUNDATION

Products of prime numbers

Every positive integer can be written as a **product** of prime numbers.

For example, $5472 = 2 \times 2 \times 2 \times 2 \times 2 \times 3 \times 3 \times 19$.

We write this more concisely as $5472 = 2^5 \times 3^2 \times 19$.

2^5 means $2 \times 2 \times 2 \times 2 \times 2$ and 3^2 means 3×3.

EXAMPLE 2

Write 702 as a product of prime numbers.

Keep dividing by prime numbers, starting with the lowest.

702 is even so it is divisible by 2:

$$2\overline{)702}$$
$$351$$

351 is not divisible by 2 but it is divisible by 3.

$$3\overline{)351}$$
$$117$$

Divide by 3 again:

$$3\overline{)117}$$
$$39$$

> **HINTS AND TIPS**
>
> Remember the prime numbers are 2, 3, 5, 7, 11, 13, 17 …

Divide by 3 again:

$$3\overline{)\ 39}$$
$$13$$

13 is prime so we stop there.

These calculations can be written more concisely like this:

$$2\overline{)702}$$
$$3\overline{)351}$$
$$3\overline{)117}$$
$$3\overline{)\ 39}$$
$$13$$

Now just write down all the prime numbers shown:

$702 = 2 \times 3 \times 3 \times 3 \times 13$ (Check that this is correct with a calculator)

We can write this more concisely as $702 = 2 \times 3^3 \times 13$.

EXERCISE 1F

1 Calculate the following:

a $2^4 \times 3$

b $3^3 \times 7^2$

c $2^5 \times 5^5$

d $3^5 \times 5$

e $2^4 \times 5^4$

f $2^{10} \times 3^4$

> **HINTS AND TIPS**
>
> $2^4 \times 3 = 2 \times 2 \times 2 \times 2 \times$

2 Write each of these as a product of prime numbers:

a 72 **b** 100 **c** 252 **d** 560 **e** 285

f 729 **g** 444 **h** 896 **i** 675 **j** 1323

3 **a** Choose any 3 digit number.

Multiply it by 7, multiply the answer by 11 and then multiply that answer by 13. What happens?

b Does what happened in part **a** happen with any three digit number? Why?

1.6 **HCF and LCM**

A common factor is a factor common to two or more numbers.

The numbers 60 and 72 have a number of common factors, including 2 and 3:

$60 = 2^2 \times 3 \times 5$
$72 = 2^3 \times 3^2$

The prime factors *common to both* are 2^2 and 3.

$$\boxed{2} \times \boxed{2} \times \boxed{3} \times 5$$
$$2 \times \boxed{2} \times \boxed{2} \times \boxed{3} \times 3$$

Multiply these together to find the **highest common factor (HCF)**.

The HCF of 60 and 72 is $2^2 \times 3 = 12$

This is the highest number that is a factor of 60 and 72.

Multiples of 60 are 60, 120, 180, …

Multiples of 72 are 72, 144, 216, …

They will have a number of common multiples.

We can use prime factors to find the **lowest common multiple (LCM)**:

$60 = 2^2 \times 3 \times 5$
$72 = 2^3 \times 3^2$

Any common multiple must contain all the factors of both numbers.

It must contain 2^3 and 3^2 and 5.

The LCM of 60 and 72 = $2^3 \times 3^2 \times 5 = 360$

> **HINTS AND TIPS**
>
> Choose the highest power of each number in either list, e.g. 2^3 not 2^2

You can use the same method to find the HCF or LCM of more than two numbers.

EXAMPLE 3

Here are three numbers: 150   225   180

Find   **a** the HCF   **b** the LCM

Write each number as a product of prime numbers

$150 = 2 \times 75 = 2 \times 3 \times 25$ $= 2 \times 3 \times 5^2$

$225 = 3 \times 75 = 3 \times 3 \times 25$ $= 3^2 \times 5^2$

$180 = 2 \times 90 = 2 \times 2 \times 45 = 2 \times 2 \times 3 \times 15$ $= 2^2 \times 3^2 \times 5$

a   The LCM is $3 \times 5 = 15$

b   The HCF is $2^2 \times 3^2 \times 5^2 = 900$

EXERCISE 1G

1 **a** Show that 2 is a common factor of 10 and of 20.

b Is it the highest common factor?

2 Find the highest common factor (HCF) of each of these pairs of numbers. You should be able to spot these without writing out a list of prime factors.

a 8 and 12

b 9 and 12

c 4 and 20

d 15 and 24

e 20 and 50

f 100 and 150

3 Find the highest common factor (HCF) of each of these pairs of numbers.

a 24 and 30

b 36 and 48

c 72 and 96

d 60 and 84

e 108 and 63

f 66 and 78

g 84 and 140

h 165 and 385

> **HINTS AND TIPS**
>
> Write each number as a product of prime factors.

4 **a** Show that 60 is a common multiple of 2 and of 3.

    b Is it the lowest common multiple?

5 Find the lowest common multiple (LCM) of these pairs of numbers.

    a 2 and 5

    b 2 and 7

    c 3 and 5

    d 3 and 7

6 Write each of these pairs of numbers as a product of prime factors. Hence find the LCM.

    a 12 and 15

    b 16 and 24

    c 12 and 14

    d 25 and 40

    e 18 and 21

    f 60 and 80

    g 32 and 48

    h 70 and 55

7 Find the highest common factor of 96, 168 and 240

8 Find the lowest common multiple of 35, 30 and 12

9 Paola has a large number of $1 coins.

She can divide them if 3, 4, 5 or 6 equal piles.

Find the smallest possible number of coins that Paola has.

Why this chapter matters

We use percentages and fractions in many situations in our everyday lives.

Why use fractions and percentages?

Because:

- basic percentages and simple fractions are easy to understand
- they are a good way of comparing quantities
- fractions and percentages are used a lot in everyday life.

Who uses them?

Here are some examples of what you might see:

- Shops and businesses
    - Everything at half price in the sales!
    - Special offer — 10% off!

- Banks
    - Interest rates on loans 6.25%.
    - Interest rates on savings 2.5%.

- Salespeople
    - Earn 7.5% commission on sales.

- Government
    - Half of government workers are over 55.
    - Unemployment has fallen by 1%.

- Workers
    - My pay rise is 2.3%.
    - My income tax is 20%.

- Teachers
    - Your test result is 67%.
    - Three-fifths of our students gain a grade C or above in IGCSE mathematics.

Can you think of other examples?

You will find many everyday uses of fractions and percentages in this chapter.

2

Fractions and percentages

ics		Level		Key words
	Equivalent fractions	**FOUNDATION**		numerator, denominator, cancel, lowest terms, simplest form, proper fraction, vulgar fraction, mixed number, top-heavy fraction
	Fractions and decimals	**FOUNDATION**		decimal, fraction, recurring decimal, terminating decimal
	Recurring decimals	**HIGHER**		
	Percentages, fractions and decimals	**FOUNDATION**		percentage, decimal equivalent
	Calculating a percentage	**FOUNDATION**		quantity, multiplier
	Increasing or decreasing quantities by a percentage	**FOUNDATION**		
	Expressing one quantity as a percentage of another	**FOUNDATION**		percentage change, percentage increase, percentage decrease, percentage profit, percentage loss
	Reverse percentage	**FOUNDATION**		unitary method
	Interest and depreciation	**FOUNDATION**		compound interest, depreciation
	Compound interest problems	**HIGHER**		
	Repeated percentage change	**HIGHER**		

What you need to be able to do in the examinations:

FOUNDATION	HIGHER
Understand and use equivalent fractions, mixed numbers and vulgar fractions and simplify a fraction by cancelling common factors. Express a given number as a fraction or percentage of another number. Convert a fraction to a decimal or percentage and vice versa. Recognise that a terminating decimal is a fraction. Understand percentages and their multiplicative nature as operators. Solve simple percentage problems, including percentage increase and decrease. Use reverse percentages. Use compound interest and depreciation.	• Convert recurring decimals into fractions. • Use repeated percentages. • Solve compound interest problems.

Equivalent fractions are two or more fractions that represent the same part of a whole.

EXAMPLE 1

Complete the following.

a $\dfrac{3}{4} \longrightarrow \dfrac{\times 4}{\times 4} = \dfrac{\square}{16}$

b $\dfrac{2}{5} = \dfrac{\square}{15}$

a Multiplying the **numerator** by 4 gives 12. This means $\dfrac{12}{16}$ is an equivalent fraction to $\dfrac{3}{4}$.

b To convert the **denominator** from 5 to 15, you multiply by 3. Do the same thing to the numerator, which gives $2 \times 3 = 6$. So, $\dfrac{2}{5} = \dfrac{6}{15}$.

The fraction $\dfrac{3}{4}$, in Example 1a, is in its **lowest terms** or **simplest form**.

This means that the only number that is a factor of both the numerator and denominator is 1.

A fraction with the numerator (top number) smaller than the denominator (bottom number) is called a **proper fraction**. An example of a proper fraction is $\dfrac{4}{5}$.

A **vulgar fraction** has a bigger numerator (top number) than the denominator (bottom number). An example of an vulgar fraction is $\dfrac{9}{5}$. It is sometimes called a **top-heavy fraction**.

A **mixed number** is made up of a whole number and a proper fraction. An example of a mixed number is $1\dfrac{3}{4}$.

EXAMPLE 2

Convert $\dfrac{14}{5}$ into a mixed number.

$\dfrac{14}{5}$ means $14 \div 5$.

Dividing 14 by 5 gives 2 with a remainder of 4 (5 fits into 14 two times, with 4 left over).

This means that there are 2 whole ones and $\dfrac{4}{5}$ left over.

So, $\dfrac{14}{5} = \dfrac{5}{5} + \dfrac{5}{5} + \dfrac{4}{5}$

$= 2\dfrac{4}{5}$

EXAMPLE 3

What fraction of 25 is 10?

The fraction we want is $\dfrac{10}{25}$

We can simplify this:

$\dfrac{10}{25} = \dfrac{2}{5}$

because both numbers are divisible by 5.

So 10 is $\dfrac{2}{5}$ of 25

EXERCISE 2A

1 Copy and complete the following.

a $\frac{2}{5} \longrightarrow \frac{\times 4}{\times 4} = \frac{\square}{20}$

b $\frac{1}{4} \longrightarrow \frac{\times 3}{\times 3} = \frac{\square}{12}$

c $\frac{3}{8} \longrightarrow \frac{\times 5}{\times 5} = \frac{\square}{40}$

d $\frac{2}{3} \longrightarrow \frac{\times \square}{\times \square} = \frac{\square}{18}$

e $\frac{3}{4} \longrightarrow \frac{\times \square}{\times \square} = \frac{\square}{12}$

f $\frac{5}{8} \longrightarrow \frac{\times \square}{\times \square} = \frac{\square}{40}$

2 Copy and complete the following.

a $\frac{10}{15} \longrightarrow \frac{\div 5}{\div 5} = \frac{\square}{\square}$

b $\frac{12}{15} \longrightarrow \frac{\div 3}{\div 3} = \frac{\square}{\square}$

c $\frac{20}{28} \longrightarrow \frac{\div 4}{\div 4} = \frac{\square}{\square}$

d $\frac{12}{18} \longrightarrow \frac{\div \square}{\div \square} = \frac{\square}{\square}$

e $\frac{15}{25} \longrightarrow \frac{\div 5}{\div \square} = \frac{\square}{\square}$

f $\frac{21}{30} \longrightarrow \frac{\div \square}{\div \square} = \frac{\square}{\square}$

3 **Cancel** each of these fractions to its simplest form.

a $\frac{4}{6}$

b $\frac{5}{15}$

c $\frac{12}{18}$

d $\frac{6}{8}$

e $\frac{3}{9}$

f $\frac{5}{10}$

g $\frac{14}{16}$

h $\frac{28}{35}$

i $\frac{10}{20}$

j $\frac{4}{16}$

4 Put the fractions in each set in order, with the smallest first.

a $\frac{1}{2}, \frac{5}{6}, \frac{2}{3}$

b $\frac{3}{4}, \frac{1}{2}, \frac{5}{8}$

c $\frac{7}{10}, \frac{2}{5}, \frac{1}{2}$

d $\frac{2}{3}, \frac{3}{4}, \frac{7}{12}$

e $\frac{1}{6}, \frac{1}{3}, \frac{1}{4}$

f $\frac{9}{10}, \frac{3}{4}, \frac{4}{5}$

5 What fraction of 20 is:

a 10

b 5

c 4

d 15

e 6?

Write your answers in the lowest terms.

6 Write your answers to this question as simply as possible.

a What fraction of 16 is 12?

b What fraction of 45 is 30?

c What fraction of 35 is 21?

d What fraction of 48 is 16?

e What fraction of 40 is 15?

7 Convert each of these vulgar fractions into a mixed number.

a $\frac{7}{3}$

b $\frac{8}{3}$

c $\frac{9}{4}$

d $\frac{10}{7}$

e $\frac{12}{5}$

f $\frac{7}{5}$

8 Convert each of these mixed numbers into an vulgar fraction.

a $3\frac{1}{3}$

b $5\frac{5}{6}$

c $1\frac{4}{5}$

d $5\frac{2}{7}$

e $4\frac{1}{10}$

f $5\frac{2}{3}$

g $2\frac{1}{2}$

h $3\frac{1}{4}$

i $7\frac{1}{6}$

j $3\frac{5}{8}$

k $6\frac{1}{3}$

l $9\frac{8}{9}$

9 Check your answers to questions **1** and **2**, using the fraction buttons on your calculator.

10 Which of these vulgar fractions has the largest value?

$$\frac{27}{4} \qquad \frac{31}{5} \qquad \frac{13}{2}$$

Show your working to justify your answer.

11 Find a mixed number that is greater than $\frac{85}{11}$ but smaller than $\frac{79}{10}$.

2.2 Fractions and decimals

Here are three **decimals**:

0.6   0.62   0.615.

Which is the largest?

Put them in a place value table:

Units	.	Tenths	Hundredths	Thousandths
0	.	6		
0	.	6	2	
0	.	6	1	5

They all have 6 tenths. The largest is 6.2 because it has 2 hundredths. The smallest is 0.6 because it has no hundredths.

EXAMPLE 4

Express 0.32 as a **fraction**.

$$0.32 = \frac{32}{100}$$

This cancels to $\frac{8}{25}$

So, $0.32 = \frac{8}{25}$

You can convert a fraction into a decimal by dividing the numerator by the denominator.

EXAMPLE 5

a   Express $\frac{3}{8}$ as a decimal.

$\frac{3}{8}$ means $3 \div 8$. This is a division calculation.

So, $\frac{3}{8} = 3 \div 8 = 0.375$

---

b Express $\frac{5}{9}$ as a decimal.

$\frac{5}{9} = 5 \div 9 = 0.555\ldots$

The decimal expression does not stop. The dots show that the sequence of 5s could continue forever. We call this a **recurring decimal**. It can be written as $0.\dot{5}$.

0.375 is called a **terminating decimal**. The decimal expression stops after three digits in this case.

EXERCISE 2B

1 Convert each of these decimals to fractions, cancelling where possible.

a 0.7   b 0.4   c 0.5   d 0.03   e 0.06

f 0.13   g 0.25   h 0.38   i 0.55   j 0.64

2 Convert each of these fractions to decimals.

a $\frac{1}{2}$ b $\frac{3}{4}$ c $\frac{3}{5}$ d $\frac{9}{10}$

e $\frac{1}{8}$ f $\frac{5}{8}$ g $\frac{7}{8}$ h $\frac{7}{20}$

3 Put each of the following sets of numbers in order, with the smallest first.

a 0.6, 0.3, $\frac{1}{2}$

b $\frac{2}{5}$, 0.8, 0.3

c 0.35, $\frac{1}{4}$, 0.15

d $\frac{7}{10}$, 0.72, 0.71

e 0.8, $\frac{3}{4}$, 0.7

f 0.08, 0.1, $\frac{1}{20}$

g 0.55, $\frac{1}{2}$, 0.4

h $1\frac{1}{4}$, 1.2, 1.23

> **HINTS AND TIPS**
>
> Convert the fractions to decimals first.

4 Write these fractions as recurring decimals:

a $\frac{1}{3}$ b $\frac{2}{3}$ c $\frac{1}{9}$ d $\frac{4}{9}$ e $\frac{1}{11}$ f $\frac{8}{11}$

5 Say whether these fractions can be written as terminating or recurring decimals:

a $\frac{5}{8}$ b $\frac{5}{9}$ c $\frac{5}{10}$ d $\frac{5}{11}$ e $\frac{5}{12}$

6 Which is bigger, $\frac{7}{8}$ or 0.87?

Show your working.

7 Which is smaller, $\frac{2}{3}$ or 0.7?

Show your working.

Writing fractions as recurring decimals is easy.

Writing recurring decimals as fractions is more difficult.

Suppose we want to write $0.\dot{8} = 0.888...$ as a fraction:

Write $\qquad f = 0.888...$

Multiply by 10 $\qquad 10f = 8.888...$

Now subtract the top row from the bottom.

$9f = 8$ (When you subtract, the digits after the decimal point cancel ou

$\Rightarrow \quad f = \dfrac{8}{9}$

If there are two recurring digits, multiply by 100. If there are three recurring digits, multiply b
1000, and so on.

For example, to write $0.\dot{3}\dot{6} = 0.363636...$ as a fraction:

Let $\qquad f = 0.3636...$

Multiply by 100 $\qquad 100f = 36.3636$

Subtract $\qquad 99f = 36$

$\Rightarrow \quad f = \dfrac{36}{99}$ which cancels to $\dfrac{4}{11}$

So $\qquad 0.3636... = \dfrac{4}{11}$

EXERCISE 2C

1 Write $\dfrac{2}{3}$ as a recurring decimal.

2 **a** Write 0.222... as a fraction.

b Write 0.7777... as a fraction.

c Write 0.4444... as a fraction.

d Look at your answers above. What do they suggest 0.9999… is as a fraction?

3 **a** Write 0.272727... as a fraction.

b Write 0.090909... as a fraction.

c Write 0.636363... as a fraction.

d What do your answers to **a**, **b** and **c** suggest about other fractions which give recurr
decimals? Check your suggestions.

4 Write these recurring decimals as fractions:

    a $0.\dot{5}$ **b** $0.\dot{2}\dot{4}$ **c** $0.\dot{4}\dot{8}$

5   **a** You know that $\frac{1}{3} = 0.3333...$

       What fraction is $0.03333...$?

    b What fraction is $0.06666...$?

2.4 Percentages, fractions and decimals

100% means the *whole* of something. So if you want to, you can express *part* of the whole as a **percentage**.

Per cent means 'out of 100'.

So, any percentage can be converted to a fraction with denominator 100.

For example:

    $32\% = \frac{32}{100}$ which can be simplified by cancelling to $\frac{8}{25}$

Also, any percentage can be converted to a decimal by dividing the percentage number by 100. This means moving the digits two places to the right.

For example:

    $65\% = 65 \div 100 = 0.65$

Any decimal can be converted to a percentage by multiplying by 100%.

For example:

    $0.43 = 0.43 \times 100\% = 43\%$

Any fraction can be converted to a percentage by converting the denominator to 100 and taking the numerator as the percentage.

For example:

    $\frac{2}{5} = \frac{40}{100} = 40\%$

Fractions can also be converted to percentages by dividing the numerator by the denominator and multiplying by 100%.

For example:

    $\frac{2}{5} = 2 \div 5 \times 100\% = 40\%$

Knowing the percentage and **decimal equivalents** of common fractions is extremely useful.

$\frac{1}{2} = 0.5 = 50\%$ $\frac{1}{4} = 0.25 = 25\%$ $\frac{3}{4} = 0.75 = 75\%$ $\frac{1}{8} = 0.125 = 12.5\%$

$\frac{1}{10} = 0.1 = 10\%$ $\frac{1}{5} = 0.2 = 20\%$ $\frac{1}{3} = 0.33 = 33\frac{1}{3}\%$ $\frac{2}{3} = 0.67 = 67\%$

The following table shows how to convert from one to the other.

Convert from percentage to:	
Decimal	**Fraction**
Divide the percentage by 100, for example 52% = 52 ÷ 100 = 0.52	Make the percentage into a fraction with a denominator of 100 and simplify by cancelling down if possible, for example $52\% = \frac{52}{100} = \frac{13}{25}$

Convert from decimal to:	
Percentage	**Fraction**
Multiply the decimal by 100%, for example 0.65 = 0.65 × 100% = 65%	If the decimal has 1 decimal place put it over the denominator 10. If it has 2 decimal places put it over the denominator 100, etc. Then simplify by cancelling down if possible, for example $0.65 = \frac{65}{100} = \frac{13}{20}$

Convert from fraction to:	
Percentage	**Decimal**
Write the fraction as an equivalent with a demonimator of 100 if possible, then the numerator is the percentage, for example $\frac{3}{20} = \frac{15}{100} = 15\%$ or convert to a decimal and change the decimal to a percentage, for example $\frac{7}{8} = 7 ÷ 8 = 0.875 = 87.5\%$	Divide the numerator by the denominator, for example $\frac{9}{40} = 9 ÷ 40 = 0.225$

EXAMPLE 6

Convert the following to decimals: **a** 78%   **b** 35%   **c** $\frac{3}{25}$ **d** $\frac{7}{40}$.

a   78% = 78 ÷ 100 = 0.78

b   35% = 35 ÷ 100 = 0.35

c $\frac{3}{25} = 3 ÷ 25 = 0.12$

d $\frac{7}{40} = 7 ÷ 40 = 0.175$

XAMPLE 7

Convert the following to percentages: **a** 0.85 **b** 0.125 **c** $\frac{7}{20}$ **d** $\frac{3}{8}$.

a $0.85 = 0.85 \times 100\% = 85\%$

b $0.125 = 0.125 \times 100\% = 12.5\%$

c $\frac{7}{20} = \frac{35}{100} = 35\%$

d $\frac{3}{8} = 3 \div 8 \times 100\% = 0.375 \times 100\% = 37.5\%$

XAMPLE 8

Convert the following to fractions: **a** 0.45 **b** 0.4 **c** 32% **d** 15%.

a $0.45 = \frac{45}{100} = \frac{9}{20}$

b $0.4 = \frac{4}{10} = \frac{2}{5}$

c $32\% = \frac{32}{100} = \frac{8}{25}$

d $15\% = \frac{15}{100} = \frac{3}{20}$

EXERCISE 2D

1 Write each percentage as a fraction in its simplest form.

   a 8%   **b** 50%   **c** 25%

   d 35%   **e** 90%   **f** 75%

2 Write each percentage as a decimal.

   a 27%   **b** 85%   **c** 13%

   d 6%   **e** 80%   **f** 32%

3 Write each decimal as a fraction in its simplest form.

   a 0.12   **b** 0.4   **c** 0.45

   d 0.68   **e** 0.25   **f** 0.625

4 Write each decimal as a percentage.

   a 0.29   **b** 0.55   **c** 0.03

   d 0.16   **e** 0.6   **f** 1.25

5 Write each fraction as a percentage.

   a $\frac{7}{25}$ **b** $\frac{3}{10}$ **c** $\frac{19}{20}$

   d $\frac{17}{50}$ **e** $\frac{11}{40}$ **f** $\frac{7}{8}$

6 Write each fraction as a decimal.

   a $\frac{9}{15}$ **b** $\frac{3}{40}$ **c** $\frac{19}{25}$

   d $\frac{5}{16}$ **e** $\frac{1}{20}$ **f** $\frac{1}{8}$

FOUNDATION

7 **a** Convert each of the following test scores into a percentage. Give each answer to the nearest whole number.

Subject	Result	Percentage
Mathematics	38 out of 60	
English	29 out of 35	
Science	27 out of 70	
History	56 out of 90	
Technology	58 out of 75	

b If all the tests are of the same standard, which was the highest result?

8 Copy and complete the table.

Percentage	Decimal	Fraction
34%		
	0.85	
		$\frac{3}{40}$
45%		
	0.3	
		$\frac{2}{3}$
84%		
	0.45	
		$\frac{3}{8}$

2.5 Calculating a percentage

To calculate a percentage of a **quantity**, you multiply the quantity by the percentage. The percentage may be expressed as either a fraction or a decimal. When finding percentages without a calculator, base the calculation on 10% (or 1%) as these are easy to calculate.

EXAMPLE 9

Calculate: **a** 10% of 54 kg   **b** 15% of 54 kg.

a 10% is $\frac{1}{10}$ so $\frac{1}{10}$ of 54 kg = 54 kg ÷ 10 = 5.4 kg

b 15% is 10% + 5% = 5.4 kg + 2.7 kg = 8.1 kg

Using a percentage multiplier

You have already seen that percentages and decimals are equivalent so it is easier, particularly when using a calculator, to express a percentage as a decimal and use this to do the calculation.

For example, 13% is a **multiplier** of 0.13, 20% a multiplier of 0.2 (or 0.20) and so on.

EXAMPLE 10

Calculate 45% of 160 cm.

45% = 0.45, so 45% of 160 = 0.45 × 160 = 72 cm

Find 52% of $460.

52% = 0.52

So, 0.52 × 460 = 239.2

This gives $239.20

Remember to always write a money answer with 2 decimal places.

EXERCISE 2E

1 What multipliers are equivalent to these percentages?

   a 88%   **b** 30%   **c** 25%

   d 8%   **e** 115%

2 What percentages are equivalent to these multipliers?

   a 0.78   **b** 0.4   **c** 0.75

   d 0.05   **e** 1.1

3 Calculate the following:

   a 15% of $300   **b** 6% of $105

   c 23% of 560 kg   **d** 45% of 2.5 kg

   e 12% of 9 hours   **f** 21% of 180 cm

   g 4% of $3   **h** 35% of 8.4 m

   i 95% of $8   **j** 11% of 308 minutes

   k 20% of 680 kg   **l** 45% of $360

4 An estate agent charges 2% commission on every house he sells. How much commission will he earn on a house that he sells for $120 500?

5 A store had 250 employees. During one week of a flu epidemic, 14% of the store's employees were absent.

    a What percentage of the employees went into work?

    b How many of the employees went into work?

6 It is thought that about 20% of fans at a soccer match are women. For one match there were 42 600 fans. How many of these do you think were women?

7 At a Paris railway station, in one week 350 trains arrived. Of these trains, 5% arrived ea and 13% arrived late. How many arrived on time?

8 A school estimates that for a school play 60% of the students will attend. There are 1500 students in the school. The caretaker is told to put out one seat for each person expected to attend plus an extra 10% of that number in case more attend. How many seats does he need to put out?

> **HINTS AND TIPS**
>
> It is not 70% of the number of students in the school.

9 A school had 850 pupils and the attendance record in one week was:

    Monday 96%   Tuesday 98%   Wednesday 100%   Thursday 94%   Friday 88%

How many pupils were present each day?

10 Calculate the following.

    a 12.5% of $26   **b** 6.5% of 34 kg

    c 26.8% of $2100   **d** 7.75% of $84

    e 16.2% of 265 m   **f** 0.8% of $3000

11 Air consists of 80% nitrogen and 20% oxygen (by volume). A man's lungs have a capac of 600 cm^3. How much of each gas will he have in his lungs when he has just taken a deep breath?

12 A factory estimates that 1.5% of all the garments it produces will have a fault in them. One week the factory produces 850 garments. How many are likely to have a fault?

13 An insurance firm sells house insurance and the annual premiums are usually set at 0.3% of the value of the house. What will be the annual premium for a house valued at $90 00

14 Average prices in a shop went up by 3% last year and 3% this year. Did the actual average price of items this year rise by more, the same amount, or less than last year?

Explain how you decided.

Increasing or decreasing quantities by a percentage

Increasing by a percentage

There are two methods for increasing a quantity by a percentage.

Method 1

Work out the increase and add it on to the original amount.

EXAMPLE 11

Increase $6 by 5%.

Work out 5% of $6: $(5 \div 100) \times 6 = \$0.30$

Add the $0.30 to the original amount: $\$6 + \$0.30 = \$6.30$

Method 2

Use a multiplier. An increase of 6% is equivalent to the original 100% *plus* the extra 6%. This is a total of 106% and is equivalent to the multiplier 1.06

EXAMPLE 12

Increase $6.80 by 5%.

A 5% increase is a multiplier of 1.05

So $6.80 increased by 5% is $\$6.80 \times 1.05 = \$7.14$

EXERCISE 2F

1 What multiplier is used to increase a quantity by:

   a 10%   **b** 3%   **c** 20%   **d** 7%   **e** 12%?

2 Increase each of the following by the given percentage. (Use any method you like.)

   a $60 by 4%   **b** 12 kg by 8%

   c 450 g by 5%   **d** 545 m by 10%

   e $34 by 12%   **f** $75 by 20%

   g 340 kg by 15%   **h** 670 cm by 23%

   i 130 g by 95%   **j** $82 by 75%

   k 640 m by 15%   **l** $28 by 8%

FOUNDATION

3 Azwan, who was on a salary of $27 500, was given a pay rise of 7%. What is his new salary?

4 In 2005 the population of a city was 1 565 000. By 2010 it had increased by 8%. What was the population of the city in 2010?

5 A small firm made the same pay increase of 5% for all its employees.

a Calculate the new pay of each employee listed below. Each of their salaries before the increase is given.

Caretaker, $16 500   Supervisor, $19 500
Driver, $17 300   Manager, $25 300

b Explain why the actual pay increases are different for each employee.

6 A bank pays 7% interest on the money that each saver keeps in the bank for a year. Allison keeps $385 in the bank for a year. How much will she have in the bank after th year?

7 In 1980 the number of cars on the roads of a town was about 102 000. Since then it ha increased by 90%. Approximately how many cars are there on the roads of the town ne

8 An advertisement for a breakfast cereal states that a special-offer packet contains 15% more cereal for the same price as a normal 500 g packet. How much breakfast cereal i a special-offer packet?

9 A headteacher was proud to point out that, since he had arrived at the school, the num of students had increased by 35%. How many students are now in the school, if there were 680 when the headteacher started at the school?

10 At a school concert there are always about 20% more girls than boys. If at one concert there were 50 boys, how many girls were there?

11 A government adds a sales tax to the price of most goods in shops. One year it is 17.5% on all electrical equipment.

Calculate the price of the following electrical equipment when sales tax of 17.5% is adde

Equipment	Pre-sales tax price
TV set	$245
Microwave oven	$72
CD player	$115
Personal stereo	$29.50

12 A television costs $400 before sales tax at 17.5% is added.

If the rate of sales tax goes up from 17.5% to 20%, by how much will the cost of the television increase?

Decreasing by a percentage

There are two methods for decreasing by a percentage.

Method 1

Work out the decrease and subtract it from the original amount.

EXAMPLE 13

Decrease $8 by 4%.

Work out 4% of $8: (4 ÷ 100) × 8 = $0.32
Subtract the $0.32 from the original amount: $8 – $0.32 = $7.68

Method 2

Use a multiplier. A 7% decrease is equivalent to 7% less than the original 100%, so it represents 100% – 7% = 93% of the original. This is a multiplier of 0.93

EXAMPLE 14

Decrease $8.60 by 5%.

A decrease of 5% is a multiplier of 0.95

So $8.60 decreased by 5% is $8.60 × 0.95 = $8.17

EXERCISE 2G

1 What multiplier is used to decrease a quantity by:

  a 8%   **b** 15%   **c** 25%   **d** 9%   **e** 12%?

2 Decrease each of the following by the given percentage. (Use any method you like.)

  a $10 by 6%   **b** 25 kg by 8%

  c 236 g by 10%   **d** 350 m by 3%

  e $5 by 2%   **f** 45 m by 12%

  g 860 m by 15%   **h** 96 g by 13%

  i 480 cm by 25%   **j** 180 minutes by 35%

  k 86 kg by 5%   **l** $65 by 42%

3 A car valued at $6500 last year is now worth 15% less. What is its value now?

4 A new diet guarantees that you will lose 12% of your mass in the first month. What ma should the following people have after one month on the diet?

a Gracia, who started at 60 kg

b Pierre, who started at 75 kg

c Greta, who started at 52 kg

5 A motor insurance firm offers no-claims discounts off the full premium, as follows.

1 year with no claims	15% discount off the full premium
2 years with no claims	25% discount off the full premium
3 years with no claims	45% discount off the full premium
4 years with no claims	60% discount off the full premium

Mr Patel and his family are all offered motor insurance from this firm.

Mr Patel has four years' no-claims discount and the full premium would be $440.

Mrs Patel has one year's no-claims discount and the full premium would be $350.

Sandeep has three years' no-claims discount and the full premium would be $620.

Priyanka has two years' no-claims discount and the full premium would be $750.

Calculate the actual amount each member of the family has to pay for the motor insuran

6 A large factory employed 640 people. It had to streamline its workforce and lose 30% of the workers. How big is the workforce now?

7 On the last day of term, a school expects to have an absence rate of 6%. If the school population is 750 students, how many students will the school expect to see on the last day of term?

8 Most speedometers in cars have an error of about 5% from the true reading. When my speedometer says I am driving at 70 km/h,

a what is the lowest speed I could be doing?

b what is the highest speed I could be doing?

9 Kerry wants to buy a sweatshirt ($19), a tracksuit ($26) and some running shoes ($56). If she joins the store's premium club which costs $25 to join she can get 20% off the co of the goods.

Should she join or not? Use calculations to support your answer.

10 A biscuit packet normally contains 300 g of biscuits and costs $1.40.

There are two special offers.

Offer A: 20% more for the same price

Offer B: Same amount for 20% off the normal price

Which is the better offer?

a Offer A   **b** Offer B   **c** Both the same   **d** Cannot tell

Justify your choice.

Expressing one quantity as a percentage of another

You find one quantity as a percentage of another by writing the first quantity as a fraction of the second, making sure that the *units of each are the same*. Then you can convert the fraction into a percentage by multiplying by 100%.

EXAMPLE 15

Express $6 as a percentage of $40.

Set up the fraction and multiply by 100%.

$$\frac{6}{40} \times 100\% = 15\%$$

EXAMPLE 16

Express 75 cm as a percentage of 2.5 m.

First, convert 2.5 m to 250 cm to get a common unit.

So, the problem now becomes: Express 75 cm as a percentage of 250 cm.

Set up the fraction and multiply by 100%.

$$\frac{75}{250} \times 100\% = 30\%$$

Percentage change

A **percentage change** may be a **percentage increase** or a **percentage decrease**.

$$\text{Percentage change} = \frac{\text{change}}{\text{original amount}} \times 100\%$$

Use this to calculate **percentage profit** or **percentage loss** in a financial transaction.

EXAMPLE 17

Jake buys a car for $1500 and sells it for $1800. What is Jake's percentage profit?

Jake's profit is $300, so his percentage profit is:

$$\text{percentage profit} = \frac{\text{profit}}{\text{original amount}} \times 100\% = \frac{300}{1500} \times 100\% = 20\%$$

EXERCISE 2H

1. Express each of the following as a percentage. Give suitably rounded figures (see page 1? where necessary.

   a $5 of $20

   b $4 of $6.60

   c   241 kg of 520 kg

   d   3 hours of 1 day

   e   25 minutes of 1 hour

   f   12 m of 20 m

   g   125 g of 600 g

   h   12 minutes of 2 hours

   i   1 week of a year

   j   1 month of 1 year

   k   25 cm of 55 cm

   l   105 g of 1 kg

2. Liam went to school with his pocket money of $2.50. He spent 80 cents at the shop. What percentage of his pocket money had he spent?

3. In Greece, there are 3 654 000 acres of agricultural land. Olives are grown on 237 000 acres of this land. What percentage of the agricultural land is used for olives?

4. During one year, it rained in Detroit on 123 days of the year. What percentage of days were wet?

5. Find the percentage profit on the following. Give your answers to one decimal place.

Item	Retail price (selling price)	Wholesale price (price the shop paid)
a CD player	$89.50	$60
b TV set	$345.50	$210
c Computer	$829.50	$750

6. Before Anton started to diet, his mass was 95 kg. His mass is now 78 kg. What percenta of his original mass has he lost?

7. In 2009 a city raised $14 870 000 in local tax. In 2010 it raised $15 597 000 in tax. What was the percentage increase?

8. When Ziad's team won the soccer league in 1995, they lost only four of their 42 league games. What percentage of games did they *not* lose?

9. In one year Britain's imports were as follows.

British Commonwealth	$109 530 000
USA	$138 790 000
France	$53 620 000
Other countries	$221 140 000

   a   What percentage of the total imports came from each source? Give your answers to 1 decimal place.

   b   Add up your answers to part **a**. What do you notice? Explain your answer.

10 Imran and Nadia take the same tests. Both tests are out of the same mark.

Here are their results.

	Test A	Test B
Imran	12	17
Nadia	14	20

Whose result has the greater percentage increase from test A to test B? Show your working.

11 A supermarket advertises its cat food as shown.

A government inspector is checking the claim.

She observes that over one hour, 46 people buy cat food and 38 buy the store's own brand.

Based on these figures, is the store's claim correct?

8 out of 10 cat owners choose our cat food.

12 Aya buys antiques and then sells them on the internet.

Find her percentage profit or loss on each of these items:

Item	Aya bought for:	Aya sold for:
Vase	$105	$84
Radio	$72	$90
Doll	$15	$41.25
Toy train	$50	$18

2.8 Reverse percentage

Reverse percentage questions involve working backwards from the final amount to find the original amount when you know, or can work out, the final amount as a percentage of the original amount.

Method 1: The unitary method

The **unitary method** has three steps.

Step 1: Equate the final percentage to the final value.

Step 2: Use this to calculate the value of 1%.

Step 3: Multiply by 100 to work out 100% (the original value).

EXAMPLE 18

The price of a car increased by 6% to $9116. Work out the price before the increase.

106% represents $9116.

Divide by 106.   1% represents $9116 ÷ 106

Multiply by 100.   100% represents original price: $9116 ÷ 106 × 100 = $8600

So the price before the increase was $8600.

Method 2: The multiplier method

The multiplier method involves fewer steps.

Step 1: Write down the multiplier.

Step 2: Divide the final value by the multiplier to give the original value.

EXAMPLE 19

In a sale the price of a freezer is reduced by 12%. The sale price is $220.
What was the price before the sale?

A decrease of 12% gives a multiplier of 0.88

Dividing the sale price by the multiplier gives $220 ÷ 0.88 = $250

So the price before the sale was $250.

EXERCISE 2I

1 Find what 100% represents in these situations.

   a 40% represents 320 g   **b** 14% represents 35 m

   c 45% represents 27 cm   **d** 4% represents $123

   e 2.5% represents $5   **f** 8.5% represents $34

2 A group of students go on a training course. Only 28 complete the course. This represented 35% of the original group. How large was the original group?

3 Sales tax is added to goods and services. With sales tax at 17.5%, what is the pre-sales price of the following goods?

| T-shirt | $9.87 | Tights | $1.41 | Shorts | $6.11 |
| Sweater | $12.62 | Trainers | $29.14 | Boots | $38.07 |

4 Howard spends $200 a month on food. This is 24% of his monthly pay. How much is his monthly pay?

5 Tina's weekly pay is increased by 5% to $315. What was Tina's pay before the increase?

6 The number of workers in a factory fell by 5% to 228. How many workers were there originally?

7 In a sale the price of a TV is reduced to $500. This is a 7% reduction on the original price. What was the original price?

8 If 38% of plastic bottles in a production line are blue and the remaining 7750 plastic bottles are brown, how many plastic bottles are blue?

9 I pay $385 sales tax on a car. Sales tax is 17.5% of the purchase price. How much did I pay for the car?

10 A company asks their workers to take a 10% pay cut.

Rob works out that his pay will be $1296 per month after the cut. How much is his pay now?

11 Manza buys a car and sells it for $2940. He made a profit of 20%.

What was the original price of the car?

12 When a suit is sold in a shop the selling price is $171 and the profit is 80%.

What was the original price?

13 Oliver buys a chair. He sells it for $63 in an auction and makes a loss of 55%.

What did he pay for the chair?

14 A woman's salary increased by 5% in one year. Her new salary was $19 845.

How much was the increase in dollars?

15 After an 8% increase, the monthly salary of a chef was $1431. What was the original monthly salary?

16 Cassie invested some money at 4% compound interest per annum for two years. After two years, she had $1406.08 in the bank. How much did she invest originally?

17 A teacher asked her class to work out the original price of a cooker for which, after a 12% increase, the price was 291.20 dollars.

This is Lee's answer: 12% of 291.20 = 34.94 dollars

Original price = 291.2 − 34.94 = 256.26 ≈ 260 dollars

When the teacher read out the answer Lee ticked his work as correct.
What errors has he made?

2.9 Interest and depreciation

When you put money in a bank you are paid interest each year.

EXAMPLE 20

Boris puts $600 in a bank and leaves it there for 2 years.

He is paid 5% interest every year.

How much does he have after   **a** one year   **b** two years?

It is best to use the multiplier method. To increase by 5% you multiply by 1.05

a   After one year he has $600 × 1.05 = $630

b   After two years he has $630 × 1.05 = $661.50

In the example Boris was paid $30 interest in the first year and $31.50 in the second year.

The amount of interest increases each year as his money increases.

This is an example of **compound interest**.

If you buy a new car or a computer or a washing machine, the value goes down each year. This is called **depreciation**.

Depreciation is often expressed as a percentage.

EXAMPLE 21

A businessman buys new machinery for $9000.

The value goes down by 20% in the first year and by 10% in the second year.

What is the value after 2 years?

a   To decrease by 20% the multiplier is 0.8
After 1 year the value of the machinery is $9000 × 0.8 = $7200

b   To decrease by 10% the multiplier is 0.9
After 2 years the value is $7200 × 0.9 = $6480

EXERCISE 2J

1. Samir puts $2000 in a bank and leaves it there for 2 years.

   She is paid 3% per year interest.

   Work out how much she has after   a 1 year   b 2 years

2. Repeat question 1 if the rate of interest is 6%

3 Luis puts $750 in a bank.

He is paid 5% interest in the first year and 4% in the second year.

a Work out how much he has after two years.

b How much interest is paid to him?

4 Carla puts $6500 in a bank.

She is paid 2% interest each year for three years.

a How much does she have after 3 years?

b How much interest does she receive?

5 Hamid buys a car for $15 000.

The value depreciates by 25% in the first year and 15% in the second year.

Work out the value of the car after **a** 1 year **b** 2 years.

6 A factory owner buys a machine for $35 000.

The value decreases by 12% each year.

Find the value after **a** 1 year **b** 2 years **c** 3 years.

7 Marta puts $10 000 in a bank. She is paid interest of 10% a year.

a How much does she have after **i** 1 year **ii** 2 years **iii** 3 years?

b Work out the interest she is she paid **i** in the first year **ii** in the second year **iii** in the third year.

8 Eric buys a car for $25 000.

The value of the car depreciates by 20% a year.

a Find the value of the car after **i** 1 year **ii** 2 years **iii** 3 years.

b Work out the fall in value **i** in the first year **ii** in the second year **iii** in the third year.

9 Yasmin has $5000 to put in a bank for 2 years.

Axel Bank offers 2% interest in the first year and 12% in the second year.

Barco Bank offers 7% interest each year.

Yasmin says "2 + 12 = 14 and 7 + 7 = 14 so they will both give the same amount of interest".

Is Yasmin correct? Give a reason for your answer.

10 Rory buys a boat for $6000.

The value depreciates by 25% a year.

Rory says "After 2 years it will be worth $3000". Is Rory correct? Give a reason for your answer.

2.10 Compound interest problems

When you are working out compound interest for a number of years, you can use the power button on your calculator to find the answer more efficiently.

EXAMPLE 22

Elspeth put $4000 in a bank account. She is paid 6% compound interest.
Work out how much interest she has after 5 years.

The multiplier for an increase of 6% is 1.06
After 5 years she has $4000 × 1.06 × 1.06 × 1.06 × 1.06 × 1.06
You can write this as $4000 × 1.06^5 = $5352.90
You should be able to use your calculator to find this.
Round your answer to 2 decimal places.
The interest is $5352 − $4000 = $1352

EXERCISE 2K

1 Aglaya has $650 in a bank account. The rate of interest is 8%.

Work out how much she has after 4 years.

2 Rahul puts $3000 in a bank. He is paid compound interest of 4%.

Work out:

a the amount he has after 5 years   b the interest he is paid

3 Lee put $1000 in a bank. Find the total value in the following cases:

a 3% interest is paid for 7 years

b 7% interest is paid for 3 years

c 5% interest is paid for 5 years

4 Rania puts $2000 in a bank where she is given 5% interest. Work out the interest she h

a after 1 year   b after 5 years   c after 10 years

5 A man puts $500 in a bank for 4 years. How much is it worth if the interest rate is

a 2.5%   b 5%   c 7.5%   d 10%?

6 A woman puts $4000 in a bank account. The rate of interest is 5%.

How many years will it be until she has more than $6000?

Repeated percentage change

Repeated percentage changes include compound interest and depreciation but there are many other examples.

XAMPLE 23

In one year the population of a town increases by 5%.

The next year the population increases by 10%.

Work out the overall percentage change.

You have not been told the initial population of the town. You can work out the overall percentage change without it.

The multiplier for a 5% increase is 1.05

The multiplier for a 10% increase is 1.1

The combined multiplier is $1.05 \times 1.1 = 1.155$

This is the multiplier for an increase of 15.5%.

If you want to check that is correct you can choose any population to start with.

Suppose the population is 10 000.

After one year it is $10\,000 \times 1.05 = 10\,500$

After 2 years it is $10\,500 \times 1.1 = 11\,550$

The increase is 1550 and the percentage increase is $\frac{1550}{10\,000} \times 100\% = 15.5\%$

This is the same answer. It is quicker just to use the multipliers.

XAMPLE 24

The number of birds of one species in a wood increases by 12% in one year.

The next year it decreases by 15%.

Find the overall percentage change.

The multipliers are 1.12 and 0.85

The combined multiplier is $1.12 \times 0.85 = 0.952$

$1 - 0.952 = 0.048$ so this is the multiplier for a 4.8% decrease.

EXERCISE 2L

1. The mass of a baby is 3.00 kg.

   One month the mass increases by 5%. The next month the mass increases by 8%.

   a Work out the mass of the baby after two months.

   b Work out the overall percentage increase in mass.

HIGHER

HIGHER

2 One year the average attendance at a football ground is 8000.

The next year the attendance increases by 20%.

The year after that the attendance increases by 30%.

Find the total percentage increase over the two years.

3 Marsha puts some money in a bank.

The money earns 6% interest for three years.

What is the overall percentage increase in the value of her money?

4 The price of a coat is $750.

In a sale the price is reduced by 30% and then by a further 30%.

  a Work out the price of the coat after the two reductions.

  b Work out the total percentage decrease.

5 Mathias gets a 15% pay increase every year for three years.

Find his total percentage pay increase.

6 The value of a car decreases by 25% in its first year and then by 20% in each subsequent year.

Work out the total percentage decrease in value after

  a 2 years   **b** 4 years

7 A newspaper says the price of property in the centre of a city is increasing by 15% each year.

Work out the total percentage increase after

  a 2 years   **b** 5 years

8 A tree is 2 m tall and the height is increasing by 10% a year.

Show that if the tree continues to grow at this rate it will be over 8 m tall in 15 years' ti

9 From 1920 to 1960 the population of a town increased by 31%.

From 1960 to 2000 the population decreased by 17%.

Work out the percentage change from 1920 to 2000.

10 **a** Show that a 20% increase followed by another 20% increase is equivalent to a total increase of 44%.

  b Show that a 20% decrease followed by another 20% decrease is equivalent to a total decrease of 36%.

  c Find the total percentage change after an increase of 20% followed by a decrease of 20%.

11 Find the total percentage change in the following cases:

   a An increase of 10% followed by a decrease of 10%

   b An increase of 25% followed by a decrease of 25%

   c An increase of 75% followed by a decrease of 75%

12 Look at these two changes:

   X a decrease of 13% followed by an increase of 42%

   Y an increase of 42% followed by a decrease of 13%

   Which has the greater total percentage change? Give a reason for your answer.

HIGHER

Most jobs will require you to use some mathematics every day. Having good number skills will help you to be more successful in your job.

The mathematics used in jobs ranges from simple calculations such as addition, subtraction, multiplication and division, to more complex calculations involving negative numbers and approximation. You will need to select the right mathematics for the job.

Jobs using mathematics

How many jobs can you think of that require some mathematics?

Here are a few ideas.

Engineer – What measurement do I need to take? How much of each type of material will be needed?

Pilot – How much fuel do I need?

Accountant – How much profit have they made?

Delivery driver – What is the best route?

Cashier – What coins do I need to give as change? What is the best price to sell my goods at?

Doctor – How much medicine should I prescribe?

Sports commentator – How many minutes are left in the game? What is his batting average?

Baker – What quantity of flour should I order?

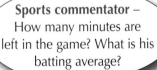

If you already know what job you would like to do, think about what mathematics you might need for it.

3

The four rules

ics	Level	Key words
Order of operations	**FOUNDATION**	operation, brackets, order
Choosing the correct operation	**FOUNDATION**	
Finding a fraction of a quantity	**FOUNDATION**	quantity, fraction
Adding and subtracting fractions	**FOUNDATION**	proper fraction, vulgar fraction, lowest terms, simplest form, denominator, mixed number, equivalent fraction
Multiplying and dividing fractions	**FOUNDATION**	numerator, reciprocal

hat you need to be able to do in the examinations:

FOUNDATION

Use the four rules of addition, subtraction, multiplication and division.

Use brackets and the hierarchy of operations.

Calculate a given fraction of a given quantity, expressing the answer as a fraction.

Use common denominators to order, add and subtract fractions.

Understand and use unit fractions as multiplicative inverses.

Multiply and divide a given fraction by an integer, by a unit fraction and by a general fraction.

Suppose you have to work out the answer to 4 + 5 × 2. You may say the answer is 18, but the correct answer is 14.

There is an **order** of **operations** which you must follow when working out calculations like this. The × is always done *before* the +.

In 4 + 5 × 2 this gives 4 + 10 = 14.

Now suppose you have to work out the answer to (3 + 2) × (9 − 5). The correct answer is 20.

You have probably realised that the parts in the **brackets** have to be done *first*, giving 5 × 4 = 2

So, how do you work out a problem such as 9 ÷ 3 + 4 × 2?

To answer questions like this, you *must* follow the BIDMAS (or BODMAS) rule. This tells you th order in which you *must* do the operations.

B	Brackets		**B**	Brackets
I	Indices (Powers)		**O**	pOwers or Order
D	Division		**D**	Division
M	Multiplication		**M**	Multiplication
A	Addition		**A**	Addition
S	Subtraction		**S**	Subtraction

For example, to work out 9 ÷ 3 + 4 × 2:

First divide:	$9 \div 3 = 3$	giving	$3 + 4 \times 2$
Then multiply:	$4 \times 2 = 8$	giving	$3 + 8$
Then add:	$3 + 8 = 11$		

And to work out $60 - 5 \times 3^2 + (4 \times 2)$:

First, work out the brackets:	$(4 \times 2) = 8$	giving	$60 - 5 \times 3^2 + 8$
Then the index (power):	$3^2 = 9$	giving	$60 - 5 \times 9 + 8$
Then multiply:	$5 \times 9 = 45$	giving	$60 - 45 + 8$
Then add:	$60 + 8 = 68$	giving	$68 - 45$
Finally, subtract:	$68 - 45 = 23$		

EXERCISE 3A

1 Work out each of these.

a 2 × 3 + 5 =   **b** 6 ÷ 3 + 4 =   **c** 5 + 7 − 2 =

d 4 × 6 ÷ 2 =   **e** 2 × 8 − 5 =   **f** 3 × 4 + 1 =

g 3 × 4 − 1 =   **h** 3 × 4 ÷ 1 =   **i** 12 ÷ 2 + 6 =

j 12 ÷ 6 + 2 =   **k** 3 + 5 × 2 =   **l** 12 − 3 × 3 =

2 Work out each of the following. Remember: first work out the bracket.

a $2 \times (3 + 5) =$ **b** $6 \div (2 + 1) =$ **c** $(5 + 7) - 2 =$

d $5 + (7 - 2) =$ **e** $3 \times (4 \div 2) =$ **f** $3 \times (4 + 2) =$

g $2 \times (8 - 5) =$ **h** $3 \times (4 + 1) =$ **i** $3 \times (4 - 1) =$

j $3 \times (4 \div 1) =$ **k** $12 \div (2 + 2) =$ **l** $(12 \div 2) + 2 =$

3 Copy each of these and then put in brackets where necessary to make each answer true.

a $3 \times 4 + 1 = 15$ **b** $6 \div 2 + 1 = 4$ **c** $6 \div 2 + 1 = 2$

d $4 + 4 \div 4 = 5$ **e** $4 + 4 \div 4 = 2$ **f** $16 - 4 \div 3 = 4$

g $3 \times 4 + 1 = 13$ **h** $16 - 6 \div 3 = 14$ **i** $20 - 10 \div 2 = 5$

j $20 - 10 \div 2 = 15$ **k** $3 \times 5 + 5 = 30$ **l** $6 \times 4 + 2 = 36$

m $15 - 5 \times 2 = 20$ **n** $4 \times 7 - 2 = 20$ **o** $12 \div 3 + 3 = 2$

p $12 \div 3 + 3 = 7$ **q** $24 \div 8 - 2 = 1$ **r** $24 \div 8 - 2 = 4$

4 Ravi says that $5 + 6 \times 7$ is equal to 77.

Is he correct?

Explain your answer.

5 Three different dice give scores of 2, 3, 5. Add $\div$, $\times$, $+$ or $-$ signs, and brackets where necessary, to make each calculation work.

a $2 \quad 3 \quad 5 = 11$ **b** $2 \quad 3 \quad 5 = 16$ **c** $2 \quad 3 \quad 5 = 17$

d $5 \quad 3 \quad 2 = 4$ **e** $5 \quad 3 \quad 2 = 13$ **f** $5 \quad 3 \quad 2 = 30$

6 Which is smaller?

$4 + 5 \times 3$ or $(4 + 5) \times 3$

Show your working.

7 Here is a list of numbers, some signs and one pair of brackets.

$$2 \quad 5 \quad 6 \quad 18 \quad - \quad \times \quad = \quad (\quad)$$

Use **all** of them to make a correct calculation.

8 Here is a list of numbers, some signs and one pair of brackets.

$$3 \quad 4 \quad 5 \quad 8 \quad - \quad \div \quad = \quad (\quad)$$

Use **all** of them to make a correct calculation.

FOUNDATION

Choosing the correct operation

When a problem is given in words you will need to decide the correct operation to use. Should you add, subtract, multiply or divide?

EXAMPLE 1

A party of 613 children and 59 adults are going on a day out to a theme park.

a How many coaches, each holding 53 people, will be needed?

b One adult gets into the theme park free for every 15 children. How many adults will have to pay to get in?

a Altogether there are 613 + 59 = 672 people.

So the number of coaches needed is 672 ÷ 53 (number of seats on each coach)
= 12.67 …

13 coaches are needed (12 will not be enough).

b This is also a division, 613 ÷ 15 = 40.86 …

40 adults will get in free.

59 − 40 = 19 will have to pay.

EXERCISE 3B

1 There are 48 cans of soup in a crate. A shop had a delivery of 125 crates of soup.

a How many cans of soup were in this delivery?

b The shop is running a promotion on soup. If you buy five cans you get one free. Each can costs 39 cents. How much will it cost to get 32 cans of soup?

2 A school has 12 classes, each of which has 24 students.

a How many students are there at the school?

b The student–teacher ratio is 18 to 1. That means there is one teacher for every 18 student How many teachers are there at the school?

3 A football club is organising travel for an away game. 1300 adults and 500 children wan go. Each coach holds 48 people and costs $320 to hire.
Tickets to the match are $18 for adults and $10 for children.

a How many coaches will be needed?

b The club is charging adults $26 and children $14 for travel and a ticket. How much profit does the club make out of the trip?

4 A large letter costs 39 cents to post and a small letter costs 30 cents. How many dollars it cost to send 20 large and 90 small letters?

5 Kirsty collects small models of animals. Each one costs 45 cents. She saves enough to buy 23 models but when she goes to the shop she finds that the price has gone up to 55 cents. How many can she buy now?

6 Michelle wants to save up for a bike that costs $250. She baby-sits each week for 6 hours for $2.75 an hour, and does a Saturday job that pays $27.50. She saves three-quarters of her weekly earnings. How many weeks will it take her to save enough to buy the bike?

7 The magazine *Teen Dance* comes out every month. In a newsagent the magazine costs $2.45. The annual subscription for the magazine is $21. How much cheaper is each magazine when bought on subscription?

8 Paula buys a sofa. She pays a deposit of 10% of the cash price and then 36 monthly payments of $12.50. In total she pays $495. How much was the cash price of the sofa?

9 There are 125 people at a wedding. They need to get to the reception.

52 people are going by coach and the rest are travelling in cars. Each car can take up to five people.

What is the least number of cars needed to take everyone to the reception?

10 Gavin's car does 8 kilometres to each litre of fuel. He drives 12 600 kilometres a year of which 4 600 is on company business.

Fuel costs 95 cents per litre.

Insurance and servicing costs $800 a year.

Gavin's company gives him 40 cents for each kilometre he drives on company business.

How much does Gavin pay from his own money towards running his car each year?

3.3 Finding a fraction of a quantity

To do this, you simply multiply the **fraction** by the **quantity**, for example, $\frac{1}{2}$ of 30 is the same as $\frac{1}{2} \times 30$.

Remember: In mathematics 'of' is interpreted as ×.

For example, two lots of three is the same as 2 × 3.

EXAMPLE 2

Find $\frac{3}{4}$ of $196.

First, find $\frac{1}{4}$ by dividing by 4. Then find $\frac{3}{4}$ by multiplying your answer by 3.

196 ÷ 4 = 49   then   49 × 3 = 147

The answer is $147.

EXERCISE 3C

1 Calculate each of these.

a $\frac{3}{5}$ of 30   **b** $\frac{2}{7}$ of 35   **c** $\frac{3}{8}$ of 48   **d** $\frac{7}{10}$ of 40

2 Calculate each of these quantities.

a $\frac{3}{4}$ of $2400   **b** $\frac{2}{5}$ of 320 grams   **c** $\frac{5}{8}$ of 256 kilograms

d $\frac{2}{3}$ of $174   **e** $\frac{5}{6}$ of 78 litres   **f** $\frac{3}{4}$ of 120 minutes

3 For each pair, which is the larger number?

a $\frac{2}{5}$ of 60 or $\frac{5}{8}$ of 40   **b** $\frac{3}{4}$ of 280 or $\frac{7}{10}$ of 290

c $\frac{2}{3}$ of 78 or $\frac{4}{5}$ of 70   **d** $\frac{5}{6}$ of 72 or $\frac{11}{12}$ of 60

4 A director receives $\frac{2}{15}$ of his company's profits. The company made a profit of $45 600 in one year. How much did the director receive?

5 A woman left $84 000 in her will.

She left $\frac{3}{8}$ of the money to charity.

How much did she leave to charity?

6 $\frac{2}{3}$ of a person's mass is water. Paul has a mass of 78 kg. How much of his body mass is water?

7 **a** Information from the first census in Singapore showed that $\frac{2}{25}$ of the population were Indian. The total population was 10 700. How many people were Indian?

b By 1990 the population of Singapore had grown to 3 002 800. Only $\frac{1}{16}$ of this populat were Indian. How many Indians were living in Singapore in 1990?

8 Mark normally earns $500 a week. One week he is given a bonus of $\frac{1}{10}$ of his wage.

a Find $\frac{1}{10}$ of $500.

b How much does he earn altogether for this week?

9 The price of a new TV costing $360 is reduced by $\frac{1}{3}$ in a sale.

a Find $\frac{1}{3}$ of $360.

b How much does the TV cost in the sale?

10 A car is advertised at Lion Autos at $9000 including extras but with a special offer of $\frac{1}{5}$ off this price.

The same car is advertised at Tiger Motors for $6000 but the extras add $\frac{1}{4}$ to this price.

Which garage is the cheaper?

11 A jar of coffee normally contains 200 g and costs $2.

There are two special offers on a jar of coffee.

   Offer A: $\frac{1}{4}$ extra for the same price.

   Offer B: Same mass for $\frac{3}{4}$ of the original price.

Which offer is the best value?

In the last exercise the answer was always a whole number. When that is not the case and we want the answer as a fraction, it is easier to change the order of multiplying and dividing.

XAMPLE 3

Find $\frac{2}{3}$ of 17, giving the answer as a fraction.

We want to find $17 \div 3 \times 2$.

Because $17 \div 3$ is not a whole number, it is easier to change the order to:

$17 \times 2 \div 3 = 34 \div 3 = 11\frac{1}{3}$ ($34 \div 3 = 11$ with 1 remainder)

EXERCISE 3D

1 To find $\frac{3}{5}$ of 75 you could calculate $75 \div 5 \times 3$ or $75 \times 3 \div 5$.

Show that both give the same answer.

2 Calculate the following, giving your answers as fractions:

   a $\frac{2}{3}$ of 8   **b** $\frac{3}{4}$ of 7   **c** $\frac{3}{4}$ of 13

   d $\frac{2}{5}$ of 4   **e** $\frac{3}{5}$ of 6   **f** $\frac{5}{8}$ of 5

3 Calculate the following, giving the answers as fractions:

   a $\frac{3}{10}$ of 3   **b** $\frac{3}{20}$ of 6   **c** $\frac{4}{15}$ of 2

4 Copy and complete this table:

	$\frac{2}{3}$	$\frac{3}{4}$	$\frac{5}{6}$
10	$6\frac{2}{3}$		
20			

5 Calculate the following, giving the answers as fractions:

a $\frac{3}{4}$ of 25

b $\frac{2}{3}$ of 40

c $\frac{2}{5}$ of 24

d $\frac{3}{20}$ of 34

e $\frac{5}{8}$ of 29

f $\frac{7}{8}$ of 30

3.4 Adding and subtracting fractions

When you add two fractions with the same **denominator**, you get one of the following:

● a **proper fraction** that cannot be simplified, for example:

$$\frac{1}{5} + \frac{2}{5} = \frac{3}{5}$$

● a proper fraction that can be simplified to its **lowest terms** or **simplest form**, for example:

$$\frac{1}{8} + \frac{3}{8} = \frac{4}{8} = \frac{1}{2}$$

● a **vulgar fraction** that cannot be simplified, so it is converted to a **mixed number**, for exam

$$\frac{6}{7} + \frac{2}{7} = \frac{8}{7} = 1\frac{1}{7}$$

● a vulgar fraction that *can* be simplified before it is converted to a mixed number, for example:

$$\frac{5}{8} + \frac{7}{8} = \frac{12}{8} = \frac{3}{2} = 1\frac{1}{2}$$

When you subtract two fractions with the same denominator, you get one of the following:

● a proper fraction that cannot be simplified, for example:

$$\frac{3}{5} - \frac{1}{5} = \frac{2}{5}$$

● a proper fraction that can be simplified, for example:

$$\frac{1}{2} - \frac{1}{10} = \frac{5}{10} - \frac{1}{10} = \frac{4}{10} = \frac{2}{5}$$

Note: You must *always* simplify fractions by cancelling if possible.

XAMPLE 4

Find $\frac{1}{2} + \frac{5}{8}$

These fractions do not have the same denominator.

However $\frac{1}{2} = \frac{4}{8}$ so we can write:

$$\frac{1}{2} + \frac{5}{8} = \frac{4}{8} + \frac{5}{8} = \frac{9}{8} = 1\frac{1}{8}$$

EXERCISE 3E

1 Work out:

a $\frac{3}{7} + \frac{2}{7}$

b $\frac{5}{9} + \frac{2}{9}$

c $\frac{3}{5} + \frac{1}{5}$

d $\frac{3}{7} + \frac{3}{7}$

2 Work out:

a $\frac{4}{7} - \frac{1}{7}$

b $\frac{5}{9} - \frac{4}{9}$

c $\frac{7}{11} - \frac{3}{11}$

d $\frac{9}{13} - \frac{2}{13}$

3 Work out:

a $\frac{5}{8} + \frac{1}{8}$

b $\frac{3}{10} + \frac{1}{10}$

c $\frac{2}{9} + \frac{4}{9}$

d $\frac{1}{4} + \frac{1}{4}$

4 Work out:

a $\frac{7}{8} - \frac{3}{8}$

b $\frac{7}{10} - \frac{3}{10}$

c $\frac{5}{6} - \frac{1}{6}$

d $\frac{9}{10} - \frac{1}{10}$

5 Work out each of these additions. Use **equivalent fractions** to make the denominators the same. Show your working.

a $\frac{1}{2} + \frac{7}{10}$

b $\frac{1}{2} + \frac{5}{8}$

c $\frac{3}{4} + \frac{3}{8}$

d $\frac{3}{4} + \frac{7}{8}$

e $\frac{1}{2} + \frac{7}{8}$

f $\frac{1}{3} + \frac{5}{6}$

g $\frac{5}{6} + \frac{2}{3}$

h $\frac{3}{4} + \frac{1}{2}$

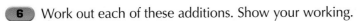

FOUNDATION

6 Work out each of these additions. Show your working.

a $\dfrac{3}{8} + \dfrac{7}{8}$

b $\dfrac{3}{4} + \dfrac{3}{4}$

c $\dfrac{2}{5} + \dfrac{3}{5}$

d $\dfrac{7}{10} + \dfrac{9}{10}$

7 Work out each of these subtractions. Use equivalent fractions to make the denominators same. Show your working.

a $\dfrac{7}{8} - \dfrac{1}{4}$

b $\dfrac{7}{10} - \dfrac{1}{5}$

c $\dfrac{3}{4} - \dfrac{1}{2}$

d $\dfrac{5}{8} - \dfrac{1}{4}$

e $\dfrac{1}{2} - \dfrac{1}{4}$

f $\dfrac{7}{8} - \dfrac{1}{2}$

g $\dfrac{9}{10} - \dfrac{1}{2}$

h $\dfrac{11}{16} - \dfrac{3}{8}$

Fractions with different denominators can only be added or subtracted after you have convert them to equivalent fractions with the same denominator.

EXAMPLE 5

i Find $\frac{2}{3} + \frac{1}{5}$

Note you can write both fractions as equivalent fractions with a denominator of 15. This is the lowest common multiple of 3 and 5.

This then becomes:

$$\frac{2 \times 5}{3 \times 5} + \frac{1 \times 3}{5 \times 3} = \frac{10}{15} + \frac{3}{15} = \frac{13}{15}$$

ii Find $2\frac{3}{4} - 1\frac{5}{6}$

Split the calculation into $\left(2 + \dfrac{3}{4}\right) - \left(1 + \dfrac{5}{6}\right)$.

This then becomes:

$$2 - 1 + \frac{3}{4} - \frac{5}{6}$$

Note you can write both fractions as equivalent fractions with a denominator of 12.

$$= 1 + \frac{9}{12} - \frac{10}{12} = 1 - \frac{1}{12}$$

$$= \frac{11}{12}$$

EXERCISE 3F

1 Work out the following. Show your working.

a $\dfrac{1}{3} + \dfrac{1}{5}$

b $\dfrac{1}{3} + \dfrac{1}{4}$

c $\dfrac{1}{5} + \dfrac{1}{10}$

d $\dfrac{2}{3} + \dfrac{1}{4}$

e $\dfrac{3}{4} + \dfrac{1}{8}$

f $\dfrac{1}{3} + \dfrac{1}{6}$

g $\dfrac{1}{2} - \dfrac{1}{3}$

h $\dfrac{1}{4} - \dfrac{1}{5}$

i $\dfrac{1}{5} - \dfrac{1}{10}$

j $\dfrac{7}{8} - \dfrac{3}{4}$

k $\dfrac{5}{6} - \dfrac{3}{4}$

l $\dfrac{5}{6} - \dfrac{1}{2}$

m $\dfrac{5}{12} - \dfrac{1}{4}$

n $\dfrac{1}{3} + \dfrac{4}{9}$

o $\dfrac{1}{4} + \dfrac{3}{8}$

p $\dfrac{7}{8} - \dfrac{1}{2}$

q $\dfrac{3}{5} - \dfrac{8}{15}$

r $\dfrac{11}{12} + \dfrac{5}{8}$

s $\dfrac{7}{16} + \dfrac{3}{10}$

t $\dfrac{4}{9} - \dfrac{2}{21}$

u $\dfrac{5}{6} - \dfrac{4}{27}$

2 Work out the following. Show your working.

a $2\dfrac{1}{7} + 1\dfrac{3}{14}$

b $6\dfrac{3}{10} + 1\dfrac{4}{5} + 2\dfrac{1}{2}$

c $3\dfrac{1}{2} - 1\dfrac{1}{3}$

d $1\dfrac{7}{18} + 2\dfrac{3}{10}$

e $3\dfrac{2}{6} + 1\dfrac{9}{20}$

f $1\dfrac{1}{8} - \dfrac{5}{9}$

g $1\dfrac{3}{16} - \dfrac{7}{12}$

h $\dfrac{5}{6} + \dfrac{7}{16} + \dfrac{5}{8}$

i $\dfrac{7}{10} + \dfrac{3}{8} + \dfrac{5}{6}$

j $1\dfrac{1}{3} + \dfrac{7}{10} - \dfrac{4}{15}$

k $\dfrac{5}{14} + 1\dfrac{3}{7} - \dfrac{5}{12}$

3 In a class of children, $\dfrac{3}{4}$ are Chinese, $\dfrac{1}{5}$ are Malay and the rest are Indian. What fraction of the class are Indian?

4 **a** In a class election, $\dfrac{1}{2}$ the class voted for Aminah, $\dfrac{1}{3}$ voted for Reshma and the rest voted for Peter. What fraction of the class voted for Peter?

b One of the following is the number of people in the class.

25   28   30   32

How many people are in the class?

What is $\frac{1}{2}$ of $\frac{1}{4}$? The diagram shows the answer is $\frac{1}{8}$.

In mathematics, you always write $\frac{1}{2}$ of $\frac{1}{4}$ as $\frac{1}{2} \times \frac{1}{4}$

So you know that $\frac{1}{2} \times \frac{1}{4} = \frac{1}{8}$

To multiply fractions, you multiply the **numerators** together and you multiply the denominators together.

EXAMPLE 6

Work out $\frac{1}{4} \times \frac{2}{5}$.

$$\frac{1}{4} \times \frac{2}{5} = \frac{1 \times 2}{4 \times 5} = \frac{2}{20} = \frac{1}{10}$$

Sometimes you can simplify by cancelling *before* you multiply.

EXAMPLE 7

Find $\frac{3}{8} \times \frac{5}{9}$

$$\frac{3}{8} \times \frac{5}{9} = \frac{\overset{1}{\cancel{3}}}{8} \times \frac{5}{\underset{3}{\cancel{9}}} \qquad \text{(3 is a factor of 3 and 9.)}$$

$$= \frac{5}{24} \qquad \begin{array}{l} (5 = 1 \times 5) \\ (24 = 8 \times 3) \end{array}$$

To multiply mixed numbers, first write them as vulgar fractions.

EXAMPLE 8

Find $1\frac{3}{4} \times 2\frac{1}{2}$

$$1\frac{3}{4} \times 2\frac{1}{2} = \frac{7}{4} \times \frac{5}{2}$$

$$= \frac{35}{8}$$

$$= 4\frac{3}{8}$$

EXERCISE 3G

1 Work out the following, leaving each answer in its simplest form. Show your working.

a $\dfrac{1}{2} \times \dfrac{1}{3}$ **b** $\dfrac{1}{4} \times \dfrac{2}{5}$

c $\dfrac{3}{4} \times \dfrac{1}{2}$ **d** $\dfrac{3}{7} \times \dfrac{1}{2}$

e $\dfrac{2}{3} \times \dfrac{4}{5}$ **f** $\dfrac{1}{3} \times \dfrac{3}{5}$

g $\dfrac{1}{3} \times \dfrac{6}{7}$ **h** $\dfrac{3}{4} \times \dfrac{2}{5}$

i $\dfrac{2}{3} \times \dfrac{3}{4}$ **j** $\dfrac{1}{2} \times \dfrac{4}{5}$

2 Work out the following, leaving each answer in its simplest form. Show your working.

a $\dfrac{5}{16} \times \dfrac{3}{10}$ **b** $\dfrac{9}{10} \times \dfrac{5}{12}$

c $\dfrac{14}{15} \times \dfrac{3}{8}$ **d** $\dfrac{8}{9} \times \dfrac{6}{15}$

e $\dfrac{6}{7} \times \dfrac{21}{30}$ **f** $\dfrac{9}{14} \times \dfrac{35}{36}$

3 $\dfrac{1}{4}$ of Lee's stamp collection was given to him by his sister. Unfortunately $\dfrac{2}{3}$ of these were torn. What fraction of his collection was given to him by his sister and were not torn?

4 Bilal eats $\dfrac{1}{4}$ of a cake, and then $\dfrac{1}{2}$ of what is left. How much cake is left uneaten?

5 Work out the following, giving each answer as a mixed number where possible. Show your working.

a $1\dfrac{1}{4} \times \dfrac{1}{3}$ **b** $1\dfrac{2}{3} \times 1\dfrac{1}{4}$

c $2\dfrac{1}{2} \times 2\dfrac{1}{2}$ **d** $1\dfrac{3}{4} \times 1\dfrac{2}{3}$

e $3\dfrac{1}{4} \times 1\dfrac{1}{5}$ **f** $1\dfrac{1}{4} \times 2\dfrac{2}{3}$

g $2\dfrac{1}{2} \times 5$ **h** $7\dfrac{1}{2} \times 4$

6 Which is larger, $\dfrac{3}{4}$ of $2\dfrac{1}{2}$ or $\dfrac{2}{5}$ of $6\dfrac{1}{2}$?

Dividing fractions

Look at the problem $3 \div \frac{3}{4}$

This is equivalent to asking, 'How many $\frac{3}{4}$s are there in 3?'

Look at the diagram.

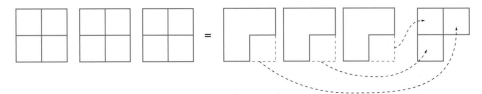

Each of the three whole shapes is divided into quarters. What is the total number of quarters divided by 3?

Can you see that you could fit the four shapes on the right-hand side of the = sign into the three shapes on the left-hand side?

i.e. $3 \div \frac{3}{4} = 4$

or $3 \div \frac{3}{4} = 3 \times \frac{4}{3} = \frac{3 \times 4}{3} = \frac{12}{3} = 4$

So, to divide by a fraction, you turn the fraction upside down (finding its **reciprocal**), and then multiply.

EXAMPLE 9

Find $2\frac{1}{2} \div \frac{3}{4}$

$2\frac{1}{2} \div \frac{3}{4} = \frac{5}{2} \times \frac{4}{3}$ (write $2\frac{1}{2}$ as a vulgar fraction)

$= \frac{5}{\cancel{2}_1} \times \frac{\cancel{4}^2}{3}$ (2 and 4 have 2 as a common factor)

$= \frac{10}{3}$

$= 3\frac{1}{3}$

This means that $3\frac{1}{3} \times \frac{3}{4} = 2\frac{1}{2}$

EXERCISE 3H

1 Work out the following, giving your answer as a mixed number where possible. Show your working.

a $\dfrac{1}{4} \div \dfrac{1}{3}$ **b** $\dfrac{2}{5} \div \dfrac{2}{7}$

c $\dfrac{4}{5} \div \dfrac{3}{4}$ **d** $\dfrac{3}{7} \div \dfrac{2}{5}$

e $5 \div 1\dfrac{1}{4}$ **f** $6 \div 1\dfrac{1}{2}$

g $7\dfrac{1}{2} \div 1\dfrac{1}{2}$ **h** $3 \div 1\dfrac{3}{4}$

i $1\dfrac{5}{12} \div 3\dfrac{3}{16}$ **j** $3\dfrac{3}{5} \div 2\dfrac{1}{4}$

2 A grain merchant has only $13\frac{1}{2}$ tonnes in stock. He has several customers who are all ordering $\frac{3}{4}$ of a tonne. How many customers can he supply?

3 For a party, Zahar made $12\frac{1}{2}$ litres of lemonade. His glasses could each hold $\frac{5}{16}$ of a litre. How many of the glasses could he fill from the $12\frac{1}{2}$ litres of lemonade?

4 How many strips of ribbon, each $3\frac{1}{2}$ centimetres long, can I cut from a roll of ribbon that is $52\frac{1}{2}$ centimetres long?

5 Joe's stride is $\frac{3}{4}$ of a metre long. How many strides does he take to walk the length of a bus 12 metres long?

6 Work out the following, giving your answers as a mixed number where possible.

a $2\dfrac{2}{9} \times 2\dfrac{1}{10} \times \dfrac{16}{35}$ **b** $3\dfrac{1}{5} \times 2\dfrac{1}{2} \times 4\dfrac{3}{4}$

c $1\dfrac{1}{4} \times 1\dfrac{2}{7} \times 1\dfrac{1}{6}$ **d** $\dfrac{18}{25} \times \dfrac{15}{16} \div 2\dfrac{2}{5}$

e $\left(\dfrac{2}{5} \times \dfrac{2}{5}\right) \times \left(\dfrac{5}{6} \times \dfrac{5}{6}\right) \times \left(\dfrac{3}{4} \times \dfrac{3}{4}\right)$ **f** $\left(\dfrac{4}{5} \times \dfrac{4}{5}\right) \div \left(1\dfrac{1}{4} \times 1\dfrac{1}{4}\right)$

Why this chapter matters

Life is full of opposites: up and down, hot and cold, left and right, light and dark, rough and smooth, to name a few. One important pair of opposites in maths is positive and negative.

So far you have worked mostly with positive numbers, carrying out calculations with them and using them in real life problems. But negative numbers are also important, both in maths and everyday life. Positive and negative numbers are called directed numbers. You can think of the + and – showing which direction they move from 0. This is important in many situations:

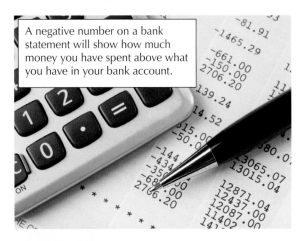

A negative number on a bank statement will show how much money you have spent above what you have in your bank account.

On the Celsius temperature scale zero is known as 'freezing point'. In many places temperatures fall below freezing point. We need negative numbers to represent these temperatures.

Jet pilots experience g-forces when their aircraft accelerates or decelerates quickly. Negative g-forces can be felt when an object accelerates downwards very quickly and they are represented by negative numbers.

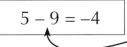
$$5 - 9 = -4$$

When a bigger number is taken from a smaller one, the result is a negative number.

In lifts, negative numbers are used to represent floors below ground level.

Sea level can be given the value 'zero'. Mountains are described as being 'above sea level' and ocean floors as 'below sea level'. This means that depths under the sea are given using negative numbers.

As you can see, negative numbers are just as important as positive numbers and you will encounter them in your everyday life.

Directed numbers

Topics	Level	Key words
1 Introduction to directed numbers	FOUNDATION	negative, positive, directed numbers
2 Everyday use of directed numbers	FOUNDATION	profit, loss
3 The number line	FOUNDATION	number line, less than, more than, greater than
4 Adding and subtracting directed numbers	FOUNDATION	
5 Multiplying and dividing directed numbers	FOUNDATION	

What you need to be able to do in the examinations:

FOUNDATION

- Use directed numbers in practical situations.
- Understand and use integers (positive, negative and zero) both as positions and translations on a number line.

Introduction to directed numbers

4.1

Negative numbers are numbers *below* zero. You use negative numbers when the temperature falls below freezing (0 °C).

The diagram below shows a thermometer with negative temperatures. The temperature is –3 °C. This means the temperature is three degrees below zero.

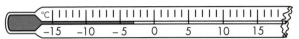

The number line below shows **positive** and negative numbers.

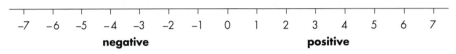

Positive and negative numbers together are called **directed numbers**.

EXERCISE 4A

FOUNDATION

1 Write down the temperature for each thermometer.

a

b

c

d

e

2

Edinburgh –3 °C

London +8 °C

Cardiff –1 °C

a How much colder is it in Edinburgh than in London?

b How much warmer is it in London than in Cardiff?

3 The instructions on a bottle of de-icer say that it will stop water freezing at temperatures down to –12 °C. The temperature is –4 °C.

How many more degrees does the temperature need to fall before the de-icer stops working?

4.2 Everyday use of directed numbers

There are many other situations where directed numbers are used. Here are three examples.

- When +15 m means 15 metres *above* sea level, then –15 m means 15 metres *below* sea level.

- When +2 h means 2 hours *after* midday, then –2 h means 2 hours *before* midday.

- When +$60 means a **profit** of $60, then –$60 means a **loss** of $60.

You also meet negative numbers on graphs, and you may already have plotted coordinates with negative numbers.

On bank statements and bills a negative number means you owe money. A positive number means they owe you money.

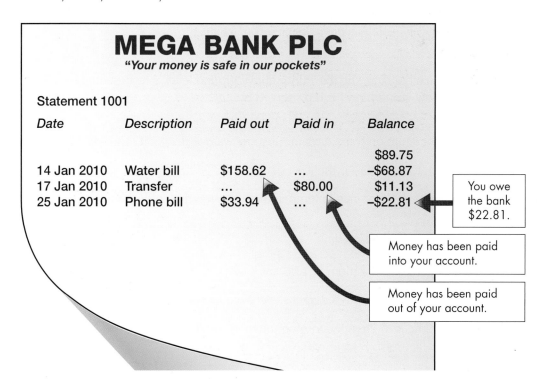

MEGA BANK PLC
"Your money is safe in our pockets"

Statement 1001

Date	Description	Paid out	Paid in	Balance
				$89.75
14 Jan 2010	Water bill	$158.62	...	–$68.87
17 Jan 2010	Transfer	...	$80.00	$11.13
25 Jan 2010	Phone bill	$33.94	...	–$22.81

You owe the bank $22.81.

Money has been paid into your account.

Money has been paid out of your account.

EXERCISE 4B

Copy and complete each of the following.

1 If +$5 means a profit of five dollars, then …… means a loss of five dollars.

2 If +200 m means 200 metres above sea level, then …… means 200 metres below sea leve

3 If –100 m means one hundred metres below sea level, then +100 m means one hundred metres …… sea level.

4 If +5 h means 5 hours after midday, then …… means 5 hours before midday.

5 If +2 °C means two degrees above freezing point, then …… means two degrees below freezing point.

6 If +70 km means 70 kilometres north of the equator, then …… means 70 kilometres south of the equator.

7 If 10 minutes before midnight is represented by –10 minutes, then five minutes after midnight is represented by …… .

8 If a car moving forwards at 10 kilometres per hour is represented by +10 km/h, then a car moving backwards at 5 kilometres per hour is represented by …… .

9 In an office building, the third floor above ground level is represented by +3. So, the secor floor below ground level is represented by …… .

10 The temperature on three days in Moscow was –7 °C, –5 °C and –11 °C.

  a Which temperature is the lowest?

  b What is the difference in temperature between the coldest and the warmest days?

11 A thermostat is set at 16 °C.

The temperature in a room at 1.00 am is –2 °C.

The temperature rises two degrees every 6 minutes.

At what time is the temperature on the thermostat reached?

The number line

Look at the **number line**.

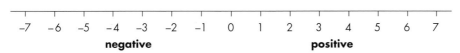

Notice that the negative numbers are to the left of 0 and the positive numbers are to the right of 0.

Numbers to the right of any number on the number line are always bigger than that number.

Numbers to the left of any number on the number line are always smaller than that number.

So, for example, you can see from a number line that:

2 is *smaller* than 5 because 2 is to the left of 5 on the number line.

You can write this as 2 < 5.

–3 is *smaller* than 2 because –3 is to the *left* of 2 on the number line.

You can write this as –3 < 2.

7 is *bigger* than 3 because 7 is to the *right* of 3 on the number line.

You can write this as 7 > 3.

–1 is *bigger* than –4 because –1 is to the *right* of –4 on the number line.

You can write this as –1 > –4.

Reminder: The inequality signs:

    < means 'is **less than**'

    > means 'is **greater than**' or 'is **more than**'

> **HINTS AND TIPS**
>
> The point of the sign points towards the smaller numbers.

EXERCISE 4C

1 Copy each of these and put the correct symbol (< or >) in each space.

  a –1 3   **b** 3 2   **c** –4 ...... –1   **d** –5 ...... –4

  e 1 ...... –6   **f** –3 0   **g** –2 ...... –1   **h** 2 ...... –3

  i 5 ...... –6   **j** 3 4   **k** –7 ...... –5   **l** –2 ...... –4

2

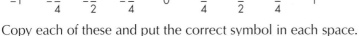

Copy each of these and put the correct symbol in each space.

  a $\frac{1}{4}$ ...... $\frac{3}{4}$ **b** $-\frac{1}{2}$ 0   **c** $-\frac{3}{4}$ ...... $\frac{3}{4}$

  d $\frac{1}{4}$ ...... $-\frac{1}{2}$ **e** –1 ...... $\frac{3}{4}$ **f** $\frac{1}{2}$ 1

3 Copy these number lines and fill in the missing numbers.

a
−5   −2   0   1   3   5

b
−20   −10   0   5   15

c
−8   −4   0   2   6

d
−30   −10   0   10   20

4 Here are some temperatures.

2 °C   −2 °C   −4 °C   6 °C

Copy and complete the weather report, using these temperatures.

> The hottest place today is Barnsley with a temperature of ____,
> while in Eastbourne a ground frost has left the temperature just
> below zero at ____. In Bristol it is even colder at ____. Finally,
> in Tenby the temperature is just above freezing at ____.

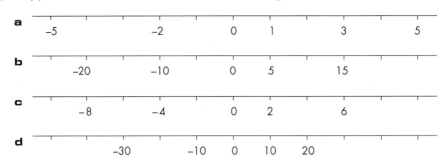

4.4   Adding and subtracting directed numbers

Adding and subtracting positive numbers

These two operations can be illustrated on a thermometer scale.

● *Adding* a positive number moves the marker up the thermometer scale.

For example,

−2 + 6 = 4

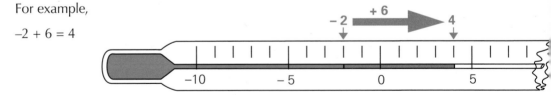

● *Subtracting* a positive number moves the marker *down* the thermometer scale.

For example,

3 − 5 = −2

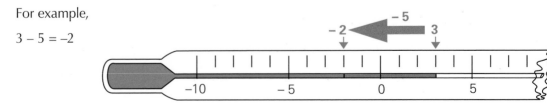

EXAMPLE 1

The temperature at midnight was 2 °C but then it fell by five degrees. What was the new temperature?

Falling five degrees means the calculation is 2 – 5, which is equal to –3. So, the new temperature is –3 °C.

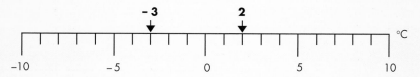

EXERCISE 4D

1 Find the answer to each of the following.

a $2° – 4° =$	**b** $4° – 7° =$	**c** $3° – 5° =$	**d** $1° – 4° =$
e $6° – 8° =$	**f** $5° – 8° =$	**g** $–2 + 5 =$	**h** $–1 + 4 =$
i $–4 + 3 =$	**j** $–6 + 5 =$	**k** $–3 + 5 =$	**l** $–5 + 2 =$
m $–1 – 3 =$	**n** $–2 – 4 =$	**o** $–5 – 1 =$	**p** $3 – 4 =$
q $2 – 7 =$	**r** $1 – 5 =$	**s** $–3 + 7 =$	**t** $5 – 6 =$
u $–2 – 3 =$	**v** $2 – 6 =$	**w** $–8 + 3 =$	**x** $4 – 9 =$

2 At 5 am the temperature in Lisbon was –4 °C. At 11 am the temperature was 3 °C.

  a By how many degrees did the temperature rise?

  b The temperature in Madrid was two degrees lower than in Lisbon at 5 am. What was the temperature in Madrid at 5 am?

3 Here are five numbers.

  4   7   8   2   5

  a Use two of the numbers to make a calculation with an answer of –6.

  b Use three of the numbers to make a calculation with an answer of –1.

  c Use four of the numbers to make a calculation with an answer of –18.

  d Use all five of the numbers to make a calculation with an answer of –12.

4 A submarine is 600 metres below sea level.

A radar system can detect submarines down to 300 metres below sea level.

To safely avoid detection, the submarine captain keeps the submarine 50 metres below the level of detection.

How many metres can the submarine climb to be safe from detection?

Adding and subtracting negative numbers

To *add a negative number* …

… treat the + – as a –

For example: $3 + (-5) = 3 - 5 = -2$

is equivalent to:

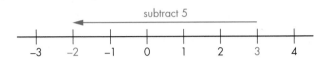

To *subtract a negative number* …

… treat the – – as a +

For example: $3 - (-5) = 3 + 5 = 8$

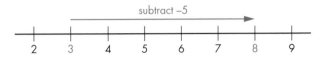

is equivalent to:

Using your calculator

Calculations involving negative numbers can be done by using the (–) key.

EXAMPLE 2

Work out $-6 - -2$.

Press (–) 6 – (–) 2 =

The answer should be –4.

EXERCISE 4E

1 Write down the answer to each of the following, then check your answers on a calculator.

a $-3 - 5 =$	**b** $-2 - 8 =$	**c** $-5 - 6 =$	**d** $6 - 9 =$	**e** $5 - 3 =$
f $3 - 8 =$	**g** $-4 + 5 =$	**h** $-3 + 7 =$	**i** $-2 + 9 =$	**j** $-6 + -2 =$
k $-1 + -4 =$	**l** $-8 + -3 =$	**m** $5 - -6 =$	**n** $3 - -3 =$	**o** $6 - -2 =$
p $3 - -5 =$	**q** $-5 - -3 =$	**r** $-2 - -1 =$	**s** $-4 - 5 =$	**t** $2 - 7 =$

2 What is the *difference* between the following temperatures?

a 4 °C and –6 °C **b** –2 °C and –9 °C **c** –3 °C and 6 °C

3 Find what you have to *add to* 5 to get:

a 7 **b** 2 **c** 0 **d** –2 **e** –5 **f** –15

4 Find what you have to *subtract from* 4 to get:

a 2 **b** 0 **c** 5 **d** 9 **e** 15 **f** –4

5 Find what you have to *add to* –5 to get:

a 8 **b** –3 **c** 0 **d** –1 **e** 6 **f** –7

6 Find what you have to *subtract from* –3 to get:

a 7 **b** 2 **c** –1 **d** –7 **e** –10 **f** 1

7 You have the following cards.

a Which card should you choose to make the answer to the following sum as large as possible? What is the answer?

 + ☐ = ...

b Which card should you choose to make the answer to part **a** as small as possible? What is the answer?

c Which card should you choose to make the answer to the following subtraction as large as possible? What is the answer?

+6 – ☐ = ...

d Which card should you choose to make the answer to part **c** as small as possible? What is the answer?

8 The thermometer in a car is inaccurate by up to two degrees.

An ice alert warning comes on at 3 °C, according to the thermometer temperature.

If the actual temperature is 2 °C, will the alert come on?

Explain how you decide.

9 Two numbers have a sum of 5.

One of the numbers is negative.

The other is a positive even number.

What are the two numbers if the even number is as small as possible?

The rules for multiplying and dividing two directed numbers are very easy.

- When the signs of the two numbers are the *same*, the answer is *positive*.

- When the signs of the two numbers are *different*, the answer is *negative*.

Here are some examples.

$$2 \times 4 = 8 \qquad 12 \div -3 = -4 \qquad -2 \times -3 = 6 \qquad -12 \div -3 = 4$$

A common error is to confuse, for example, -3^2 and $(-3)^2$.

$$-3^2 = -3 \times 3 = -9$$

but,

$$(-3)^2 = -3 \times -3 = +9.$$

So, this means that if we use a variable, for example, $a = -5$, the calculation would be as follows:

$$a^2 = -5 \times -5 = +25$$

EXAMPLE 3

$a = -2$ and $b = -6$

Work out the following:

a a^2 **b** $a^2 + b^2$ **c** $b^2 - a^2$ **d** $(a - b)^2$

a a^2 $= -2 \times -2 = +4$

b $a^2 + b^2$ $= +4 + -6 \times -6 = 4 + 36 = 40$

c $b^2 - a^2$ $= 36 - 4 = 32$

d $(a - b)^2$ $= (-2 - -6)^2 = (-2 + 6)^2 = (4)^2 = 16$

EXERCISE 4F

1 Write down the answers to the following.

a -3×5 **b** -2×7 **c** -4×6

d -2×-3 **e** -7×-2 **f** $-12 \div -6$

g $-16 \div 8$ **h** $24 \div -3$ **i** $16 \div -4$

j $-6 \div -2$ **k** 4×-6 **l** 5×-2

m 6×-3 **n** -2×-8 **o** -9×-4

p $24 \div -6$ **q** $12 \div -1$ **r** $-36 \div 9$

s $-14 \div -2$ **t** $100 \div 4$ **u** -2×-9

2 Write down the answers to the following.

a	−3 + −6	**b**	−2 × −8	**c**	2 + −5
d	8 × −4	**e**	−36 ÷ −2	**f**	−3 × −6
g	−3 − −9	**h**	48 ÷ −12	**i**	−5 × −4
j	7 − −9	**k**	−40 ÷ −5	**l**	−40 + −8
m	4 − −9	**n**	5 − 18	**o**	72 ÷ −9
p	−7 − −7	**q**	8 − −8	**r**	6 × −7

3 What number do you multiply by −3 to get the following?

a 6 **b** −90 **c** −45

d 81 **e** 21

4 What number do you divide −36 by to get the following?

a −9 **b** 4 **c** 12

d −6 **e** 9

5 Evaluate the following.

a −6 + (4 − 7) **b** −3 − (−9 − −3) **c** 8 + (2 − 9)

6 Evaluate the following.

a 4 × (−8 ÷ −2) **b** −8 − (3 × −2) **c** −1 × (8 − −4)

7 What do you get if you divide −48 by the following?

a −2 **b** −8

c 12 **d** 24

8 Write down six different multiplications that give the answer −12.

9 Write down six different divisions that give the answer −4.

10 Put these calculations in order from the lowest to the highest answer.

−5 × 4   −20 ÷ 2   −16 ÷ −4   3 × −6

11 $x = -2$, $y = -3$ and $z = -4$. Work out the following:

a x^2 **b** $y^2 + z^2$

c $z^2 - x^2$ **d** $(x - y)^2$

Why this chapter matters

The squares and square roots of numbers are important tools in mathematics and mathematicians have helped us to use them by inventing notation.

We often need to multiply a number by itself two or three times, for example when finding the area of a square or the volume of a cube.

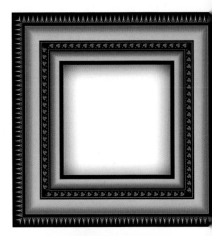

You have seen in Chapter One how we can write 5^2 instead of 5×5. We can also show $5 \times 5 \times 5$ by using 5^3. This short cut is called index notation (see Chapter 18).

The notation for the square root of a number (e.g. $\sqrt{25}$) is even more convenient. Without this, we would have to write 'the number which multiplies by itself to make 25'. The sign for the cube root of a number is $\sqrt[3]{}$.

The root signs are especially convenient when the roots are hard to work out and difficult to express accurately. Square numbers such as 4, 9, 16 and 25 have whole numbers as their square roots but most numbers have fractions. If the fractions are expressed as decimals they are sometimes recurring (that is, they never end) which means that they can never be written down accurately.

The notation we now use was only introduced in the sixteenth century (CE). One of the first people to use it in print was a German mathematician called Christoph Rudolff. The notation was simple and easy to understand and was soon widely used.

Ways of working out square roots have been developed by mathematicians over the centuries. This Babylonian tablet showing how to calculate the square root of two is 2500 years ago.

Notation can be used in calculations to represent the square roots of numbers without actually having to work them out or write them down. This makes it possible to carry out more sophisticated and accurate calculations.

Squares, cubes and roots

pics	Level	Key words
1 Squares and square roots	**FOUNDATION**	square, square root
Cubes and cube roots	**FOUNDATION**	cube, cube root
3 Surds	**HIGHER**	surd, rationalise

What you need to be able to do in the examinations:

FOUNDATION	HIGHER
• Identify square numbers and cube numbers. • Calculate squares, square roots, cubes and cube roots.	• Understand the meaning of surds. • Manipulate surds, including rationalising the denominator where the denominator is a pure surd.

Squares and square roots

Take $6 \times 6 = 36$

This can also be written as 6^2, which we say as 'six squared'.

Another way to describe this is to say: '6 is the **square root** of 36'.

This can be shown as $\sqrt{36}$.

In fact 36 has *two* square roots:

$$6 \times 6 = 36$$
$$\text{and} \quad -6 \times -6 = 36$$

So the square roots of 36 are 6 and -6 or $\sqrt{36}$ and $-\sqrt{36}$.

We can **square** decimals as well as whole numbers:

$$2.5^2 = 6.25 \text{ so } \sqrt{6.25} = 2.5$$

Most calculators have a 'square' button x^2 and a 'square root' button $\sqrt{}$. Check that you know how to use them.

The calculator will show the positive square root, e.g. $\sqrt{36} = 6$

EXERCISE 5A

FOUNDATION

1 Find the value of:

a 7^2 b 10^2 c 1.2^2 d 2.5^2

e 16^2 f 20^2 g 3.1^2 h 4.5^2

i $(-3)^2$ j $(-8)^2$ k 0.5^2 l $(-0.5)^2$

2 Write down the two square roots of:

a 9 b 100 c 121 d 1.44

e 400 f 12.25 g 1 h $10\,000$

3 Find:

a $\sqrt{25}$ b $\sqrt{36}$ c $\sqrt{100}$ d $\sqrt{49}$

e $\sqrt{64}$ f $\sqrt{2.25}$ g $\sqrt{30.25}$ h $\sqrt{1.44}$

i $\sqrt{400}$ j $\sqrt{0.25}$

4 Write down the value of each of these. You will need to use your calculator for some of them. Look for the key.

   a 9^2 **b** $\sqrt{1600}$ **c** 10^2 **d** $\sqrt{196}$

   e 6^2 **f** $\sqrt{225}$ **g** 7^2 **h** $\sqrt{144}$

   i 5^2 **j** $\sqrt{441}$

5 Write down each of the following. You will need to use your calculator.

   a $\sqrt{576}$ **b** $\sqrt{961}$ **c** $\sqrt{2025}$ **d** $\sqrt{1600}$

   e $\sqrt{4489}$ **f** $\sqrt{10\,201}$ **g** $\sqrt{12.96}$ **h** $\sqrt{42.25}$

   i $\sqrt{193.21}$ **j** $\sqrt{492.84}$

6 Put these in order, starting with the smallest value.

   3^2 $\sqrt{90}$ $\sqrt{50}$ 4^2

7 **a** Explain how you know that $\sqrt{40}$ is between 6 and 7.

   b Use your calculator to find $\sqrt{40}$. Write down all the decimal places on your display.

8 Between which two consecutive whole numbers does the square root of 20 lie?

9 Find two consecutive integers between which each of these square roots lie:

   a $\sqrt{68}$ **b** $\sqrt{96}$ **c** $\sqrt{155}$ **d** $\sqrt{250}$

10 Use these number cards to make this calculation correct.

   1   **2**   **3**   **4**   **8**

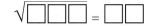

11 A square wall in a kitchen is being tiled.

Altogether it needs 225 square tiles.
How many tiles are there in each row?

Cubes and cube roots

Take $6 \times 6 \times 6 = 216$

216 is 'six cubed'. A **cube** of a number is formed when you multiply the number by itself and then by itself again. This can also be written as 6^3:

   $6 \times 6 \times 6 = 6^3$ or 'six cubed'

Another way to describe this is to say: '6 is the **cube root** of 216'.

The symbol for a cube root is $\sqrt[3]{}$ so this can also be shown as $\sqrt[3]{216}$.

Many calculators have a button for cubes $\boxed{x^3}$ and cube roots $\boxed{\sqrt[3]{}}$. Check that you know how to use them.

EXAMPLE 1

Find the cube roots of 64 and –64.

$4 \times 4 \times 4 = 64$ so $\sqrt[3]{64} = 4$

$-4 \times -4 \times -4 = -64$ so $\sqrt[3]{-64} = -4$

Notice that 64 and –64 have just one cube root each.

EXERCISE 5B

1 Find the following cubes:

a 2^3 b 3^3 c 8^3 d 10^3

e 1.1^3 f 2.5^3 g $(-3)^3$ h $(-5)^3$

i 20^3 j 4.1^3 k $(-4.1)^3$

2 Find these cube roots:

a $\sqrt[3]{8}$ b $\sqrt[3]{125}$ c $\sqrt[3]{729}$ d $\sqrt[3]{1}$

e $\sqrt[3]{27}$ f $\sqrt[3]{-27}$ g $\sqrt[3]{1000}$ h $\sqrt[3]{3.375}$

i $\sqrt[3]{91.125}$ j $\sqrt[3]{0.125}$

3 $4^3 = 64$ and $5^3 = 125$ so $\sqrt[3]{100}$ is between 4 and 5.

Find two consecutive integers between which these cube roots lie:

a $\sqrt[3]{200}$ b $\sqrt[3]{300}$ c $\sqrt[3]{500}$ d $\sqrt[3]{-500}$

4 Here are four numbers:

   2^3 3^2 $\sqrt{81}$ $\sqrt[3]{729}$

Which is the odd one out and why?

5 Can you find two different positive integers, A and B, such that $A^2 = B^3$?

6 Put these numbers in order, smallest first:

2.5^3 $\sqrt{225}$ 4^2 $\sqrt[3]{2000}$

7 Copy and complete this table:

Number	Square	Cube
	100	
5		
		64
11		
	81	

8 Write these numbers in order with the smallest first:

$\sqrt{0.8}$ $\sqrt[3]{0.8}$ 0.8^2 0.8^3

5.3   Surds

It is useful to be able to work with **surds**, which are roots of whole numbers written as, for example:

$\sqrt{2}$ $\sqrt{5}$ $\sqrt{15}$ $\sqrt{6}$ $\sqrt{3}$ $\sqrt{10}$

Here are four general rules for simplifying surds.

You can check that these rules work by taking numerical examples.

$$\sqrt{a} \times \sqrt{b} = \sqrt{ab} \qquad\qquad C\sqrt{a} \times D\sqrt{b} = CD\sqrt{ab}$$

$$\sqrt{a} \div \sqrt{b} = \sqrt{\frac{a}{b}} \qquad\qquad C\sqrt{a} \div D\sqrt{b} = \frac{C}{D}\sqrt{\frac{a}{b}}$$

For example:

$\sqrt{2} \times \sqrt{2} = \sqrt{4} = 2 \qquad\qquad \sqrt{2} \times \sqrt{10} = \sqrt{20} = \sqrt{(4 \times 5)} = \sqrt{4} \times \sqrt{5} = 2\sqrt{5}$

$\sqrt{2} \times \sqrt{3} = \sqrt{6} \qquad\qquad\qquad \sqrt{6} \times \sqrt{15} = \sqrt{90} = \sqrt{9} \times \sqrt{10} = 3\sqrt{10}$

$\sqrt{2} \times \sqrt{8} = \sqrt{16} = 4 \qquad\qquad 3\sqrt{5} \times 4\sqrt{3} = 12\sqrt{15}$

$\sqrt{75} = \sqrt{25 \times 3} = \sqrt{25} \times \sqrt{3} = 5\sqrt{3} \qquad \sqrt{8} = \sqrt{4 \times 2} = 2\sqrt{2}$

If you type into your calculator the expressions before the equals sign in these problems, the display will give you the answer. However, you need to be able to manipulate surds without a calculator. This will help you understand them better. Make sure you understand the above rules.

EXERCISE 5C

Do this exercise without a calculator.

1 Simplify each of the following. Leave your answers in surd form if necessary.

    a $\sqrt{2} \times \sqrt{3}$ **b** $\sqrt{5} \times \sqrt{3}$ **c** $\sqrt{2} \times \sqrt{2}$ **d** $\sqrt{2} \times \sqrt{8}$

    e $\sqrt{2} \times \sqrt{7}$ **f** $\sqrt{2} \times \sqrt{18}$ **g** $\sqrt{6} \times \sqrt{6}$ **h** $\sqrt{5} \times \sqrt{6}$

2 Simplify each of the following. Leave your answers in surd form if necessary.

    a $\sqrt{12} \div \sqrt{3}$ **b** $\sqrt{15} \div \sqrt{3}$ **c** $\sqrt{12} \div \sqrt{2}$ **d** $\sqrt{24} \div \sqrt{8}$

    e $\sqrt{28} \div \sqrt{7}$ **f** $\sqrt{48} \div \sqrt{8}$ **g** $\sqrt{6} \div \sqrt{6}$ **h** $\sqrt{54} \div \sqrt{6}$

3 Simplify each of the following. Leave your answers in surd form if necessary.

    a $\sqrt{2} \times \sqrt{3} \times \sqrt{2}$ **b** $\sqrt{5} \times \sqrt{3} \times \sqrt{15}$ **c** $\sqrt{2} \times \sqrt{2} \times \sqrt{8}$ **d** $\sqrt{2} \times \sqrt{8} \times \sqrt{3}$

    e $\sqrt{2} \times \sqrt{7} \times \sqrt{2}$ **f** $\sqrt{2} \times \sqrt{18} \times \sqrt{5}$ **g** $\sqrt{6} \times \sqrt{6} \times \sqrt{3}$ **h** $\sqrt{5} \times \sqrt{6} \times \sqrt{30}$

4 Simplify each of the following. Leave your answers in surd form.

    a $\sqrt{2} \times \sqrt{3} \div \sqrt{2}$ **b** $\sqrt{5} \times \sqrt{3} \div \sqrt{15}$ **c** $\sqrt{32} \times \sqrt{2} \div \sqrt{8}$ **d** $\sqrt{2} \times \sqrt{8} \div \sqrt{8}$

    e $\sqrt{5} \times \sqrt{8} \div \sqrt{8}$ **f** $\sqrt{3} \times \sqrt{3} \div \sqrt{3}$ **g** $\sqrt{8} \times \sqrt{12} \div \sqrt{48}$ **h** $\sqrt{7} \times \sqrt{3} \div \sqrt{3}$

    i $\sqrt{2} \times \sqrt{7} \div \sqrt{2}$ **j** $\sqrt{2} \times \sqrt{18} \div \sqrt{3}$ **k** $\sqrt{6} \times \sqrt{6} \div \sqrt{3}$ **l** $\sqrt{5} \times \sqrt{6} \div \sqrt{30}$

5 Simplify each of these expressions.

    a $\sqrt{a} \times \sqrt{a}$ **b** $\sqrt{a} \div \sqrt{a}$ **c** $\sqrt{a} \times \sqrt{a} \div \sqrt{a}$

6 Simplify each of the following surds into the form $a\sqrt{b}$.

    a $\sqrt{18}$ **b** $\sqrt{24}$ **c** $\sqrt{12}$ **d** $\sqrt{50}$ **e** $\sqrt{8}$ **f** $\sqrt{27}$

    g $\sqrt{48}$ **h** $\sqrt{75}$ **i** $\sqrt{45}$ **j** $\sqrt{63}$ **k** $\sqrt{32}$ **l** $\sqrt{200}$

7 Simplify each of these.

    a $2\sqrt{18} \times 3\sqrt{2}$ **b** $4\sqrt{24} \times 2\sqrt{5}$ **c** $3\sqrt{12} \times 3\sqrt{3}$ **d** $2\sqrt{8} \times 2\sqrt{8}$

    e $2\sqrt{27} \times 4\sqrt{8}$ **f** $2\sqrt{48} \times 3\sqrt{8}$ **g** $2\sqrt{45} \times 3\sqrt{3}$ **h** $2\sqrt{63} \times 2\sqrt{7}$

    i $2\sqrt{32} \times 4\sqrt{2}$ **j** $\sqrt{1000} \times \sqrt{10}$ **k** $\sqrt{250} \times \sqrt{10}$ **l** $2\sqrt{98} \times 2\sqrt{2}$

8 Simplify each of these.

    a $4\sqrt{2} \times 5\sqrt{3}$ **b** $2\sqrt{5} \times 3\sqrt{3}$ **c** $4\sqrt{2} \times 3\sqrt{2}$ **d** $2\sqrt{2} \times 2\sqrt{8}$

    e $2\sqrt{5} \times 3\sqrt{8}$ **f** $3\sqrt{3} \times 2\sqrt{3}$ **g** $2\sqrt{6} \times 5\sqrt{2}$ **h** $5\sqrt{7} \times 2\sqrt{3}$

    i $2\sqrt{2} \times 3\sqrt{7}$ **j** $2\sqrt{2} \times 3\sqrt{18}$ **k** $2\sqrt{6} \times 2\sqrt{6}$ **l** $4\sqrt{5} \times 3\sqrt{6}$

9 Simplify each of these.

    a $6\sqrt{12} \div 2\sqrt{3}$ **b** $3\sqrt{15} \div \sqrt{3}$ **c** $6\sqrt{12} \div \sqrt{2}$ **d** $4\sqrt{24} \div 2\sqrt{8}$

    e $12\sqrt{40} \div 3\sqrt{8}$ **f** $5\sqrt{3} \div \sqrt{3}$ **g** $14\sqrt{6} \div 2\sqrt{2}$ **h** $4\sqrt{21} \div 2\sqrt{3}$

    i $9\sqrt{28} \div 3\sqrt{7}$ **j** $12\sqrt{56} \div 6\sqrt{8}$ **k** $25\sqrt{6} \div 5\sqrt{6}$ **l** $32\sqrt{54} \div 4\sqrt{6}$

10 Simplify each of these.

a $4\sqrt{2} \times \sqrt{3} \div 2\sqrt{2}$

b $4\sqrt{5} \times \sqrt{3} \div \sqrt{15}$

c $2\sqrt{32} \times 3\sqrt{2} \div 2\sqrt{8}$

d $6\sqrt{2} \times 2\sqrt{8} \div 3\sqrt{8}$

e $3\sqrt{5} \times 4\sqrt{8} \div 2\sqrt{8}$

f $12\sqrt{3} \times 4\sqrt{3} \div 2\sqrt{3}$

g $3\sqrt{8} \times 3\sqrt{12} \div 3\sqrt{48}$

h $4\sqrt{7} \times 2\sqrt{3} \div 8\sqrt{3}$

i $15\sqrt{2} \times 2\sqrt{7} \div 3\sqrt{2}$

j $8\sqrt{2} \times 2\sqrt{18} \div 4\sqrt{3}$

k $5\sqrt{6} \times 5\sqrt{6} \div 5\sqrt{3}$

l $2\sqrt{5} \times 3\sqrt{6} \div \sqrt{30}$

11 Simplify each of these expressions.

a $a\sqrt{b} \times c\sqrt{b}$

b $a\sqrt{b} \div c\sqrt{b}$

c $a\sqrt{b} \times c\sqrt{b} \div a\sqrt{b}$

12 Find the value of a that makes each of these surds true.

a $\sqrt{5} \times \sqrt{a} = 10$

b $\sqrt{6} \times \sqrt{a} = 12$

c $\sqrt{10} \times 2\sqrt{a} = 20$

d $2\sqrt{6} \times 3\sqrt{a} = 72$

e $2\sqrt{a} \times \sqrt{a} = 6$

f $3\sqrt{a} \times 3\sqrt{a} = 54$

13 Simplify the following.

a $\left(\dfrac{\sqrt{3}}{2}\right)^2$

b $\left(\dfrac{5}{\sqrt{3}}\right)^2$

c $\left(\dfrac{\sqrt{5}}{4}\right)^2$

d $\left(\dfrac{6}{\sqrt{3}}\right)^2$

e $\left(\dfrac{\sqrt{8}}{2}\right)^2$

14 Decide whether each statement is true or false.

Show your working.

a $\sqrt{(a + b)} = \sqrt{a} + \sqrt{b}$

b $\sqrt{(a - b)} = \sqrt{a} - \sqrt{b}$

15 Write down a product of two different surds which has an integer answer.

Calculating with surds

The following examples show how surds can be used in calculations.

EXAMPLE 2

Calculate the area of a square with a side of $2 + \sqrt{3}$ cm.

Give your answer in the form $a + b\sqrt{3}$.

Area in cm^2 = $(2 + \sqrt{3})^2$

$= (2 + \sqrt{3})(2 + \sqrt{3})$

$= 2(2 + \sqrt{3}) + \sqrt{3}\,(2 + \sqrt{3})$

$= 4 + 2\sqrt{3} + 2\sqrt{3} + 3$

$= 7 + 4\sqrt{3}$

> **HINTS AND TIPS**
>
> First multiply the second bracket by the 2 from the first bracket, then by the $\sqrt{3}$.

EXAMPLE 3

Simplify $(4 - \sqrt{5})(2 + \sqrt{5})$

$$(4 - \sqrt{5})(2 + \sqrt{5}) = 4(2 + \sqrt{5}) - \sqrt{5}\,(2 + \sqrt{5})$$
$$= 8 + 4\sqrt{5} - 2\sqrt{5} - 5$$
$$= 3 + 2\sqrt{5}$$

Rationalising a denominator

When surds are written as fractions in answers they are usually given with a rational denominato which means that it is a whole number.

Multiplying the numerator and denominator by an appropriate surd will **rationalise** the denominato

EXAMPLE 4

Rationalise the denominator of: **a** $\dfrac{1}{\sqrt{3}}$ and   **b** $\dfrac{2\sqrt{3}}{\sqrt{8}}$

a Multiply the numerator and denominator by $\sqrt{3}$.

$$\frac{1 \times \sqrt{3}}{\sqrt{3} \times \sqrt{3}} = \frac{\sqrt{3}}{3}$$

b Multiply the numerator and denominator by $\sqrt{8}$.

$$\frac{2\sqrt{3} \times \sqrt{8}}{\sqrt{8} \times \sqrt{8}} = \frac{2\sqrt{24}}{8} = \frac{4\sqrt{6}}{8} = \frac{\sqrt{6}}{2}$$

or

$$\sqrt{8} = 2\sqrt{2}$$

So $\dfrac{2\sqrt{3}}{\sqrt{8}} = \dfrac{2\sqrt{3}}{2\sqrt{2}} = \dfrac{\sqrt{3}}{\sqrt{2}}$

Multiplying the numerator and denominator by $\sqrt{2}$:

$$\frac{\sqrt{3} \times \sqrt{2}}{\sqrt{2} \times \sqrt{2}} = \frac{\sqrt{6}}{2}$$

EXAMPLE 5

Rationalise the denominator of   **a** $\dfrac{1}{\sqrt{3}+1}$ **b** $\dfrac{\sqrt{2}+1}{\sqrt{2}-1}$

a Multiply the numerator and denominator by $\sqrt{3} - 1$

$$\frac{1}{\sqrt{3}+1} = \frac{\sqrt{3}-1}{(\sqrt{3}+1)(\sqrt{3}-1)} = \frac{\sqrt{3}-1}{3-1} = \frac{1}{2}(\sqrt{3}-1)$$

This works because $(\sqrt{3}+1)(\sqrt{3}-1)=(\sqrt{3})^2+\sqrt{3}-\sqrt{3}-1 = 3 - 1 = 2$ and this rationalises the denominator.

b Multiply the numerator and denominator by $\sqrt{2}+1$

$$\frac{\sqrt{2}+1}{\sqrt{2}-1} = \frac{(\sqrt{2}+1)(\sqrt{2}+1)}{(\sqrt{2}-1)(\sqrt{2}+1)} = \frac{2+\sqrt{2}+\sqrt{2}+1}{2-1} = 3 + 2\sqrt{2}$$

This time the denominator is $(\sqrt{2}-1)(\sqrt{2}+1)=(\sqrt{2})^2+\sqrt{2}-\sqrt{2}-1 = 2 - 1 = 1$

EXERCISE 5D

Do this exercise without a calculator.

1 Show that:

a $(2 + \sqrt{3})(1 + \sqrt{3}) = 5 + 3\sqrt{3}$ **b** $(1 + \sqrt{2})(2 + \sqrt{3}) = 2 + 2\sqrt{2} + \sqrt{3} + \sqrt{6}$

c $(4 - \sqrt{3})(4 + \sqrt{3}) = 13$

2 Expand and simplify where possible.

a $\sqrt{3}(2 - \sqrt{3})$
b $\sqrt{2}(3 - 4\sqrt{2})$
c $\sqrt{5}(2\sqrt{5} + 4)$

d $3\sqrt{7}(4 - 2\sqrt{7})$
e $3\sqrt{2}(5 - 2\sqrt{8})$
f $\sqrt{3}(\sqrt{27} - 1)$

3 Expand and simplify where possible.

a $(1 + \sqrt{3})(3 - \sqrt{3})$
b $(2 + \sqrt{5})(3 - \sqrt{5})$
c $(1 - \sqrt{2})(3 + 2\sqrt{2})$

d $(3 - 2\sqrt{7})(4 + 3\sqrt{7})$
e $(1 - \sqrt{2})^2$
f $(3 + \sqrt{2})^2$

4 Calculate the area of each of these rectangles, simplifying your answers where possible.
(The area of a rectangle with length l and width w is $A = l \times w$.)

a

$1 + \sqrt{3}$ cm

$2 - \sqrt{3}$ cm

b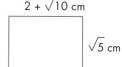

$2 + \sqrt{10}$ cm

$\sqrt{5}$ cm

c

$2\sqrt{3}$ cm

$1 + \sqrt{27}$ cm

5 Rationalise the denominators of these expressions.

a $\dfrac{1}{\sqrt{3}}$
b $\dfrac{1}{\sqrt{2}}$
c $\dfrac{1}{\sqrt{5}}$
d $\dfrac{1}{2\sqrt{3}}$
e $\dfrac{3}{\sqrt{3}}$
f $\dfrac{5}{\sqrt{2}}$

g $\dfrac{3\sqrt{2}}{\sqrt{8}}$
h $\dfrac{5\sqrt{3}}{\sqrt{6}}$
i $\dfrac{\sqrt{7}}{\sqrt{3}}$
j $\dfrac{1 + \sqrt{2}}{\sqrt{2}}$
k $\dfrac{2 - \sqrt{3}}{\sqrt{3}}$
l $\dfrac{5 + 2\sqrt{3}}{\sqrt{3}}$

6 **a** Expand and simplify the following.

i $(2 + \sqrt{3})(2 - \sqrt{3})$
ii $(1 - \sqrt{5})(1 + \sqrt{5})$
iii $(\sqrt{3} - 1)(\sqrt{3} + 1)$

iv $(3\sqrt{2} + 1)(3\sqrt{2} - 1)$
v $(2 - 4\sqrt{3})(2 + 4\sqrt{3})$

b What happens in the answers to part **a**? Why?

7 Simplify the following.

a $(3 + \sqrt{2})(3 - \sqrt{2})$
b $(2 - \sqrt{3})(2 + \sqrt{3})$
c $(\sqrt{6} + 2)(\sqrt{6} - 2)$

d $(1 + \sqrt{2})(1 - \sqrt{2})$
e $(2 - \sqrt{7})(2 + \sqrt{7})$
f $(\sqrt{20} + 4)(\sqrt{20} - 4)$

8 Rationalise the denominator.

a $\dfrac{2}{\sqrt{3} - 1}$
b $\dfrac{4}{2 + \sqrt{3}}$
c $\dfrac{8}{5 - \sqrt{3}}$
d $\dfrac{\sqrt{2}}{\sqrt{2} + 1}$
e $\dfrac{\sqrt{5}}{\sqrt{5} - 2}$
f $\dfrac{1 + \sqrt{3}}{2 - \sqrt{3}}$

9 The reciprocal of $\dfrac{3 - \sqrt{3}}{12}$ is $\dfrac{12}{3 - \sqrt{3}}$

Rationalise the denominator of the reciprocal.

10 The number $r = \dfrac{\sqrt{5} + 1}{2}$ is called the golden ratio. Show that the reciprocal of r is equal to $r - 1$.

Why this chapter matters

Sets are collections of objects. Set notation gives us a way of seeing the logical connection between sets. The mathematics of sets is very useful in designing computer circuits and electronic components.

Alice in Wonderland

You have probably heard of Alice in Wonderland. Did you know that the author, Lewis Carroll, was actually a lecturer in mathematics at the University of Oxford, in the nineteenth century? His real name was Charles Dodgson.

He also wrote a mathematics book called 'Symbolic Logic'. Here is a problem from it:

1. All humming birds are richly coloured.
2. No large birds can live on honey.
3. Birds that do not live on honey are dull in colour.

What conclusion follows?

A Venn diagram in glass

'Symbolic Logic' was about how to write sentences in symbols so that conclusions could be seen more easily. You can use set notation and Venn diagrams to do this and you will learn about them in this chapter.

Venn diagrams were invented by the logician John Venn and may be the only mathematical invention to be celebrated in a stained glass window!

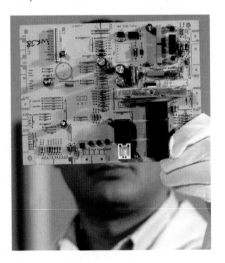

Symbolic logic has a very modern application in the design of computer circuits and the construction of electronic components. When engineers talk about NAND and NOR gates they are using ideas which were first developed in the nineteenth century for very different purposes.

Set language and notation

Topics	Level	Key words
1 Inequalities	**FOUNDATION**	greater than, less than
2 Sets	**FOUNDATION**	universal set, empty set, union, intersection, element
3 Venn diagrams	**FOUNDATION**	complement, Venn diagram
4 More notation	**HIGHER**	subset
5 Practical problems	**HIGHER**	

What you need to be able to do in the examinations:

FOUNDATION	HIGHER
Understand and use the symbols $>$, $<$, $\geq$, $\leq$.	• Understand sets defined in algebraic terms.
Understand the definition of a set.	• Understand and use the complement of a set.
Use the set notation $\cup$, $\cap$, $\in$, $\notin$	• Use the notation $n(A)$ for the number of elements in the set A.
Understand the concept of the Universal Set and the Empty Set and the symbols for these sets.	• Use sets in practical situations.
Understand and use subsets.	
Use Venn diagrams to represent sets and the number of elements in sets.	

Inequalities

You need to know the meaning of the following symbols:

$=$ equals

$\neq$ does not equal

$>$ is **greater than**

$<$ is **less than**

$\geqslant$ is equal to or greater than

$\leqslant$ is equal to or less than

EXAMPLE 1

n is a positive integer and $n < 6$. What are the possible values of n?

Because n is positive the possible values are 1, 2, 3, 4 and 5. 6 is not included because the sign says 'less than'. If the question said $n \leqslant 6$ then 6 would be included.

Sometimes two inequalities are used together.

EXAMPLE 2

Find the possible values of x if x is an integer and $-3 < x \leqslant 3$.

This means that x is a whole number between -3 and 3.

The possible values of x are -2, -1, 0, 1, 2, 3. Note that -3 is not in the list but 3 is included because $\leqslant 3$ means equal to as well as less than 3.

EXERCISE 6A

1   Here are three symbols: $< = >$.

Put the correct symbol between each pair of numbers:

a   3.5 ... 3.15   **b**   180 cm ... 2 m

c 5×7 ... 6×6 **d**   5 km ... 5000 m

e $\frac{1}{3}$ of 27 ... $\frac{3}{4}$ of 12   **f** 4^2 ... 8

g 10×10 ... 8×12 **h** $\sqrt{64}$ ... 3^2

2   Here are three fractions: $\frac{1}{3}$, $\frac{3}{5}$, $\frac{1}{2}$.

Fill in the gaps below to list them in order, smallest first:

...$<$...$<$...

3 d is the number scored when a normal six-sided dice is thrown.

List the possible values of d in the following cases:

a $d \geqslant 4$ **b** $d < 3$

c $d > 5$ **d** $d \leqslant 5$

e $2 \leqslant d \leqslant 4$ **f** $3 < d < 6$

g $1 \leqslant d < 4$ **h** $5 < d \leqslant 6$

4 The table shows whether babies of a particular age are underweight or overweight.

Mass < 6.5 kg	Underweight
6.5 kg $\leqslant$ mass $\leqslant 8.5$ kg	Normal
Mass > 8.5 kg	Overweight

Are babies of the following masses underweight, overweight or normal?

a 6.3 kg   **b** 9.3 kg

c 7.8 kg   **d** 8.5 kg

5 e is an even number and $20 \leqslant e < 30$ and $e \neq 24$. List the possible values of e.

6 In a lottery each ball has a number n where $1 \leqslant n \leqslant 49$.

a What is the largest number on a ball?

b The first ball is a multiple of 5 and $n > 40$. What are the possible values of n?

c The second ball is a multiple of 3 and $n \leqslant 10$. What are the possible values of n?

d On the third ball $15 < n \leqslant 20$. What are the possible values of n?

7 True or false? State which in each case.

a $3 \neq -3$ **b** $-3 < -5$

c $1.99 < 2$ **d** $2 < \sqrt{5} < 3$

e $20^2 \leqslant 300$ **f** 200 minutes $\geqslant$ 3 hours

8 List all the possible values for an integer x in the following cases.

a $5 < x < 9$ **b** $26 \leqslant x \leqslant 28$

c $-8 < x \leqslant -4$ **d** $-2 \leqslant x < 2$

e $17 < x < 18$ **f** $32.5 \leqslant x < 33.5$

A set is any collection of items. They could be numbers, objects or symbols.

The items in a set are called **elements**. The elements in a set can be listed or described inside brackets like this { }.

The elements are separated by commas, e.g. {vowels in English} = {a, e, i, o, u}.

We usually use capital letters to stand for sets, e.g. $N = \{1, 3, 5, 7, 9\}$.

Dots show when we cannot list all the elements, e.g. {even numbers} = {2, 4, 6,…}.

You can describe sets using inequality symbols e.g. {integer $x : 5 \leqslant x \leqslant 10$} = {5, 6, 7, 8, 9, 1(} can be read as 'the set of integers between 5 and 10 inclusive'.

Describing elements

HINTS AND TIPS

Only write each letter once.

Suppose **M** = {letters in the word "mathematics"}
= {a, c, e, h, i, m, s, t}

L = {letters in "language"} = {a, e, g, l, n, u}

S = {letters in "science"} = {c, e, i, n, s}

A = {letters in "art"} = {a, r, t}

HINTS AND TIPS

The letters can be given in any order.

We use $\in$ to mean "*is a member of*" and $\notin$ to mean "*is not a member of*".

So $a \in$ **M** but $a \notin$ **S**

$g \in$ **L** but $g \notin$ **M**

Some letters are in more than one set.

For example, a and e are in **M** and **L**. We say that {a, e} is the **intersection** of **M** and **L** and we write **M** $\cap$ **L** = {a, e}. They are the only letters in both **M** and **L**.

Check that **M** $\cap$ **S** = {c, e, i, s} and **L** $\cap$ **S** = {e, n}.

There are no letters which occur in both **S** and **A**.

We write **S** $\cap$ **A** = { } and this is called the **empty set**. It has no elements.

Sometimes the symbol $\varnothing$ is used for the empty set so we could write **S** $\cap$ **A** = $\varnothing$. If we list all the letters in **M** or **L** (or both), this is called **union** of **M** and **L**.

We write **M** $\cup$ **L** = {$a, c, e, g, h, i, l, m, n, u, s, t$}.

Check that **L** $\cup$ **S** has 9 elements.

HINTS AND TIPS

 $\cup$ means **U**nion
$\cap$ means **in**tersection

Throughout this example we have been looking at the letters of the alphabet. That is the **universal set** for this example.

We write $\mathscr{E}$ = {letters of the alphabet}. The elements of **M**, **L**, **S**, and **A** are all taken from $\mathscr{E}$.

EXERCISE 6B

1 Describe these sets in words.

    a {Monday, Tuesday, Wednesday...}   **b** {1, 3, 5, 7...}

    c {Mercury, Venus, Earth, Mars...}   **d** {North, South, East, West}

    e {1, 2, 3, 4, 5, 6}

2 List the elements of the following sets.

    a A = {integer x: $5 < x < 10$}   **b** B = {factors of 12}

    c C = {prime numbers between 20 and 30}   **d** D = { x: $x^2 = 9$}

In questions **3** to **5**:

    $\mathscr{E}$ = {numbers on a clock face}

    E = {even numbers}, O = {odd numbers}, T = {multiples of 3}, P = {prime numbers}

3 Say whether these statements are true or false.

    a $3 \in E$ **b** $3 \in O$ **c** $3 \in T$ **d** $3 \in P$

    e $5 \notin E$ **f** $5 \notin O$ **g** $5 \notin T$ **h** $5 \notin P$

4 List the elements of the following sets:

    a $E \cap T$ **b** $E \cap P$ **c** $E \cap O$ **d** $O \cap T$

5 List the elements of:

    a $T \cup P$ **b** $T \cup E$ **c** $E \cup O$

In questions **6** to **8**:

    I = {letters in the word india}

    E = {letters in the word europe}

    A = {letters in the word america}

6 Suggest a universal set for **I**, **E**, and **A**.

7 List the elements of:

    a $I \cap E$ **b** $I \cap A$ **c** $A \cap E$

8 How many elements are there in:

    a $E \cup I$ **b** $E \cup A$ **c** $I \cup A$

9 **X** = {1, 3, 4, 6, 8}   **X** $\cup$ **Y** = {1, 2, 3, 4, 5, 6, 7, 8}   **X** $\cap$ **Y** = {4, 8}

List the elements of **Y**.

Suppose $\mathcal{E}$ = {1, 2, 3, 4, 5, 6, 7, 8, 9}.

A = {x: x is even}

B = {x: $x \leqslant 5$)

We can show these sets in a **Venn diagram**.

Elements in $A \cap B$ are placed where the two circles overlap.

$A \cap B$ = {2, 4}

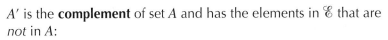

Each element of the universal set is in A, in B, in both, or in neither.

A' is the **complement** of set A and has the elements in $\mathcal{E}$ that are *not* in A:

A' = {1, 3, 5, 7, 9}

$A \cup B$ = {1, 2, 3, 4, 5, 6, 8}

$(A \cup B)'$ is the complement of $A \cup B$ and contains elements which are *not* in A and B togethe

$(A \cup B)'$ = {7, 9}

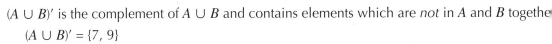

We can also use Venn diagrams with three sets.

Suppose we add C = {x: $3 \leqslant x \leqslant 7$} to our example.

C = {3, 4, 5, 6, 7}

$A \cup B$ = {1, 2, 3, 4, 5, 6, 8}

$(A \cup B) \cap C$ = {3, 4, 5, 6}

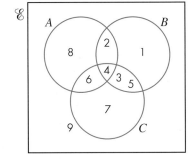

We can illustrate this by shading:

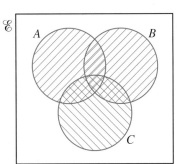

$A \cup B$

C

$(A \cup B) \cap C$

EXERCISE 6C

1

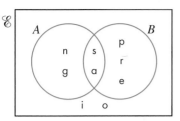

a List the elements of:

   i A **ii** B **iii** $A \cap B$ **iv** $A \cup B$ **v** A'

   vi $(A \cup B)'$ **vii** $A' \cap B$ **viii** $A \cap B'$

b $\mathcal{E}$ = {letters in the name of a city}. What is the city?

2

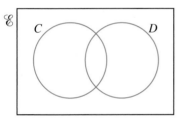

$\mathcal{E}$ = {f, r, a, c, t, i, o, n, s}
C = {r, a, t, i, o}
D = {f, i, r, s, t}

a Copy the diagram and put the elements in the correct places.

b List the elements of:

   i $(C \cup D)'$ **ii** C' **iii** $C' \cup D$

3 $\mathcal{E}$ = {integer x: $1 \leqslant x \leqslant 12$}
A = {multiples of 2}
B = {multiples of 3}

a Show these sets on a Venn diagram.

b Describe set $A \cap B$.

FOUNDATION

4

a List the elements of these sets:

i A

ii B

iii $A \cap B$

iv $A \cap C$

v $B \cup C$

vi C'

vii $A \cap B \cap C$

viii $(A \cup B) \cap C$

ix $(B \cup C)'$

x $(B \cup C)' \cap A$

b Complete these descriptions:

i $\mathscr{E} = \{$integer $x: ... \leqslant x \leqslant ...\}$

ii $A = \{$factors of $...\}$

iii $C = \{...$numbers$\}$

5 On copies of this Venn diagram shade:

a $A \cap B$

b $A' \cap B$

c $(A \cap B)'$

d $(A \cup B)'$

e $A' \cup B'$

f $A' \cap B'$

6 Which of your answers in question 5 are identical?

7 On copies of this Venn diagram shade:

a $(A \cup B) \cap C$

b $A' \cap (B \cup C)$

c $(A \cup B) \cap (A \cup C)$

d $(A \cup B \cup C)'$

e $(B \cup C)' \cap A$

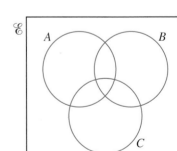

Suppose $R = \{10, 12, 14, 16, 18, 20, 22, 24\}$ and $S = \{9, 12, 15, 18, 21, 24\}$ and $T = \{10, 20\}$

$n(A)$ means "the number of elements in set A".

Check that $n(R) = 8$, $n(T) = 2$ and $n(R \cap S) = 3$

If all the elements of set X are in set Y we say that X is a **subset** of Y.

Check that $T \subset R$ and $\{9, 15, 21\} \subset S$ and $\{10\} \subset T$.

EXERCISE 6D

1

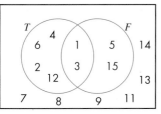

Write down:

a $n(T)$ **b** $n(F)$ **c** $n(\xi)$

d $n(T \cap F)$ **e** $n(T \cup F)$ **f** $n(T')$

g $n((T \cup F)')$ **h** $n((T \cap F)')$

2 In this Venn diagram

$\mathcal{E} = \{\text{positive integers}\}$

$P = \{\text{prime numbers}\}$

$E = \{\text{even numbers}\}$

$S = \{\text{square numbers}\}$

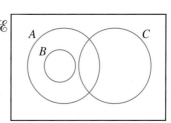

a Explain why $n(P \cap E) = 1$.

b Explain why $P \cap S = \varnothing$.

c Describe the elements of $E' \cap S$.

d Write down a number x such that $x \notin P$ and $x \notin E$ and $x \notin S$.

3 Look at this Venn diagram.

Say whether these statements are true or false:

a $A \subset B$ **b** $B \subset A$

c $A \cap B = B$ **d** $A \cup B = B$

e $B \cap C = \varnothing$ **f** $B \cup C = \mathcal{E}$

4

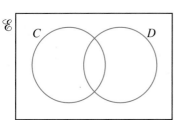

$n(A) = 50$ $n(B) = 40$ $n(A \cap B) = 27$

What is $n(A \cup B)$?

5 *A* and *B* are two sets. Write as simply as possible:

a $A \cup A'$ **b** $B \cap B'$ **c** $((A \cup B)')'$ **d** $(A \cap B) \cap B$ **e** $(A \cup B)$

6 $\mathscr{E} = \{x: x \text{ is a positive integer}\}$
$E = \{x: x \text{ is a multiple of 2}\}$
$T = \{x: x \text{ is a multiple of 3}\}$
$F = \{x: x \text{ is a multiple of 5}\}$

a Draw a Venn diagram and place these numbers in it: 10, 11, 12, 13, 14, 15.

b Complete this description:
$E \cap T = \{x: x \text{ is a multiple of } .....\}$

c Write a description of $T \cap F$.

d Find an element of $E \cap T \cap F$.

e Write a description of $E \cap T \cap F$.

6.5 Practical problems

We can use Venn diagrams to help solve practical problems.

EXAMPLE 3

There are 20 students in a class.

12 play football. 14 swim regularly. 8 do both.

How many do neither?

Draw a Venn diagram with sets for football (F) and swimming (S).

We know $n(\mathscr{E}) = 20$

$n(F) = 12$

$n(S) = 14$

$n(F \cap S) = 8$

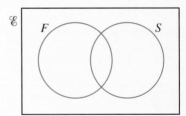

We can use this information to write the number of students in each region:

2 students do neither.

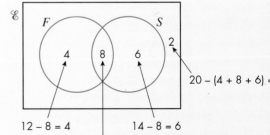

$12 - 8 = 4$ $14 - 8 = 6$
Start with 8

$20 - (4 + 8 + 6) =$

EXERCISE 6E

HIGHER

1 A group of people were asked if they had visited the theatre (*T*) or the cinema (*C*) recently. The numbers are shown in the Venn diagram.

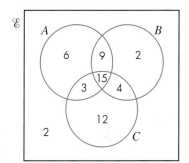

   a How many had been to the theatre?

   b How many had been to the cinema but not the theatre?

   c How many had done neither?

   d How many were in the survey?

2 A panel were asked to try three new flavours of fruit drink (*A*, *B*, and *C*). The Venn diagram shows how many people liked each one.

   a How many liked *A*?

   b How many liked *B* and *C*?

   c How many liked two drinks but not the third one?

   d How many did not like *B*?

3 24 people were asked if they enjoyed two TV programmes, *X* and *Y*.

10 said they enjoyed *A*, 17 enjoyed *B*, and 8 enjoyed both.

   a Show this information on a Venn diagram

   b How many people enjoyed neither?

4 In a group of people, 15 wear glasses and there are 7 females.

5 males and 4 females do not wear glasses.

How many people are there in the group?

5 A group of 30 people were asked if they liked jazz or classical music.

12 said jazz, 17 said classical and 8 said neither.

How many said both?

6 A class of 28 students are all learning at least one language out of English, Chinese and Russian.

   5 are learning English and Chinese

   3 are learning Russian and Chinese

   7 are learning English and Russian

   2 are learning all three languages.

How many are learning just one language?

Why this chapter matters

We use ratio, proportion and speed in our everyday lives to help us compare two or more pieces of information.

Ratio and proportions are often used to compare sizes; speed is used to compare distances with the time taken to travel them.

A 100-m sprinter

Speed

When is a speed fast?

On 16 August 2009 Usain Bolt set a new world record for the 100-m sprint of 9.58 seconds. This is an average speed of 37.6 km/h.

The sailfish is the fastest fish and can swim at 110 km/h.

The cheetah is the fastest land animal and can travel at 121 km/h.

The fastest bird is the swift which can travel at 170 km/h.

Sailfish

Cheetah

Swift

Ratio and proportion facts

- Russia is the largest country.
- Vatican City is the smallest country.
- The area of Russia is nearly 39 million times the area of Vatican City.
- Monaco has the most people per square mile.
- Mongolia has the least people per square mile.

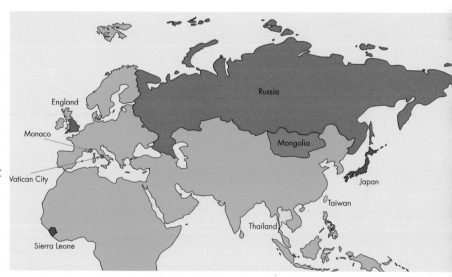

- The number of people per square mile in Monaco to the number of people in Mongolia is in the ratio 10 800 : 1.
- Japan has the highest life expectancy.
- Sierra Leone has the lowest life expectancy.
- On average people in Japan live over twice as long as people in Sierra Leone.
- Taiwan has the most mobile phones per 100 people (106.5).
- This is approximately four times that of Thailand (26.04).

7

Ratio, proportion and speed

Topics	Level	Key words
1 Ratio	**FOUNDATION**	ratio, cancel, simplest form, map scale
2 Speed	**FOUNDATION**	average, speed, distance, time
3 Density and pressure	**FOUNDATION**	density, pressure, Pascals
4 Direct proportion	**FOUNDATION**	unitary method, direct proportion, single unit value
5 Proportional variables	**FOUNDATION**	multiplier

What you need to be able to do in the examinations:

FOUNDATION

- Use ratio notation, including reduction to its simplest form and its various links to fraction notation.
- Divide a quantity in a given ratio or ratios.
- Use compound measures such as speed, density and pressure.
- Use the process of proportionality to evaluate unknown quantities.
- Calculate an unknown quantity from quantities that vary in direct proportion.
- Solve word problems about ratio and proportion.
- Understand and use the relationship between average speed, distance and time.
- Use and interpret maps.

A **ratio** is a way of comparing the sizes of two or more quantities.

A ratio can be expressed in a number of ways. For example, if Tasnim is five years old and Ziad is 20 years old, the ratio of their ages is:

Tasnim's age : Ziad's age

which is:	5 : 20
which simplifies to:	1 : 4 (dividing both sides by 5)

A ratio is usually given in one of these three ways.

Tasnim's age : Ziad's age	or	5 : 20	or	1 : 4
Tasnim's age to Ziad's age	or	5 to 20	or	1 to 4
$\dfrac{\text{Tasnim's age}}{\text{Ziad's age}}$	or	$\dfrac{5}{20}$	or	$\dfrac{1}{4}$

Common units

When working with a ratio involving different units, *always convert them to the same units*. A ratio can be simplified only when the units of each quantity are the *same*, because the ratio itself has no units. Once the units are the same, the ratio can be simplified or **cancelled** like a fraction.

For example, the ratio of 125 g to 2 kg must be converted to the ratio of 125 g to 2000 g, so that you can simplify it.

125 : 2000

Divide both sides by 25:	5 : 80
Divide both sides by 5:	1 : 16

The ratio 125 : 2000 can be simplified to 1 : 16.

EXAMPLE 1

Express 25 minutes : 1 hour as a ratio in its simplest form.

The units must be the same, so write 1 hour as 60 minutes.

25 minutes : 1 hour	=	25 minutes : 60 minutes	
	=	25 : 60	Cancel the units (minutes)
	=	5 : 12	Divide both sides by 5

So, 25 minutes : 1 hour simplifies to 5 : 12

Ratios as fractions

A ratio in its **simplest form** can be expressed as portions of a quantity by expressing the whole numbers in the ratio as fractions with the same denominator (bottom number).

XAMPLE 2

A garden is divided into 5 equal parts: 3 parts lawn and 2 parts shrubs.

We say it is divided into lawn and shrubs in the ratio 3 : 2.

What fraction of the garden is covered by **a** lawn, **b** shrubs?

The denominator (bottom number) of the fraction comes from adding the numbers in the ratio (that is, $2 + 3 = 5$).

a the lawn covers $\frac{3}{5}$ of the garden

b and the shrubs cover $\frac{2}{5}$ of the garden.

EXERCISE 7A

1 Express each of the following ratios in its simplest form.

    a 6 : 18   **b** 5 : 20   **c** 16 : 24   **d** 24 : 12

    e 20 : 50   **f** 12 : 30   **g** 25 : 40   **h** 150 : 30

2 Write each of the following ratios of quantities in its simplest form. (Remember to always express both parts in a common unit before you simplify.)

    a 40 minutes : 5 minutes   **b** 3 kg : 250 g

    c 50 minutes to 1 hour   **d** 1 hour to 1 day

    e 12 cm to 2.5 mm   **f** 1.25 kg : 500 g

    g 75 cents : $2   **h** 400 m: 2 km

3 A length of wood is cut into two pieces in the ratio 3 : 7

What fraction of the original length is the longer piece?

4 Tareq and Hassan find a bag of marbles that they share between them in the ratio of their ages. Tareq is 10 years old and Hassan is 15 years old. What fraction of the marbles did Tareq get?

5 Mona and Petra share a pizza in the ratio 2 : 3

They eat it all.

    a What fraction of the pizza did Mona eat?

    b What fraction of the pizza did Petra eat?

FOUNDATION

6 A camp site allocates space to caravans and tents in the ratio 7 : 3. What fraction of the total space is given to:

a the caravans

b the tents?

7 In a safari park at feeding time, the elephants, the lions and the chimpanzees are given food in the ratio 10 : 7 : 3. What fraction of the total food is given to:

a the elephants

b the lions

c the chimpanzees?

8 Paula wins $\frac{3}{4}$ of her tennis matches. She loses the rest.

What is the ratio of wins to losses?

9 Three brothers share some cash.

The ratio of Marco's and Dani's share is 1 : 2

The ratio of Dani's and Paulo's share is 1 : 2

What is the ratio of Marco's share to Paulo's share?

Dividing amounts in a given ratio

To divide an amount in a given ratio, you first look at the ratio to see how many parts there are altogether.

For example, 4 : 3 has 4 parts and 3 parts giving 7 parts altogether.

7 parts is the whole amount.

1 part can then be found by dividing the whole amount by 7.

3 parts and 4 parts can then be worked out from 1 part.

EXAMPLE 3

Divide $28 in the ratio 4 : 3

4 + 3 = 7 parts altogether

So 7 parts = $28

Dividing by 7:

1 part = $4

4 parts = 4 × $4 = $16 and 3 parts = 3 × $4 = $12

So $28 divided in the ratio 4 : 3 = $16 : $12

Map scales

Map scales are often given as ratios in the form $1 : n$

XAMPLE 4

A map of New Zealand has a scale of $1 : 900\,000$.

The distance on the map from Auckland to Hamilton is 11.5 centimetres.

What is the actual distance?

1 cm on the map = 900 000 centimetres on the ground.

$\qquad$ = 9000 metres (100 centimetres = 1 metre)

$\qquad$ = 9 kilometres (1000 metres = 1 kilometre).

The distance is 11.5 × 9 kilometres

$\qquad$ = 103.5 kilometres.

EXERCISE 7B

1 Divide the following amounts according to the given ratios.

a 400 g in the ratio 2 : 3   **b** 280 kg in the ratio 2 : 5

c 500 in the ratio 3 : 7   **d** 1 km in the ratio 19 : 1

e 5 hours in the ratio 7 : 5   **f** $100 in the ratio 2 : 3 : 5

g $240 in the ratio 3 : 5 : 12   **h** 600 g in the ratio 1 : 5 : 6

i $5 in the ratio 7 : 10 : 8   **j** 200 kg in the ratio 15 : 9 : 1

2 The ratio of female to male members of a sports club is 7 : 3
The total number of members of the group is 250.

a How many members are female?

b What percentage of members are male?

3 A store sells small and large TV sets.

The ratio of small : large = 2 : 3

The total stock is 70 sets.

a How many small sets are in stock?

b How many large sets?

4 The Illinois Department of Health reported that, for the years 1981 to 1992 when they tested a total of 357 horses for rabies, the ratio of horses with rabies to those without was 1 : 16

How many of these horses had rabies?

FOUNDATION

5 Joshua, Aicha and Mariam invest $10 000 in a company.

The ratio of the amount they invest is:

Joshua : Aicha : Mariam = 5 : 7 : 8

How much does each of them invest?

6 Rewrite the following scales as ratios in the form 1 : n

a 1 cm to 4 km

b 4 cm to 5 km

c 2 cm to 5 km

d 4 cm to 1 km

e 5 cm to 1 km

f 2.5 cm to 1 km

g 8 cm to 5 km

h 10 cm to 1 km

i 5 cm to 3 km

7 A map has a scale of 1 cm to 10 km.

a Rewrite the scale as a ratio in its simplest form.

b What is the actual length of a lake that is 4.7 cm long on the map?

c How long will a road be on the map if its actual length is 8 km?

> **HINTS AND TIPS**
>
> 1 km = 1000 m
>   = 100 000 cm

8 A map has a scale of 2 cm to 5 km.

a Rewrite the scale as a ratio in its simplest form.

b How long is a path that measures 0.8 cm on the map?

c How long should a 12 km road be on the map?

9 The scale of a map is 5 cm to 1 km.

a Rewrite the scale as a ratio in the form 1 : n.

b How long is a wall that is shown as 2.7 cm on the map?

c The distance between two points is 8 km. How far will this be on the map?

10 You can simplify a ratio by writing it in the form 1 : n. For example, 5 : 7 can be rewritten

$$\frac{5}{5} : \frac{7}{5} = 1 : 1.4$$

Rewrite each of the following ratios in the form 1 : n

a 5 : 8

b 4 : 13

c 8 : 9

d 25 : 36

e 5 : 27

f 12 : 18

g 5 hours : 1 day

h 4 hours : 1 week

i £4 : £5

Calculating with ratios when only part of the information is known

XAMPLE 5

A fruit drink is made by mixing orange squash with water in the ratio 2 : 3
How much water needs to be added to 5 litres of orange squash to make the drink?

2 parts is 5 litres.

Dividing by 2:

1 part is 2.5 litres

3 parts = 2.5 litres × 3 = 7.5 litres

So 7.5 litres of water is needed to make the drink.

XAMPLE 6

Two business partners, Lubna and Adama, divided their total profit in the ratio 3 : 5
Lubna received $2100. How much did Adama get?

Lubna's $2100 was $\frac{3}{8}$ of the total profit. (Check that you know why.)
$\frac{1}{8}$ of the total profit = $2100 ÷ 3 = $700

So Adama's share, which was $\frac{5}{8}$, amounted to $700 × 5 = $3500.

EXERCISE 7C

1. Sean, aged 15, and Ricki, aged 10, shared some sweets in the same ratio as their ages. Sean had 48 sweets.

   a Simplify the ratio of their ages.

   b How many sweets did Ricki have?

   c How many sweets did they share altogether?

2. A blend of tea is made by mixing Lapsang with Assam in the ratio 3 : 5. I have a lot of Assam tea but only 600 g of Lapsang. How much Assam do I need to make the blend using all the Lapsang?

3. The ratio of male to female spectators at a hockey game is 4 : 5. 4500 men watched the match. What was the total attendance at the game?

4. A teacher always arranged the content of each of his lessons as 'teaching' and 'practising skills' in the ratio 2 : 3

   a If a lesson lasted 35 minutes, how much teaching would he do?

   b If he decided to teach for 30 minutes, how long would the lesson be?

FOUNDATION

5 A 'good' children's book has pictures and text in the ratio 17 : 8
In a book I have just looked at, the pictures occupy 23 pages.

a Approximately how many pages of text should this book have to be a 'good' children's book?

b What percentage of a 'good' children's book will be text?

6 Three business partners, Ren, Shota and Fatima, put money into a business in the ratio 3 : 4 : 5. They shared any profits in the same ratio. Last year, Fatima made $3400 out o the profits. How much did Ren and Shota make last year?

7 **a** Iqra is making a drink from lemonade, orange and ginger ale in the ratio 40 : 9 : 1
If Iqra has only 4.5 litres of orange, how much of the other two ingredients does she need to make the drink?

b Another drink made from lemonade, orange and ginger ale uses the ratio 10 : 2 : 1

Which drink has a larger proportion of ginger ale, Iqra's or this one? Show how you work out your answer.

8 There is a group of boys and girls waiting for school buses. 25 girls get on the first bus. The ratio of boys to girls at the stop is now 3 : 2. 15 boys get on the second bus. There now the same number of boys and girls at the bus stop. How many students altogether were originally at the bus stop?

9 A jar contains 100 cm^3 of a mixture of oil and water in the ratio 1 : 4. Enough oil is ad to make the ratio of oil to water 1 : 2. How much water must be added to make the ra of oil to water 1 : 3?

7.2 Speed

The relationship between **speed, time** and **distance** can be expressed in three ways:

$$\text{speed} = \frac{\text{distance}}{\text{time}} \qquad \text{distance} = \text{speed} \times \text{time} \qquad \text{time} = \frac{\text{distance}}{\text{speed}}$$

In problems relating to speed, you usually mean **average** speed, as it would be unusual to maintain one constant speed for the whole of a journey.

This diagram will help you remember the relationships between distance (D), time (T) and speed (S).

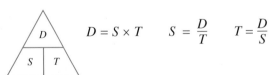

$$D = S \times T \qquad S = \frac{D}{T} \qquad T = \frac{D}{S}$$

Units for speed include km/h (kilometres per hour, or 'the number of kilometres travelled in hour') and m/s (metres per second).

XAMPLE 7

Paula drove a distance of 270 kilometres in 5 hours. What was her average speed?

Paula's average speed $= \dfrac{\text{distance she drove}}{\text{time she took}} = \dfrac{270}{5} = 54$ kilometres per hour (km/h)

XAMPLE 8

Renata drove to Frankfurt in $3\frac{1}{2}$ hours at an average speed of 60 km/h.
How far is it to Frankfurt?

Since:

distance = speed × time

the distance to Frankfurt is given by:

60 × 3.5 = 210 kilometres

Note: You need to change the time to a decimal number and use 3.5 (*not 3.30*).

XAMPLE 9

Maria is going to drive to Rome, a distance of 190 kilometres. She estimates that she will drive at an average speed of 50 km/h. How long will it take her?

Maria's time $= \dfrac{\text{distance she covers}}{\text{her average speed}} = \dfrac{190}{50} = 3.8$ hours

Write the 0.8 hour as minutes by multiplying by 60, to give 48 minutes.

So, the time for Maria's journey will be 3 hours 48 minutes.

Remember: When you calculate a time and get a decimal answer, as in Example 9, *do not mistake* the decimal part for minutes. You must either:

● leave the time as a decimal number and give the unit as hours, or

● write the decimal part as minutes by multiplying it by 60 (1 hour = 60 minutes) and give the answer in hours and minutes.

EXERCISE 7D

1 A cyclist travels a distance of 90 kilometres in 5 hours.

What was her average speed?

2 How far along a road would you travel if you drove at 110 km/h for 4 hours?

HINTS AND TIPS

Remember to convert time to a decimal if you are using a calculator, for example, 8 hours 30 minutes is 8.5 hours.

FOUNDATION

3 I drive to see my aunt in about 6 hours. The distance is 315 kilometres.

What is my average speed?

4 The distance from Leeds to London is 350 kilometres.

The train travels at an average speed of 150 km/h.

If I catch the 9.30 am train in London, at what time should I expect to arrive in Leeds?

HINTS AND TIPS

km/h means kilometres per hour.
m/s means metres per second.

5 How long will an athlete take to run 2000 metres at an average speed of 4 metres per second?

6 Copy and complete the following table.

	Distance travelled	Time taken	Average speed
a	150 km	2 hours	
b	260 km		40 km/h
c		5 hours	35 km/h
d		3 hours	80 km/h
e	544 km	8 hours 30 minutes	
f		3 hours 15 minutes	100 km/h
g	215 km		50 km/h

7 Eliot drove a distance of 660 kilometres, in 7 hours 45 minutes.

　a Write the time 7 hours 45 minutes as a decimal.

　b What was the average speed of the journey? Round your answer to 1 decimal place

8 Johan drives home from his son's house in 2 hours 15 minutes. He says that he drives a an average speed of 70 km/h.

　a Write the 2 hours 15 minutes as a decimal.

　b How far is it from Johan's home to his son's house?

9 The distance between Paris and Le Mans is 200 km. The express train between Paris ar Le Mans travels at an average speed of 160 km/h.

　a Calculate the time taken for the journey from Paris to Le Mans, giving your answer decimal number of hours.

　b Write your answer to part **a** as hours and minutes.

10 The distance between two cities is 420 kilometres.

　a What is the average speed of a journey from one to the other that takes 8 hours 45 minutes?

　b If Sam covered the distance at an average speed of 63 km/h, how long would it take

11 A train travels at 50 km/h for 2 hours, then slows down to do the last 30 minutes of its journey at 40 km/h.

    a What is the total distance of this journey?

    b What is the average speed of the train over the whole journey?

12 Suni runs and walks the 6 kilometres from home to work each day. She runs the first 4 kilometres at a speed of 16 km/h, then walks the next 2 kilometres at a steady 8 km/h.

    a How long does it take Suni to get to work?

    b What is her average speed?

> **HINTS AND TIPS**
>
> **Remember** that there are 3600 seconds in an hour and 1000 metres in a kilometre. So to change from km/h to m/s multiply by 1000 and divide by 3600.

13 Write the following speeds as metres per second.

    a 36 km/h   **b** 12 km/h

    c 60 km/h   **d** 150 km/h

    e 75 km/h

14 Write the following speeds as kilometres per hour.

    a 25 m/s   **b** 12 m/s

    c 4 m/s   **d** 30 m/s

    e 0.5 m/s

> **HINTS AND TIPS**
>
> To change from m/s to km/h multiply by 3600 and divide by 1000.

15 A train travels at an average speed of 18 m/s.

    a Express its average speed in km/h.

    b Find the approximate time the train would take to travel 500 m.

    c The train set off at 7.30 on a 40 km journey. At approximately what time will it reach its destination?

> **HINTS AND TIPS**
>
> To convert a decimal fraction of an hour to minutes, just multiply by 60.

16 A cyclist is travelling at an average speed of 24 km/h.

    a What is this speed in metres per second?

    b What distance does he travel in 2 hours 45 minutes?

    c How long does it take him to travel 2 km?

    d How far does he travel in 20 seconds?

17 How much longer does it take to travel 100 kilometres at 65 km/h than at 70 km/h?

Density

Density is the mass of a substance per unit of volume, usually expressed in grams per cm³. The relationship between the three quantities is:

$$\text{density} = \frac{\text{mass}}{\text{volume}}$$

You can remember this with a triangle similar to that for distance, speed and time.

$M = DV$ mass = density × volume

$D = \dfrac{M}{V}$ density = mass ÷ volume

$V = \dfrac{M}{D}$ volume = mass ÷ density

EXAMPLE 10

A piece of metal has a mass of 30 g and a volume of 4 cm³. What is the density of the metal?

$$\text{Density} = \frac{\text{mass}}{\text{volume}}$$
$$= \frac{30}{4}$$
$$= 7.5 \text{ g/cm}^3$$

EXAMPLE 11

What is the mass of a piece of rock that has a volume of 34 cm³ and a density of 2.25 g/cm³?

$$\text{mass} = \text{density} \times \text{volume}$$
$$= 2.25 \times 34$$
$$= 76.5 \text{ g}$$

Pressure

When you put air in the tyre of a car, you inflate it to a particular **pressure**. As you put in mo air the air pressure increases and the tyre feels harder.

The force of the compressed air in the tyre is spread over an area of the surface of the tyre.

When a force is spread over an area we define the pressure as the force divided by the area.

If the force is in Newtons and the area is in m² then the pressure is in **Pascals** (Pa).

The relationship between pressure (P), force (F) and area (A) is:

$$\text{pressure} = \frac{\text{force}}{\text{area}}$$

You can remember this with a triangle similar to that for distance, speed and time.

$P = \dfrac{F}{A}$ pressure = force ÷ area

$F = PA$ force = pressure × area

$A = \dfrac{F}{P}$ area = force ÷ pressure

EXAMPLE 12

A force of 200 Newtons is applied to a flat surface.
Calculate the pressure if the flat surface is:

a a square of side 50 cm   **b** a square of side 5 cm

a You need to know the area in m². 50 cm = 0.5 m

The area of the square is 0.5 × 0.5 = 0.25 m²

$\text{Pressure} = \dfrac{\text{Force}}{\text{Area}} = \dfrac{200}{0.25} = 800 \text{ Pa}$

b 5 cm = 0.05 m so the area is 0.05 × 0.05 = 0.0025 m²

$\text{Pressure} = \dfrac{\text{Force}}{\text{Area}} = \dfrac{200}{0.0025} = 80\,000 \text{ Pa}$

Notice that the smaller the area is, the larger the pressure will be.

As an example of pressure, think about pushing your thumb onto a piece of wood. Not much happens.

Now use your thumb to push a drawing pin onto the wood. The drawing pin will penetrate the wood.

The force is the same but the area it is applied to is much smaller so the pressure applied is much greater.

Note: If an object has a mass of x kg, then it exerts a downward force, due to gravity of xg newtons, where g is the acceleration due to gravity.

Under normal conditions, $g = 9.81$ m/s² so a mass of 1 kg exerts a force of 9.81 newtons. Usually, to make calculations easier, you take g as 10 m/s².

EXAMPLE 13

a When does a woman exert the greater pressure on the floor: when she is wearing walking boots or high-heeled shoes?

Explain your answer.

b A woman has a mass of 50 kg. She is wearing a pair of high-heeled shoes. Each shoe has an area of 40 cm² for the sole and 1 cm² for the heel.

Take $g = 10$ m/s².

i When she is standing on both shoes with the heel down, what is the average pressure exerted on the ground?

ii She swivels round on the heel of one shoe only. How much pressure, in pascals, is exerted on the ground?

You are given that 1 cm² = 0.0001 m².

a A woman exerts more pressure on the floor when she is wearing high-heeled shoes as they have a much smaller contact area with the floor than walking boots.

b i Force is $50 \times 10 = 500$ N

Area is 82×0.0001 m^2 = 0.0082 m^2

pressure = force ÷ area

So average pressure is $500 \div 0.0082 \approx 61\,000$ Pa.

ii On one heel, the pressure is $500 \div 0.0001 = 5\,000\,000$ Pa.

EXERCISE 7E

FOUNDATION

1 Find the density of a piece of wood with a mass of 6 g and a volume of 8 cm^3.

2 A force of 20 N acts over an area of 5 m^2. What is the pressure?

3 Calculate the density of a metal if 12 cm^3 of it has a mass of 100 g.

4 A pressure of 5 Pa acts on an area of $\frac{1}{2}$ m^2. What force is exerted?

5 Calculate the mass of a piece of plastic, 20 cm^3 in volume, if its density is 1.6 g/cm^3.

6 A crate weighs 200 N and exerts a pressure of 40 Pa on the ground. What is its area?

7 Calculate the volume of a piece of wood which has a mass of 102 g and a density of 0.85 g/cm^3.

8 Find the mass of a marble model, 56 cm^3 in volume, if the density of marble is 2.8 g/cr

9 Two statues look identical and both appear to be made out of gold. One of them is a fa

The density of gold is 19.3 g/cm^3.

The statues each have a volume of approximately 200 cm^3.

The first statue has a mass of 5.2 kg.

The second statue has a mass of 3.8 kg.

Which one is the fake?

10 A piece of metal has a mass of 345 g and a volume of 15 cm^3.

A different piece of metal has a mass of 400 g and a density of 25 g/cm^3.

Which piece of metal has the bigger volume and by how much?

11 Two pieces of scrap metal are melted down to make a single piece of metal.

The first piece has a mass of 1.5 tonnes and a density of 7000 kg/m^3.

The second piece has a mass of 1 tonne and a density of 8000 kg/m^3.

Work out the total volume of the new piece of metal.

Suppose you buy 12 items which each cost the *same*. The total amount you spend is 12 times the cost of one item.

That is, the total cost is said to be in **direct proportion** to the number of items bought. The cost of a single item (the unit cost) is the constant factor that links the two quantities.

Direct proportion is not only concerned with costs. Any two related quantities can be in direct proportion to each other.

The best way to solve all problems involving direct proportion is to start by finding the **single unit value**. This method is called the **unitary method**, because it refers to a single unit value. Work through Examples 14 and 15 to see how it is done.

Remember: Before solving a direct proportion problem, think about it carefully to make sure that you know how to find the required single unit value.

EXAMPLE 14

If eight pens cost $2.64, what is the cost of five pens?

First, find the cost of one pen. This is $2.64 ÷ 8 = $0.33

So, the cost of five pens is $0.33 × 5 = $1.65

EXAMPLE 15

Eight loaves of bread will make packed lunches for 18 people. How many packed lunches can be made from 20 loaves?

First, find how many lunches one loaf will make.

One loaf will make 18 ÷ 8 = 2.25 lunches.

So, 20 loaves will make 2.25 × 20 = 45 lunches.

EXERCISE 7F

1 If 30 matches have a mass of 45 g, what is the mass of 40 matches?

2 Five bars of chocolate cost $2.90. Find the cost of nine bars.

3 Eight men can chop down 18 trees in a day. How many trees can 20 men chop down in a day?

4 Find the cost of 48 eggs when 15 eggs can be bought for $2.10.

FOUNDATION

5 Seventy maths textbooks cost $875.

  a How much will 25 maths textbooks cost?

  b How many maths textbooks can you buy for $100?

6 A lorry uses 80 litres of fuel on a trip of 280 kilometres.

  a How much fuel would the same lorry use on a trip of 196 kilometres?

  b How far would the lorry get on a full tank of 100 litres of fuel?

7 During the winter, I find that 200 kg of coal keeps my fire burning for 12 weeks.

  a If I want a fire all through the winter (18 weeks), how much coal will I need to buy?

  b Last year I bought 150 kg of coal. For how many weeks did I have a fire?

8 It takes a photocopier 16 seconds to produce 12 copies. How long will it take to produce 30 copies?

9 A recipe for 12 biscuits uses:

200 g margarine	400 g sugar
500 g flour	300 g ground rice

  a What quantities are needed for:

  i 6 biscuits   **ii** 9 biscuits   **iii** 15 biscuits?

  b What is the maximum number of biscuits I could make if I had just 1 kg of each ingredient?

10 Peter the baker sells bread rolls in packs of 6 for $2.30

  Paul the baker sells bread rolls in packs of 10 for $3.50

  I have $10 to spend on bread rolls.

  If I want to buy as many bread rolls as possible from one shop, which shop should I use? Show your working.

7.5 Proportional variables

Suppose a and b are two variables in direct proportion.

Here is a table of values.

a	5	12	y
b	8	x	36

We can find the missing values x and y by finding a **multiplier** m.

$5 \times m = 8$

$\Rightarrow \quad m = 8 \div 5 = 1.6$

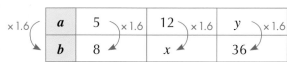

$x = 12 \times 1.6 = 19.2$

$y \times 1.6 = 36 \quad \Rightarrow \quad y = 36 \div 1.6 = 22.5$

EXERCISE 7G

1 s varies directly with t.

s	12	20
t	30	

Calculate the missing value.

2 p varies directly with q.

p	4	5	9
q	24		

Calculate the two missing values.

3 a and b are in direct proportion.

a	2.8	3.1	4.4
b	19.6		

Calculate the two missing values.

4 x and y are in direct proportion.

x	8	10	12	
y	12			24

Find the missing values.

5 The length (L) and the mass (M) of metal wire are in direct proportion.

L	150	200
M	270	

Find the missing value of M.

6

r	1.6	2.2	3.4
s	5.6	7.7	15.3

Does s vary directly with r?

Give a reason for your answer.

7

c	d
12.0	4.8
	6.0
32.8	

d varies directly with c.

Calculate the missing values.

8 The cost of electricity (C) is proportional to the number of units used.

U	C
720	432
1200	
	1200

Find the missing values in this table.

FOUNDATION

How accurate are we?

In real life it is not always sensible to use exact values. Sometimes it would be impossible to have exact measurements. People often round values without realising it. Rounding is done so that values are sensible.

Is it exactly 23 km to Utrecht and exactly 54 km to Amsterdam?

Does this box contain exactly 750 g of rice when full?

Was her time exactly 13.4 seconds?

Does the school have exactly 1500 stude

Imagine if people tried to use exact values all the time. Would life seem strange?

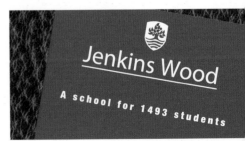

8

Approximation and limits of accuracy

Rounding whole numbers   **FOUNDATION**   approximation, rounded up, rounded down

Rounding decimals   **FOUNDATION**   round, digit, decimal place

Rounding to significant figures   **FOUNDATION**   significant figure

Approximation of calculations   **FOUNDATION**

Upper and lower bounds   **FOUNDATION**   upper bound, lower bound, limits of accuracy

Upper and lower bounds for calculations   **HIGHER**

hat you need to be able to do in the examinations:

FOUNDATION	HIGHER
Round integers to a given power of 10.	• Solve problems using upper and lower bounds where values are given to a degree of accuracy.
Round to a given number of significant figures or decimal places.	
Identify upper and lower bounds where values are given to a degree of accuracy.	
Use estimation to evaluate approximations to numerical calculations.	

You use rounded information, or **approximations** all the time. Look at the examples on the right. Each actual figure is either above or below the approximation shown here.

How do you round numbers up or down?

If you want to round a number to the nearest multiple of ten, you round it up if it ends in 5 or above, and round it down if it ends in less than 5. For example:

- 25, 26, 27, 28 and 29 are **rounded up** to 30.

- 24, 23, 22 and 21 are **rounded down** to 20.

So a box with approximately 30 matches could contain any number from 25 to 34.

You can round numbers to the nearest multiple of 10, 100, 1000 and so on. The number of runners (23 000) in the report on the marathon is rounded to the nearest 1000:

- the smallest number of people actually running would be 22 500 (22 500–22 999 are rounded up to 23 000).

- the largest number of people running would be 23 499 (23 500 would be rounded up to 24 000).

So, there could actually be from 22 500 to 23 499 people in the marathon.

EXERCISE 8A

1 Round each of these numbers to the nearest 10.

| a | 24 | b | 57 | c | 78 | d | 54 | e | 96 |
| f | 21 | g | 88 | h | 66 | i | 14 | j | 26 |

2 Round each of these numbers to the nearest 100.

| a | 240 | b | 570 | c | 780 | d | 504 | e | 967 |
| f | 112 | g | 645 | h | 358 | i | 998 | j | 1050 |

3 Round each of these numbers to the nearest 1000.

| a | 2400 | b | 5700 | c | 7806 | d | 5040 | e | 9670 |
| f | 1120 | g | 6450 | h | 3499 | i | 9098 | j | 1500 |

4

Welcome to Elsecar	Welcome to Hoyland	Welcome to Jump
Population 800 (to the nearest 100)	**Population 1200** (to the nearest 100)	**Population 600** (to the nearest 100)

Which of these sentences could be true and which must be false?

a There are 789 people living in Elsecar.　**b** There are 1278 people living in Hoyland.

c There are 550 people living in Jump.　**d** There are 843 people living in Elsecar.

e There are 1205 people living in Hoyland.　**f** There are 650 people living in Jump.

5 Here is the average attendance in four football leagues in 2008–9:

England Premier League　　35 600

Germany Bundesliga　　　　42 565

Italy Serie A　　　　　　　25 303

Spain La Liga　　　　　　　29 124

a Which were the highest and lowest?

b Round each number to the nearest thousand.

c The figure for Ligue 1 in France to the nearest thousand was 25 000. What were the largest and smallest actual values for Ligue 1?

6 Matthew and Viki are playing a game with whole numbers.

a What is the smallest number Matthew could be thinking of?

I am thinking of a number. Rounded to the nearest 10 it is 380.

I am thinking of a different number. Rounded to the nearest 100 it is 400.

b Matthew's number is the smallest possible. How many possible values are there for Vicki's number?

7 The number of adults attending a comedy show is 80 to the nearest 10.

The number of children attending is 50 to the nearest 10.

Katie says that 130 adults and children attended the comedy show.

Give an example to show that she may **not** be correct.

Rounding decimals

Decimal places

When a number is written in decimal form, the **digits** to the right of the decimal point are cal
decimal places. For example:

79.4 is written 'with one decimal place'

6.83 is written 'with two decimal places'

0.526 is written 'with three decimal places'.

To **round** a decimal number to a particular number of decimal places, take these steps:

● count along the decimal places from the decimal point and look at the first digit to be removed.

● when the value of this digit is less than five, just remove the unwanted places.

● when the value of this digit is five or more, add 1 onto the digit in the last decimal place t remove the unwanted places.

Here are some examples.

5.852 rounds to 5.85 to two decimal places

7.156 rounds to 7.16 to two decimal places

0.274 rounds to 0.3 to one decimal place

15.3518 rounds to 15.4 to one decimal place

EXERCISE 8B

1 Round each of the following numbers to one decimal place.

a 4.83   **b** 3.79   **c** 2.16   **d** 8.25

e 3.673   **f** 46.935   **g** 23.883   **h** 9.549

i 11.08   **j** 33.509

> **HINTS AND TIPS**
>
> Just look at the value of the digit in the second decimal place.

2 Round each of the following numbers to two decimal places.

a 5.783   **b** 2.358   **c** 0.977   **d** 33.085   **e** 6.007

f 23.5652   **g** 91.7895   **h** 7.995   **i** 2.3076   **j** 23.915

3 Round each of the following to the number of decimal places (dp) indicated.

a 4.568 (1 dp)   **b** 0.0832 (2 dp)   **c** 45.715 93 (3 dp)   **d** 94.8531 (2

e 602.099 (1 dp)   **f** 671.7629 (2 dp)   **g** 7.1124 (1 dp)   **h** 6.903 54 (3

i 13.7809 (2 dp)   **j** 0.075 11 (1 dp)

4 Round each of the following to the nearest whole number.

a 7.82   **b** 3.19   **c** 7.55   **d** 6.172   **e** 3.961

f 7.388   **g** 1.514   **h** 46.78   **i** 23.19   **j** 96.45

5 Anna puts the following items in her shopping basket: bread $3.20, meat $8.95, cheese $6.16 and butter $3.90

By rounding each price to the nearest dollar, work out an estimate of the total cost of the items.

6 Which of the following are correct roundings of the number 3.456?

3   3.0   3.4   3.40   3.45   3.46   3.47   3.5   3.50

7 When a number is rounded to three decimal places the answer is 4.728

Which of these could be the number?

4.71   4.7275   4.7282   4.73

8.3 Rounding to significant figures

We often use **significant figures** (sf) when we want to round a number with a lot of digits in it. We often use this technique with calculator answers.

Look at this table which shows some numbers *rounded* to one, two and three significant figures.

One sf	8	50	200	90 000	0.000 07	0.003	0.4
Two sf	67	4.8	0.76	45 000	730	0.0067	0.40
Three sf	312	65.9	40.3	0.0761	7.05	0.003 01	0.400

The steps taken to round a number to a given number of significant figures are very similar to those used for rounding to a given number of decimal places:

- from the left, count the digits. If you are rounding to 2 sf, count two digits, for 3 sf count three digits, and so on. When the original number is less than 1, start counting from the first non-zero digit.

- look at the next digit to the right. When the value of this next digit is less than 5, leave the digit you counted to the same. However if the value of this next digit is equal to or greater than 5, add 1 to the digit you counted to.

- ignore all the other digits, but put in enough zeros to keep the number the right size (value).

For example, look at the following table, which shows some numbers rounded to 1, 2 and 3 significant figures, respectively.

Number	Rounded to 1 sf	Rounded to 2 sf	Rounded to 3 sf
45 281	50 000	45 000	45 300
568.54	600	570	569
7.3782	7	7.4	7.38
8054	8000	8100	8050
99.8721	100	100	99.9
0.7002	0.7	0.70	0.700

FOUNDATION

1 Round each of the following numbers to 1 significant figure.

   a 46 313   **b** 57 123   **c** 30 569   **d** 94 558   **e** 85 299

   f 0.5388   **g** 0.2823   **h** 0.005 84   **i** 0.047 85   **j** 0.000 876

   k 9.9   **l** 89.5   **m** 90.78   **n** 199   **o** 999.99

2 Round each of the following numbers to 2 significant figures.

   a 56 147   **b** 26 813   **c** 79 611   **d** 30 578   **e** 14 009

   f 1.689   **g** 4.0854   **h** 2.658   **i** 8.0089   **j** 41.564

   k 0.8006   **l** 0.458   **m** 0.0658   **n** 0.9996   **o** 0.009 82

3 Round each of the following to the number of significant figures (sf) indicated.

   a 57 402 (1 sf)   **b** 5288 (2 sf)   **c** 89.67 (3 sf)

   d 105.6 (2 sf)   **e** 8.69 (1 sf)   **f** 1.087 (2 sf)

   g 0.261 (1 sf)   **h** 0.732 (1 sf)   **i** 0.42 (1 sf)

   j 0.758 (1 sf)   **k** 0.185 (1 sf)   **l** 0.682 (1 sf)

4 What are the lowest and the highest numbers of sweets that can be found in these jars?

   a   **b**   **c**

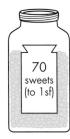
70 sweets (to 1 sf)

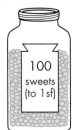
100 sweets (to 1 sf)

1000 sweets (to 1 sf)

5 What are the least and the greatest numbers of people that live in these towns?

   Satora   population 800 (to 1 significant figure)

   Nimral   population 1200 (to 2 significant figures)

   Korput   population 165 000 (to 3 significant figures)

6 There are 500 fish in a pond, to 1 sf. What is the least possible number of fish that could be taken from the pond so that there are 400 fish in the pond to 1 sf?

7 Rani says that the population of Bikran is 132 000 to the nearest thousand. Vashti says that the population of Bikran is 130 000. Explain why Vashti could also be correct.

Approximation of calculations

How could you find an approximate value of a calculation, such as 35.1 × 6.58?

One way is to round each number to 1 significant figure, and then complete the calculation. The approximation is:

$35.1 \times 6.58 \approx 40 \times 7 = 280$

Note the symbol ≈ which means 'approximately equal to'.

For the division 89.1 ÷ 2.98, the approximate answer is 90 ÷ 3 = 30.

Sometimes when dividing it can be sensible to round to 2 sf instead of 1 sf. For example,

24.3 ÷ 3.87 using 24 ÷ 4 gives an approximate answer of 6

whereas

24.3 ÷ 3.87 using 20 ÷ 4 gives an approximate answer of 5.

Both of these are sensible answers, but generally rounding to one significant figure is easier.

Finding an approximate value is always a great help in any calculation since it often stops you giving a silly answer.

EXERCISE 8D

1 Find approximate answers to the following.

    a 5435 × 7.31   **b** 5280 × 3.211   **c** 63.24 × 3.514 × 4.2

    d 3508 × 2.79   **e** 72.1 × 3.225 × 5.23   **f** 470 × 7.85 × 0.99

    g 354 ÷ 79.8   **h** 36.8 ÷ 1.876   **i** 5974 ÷ 5.29

    Check your answers on a calculator to see how close you were.

2 Find the approximate monthly pay of the following people whose annual salaries are given.

    a Paul $35 200   **b** Michael $25 600   **c** Jennifer $18 125   **d** Ross $8420

3 Find the approximate annual pay of the following people who earn:

    a Kevin $270 a week   **b** Malcolm $1528 a month   **c** David $347 a week

4 A farmer bought 2713 kg of seed at a cost of $7.34 per kg. Find the approximate total cost of this seed.

5 By rounding, find an approximate answer to each of the following.

    a $\dfrac{573 + 783}{107}$ **b** $\dfrac{783 - 572}{24}$ **c** $\dfrac{352 + 657}{999}$ **d** $\dfrac{1123 - 689}{354}$

    e $\dfrac{589 + 773}{658 - 351}$ **f** $\dfrac{793 - 569}{998 - 667}$ **g** $\dfrac{352 + 657}{997 - 656}$ **h** $\dfrac{1123 - 689}{355 + 570}$

    i $\dfrac{28.3 \times 19.5}{97.4}$ **j** $\dfrac{78.3 \times 22.6}{3.69}$ **k** $\dfrac{3.52 \times 7.95}{15.9}$ **l** $\dfrac{11.78 \times 77.8}{39.4}$

FOUNDATION

6 Find the approximate answer to each of the following.

 a $208 \div 0.378$ **b** $96 \div 0.48$ **c** $53.9 \div 0.58$

 d $14.74 \div 0.285$ **e** $28.7 \div 0.621$ **f** $406.9 \div 0.783$

Check your answers on a calculator to see how close you were.

7 A litre of paint will cover an area of about 8.7 m². Approximately how many litre cans will I need to buy to paint a room with a total surface area of 73 m²?

8 By rounding, find the approximate answer to each of the following.

 a $\dfrac{84.7 + 12.6}{0.483}$ **b** $\dfrac{32.8 \times 71.4}{0.812}$ **c** $\dfrac{34.9 - 27.9}{0.691}$ **d** $\dfrac{12.7 \times 38.9}{0.42}$

9 Kirsty arranges for magazines to be put into envelopes. She sorts out 178 magazines between 10.00 am and 1.00 pm. Approximately how many magazines will she be able to sort in a week in which she works for 17 hours?

10 An athlete runs 3.75 km every day. Approximately how far does he run in:

 a a week   **b** a month   **c** a year?

11 1 kg = 1000 g

A box full of magazines has a mass of 8 kg. One magazine has a mass of about 15 g. Approximately how many magazines are there in the box?

12 An apple has a mass of about 280 grams.

 a What is the approximate mass of a bag containing a dozen apples?

 b Approximately how many apples will there be in a sack with a mass of 50 kg?

8.5 Upper and lower bounds

Any recorded measurements have usually been rounded.

The true value will be somewhere between the **lower bound** and the **upper bound**.

The lower and upper bounds are sometimes known as the **limits of accuracy**.

A journey that is measured as 26 kilometres to the nearest kilometre could be anything between 25.5 and 26.5 kilometres:

 25.5 would round up to 26

 26.5 would round up to 27 but anything less would round down to 26

We say that 26.5 is the upper bound and 25.5 is the lower bound.

We can write 25.5 ⩽ distance < 26.5 which means that the distance is greater than or equal to 25.5 kilometres but less than 26.5 kilometres.

EXAMPLE 1

A stick of wood measures 32 cm, to the nearest centimetre.

What are the lower and upper bounds of the actual length of the stick?

The lower bound is 31.5 cm as this is the lowest value that rounds to 32 cm to the nearest centimetre.

The upper bound is 32.5 cm as anything lower rounds to 32 cm to the nearest centimetre. 32.5 cm would round to 33 cm.

We write:

   31.5 ⩽ length of stick in cm < 32.5

Note the use of < for the upper bound.

EXAMPLE 2

A time of 53.7 seconds is accurate to 1 decimal place.

What are the upper and lower bounds for the time?

The lower bound is 53.65 seconds and the upper bound is 53.75 seconds.

So 53.65 ⩽ time in seconds < 53.75

EXAMPLE 3

The number of people at a football match is 32 000 to the nearest thousand.

What are the upper and lower bounds of the size of the crowd?

In this example we are counting (people) not measuring, and the values can only be whole numbers.

● The lower bound is 31 500. This would round up to 32 000 to the nearest thousand. One less, 31 499, would round down to 31 000.

● The upper bound is 32 499. This would round down to 32 000. One more, 32 500, would round up to 33 000.

FOUNDATION

1 Write down the upper and lower bounds of the following.

a A length measured as 7 cm to the nearest cm.

b A mass measured as 120 g to the nearest 10 g.

c A length measured as 3400 km to the nearest 100 km.

d A speed measured as 50 km/h to the nearest km/h.

e An amount given as $6 to the nearest dollar.

f A length given as 16.8 cm to the nearest tenth of a centimetre.

g The number of people at a rally is 76 000 to the nearest thousand.

h A football crowd of 14 500 to the nearest 100.

i The number of votes for an election candidate is 29 000 to the nearest thousand.

j The population of Saudi Arabia is 24 000 000 to the nearest million.

2 Write down the upper and lower bounds for each of the following values, which are rounded to the given degree of accuracy. Use inequalities to show your answer. For example, part **a** should be 5.5 ⩽ length in cm < 6.5.

a	6 cm (1 significant figure)	**b**	17 kg (2 significant figures)
c	32 min (2 significant figures)	**d**	238 km (3 significant figures)
e	7.3 m (1 decimal place)	**f**	25.8 kg (1 decimal place)
g	3.4 h (1 decimal place)	**h**	87 g (2 significant figures)
i	4.23 mm (2 decimal places)	**j**	2.19 kg (2 decimal places)
k	12.67 min (2 decimal places)	**l**	25 m (2 significant figures)
m	40 cm (1 significant figure)	**n**	600 g (2 significant figures)
o	30 min (1 significant figure)	**p**	1000 m (2 significant figures)
q	4.0 m (1 decimal place)	**r**	7.04 kg (2 decimal places)
s	12.0 s (1 decimal place)	**t**	7.00 m (2 decimal places)

3 Write down the lower and upper bounds of each of these values, rounded to the accuracy stated.

a	8 m (1 significant figure)	**b**	26 kg (2 significant figures)
c	25 min (2 significant figures)	**d**	85 g (2 significant figures)
e	2.40 m (2 decimal places)	**f**	0.2 kg (1 decimal place)
g	0.06 s (2 decimal places)	**h**	300 g (1 significant figure)
i	0.7 m (1 decimal place)	**j**	366 d (3 significant figures)
k	170 weeks (2 significant figures)	**l**	210 g (2 significant figures)

4 A chain is 30 m long, to the nearest metre.

A chain is needed to fasten a boat to a harbour wall. The distance to the wall is also 30 m, to the nearest metre.

Which statement is definitely true? Explain your decision.

    A: The chain will be long enough.

    B: The chain will not be long enough.

    C: It is impossible to tell whether or not the chain is long enough.

5 A bag contains 2.5 kg of soil, to the nearest 100 g.

What is the least amount of soil in the bag?

Give your answer in kilograms and grams.

6 Chang has 40 identical marbles. Each marble has a mass of 65 g (to the nearest gram).

    a What is the greatest possible mass of one marble?

    b What is the least possible mass of one marble?

    c What is the greatest possible mass of all the marbles?

    d What is the least possible mass of all the marbles?

8.6 Upper and lower bounds for calculations   H

When rounded values are used for a calculation, we can find upper and lower bounds for the result of the calculation.

EXAMPLE 4

The dimensions of this rectangle are given to the nearest centimetre.

Find the lower bound for the perimeter and the upper bound for the area.

27 cm

21 cm

The upper and lower bounds for the sides are:

$$26.5 \leqslant \text{length in cm} < 27.5$$
$$\text{and} \quad 20.5 \leqslant \text{width in cm} < 21.5$$

Perimeter = 2 × (length + width)

The lower bound will be found using the lower bounds of the length and width.

Lower bound of perimeter = 2 × (26.5 + 20.5) = 94 cm

Area = length × width

The upper bound will be found using the upper bounds of the length and width.

Upper bound of area = 27.5 × 21.5 = 591.25 cm^2

EXAMPLE 5

A car travels 125 km (to the nearest km) and uses 16.1 litres of fuel (correct to one decimal place).

Find the upper and lower bounds of the fuel consumption in km/litre.

$124.5 \leqslant$ distance in kilometres < 125.5

$16.05 \leqslant$ fuel in litres < 16.15

The fuel consumption is distance ÷ fuel used.

To find the upper bound of this, use the *upper* bound of the distance ÷ *lower* bound of the fuel used:

upper bound is $125.5 \div 16.05 = 7.8193\ldots\ldots$

Lower bound of fuel consumption = *lower* bound of distance ÷ *upper* bound of fuel used:

$= 124.5 \div 16.15 = 7.7089\ldots\ldots$

So $7.709 \leqslant$ fuel consumption in km/litre < 7.819 when the answers are rounded off to three decimal places.

When solving a problem, write down the upper and lower bounds for the values given and then decide which to use to find the solution.

EXERCISE 8F

1 Boxes have a mass of 7 kg, to the nearest kilogram.

What are the upper and lower bounds for the total mass of 10 of these boxes?

2 Books each have a mass of 1200 g, to the nearest 100 g.

a What is the greatest possible mass of 10 books?
Give your answer in kilograms.

b A trolley can safely hold up to 25 kg of books.
How many books can safely be put on the trolley?

3 Jasmine says, "I am 45 kilos." Yolander says, "I am 53 kilos." Both are measured to the nearest kilogram.

What is the greatest possible difference between their masses?

Show how you worked out your answer.

4 For each of these rectangles, find the upper and lower bounds for the perimeter.
The measurements are shown to the level of accuracy indicated in brackets.

a 5 cm × 9 cm (nearest cm) **b** 4.5 cm × 8.4 cm (1 decimal place)

c 7.8 cm × 18 cm (2 significant figures)

5 Calculate the upper and lower bounds for the areas of each rectangle in question **4**.

6 A cinema screen is measured as 6 m by 15 m, to the nearest metre. Calculate the upper and lower bounds for the area of the screen.

7 The measurements, to the nearest centimetre, of a box are given as 10 cm × 7 cm × 4 cm. Calculate the upper and lower bounds for the volume of the box.

8 Mr Sparks is an electrician. He has a 50-m roll of cable, correct to the nearest metre. He uses 10 m on each job, to the nearest metre.

If he does four jobs, what is the maximum amount of cable he could have left?

9 Jon and Matt are exactly 7 kilometres apart. They are walking towards each other. Jon is walking at 4 km/h and Matt is walking at 2 km/h.

Both speeds are given to the nearest kilometre per hour.

Without doing any time calculations, decide whether it is possible for them to meet in 1 hour. Justify your answer.

10 The area of a rectangular field is given as 350 m^2, to the nearest 10 m^2. One length is given as 16 m, to the nearest metre. Find the upper and lower bounds for the other length of the field.

11 A stopwatch records the time for the winner of a 100-metre race as 14.7 seconds, measured to the nearest one-tenth of a second.

   a What are the upper and lower bounds of the winner's time?

   b The length of the 100-metre track is correct to the nearest 1 m. What are the upper and lower bounds of the length of the track?

   c What is the fastest possible average speed of the winner, with a time of 14.7 seconds in the 100-metre race?

12 A model car travels 40 m, measured to one significant figure, at a speed of 2 m/s, measured to one significant figure. Find the upper and lower bounds of the time taken.

13 The population of Japan is 127 000 000 to the nearest million.

The area of Japan is 378 000 km^2 to the nearest 1000 km^2.

The population density for any country is the total population divided by the area.

Find the upper and lower bounds for the population density of Japan in people/km^2. Round off your answers to two decimal places.

14 An engineer testing a car's CO_2 (carbon dioxide) emissions measures 26 kg of CO_2 when it is driven 150 km.

The mass is given to the nearest kg.

The distance is given to the nearest km.

Find the upper and lower bounds for the CO_2 emissions in grams/kilometre. Round off your answers to one decimal place.

Why this chapter matters

Very large and very small numbers can often be difficult to read. Scientists use standard form as a shorthand way of representing numbers.

The planets

Mercury is the closest planet to the Sun (and is very hot). It orbits 60 million km (6×10^7 km) away from the Sun.

Earth takes 365 days to orbit the Sun and 24 hours to complete a rotation.

Jupiter is made of gas. It has no solid land so visiting it is not recommended! It has a huge storm which rages across its surface. This is about 8 km high, 40 000 km long and 14 000 km wide. It looks like a red spot and is called 'the Great Red Spot'.

Uranus takes 84 days to orbit the Sun.

Pluto is the furthest planet from the Sun. Some astronomers dispute whether it can be classed as a planet. The average surface temperature on Pluto is about −230 °C.

Venus rotates the opposite way to the other planets and has a diameter of 12 100 km (1.21×10^4 km).

Mars has the largest volcano in the solar system. It is almost 600 km across and rises 24 km above the surface. This is five times bigger than the biggest volcano on Earth.

Saturn is the largest planet in the solar system. It is about 120 000 km across (1.2×10^5 km) and 1400 million km from the Sun (1.4×10^9 km).

Neptune is similar to Jupiter in that it is a gas planet and has violent storms. Winds can blow at up to 2000 km per hour, so a cloud can circle Neptune in about 16 hours.

The mass of an electron is about 0.000 000 000 000 000 000 000 000 000 000 91 kg.
This is written 9.1×10^{-31} kg.

The mass of the Earth is about 5 970 000 000 000 000 000 000 000 kg.
This is written 5.97×10^{24} kg.

Standard form

Topics	Level	Key words
1 Standard form	**FOUNDATION**	standard form, index
2 Calculating with standard form	**FOUNDATION**	
3 Solving problems	**HIGHER**	

What you need to be able to do in the examinations:

FOUNDATION	HIGHER
• Express numbers in the form $a \times 10^n$ where n is an integer and $1 \leqslant a < 10$. • Calculate with numbers in standard form.	• Solve problems involving standard form.

Powers of ten:

$$100 = 10 \times 10 = 10^2$$
$$1000 = 10 \times 10 \times 10 = 10^3$$

Extending this idea:

$$10\,000 = 10 \times 10 \times 10 \times 10 = 10^4$$
$$100\,000 = 10^5$$
$$1\,000\,000 = 10^6$$

and so on.

The power of 10 is called the **index**.

Standard form is a way of writing very large and very small numbers using powers of 10. In thi form, a number is given a value between 1 and 10 multiplied by a power of 10. That is,

$a \times 10^n$ where $1 \leqslant a < 10$, and n is a whole number.

Look at these examples to see how numbers are written in standard form.

$$52 = \qquad 5.2 \times 10 = \mathbf{5.2 \times 10^1}$$
$$73 = \qquad 7.3 \times 10 = \mathbf{7.3 \times 10^1}$$
$$625 = \quad 6.25 \times 100 = \mathbf{6.25 \times 10^2} \qquad \text{The numbers in bold are in standard form.}$$
$$389 = \quad 3.89 \times 100 = \mathbf{3.89 \times 10^2}$$
$$3147 = 3.147 \times 1000 = \mathbf{3.147 \times 10^3}$$

When writing a number in this way, you must always follow two rules.

- The first part must be a number between 1 and 10 (1 is allowed but 10 isn't).
- The second part must be a whole-number (negative or positive) power of 10. Note that you would *not normally* write the power 1.

You need to be able to use and manipulate the numbers in standard form both with a calculato and without one.

Standard form on a calculator

A number such as $123\,000\,000\,000$ is obviously difficult to key into a calculator. Instead, you enter it in standard form (assuming you are using a scientific calculator):

$$123\,000\,000\,000 = 1.23 \times 10^{11}$$

The key strokes to enter this into your calculator could be something like this:

1 **.** **2** **3** **×10ˣ** **1** **1**

Your calculator display will display the number either as an ordinary number, if there is enoug space, or in standard form. Make sure you know how to use standard form on your calculator.

Standard form of numbers less than 1

We use a negative index for numbers between 0 and 1:

$$0.1 = 10^{-1}$$
$$0.01 = 10^{-2}$$
$$0.001 = 10^{-3}$$
$$0.0001 = 10^{-4}$$

and so on.

For example:

$$0.000\,729 = 7.29 \times 0.0001$$
$$= 7.29 \times 10^{-4} \text{ in standard form}$$

These numbers are written in standard form. Make sure that you understand how they are formed.

a $0.4 = 4 \times 10^{-1}$ **b** $0.05 = 5 \times 10^{-2}$ **c** $0.007 = 7 \times 10^{-3}$

d $0.123 = 1.23 \times 10^{-1}$ **e** $0.007\,65 = 7.65 \times 10^{-3}$ **f** $0.9804 = 9.804 \times 10^{-1}$

g $0.0098 = 9.8 \times 10^{-3}$ **h** $0.000\,0078 = 7.8 \times 10^{-6}$

On a calculator you would enter 1.23×10^{-6}, for example, as:

Try entering some of the numbers in **a** to **h** (above) into your calculator for practice.

EXERCISE 9A

Do this exercise without a calculator.

1 These numbers are in standard form. Write them out in full.

     a 2.5×10^2 **b** 3.45×10 **c** 4.67×10^{-3} **d** 3.46×10

     e 2.0789×10^{-2} **f** 5.678×10^3 **g** 2.46×10^2 **h** 7.6×10^3

     i 8.97×10^5 **j** 8.65×10^{-3} **k** 6×10^7 **l** 5.67×10^{-4}

2 Write these numbers in standard form.

     a 250   **b** 0.345   **c** 46 700

     d 3 400 000 000   **e** 20 780 000 000   **f** 0.000 567 8

     g 2460   **h** 0.076   **i** 0.000 76

     j 0.0006   **k** 0.005 67   **l** 56.0045

In questions **3** to **5**, write the numbers given in each statement in standard form.

3 The population of India in 2003 was 1 065 000 000.

4 The total land area of Asia is 45 040 000 square kilometres.

FOUNDATION

5 The asteroid *Phaethon* comes within 12 980 000 miles of the Sun. The asteroid *Pholus*, at its furthest point, is a distance of 2997 million miles from the Earth. The closest an asteroid ever came to Earth was 93 000 miles from the planet.

6 How many times bigger is 3.2×10^6 than 3.2×10^4?

7 Here are the distances of some planets from the Sun:

Jupiter   778 million kilometres

Mercury   58 million kilometres

Pluto   5920 million kilometres

Write these distances in standard form.

8 Here are some facts about a bacterium:

Width 0.000 001 2 metres

Mass 0.000 000 000 000 95 grams

Write these numbers in standard form.

9.2 Calculating with standard form

Calculations involving very large or very small numbers can be done more easily using standard form.

You can enter numbers in a scientific calculator in standard form. This is done in different ways with different models. Make sure you know how to do this with your calculator.

EXAMPLE 1

A pixel on a computer screen is 2×10^{-2} cm long by 7×10^{-3} cm wide.

What is the area of the pixel?

Give your answer in standard form.

The area is given by length times width.

$$\text{Area in cm}^2 = 2 \times 10^{-2} \times 7 \times 10^{-3}$$
$$= 1.4 \times 10^{-4} \text{ in standard form.}$$

When you use a calculator you can enter the numbers directly without any rearranging. Your calculator may give you the answer in standard form.

EXERCISE 9B

Do questions 1, 2 and 3 without a calculator.

1 Write these numbers in standard form.

a 56.7×10^2

b 0.06×10^4

c 34.6×10^{-2}

d 0.07×10^{-2}

e 56×10

f $2 \times 3 \times 10^5$

g $2 \times 10^2 \times 35$

h 23 million

2 Work out the following. Give your answers in standard form.

a $2 \times 10^4 \times 5.4 \times 10^3$

b $1.6 \times 10^2 \times 3 \times 10^4$

c $2 \times 10^4 \times 6 \times 10^4$

d $2 \times 10^{-4} \times 5.4 \times 10^3$

e $1.6 \times 10^{-2} \times 4 \times 10^4$

f $2 \times 10^4 \times 6 \times 10^{-4}$

g $(5 \times 10^3)^2$

h $(2 \times 10^{-2})^3$

3 Work out the following. Give your answers in standard form.

a $(5.4 \times 10^4) \div (2 \times 10^3)$

b $(4.8 \times 10^2) \div (3 \times 10^4)$

c $(1.2 \times 10^4) \div (6 \times 10^4)$

d $(2 \times 10^{-4}) \div (5 \times 10^3)$

e $(1.8 \times 10^4) \div (9 \times 10^{-2})$

f ○ d $\sqrt{(36 \times 10^{-4})}$

4 A typical adult has about 20 000 000 000 000 red blood cells. Each red blood cell has a mass of about 0.000 000 000 1 g. Write both of these numbers in standard form and work out the total mass of red blood cells in a typical adult.

5 The Moon is a sphere with a radius of 1.74×10^3 kilometres. The formula for working out the surface area of a sphere is:

surface area $= 4\pi r^2$

Calculate the surface area of the Moon.

6 Evaluate $\dfrac{E}{M}$ when $E = 1.5 \times 10^3$ and $M = 3 \times 10^{-2}$, giving your answer in standard form.

7 Work out the value of $\dfrac{3.2 \times 10^7}{1.4 \times 10^2}$ giving your answer in standard form, correct to 2 significant figures.

8 A number is greater than 100 million and less than 1000 million.

Write down a possible value of the number, in standard form.

Problems in astronomy can use very large numbers. Problems in science can use very small numbers. It is better to use very large and very small numbers in standard form.

EXAMPLE 2

The distance from the Sun to the Earth is 150 million km.

Light travels at 3.00×10^5 km/second.

How long does light from the Sun take to reach the Earth?

Distance = 150 000 000 = 1.5×10^8 km

Time = distance ÷ speed

$= (1.5 \times 10^8) \div (3.00 \times 10^5)$

$= 500$ seconds (use a calculator to do this)

$= 8$ minutes 20 seconds

EXERCISE 9C

1 A man puts one grain of rice on the first square of a chess board, two on the second square, four on the third, eight on the fourth and so on.

a How many grains of rice will he put on the 64th square of the board?

b How many grains of rice will there be altogether?

Give your answers in standard form.

> **HINTS AND TIPS**
>
> Compare powers of 2 with the running totals. By the fourth square you have 8 grains altogether, and $2^3 = 8$.

2 The surface area of the Earth is approximately 3.2×10^8 square kilometres. The area of the Earth's surface that is covered by water is approximately 2.2×10^8 square kilometres.

a Calculate the area of the Earth's surface *not* covered by water. Give your answer in standard form.

b What percentage of the Earth's surface is not covered by water?

3 In 2009, British Airways carried 33 million passengers. Of these, 70% passed through Heathrow Airport. On average, each passenger carried 19.7 kg of luggage. Calculate the total mass of the luggage carried by these passengers.

4 In 2009 the world population was approximately 6.77×10^9. In 2010 the world population was approximately 6.85×10^9.

a By how much did the population rise? Give your answer as an ordinary number.

b What was the percentage increase?

5 Here are four numbers written in standard form.

$$1.6 \times 10^4 \qquad 4.8 \times 10^6 \qquad 3.2 \times 10^2 \qquad 6.4 \times 10^3$$

  a Work out the smallest answer when two of these numbers are multiplied together.

  b Work out the largest answer when two of these numbers are added together.
  Give your answers in standard form.

6 The mass of Saturn is 5.686×10^{26} tonnes. The mass of the Earth is 6.04×10^{21} tonnes. How many times heavier is Saturn than the Earth? Give your answer in standard form to a suitable degree of accuracy.

7 Here are the populations of some countries.

  a Which country has the largest population?

  b Which two countries have a very similar population size?

  c Find the total population of Senegal, Denmark and Jamaica, in standard form to two significant figures.

Country	Population
Tunisia	9.83×10^6
Denmark	5.36×10^6
Senegal	1.01×10^7
Jamaica	2.65×10^6
Mexico	1.03×10^8
India	1.07×10^9

  d Complete this sentence:

  The population of Mexico is approximately … times larger than the population of Denmark.

  e Complete this sentence:

  The population of India is approximately … times larger than the population of Jamaica.

8 This table shows the populations and the areas of five different countries.

Country	Population	Area
Russian Federation	1.43×10^8	1.71×10^7
Sri Lanka	1.91×10^7	6.56×10^4
Thailand	6.28×10^7	5.13×10^5
Togo	4.91×10^6	5.68×10^5
Iran	6.89×10^7	1.65×10^6

  a Which country has the smallest population?

  b Which country has the smallest area?

The population density is the population divided by the area.

  c Which country has the largest population density?

  d Which country has the smallest population density?

  e What fraction of the area of the Russian Federation is the area of Sri Lanka?
  Give your answer in the form $\frac{1}{N}$.

Technology is increasingly important in our lives. It helps us do many things more efficiently than we could without it.

Hundreds of years ago people in different countries had different systems of measurement. They were often based on the length of people's hands, arms or feet, but they all varied and all had different names.

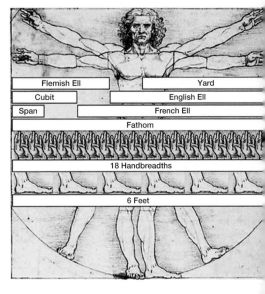

Now the world has an official standard system of measurement – the metric system. This is especially important for scientists so they can work together worldwide. It is also helpful for everyone who needs to compare lengths, masses, volumes and so on between different countries.

We also have more help now in calculating complicated measurements like volume. Calculating aids have been used for thousands of years. In about 2000 BCE the abacus was being used in Egypt and China.

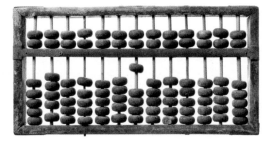

Abacuses are still widely used in China today and they were used everywhere for almost 3500 years, until John Napier devised a calculating aid called *Napier's bones*.

Napier's bones

These led to the invention of the *slide rule* by William Oughtred in 1622. This was in use until the mid-1960s. Engineers working on the first ever moon landings used slide rules to do some of their calculations.

The first *electronic computers* were produced in the mid-20th century. When the transistor was invented, the power increased and the cost and size decreased until the point where the average scientific calculator that students use in schools has more computing power than the first craft that went into space.

Applying number and using calculators

pics		Level	Key words
1	Units of measurement	FOUNDATION	metric system, length, mass, volume, capacity
2	Converting between metric units	FOUNDATION	centimetre, millimetre, metre, kilometre, gram, kilogram, tonne, litre, millilitre, centilitre
3	Reading scales	FOUNDATION	scales, division, units
4	Time	FOUNDATION	24-hour clock, 12-hour clock, timetable
5	Currency conversions	FOUNDATION	exchange rate
6	Using a calculator efficiently	FOUNDATION	

What you need to be able to do in the examinations:

FOUNDATION

- Use and apply number in everyday personal, domestic or community life.
- Carry out calculations using standard units of mass, length, area, volume and capacity.
- Convert measurements within the metric system to include linear and area units.
- Convert between units of volume within the metric system.
- Understand and carry out calculations using time.
- Calculate time intervals in terms of the 24-hour and 12-hour clock.
- Interpret scales on a range of measuring instruments.
- Carry out calculations using money, including converting between currencies.
- Use a scientific electronic calculator to determine numerical results.

The **metric system** of measurement is now in use in most countries of the world, except for the United States. Here is a list of the most common metric units.

Unit	How to estimate it
Length	
1 metre	A long stride for an average person
1 kilometre	Two and a half times round a school track
1 centimetre	The distance across a fingernail
Mass	
1 gram	A small coin has a mass of a few grams
1 kilogram	A bag of sugar
1 tonne	A saloon car
Volume/Capacity	
1 litre	A full carton of orange juice
1 centilitre	A small glass is about 10 centilitres
1 millilitre	A full teaspoon is about 5 millilitres

Volume and capacity

The term 'capacity' is normally used to refer to the volume of a liquid or a gas.

For example, when referring to the volume of petrol that a car's fuel tank will hold, people may say its capacity is 60 litres.

EXERCISE 10A

1 Decide which metric unit you would be most likely to use to measure each of the following

   a The height of your classroom
   b The distance from Athens to Vienna
   c The thickness of your little finger
   d The mass of this book
   e The amount of water in a fish tank
   f The mass of an aircraft
   g A spoonful of medicine
   h The length of a football pitch
   i The mass of your head teacher
   j The thickness of a piece of wire

2 Estimate the approximate metric length, mass or capacity of each of the following.

   a This book (both length and mass)
   b The length of your school hall
   c The capacity of a bucket
   d The diameter of a coin, and its mass
   e The mass of a cat
   f The amount of water in one raindrop
   g The dimensions of the room you are in
   h Your own height and mass

3 Bob was asked to put up some decorative bunting from the top of each lamp post in his street. He had three sets of ladders he could use: a 2 metre, a 3.5 metre and a 5 metre ladder.

He looked at the lamp posts and estimated that they were about three times his height. He is slightly below average height for an adult male.

Which of the ladders should he use? Give a reason for your choice.

10.2 Converting between metric units

You should already know the relationships between these metric units.

Length		Mass	
10 **millimetres** = 1 **centimetre**		1000 **grams** = 1 **kilogram**	
1000 millimetres = 100 centimetres		1000 kilograms = 1 **tonne**	
= 1 **metre**			
1000 metres = 1 **kilometre**			
Capacity		**Volume**	
10 **millilitres** = 1 **centilitre**		1000 litres = 1 metre^3	
1000 millilitres = 100 centilitres		1 millilitre = 1 centimetre^3	
= 1 **litre**			

Note the equivalence between the units of capacity and volume:

1 litre = 1000 cm^3 which means   1 ml = 1 cm^3

You need to be able to convert from one metric unit to another.

Since the metric system is based on powers of 10, you should be able to multiply or divide easily to change units. Work through the following examples.

EXAMPLE 1

To convert small units to larger units, always divide.

Convert:

a *732 cm to metres*
732 ÷ 100 = 7.32 m

b *840 mm to metres*
840 ÷ 1000 = 0.84 m

EXAMPLE 2

To convert *large* units to *smaller* units, always *multiply*.

Convert:

a 1.2 m to centimetres

1.2 × 100 = 120 cm

b 0.62 cm to millimetres

0.62 × 10 = 6.2 mm

EXERCISE 10B

1 Fill in the gaps, using the information in this section.

a 125 cm = … m	**b** 82 mm = … cm	**c** 550 mm = … m
d 4200 g = … kg	**e** 5750 kg = … t	**f** 85 ml = … cl
g 755 g = … kg	**h** 800 ml = … l	**i** 200 cl = … l
j 1035 l = … m^3	**k** 530 l = … m^3	**l** 34 km = … m

2 Fill in the gaps, using the information in this section.

a 3.4 m = … mm	**b** 13.5 cm = … mm	**c** 0.67 m = … cm
d 0.64 km = … m	**e** 2.4 l = … ml	**f** 5.9 l = … cl
g 3.75 t = … kg	**h** 0.94 cm^3 = … l	**i** 21.6 l = … cl
j 15.2 kg = … g	**k** 14 m^3 = … l	**l** 0.19 cm^3 = … ml

3 Sarif wanted to buy two lengths of wood, each 2 m long, and 1.5 cm by 2 cm. He went the local store where the types of wood were described as:

2000 mm × 15 mm × 20 mm

200 mm × 15 mm × 20 mm

200 mm × 150 mm × 2000 mm

1500 mm × 2000 mm × 20 000 mm

Should he choose any of these? If so, which one?

4 How many square metres are there in a square kilometre?

HINTS AND TIPS

The answer is not 1000·

You will come across **scales** in a lot of different places.

For example, there are scales on thermometers, car speedometers and weighing scales. It is important that you can read scales accurately.

There are two things to do when reading a scale. First, make sure that you know what each **division** on the scale represents. Second, make sure you read the scale in the right direction, for example some scales read from right to left.

Also, make sure you note the **units**, if given, and include them in your answer.

EXAMPLE 3

Read the values from the following scales.

a  **b**

a   The scale shows 7. This is a very straightforward scale. It reads from left to right and each division is worth 1 unit.

b   The scale shows 34 kg. The scale reads from left to right and each division is worth 2 units.

EXERCISE 10C

1 Read the values from the following scales. Remember to state the units if they are shown.

a   **b**   **c**

2 Copy (or trace) the following dials and mark on the values shown.

a 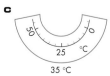   **b** 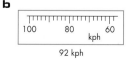   **c**

470 kg   92 kph   35 °C

3 Susie is using kitchen scales to weigh out flour.

a What is the mass of the flour shown on the scales?

b These scales can weigh items up to 400 g. Susie needs to weigh 700 g of currants using these scales. Explain how she could do this.

4 A pineapple was weighed.

A pineapple and an orange were weighed together.

a What is the mass of the pineapple? Give your answer in kilograms.

b What is the mass of the orange? Give your answer in grams.

10.4 Time

Times can be given using the **12-hour** or **24-hour clock**.

The 12-hour clock starts at midnight and runs to 12:00 at midday. After 12:59 it goes back to 1:00 and runs through to 12 again. So 7:45 could be 7:45 in the morning (am) or 7:45 in the evening (pm).

In everyday life we usually use the 12-hour clock and add 'am' or 'pm' to indicate whether we mean before or after midday.

The 24-hour clock indicates the number of hours and minutes after midnight using four digits. The first two digits are hours and the last two digits are minutes.

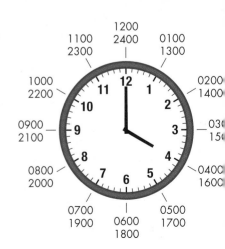

So 1:45 pm in the 12-hour clock is 1345 in the 24-hour clock, meaning 13 hours and 45 minutes after midnight. 7:30 am is 0730 but 7:30 pm is 1930 and so on.

Fifteen minutes after midnight is 0015 in the 24-hour clock and 12:15 am in the 12-hour clock.

Timetables usually use the 24-hour clock to avoid confusion.

7:35 am = 0735   1:42 pm = 1342   9:30 pm = 2130

EXAMPLE 4

A train left at 1135 and arrived at 1415.

How long did the journey take?

Do not use a calculator for this sort of question. Calculating 1415 – 1135 will not give the correct answer!

Break the journey into sections.

Here is one way:

1135 ⟶ 1200 ⟶ 1400 ⟶ 1415
   25 minutes   2 hours   15 minutes

Total time = 2 hours 40 minutes.

EXERCISE 10D

1 Here is the timetable for four trains from Rome to Naples:

Rome (depart)	0900	0927	1027	1045
Naples (arrive)	1010	1130	1236	1230

  a How long does each of the journeys take in hours and minutes?

  b Which train was the high speed express?

2 Here are the times of two trains from Rome to Venice:

Rome (depart)	0945	1036
Venice (arrive)	1333	1649

  a Write the four times as 12-hour clock times.

  b Find the length of each journey in hours and minutes.

3 A man arrived at his office at 0835 and left at 1520.

  a His journey home took 45 minutes. What time did he arrive home?

  b His journey to work in the morning took one hour and 20 minutes. What time did he leave home?

  c How long was he in the office?

4 Sunetra went on a coach trip to a forest park.

The coach left at 0830 and returned at 1855.

The journey took 2 hours and 20 minutes each way.

a What time did Sunetra arrive at the forest park?

b What time did she leave the forest park?

c How long did she spend there?

5 **a** A car left at 0845 and arrived at its destination 3 hours and 25 minutes later. What time did it arrive?

b On the return journey the car left at 1835 and arrived at 2125.

How long did this return journey take?

6 Here is the timetable for a bus journey:

Lympstone	1729
Exton	1741
Topsham	1757
Digby	1809
Sowton	1823

How long was the journey from Lympstone to:

a Exton **b** Digby **c** Sowton?

7 The express train left at 1050 and arrived at 1324.

The slow train left at 1242 and arrived at 1629.

How much shorter was the journey by express train?

8 Pierre flew from Paris to Doha on a day when the clocks in Doha were one hour ahead the clocks in Paris.

He left at 0740 and the flight took 5 hours and 35 minutes.

What was the local time when he arrived in Doha?

9 Boston is 5 hours behind London.

A flight from Boston to London left at 1935 and took 6 hours and 40 minutes.

What time did it arrive in London?

Currency conversions

Exchange rates are used to convert between one currency and another. They vary all the time depending on what happens in the world's stock exchanges.

EXAMPLE 5

If the exchange rate is 1 US dollar = 0.7775 euros:

a How many euros is 210 US dollars?

b How many dollars is 850 euros?

a 210 US dollars = 210 × 0.7775 euros
$$= 163.275 \text{ euros}$$

b 850 euros = 850 ÷ 0.7775 US dollars
$$= 1093.25 \text{ US dollars, rounding the answer to 2 decimal places.}$$

EXERCISE 10E

1 The exchange rate is 1 euro = 9.9919 Hong Kong dollars.

Change 320 euros into Hong Kong dollars, giving your answer to 2 decimal places.

2 The exchange rate is 1 Russian rouble = 0.0328 US dollars.

How many US dollars could you get for 5000 Russian roubles?

3 Copy and complete this guide for changing US dollars into euros:

$5	$10	$50	$100	$250	$500	$1000
			€77.55			

4 1 US dollar = 85.7 Pakistani rupees

If a gift costs 3686 Pakistani rupees, how many US dollars is that?

5 The exchange rate is 1 British pound to 1.2128 euros.

a A flight from London to Paris costs £185.45

How many euros is that?

b A flight from Paris to London costs €209.50

How many British pounds is that?

6 This table shows the conversion rates on one day between three currencies.

Currency	euro	US dollar	Japanese yen
1 euro =	1	1.2863	109.7406
1 US dollar =	0.7774	1	85.315
1 Japanese yen =	0.0091	0.0117	1

a Use the table to make the following conversions:

i 450 US dollars to euros.

ii 225 euros to Japanese yen. Give your answer to the nearest yen.

iii 37 000 Japanese yen to US dollars.

b Here are three amounts: 500 euros, 650 dollars, 54 000 Japanese yen.

Use the values in the table to put them in order from smallest to largest and complete this statement:

.... < .... < ....

7 On one day the exchange rate is

1 US dollar = 31.885 Taiwan dollars = 46.53 Indian rupees.

a What is 75 US dollars in Taiwan dollars?

b What is 75 US dollars in Indian rupees?

c Which is worth more, a Taiwan dollar or an Indian rupee?

d Complete this exchange rate : 1 Taiwan dollar = ... Indian rupees.

8 On 1st July 2005 the exchange rate was 1 US dollar = 8.2765 Chinese yuan.

On 1st July 2010 the exchange rate was 1 US dollar = 6.78099 Chinese yuan.

a How many *fewer* Chinese yuan could you buy for $50 on 1st July 2010 compared to 1st July 2005?

b Write the exchange rate on 1st July 2010 in the form:

1 Chinese yuan = ... US dollars.

Using a calculator efficiently

The aim of this topic is to make you aware of some of the keys on your calculator and how to use them to make calculations as efficiently as possible.

Consider the calculation $\dfrac{3.7 + 9.5}{0.38 + 0.16}$

If you just type $3.7 + 9.5 \div 0.38 + 0.16$ you will not get the correct answer.

One method is to calculate the numerator (that is, $3.7 + 9.5$) first, then the denominator and finally divide one by the other. Using the bracket keys it can all be done in one operation:

$(3.7 + 9.5) \div (0.38 + 0.16)$ gives 24.44 (to 2 decimal places)

It is also useful to check whether the answer is reasonable:

$\dfrac{3.7 + 9.5}{0.38 + 0.16} \approx \dfrac{4 + 10}{0.5} = \dfrac{14}{0.5} = 28$ so the answer seems reasonable.

The check should be a calculation you can easily do in your head.

EXERCISE 10F

Use your calculator to work out the following. Try to key in the calculation as one continuous set, without writing down any intermediate values.

1 Work out:

a $(10 - 2) \times 180 \div 10$

b $180 - (360 \div 5)$

2 Work out:

a $\dfrac{1}{2} \times (4.6 + 6.8) \times 2.2$

b $\dfrac{1}{2} \times (2.3 + 9.9) \times 4.5$

3 Work out the value of each of these, if $a = 3.4$, $b = 5.6$, and $c = 8.8$

a $2(ab + ac + bc)$

b $\dfrac{a + b}{c}$

c $\dfrac{a}{b + c}$

d $\sqrt{a + b + c}$

4 Work out the following, giving your answers to 2 decimal places.

a $\sqrt{3.2^2 + 1.6^2}$

b $\sqrt{4.8^2 + 3.6^2}$

5 Work out:

a $7.8^3 + 3 \times 7.8$

b $5.45^3 - 2 \times 5.45 - 40$

6 Do these calculations as efficiently as you can. Check that your answers are sensible.

a $\dfrac{15.89}{3.24 + 1.86}$

b $\dfrac{27}{18.1 + 17.95}$

c $\dfrac{383 + 936}{1.47 + 13.11}$

d $\dfrac{0.34^2}{0.025^2}$

e $\sqrt{3.8^2 + 9.7 \times 2.8}$

f $\sqrt{32.4^2 - 17.1^2}$

1 The table shows the distance, in kilometres, from Cairo to each of six other cities.

City	Distance from Cairo (km)
Hong Kong	8103
Jakarta	8943
London	3493
Nairobi	3518
New Delhi	4408
Singapore	8220

a Which of these cities is furthest away from Cairo? [1]

b Write the number 8103 in words. [1]

c Which number in the table is a multiple of 10? [1]

d Write the number 3518 correct to the nearest ten. [1]

Alex travels from London to Cairo.

He then travels from Cairo to Singapore.

e How far has Alex travelled? [2]

Edexcel Limited Paper 1F Q1 Jan 16

2 a Find a fraction which is equivalent to $\frac{3}{5}$ [1]

b Write $\frac{3}{5}$ as a decimal. [1]

c Write $\frac{3}{5}$ as a percentage. [1]

d Matheville School has 875 students.

$\frac{3}{5}$ of the students are girls.

i Work out $\frac{3}{5}$ of 875.

ii Work out the fraction of the students who are boys.

8% of the students were born in May.

iii Work out 8% of 875. [5]

Edexcel Limited Paper 1F Q3 Jan 14

3 Here are some numbers in a list.

2	−4	−8	5	−3

a Write the numbers in order of size.

Start with the smallest number.

b Work out

i −4 + 5

ii −8 − (−3)

iii −3 × 2

iv −8 ÷ (−4) [4]

Edexcel Limited Paper 1F Q7 May 14

4 a One morning, Lizzy went on a bus journey.
The clock shows the time that she left home.

   i Write down this time using the 12 hour clock.

Lizzy got home at ten to four in the afternoon.

   ii Write down this time using the 24-hour clock.   [2]

b On another day, Lizzy drove by car to visit her aunt.
She left home at 9:30 am.
Lizzy arrived at her aunt's house at 11:15 am.
She drove a distance of 140 km.
Work out, in km/h, Lizzy's average speed for the journey.   [3]

Edexcel Limited Paper 1F Q10 Jan 16

5 A total of 1200 passengers are booked to go on a cruise ship.

70% of the passengers will get on the ship at Southampton.

$\frac{1}{6}$ of the passengers will get on the ship at Lisbon.

The rest of the passengers will get on the ship at Venice.

a How many passengers will get on the ship at Venice?   [3]

There are 1200 passengers on the ship and 900 crew on the ship.

b Write down the ratio of the number of passengers to the number of crew.
Give your ratio in its simplest form.   [2]

Edexcel Limited Paper 1F Q11 Jan 16

6 Here is a list of the ingredients needed to make lentil soup for 6 people.

Lentil Soup (for 6 people)
120 g lentils
300 g carrots
800 ml vegetable stock
3 onions

Jenny wants to make lentil soup for 24 people.

a Work out the amount of vegetable stock she needs.   [2]

Ravi is going to make lentil soup.

He uses 450g of carrots.

b How many people is Ravi making the lentil soup for?   [2]

Edexcel Limited Paper 1F Q14 Jan 16

7 a Work out the value of 2.5^3

b Work out the value of $\dfrac{451.4}{14.1 + 10.3}$

c Work out the value of $\sqrt{7.8^2 - 7.2^2}$

Edexcel Limited Paper 1F Q14 Jan 15

8 Eloy's height was 125 cm when his age was 7 years.

His height was 153 cm when his age was 12 years.

a Work out the percentage increase in Eloy's height between the ages of 7 and 12 years. [3]

Eloy's height at the age of 12 was 85% of his height at the age of 20 years.

b Work out Eloy's height when his age was 20 years. [3]

Edexcel Limited Paper 1F Q19 Jan 15

9 The number of runners in the London Marathon on 25th April, 2010 was 37 527.

Work out an estimate for the number of these runners whose birthday was on that day.

Edexcel Limited Paper 1F Q20 May 13

PAPER 2F

1 **a**

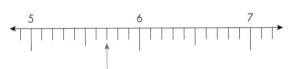

Write down the number marked with an arrow. [1]

b

3.6   3.7   3.8

i Find the number 3.76 on the number line.

Mark it with an arrow (↑)

ii Round 3.76 to the nearest whole number.

iii Write the value of the 7 in the number 3.76. [3]

c Write down the number that is exactly halfway between 3.76 and 3.77. [1]

Edexcel Limited Paper 2F Q2 May 13

2 **a** Write 0.8 as a percentage. [1]

b Write 0.023 as a fraction. [1]

c Write 5.6382 correct to 2 decimal places. [1]

d Work out $\sqrt{42.25} + 1.3^2$. [1]

Give your answer as a decimal.

e Work out $\frac{3}{8}$ of 56.8 kg. [2]

Edexcel Limited Paper 2F Q7 Jun 15

3 **a** Write these numbers in order of size.

Start with the smallest number.

$\frac{7}{9}$ $\frac{8}{11}$ 0.79   84% $\frac{4}{5}$ [3]

b Find the value of 3^5 [1]

c Find the square of −2.1 [1]

d Find the value of $\sqrt[3]{17.576}$ [1]

Edexcel Limited Paper 2F Q9 Jan 16

4 **a** Write down the square root of 100. [1]

Kwo writes down one square number and one cube number.

When she adds the two numbers together, she gets a total that is more than 80 but less than 100.

b What square number and what cube number could Kwo have written down? [3]

Edexcel Limited Paper 2F Q10 Jun 15

5

| 1 euro = 1.40 Canadian dollars |

a Alain changes 450 euros into Canadian dollars.
How many Canadian dollars should he receive? [2]

b Isabella changes 840 Canadian dollars into euros.
How many euros should she receive? [2]

| 1 euro = 1.40 Canadian dollars |

c How many cents is 1 Canadian dollar worth? [2]

Edexcel Limited Paper 2F Q11 May 13

6 Mr and Mrs Sandhu take their 3 children to a museum.

The cost of a ticket for one adult is €8.60.

The total cost of the 5 tickets is €30.40.

Work out the cost of a ticket for one child. [3]

Edexcel Limited Paper 2F Q13 Jan 16

7 ζ = {positive whole numbers less than 19}

A = {odd numbers}

B = {multiples of 5}

C = {multiples of 4}

a List the members of the set

i $A \cap B$

ii $B \cup C$ [2]

D = {prime numbers}

b Is it true that $B \cap D = \varnothing$?
Tick (✓) the appropriate box.   Yes ☐   No ☐
Explain your answer. [1]

Edexcel Limited Paper 2F Q18 Jan 15

8 The lengths of the sides of a triangle are in the ratios 2 : 6 : 7

The length of the longest side of the triangle is 24.5 cm.

Work out the perimeter of the triangle. [3]

Edexcel Limited Paper 2F Q19 Jan 16

9 Express 825 as a product of its prime factors.

Edexcel Limited Paper 2F Q20 Jan 14

PAPER 3H

1 **a** Work out the value of $\dfrac{138 \times 6.5}{7 + \sqrt{2}}$

Write down all of the figures on your calculator display. [2

b Give your answer to part **a** correct to 3 significant figures. [1

Edexcel Limited Paper 3H Q1 May 14

2 **a** $A = \{p, r, a, g, u, e\}$

$B = \{p, a, r, i, s\}$

$C = \{b, u, d, a, p, e, s, t\}$

List the members of the set

i $A \cap B$

ii $B \cup C$ [2

b $D = \{r, o, m, e\}$

$E = \{l, i, s, b, o, n\}$

$F = \{b, e, r, l, i, n\}$

Put one of the letters D, E or F in the box below to make the statement correct.

$A \cap \boxed{} = \varnothing$

Explain your answer.

Edexcel Limited Paper 3H Q7 May 14

3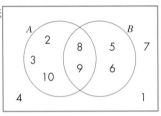

The Venn diagram shows all of the elements in sets A, B and ξ.

a Write down the elements in A' [1

b Find $n(A \cap B)'$ [1

c Find the elements in $(A \cap B) \cup (A \cup B)'$ [1

$A \cap C = \varnothing$

$B \cup C = \{5, 6, 7, 8, 9\}$

$n(C) = 3$

d Write down the elements in C. [1

Edexcel Limited Paper 3H Q19 Jan 15

4 The table shows some information about the five Great Lakes in North America.

Name	Surface area (m^2)	Volume of water (m^3)
Lake Erie	2.57×10^{10}	4.80×10^{11}
Lake Huron	6.01×10^{10}	3.52×10^{12}
Lake Michigan	5.80×10^{10}	4.87×10^{12}
Lake Ontario	1.91×10^{10}	1.64×10^{12}
Lake Superior	8.21×10^{10}	1.22×10^{13}

a Work out the total surface area of the five Great Lakes.

Give your answer in standard form. [2]

Loch Ness is the largest lake in Scotland.

The lake has a volume of water of 7.45×10^9 m^3.

The volume of water in Lake Superior is k times the volume of water in Loch Ness. [2]

b Work out the value of k.

Give your answer correct to 3 significant figures.

Edexcel Limited Paper 3H Q10 May 15

5

$3780 = 2^2 \times 3^3 \times 5 \times 7$	$3240 = 2^3 \times 3^4 \times 5$

a Find the highest common factor (HCF) of 3780 and 3240

Give your answer as a product of prime factors. [2]

b Find the lowest common multiple (LCM) of 3780 and 3240

Give your answer as a product of prime factors. [2]

Edexcel Limited Paper 3H Q11 Jan 15

6 Liam invests £8000 in a savings account for 4 years.

The savings account pays compound interest at a rate of

    4.5% for the first year;

    2.75% for all subsequent years.

a Work out the value of Liam's investment at the end of 4 years. [3]

Max invests some money in a savings bond.

The savings bond pays interest at a rate of 2% per year.

At the end of the first year, his savings bond is worth £5763.

b How much money did Max invest in his savings bond? [3]

Edexcel Limited Paper 3H Q14 Jan 16

7 **a** Correct to the nearest millimetre, the length of a side of a regular hexagon is 3.6 cm.
Calculate the upper bound for the perimeter of the regular hexagon. [2]

b Correct to 1 significant figure, the area of a rectangle is 80 cm^2
Correct to 2 significant figures, the length of the rectangle is 12 cm.
Calculate the lower bound for the width of the rectangle.
Show your working clearly. [3]

Edexcel Limited Paper 3H Q19 May 13

8

32 cm

2000 cm³   500 cm³

Zane buys mineral water in large bottles and in small bottles.
The large bottles are mathematically similar to the small bottles.
Large bottles have a height of 32 cm and a volume of 2000 cm^3
Small bottles have a volume of 500 cm^3
Work out the height of a small bottle.
Give your answer to 3 significant figures. [3]

Edexcel Limited Paper 3H Q15 Jan 15

9 A, r and T are three variables.

A is proportional to T^2

A is also proportional to r^3

$T = 47$ when $r = 0.25$

Find r when $T = 365$

Give your answer correct to 3 significant figures. [4]

Edexcel Limited Paper 3H Q22 May 15

PAPER 4H

1 x is an integer.

The Lowest Common Multiple (LCM) of x and 12 is 120.

The Highest Common Factor (HCF) of x and 12 is 4.

Work out the value of x. [2]

Edexcel Limited Paper 4H Q11 June 15

2 The pressure P, of water leaving a cylindrical pipe, is inversely proportional to the square of the radius, r, of the pipe.

$P = 22.5$ when $r = 2$

a Find a formula for P in terms of r. [3]

b Calculate the value of P when $r = 1.5$. [1]

c Calculate the value of r when $P = 10$. [2]

Edexcel Limited Paper 4H Q16 May 13

3 The Venn diagram shows a universal set ξ and 3 sets A, B and C.

2, 4, 7, 3, 6 and 10 represent **numbers** of elements.

Find

a $n(A \cup B)$ [1]

b $n(B')$ [1]

c $n(A \cap C')$ [1]

d $n(B' \cap C')$ [1]

Edexcel Limited Paper 4H Q20 May 13

4 Show that the recurring decimal $0.0\dot{1}\dot{5} = \dfrac{1}{66}$ [2]

Edexcel Limited Paper 4H Q20 Jan 14

5 **a** Write $\dfrac{1}{32}$ as a power of 2.

b Show that $(4 + \sqrt{12})(5 - \sqrt{3}) = 14 + 6\sqrt{3}$

Show each stage of your working clearly. [3]

Edexcel Limited Paper 4H Q22 Jan 15

6 Given that $\left(2^{\frac{1}{2}}\right)^n = \dfrac{2^x}{8^y}$

Express n in terms of x and y. [3]

Edexcel Limited Paper 4H Q24 May 14

7 $y = 16 \times 10^{8k}$ where k is an integer.

Find an expression, in terms of k, for $y^{\frac{5}{4}}$

Give your answer in standard form. [3]

Edexcel Limited Paper 4H Q25 Jan 16

Where is mathematics used? Is it used in:

Art   Science   Sport   Language?

In fact, it is important for them all.

Art

Mathematicians think that famous works of art are often based on the 'golden ratio'. This is the ratio of one part of the art to another. We think that human brains find the 'golden ratio' very attractive.

Science

Science needs mathematics. In 1962 a space probe went off course because someone had got a mathematical formula wrong in its programming.

Sport

Is mathematics a sport? There are national and international competitions each year that use mathematics. University students compete in the annual 'Mathematics Olympiad' and there is a world Suduko championship each year. Lots of sporting activities require maths too, such as throwing a javelin (angles).

Language

But the best description of mathematics is that it is a language

It is the only language which people in all countries understar Everyone understands the numbers on this stamp even if they not speak the language of the country.

Algebra is an important part of the language of mathematics. comes from the Arabic *al-jabr*. It was first used in a book writt in 820 CE by a Persian mathematician called al-Khwarizmi.

The use of symbols grew until the 17th century when a French mathematician called Descartes developed them into the sort algebra we use today.

Algebra and formulae

...ics	Level	Key words
The language of algebra	**FOUNDATION**	expression, symbol, variable, formula, formulae, equation, term, solve
Substitution into formulae	**FOUNDATION**	
Rearranging formulae	**FOUNDATION**	rearrange, subject, variable
More complicated formulae	**HIGHER**	

What you need to be able to do in the examinations:

FOUNDATION	HIGHER
Understand that: • Symbols may be used to represent numbers in equations or variables in expressions and formulae. • A letter may represent an unknown number or variable. • Algebraic expressions follow the generalised rules of arithmetic. Use correct notational conventions for algebraic expressions and formulae. Substitute positive and negative integers, decimals and fractions for words and letters in expressions and formulae. Use formulae from mathematics and other real-life contexts expressed initially in words or diagrammatic form and convert to letters and symbols. Derive a formula or expression. Change the subject of a formula where the subject appears once.	• Understand the process of manipulating formulae to change the subject, to include cases where the subject may appear twice or a power of the subject occurs.

Algebra is a way of expressing operations involving numbers, where one or more number is unknown. Here is an example.

Ari is buying some tickets. They cost 12 dollars each. Ari must also pay 4 dollars postage. He wants to know how to work out the cost for different numbers of tickets. He calls the number of tickets t.

The total cost is $12t + 4$ dollars.

- The letter t is a **variable**. It stands for the number of tickets which varies.
- $12t + 4$ is an **expression**. It shows how to calculate the total cost in dollars: multiply the number of tickets by 12 and add 4.
- $12t$ and 4 are both **terms** in the expression.
- If C dollars is the total cost of the tickets, we can write $C = 12t + 4$. This is a **formula**. It shows the relationship between the variables t and C.
- If Ari spends 172 dollars on tickets, we can write $12t + 4 = 172$. This is an **equation**. To **solve** this equation means finding the value of the variable t.

Algebra follows the same rules as arithmetic, and uses the same **symbols** ($+$, $-$, $\times$ and $\div$). Below are seven important algebraic rules.

- Write '4 more than x' as $4 + x$ or $x + 4$.
- Write '6 less than p' or 'p minus 6' as $p - 6$.
- Write '4 times y' as $4 \times y$ or $y \times 4$ or $4y$. The last one of these is the neatest way to write
- Write 'b divided by 2' as $b \div 2$ or $\frac{b}{2}$.
- When a number and a letter or a letter and a letter appear together, there is a hidden multiplication sign between them. So, $7x$ means $7 \times x$ and ab means $a \times b$.
- Always write '$1 \times x$' as x.
- Write 't times t' as $t \times t$ or t^2.

EXAMPLE 1

One side of this rectangle is three centimetres longer than the other.

Find a formula for the area (A) in square centimetres, and the perimeter (P) in centimetres.

$(x + 3)$ cm

x cm

Area = width × length

$A = x(x + 3)$ (We leave out the multiplication sign)

Perimeter = distance around the outside

$P = x + (x + 3) + x + (x + 3)$

We can simplify this to:

$P = 4x + 6$

EXERCISE 11A

1 Write down the algebraic expression for:

a 2 more than x

b 6 less than x

c k more than x

d x minus t

e x added to 3

f d added to m

g y taken away from b

h p added to t added to w

i 8 multiplied by x

j h multiplied by j

k x divided by 4

l 2 divided by x

m y divided by t

n w multiplied by t

o a multiplied by a

p g multiplied by itself.

2 Asha, Bernice and Charu are three sisters. Bernice is x years old. Asha is three years older than Bernice. Charu is four years younger than Bernice.

a How old is Asha?

b How old is Charu?

3 An approximation method of converting from degrees Celsius to degrees Fahrenheit is given by this rule:

Multiply by 2 and add 30.

Using C to stand for degrees Celsius and F to stand for degrees Fahrenheit, complete this formula.

$$F = \ldots\ldots$$

4 Cows have four legs. Which of these **formulae** connects the number of legs (L) and the number of cows (C)?

a $C = 4L$ **b** $L = C + 4$ **c** $L = 4C$ **d** $L + C = 4$

5 **a** Lakmini has three bags of marbles. Each bag contains n marbles. How many marbles does she have altogether?

b Rushani gives her another three marbles. How many marbles does Lakmini have now?

c Lakmini puts one of her new marbles in each bag. How many marbles are there now in each bag?

d Lakmini takes two marbles out of each bag. How many marbles are there now in each bag?

6 Lee has *n* cubes.

● Anil has twice as many cubes as Lee.

● Reza has two more than Lee.

● Dale has three fewer than Lee.

● Chen has three more than Anil.

How many cubes does each person have?

7 **a** I go shopping with $10 and spend $6. How much do I have left?

b I go shopping with $10 and spend $*x*. How much do I have left?

c I go shopping with $*y* and spend $*x*. How much do I have left?

d I go shopping with $3*x* and spend $*x*. How much do I have left?

8 Give the total cost of:

a five books at $15 each

b *x* books at $15 each

c four books at $*A* each

d *y* books at $*A* each.

9 A boy went shopping with *A* dollars. He spent *B* dollars. How much did he have left?

10 Five ties cost $*A*. What is the cost of one tie?

11 **a** My dad is 72 and I am *T* years old. How old shall we each be in *x* years' time?

b My mum is 64 years old. In two years' time she will be twice as old as I am. What age am I now?

12 I am twice as old as my son. I am *T* years old.

a How old is my son?

b How old will my son be in four years' time?

c How old was I *x* years ago?

13 What is the perimeter of each of these figures?

a

$2x$

Square

b

$4m$

Equilateral triangle

c

$3t$

Regular hexagon

14 Write down the number of marbles each student ends up with.

Student	Action	Marbles
Andrea	Start with three bags each containing n marbles and give away one marble from each bag	
Barak	Start with three bags each containing n marbles and give away one marble from one bag	
Ahmed	Start with three bags each containing n marbles and give away two marbles from each bag	
Dina	Start with three bags each containing n marbles and give away n marbles from each bag	
Emma	Start with three bags each containing n marbles and give away n marbles from one bag	
Hana	Start with three bags each containing n marbles and give away m marbles from each bag	

15 The answer to $3 \times 4m$ is $12m$.

Write down two *different* expressions for which the answer is $12m$.

1.2 Substitution into formulae

A formula describes the relationship between variables.

EXAMPLE 2

The formula for the area of a trapezium is:

$$A = \frac{(a + b)h}{2}$$

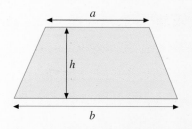

Find the area of the trapezium when $a = 5$, $b = 9$ and $h = 3$.

$$A = \frac{(5 + 9) \times 3}{2} = \frac{14 \times 3}{2} = 21$$

FOUNDATION

1 Find the value of $3x + 2$ when:

a $x = 2$ **b** $x = 5$ **c** $x = -10$

> **HINTS AND TIPS**
>
> It helps to put the numbers in brackets.
> $3(2) + 2 = 6 + 2 = 8$
> $3(5) + 2 = 15 + 2 = 17$
> etc …

2 Find the value of $4k - 1$ when:

a $k = 3.5$ **b** $k = 3$ **c** $k = 11$

3 Find the value of $5 + 2t$ when:

a $t = 8$ **b** $t = -6$ **c** $t = 3.4$

4 Evaluate $15 - 2f$ when: **a** $f = 3$ **b** $f = 5.6$ **c** $f = -4$

5 Evaluate $5m + 3$ when: **a** $m = 2$ **b** $m = \frac{1}{2}$ **c** $m = \frac{3}{4}$

6 Evaluate $3d - 2$ when: **a** $d = -6$ **b** $d = 5$ **c** $d = 4.3$

7 A taxi company uses the following rule to calculate their fares.

Fare = \$2.50 plus \$0.50 per kilometre

a How much is the fare for a journey of 3 km?

b Farook pays \$9.00 for a taxi ride. How far was the journey?

c Maisy knows that her house is 5 kilometres from town. She has \$5.50 left in her pu[r]
after a night out. Has she got enough for a taxi ride home?

8 Kaz knows that x, y and z have the values 2, 8 and 11, but she does not know which variable has which value.

a What is the maximum value that the expression $2x + 6y - 3z$ could be?

b What is the minimum value that the expression $5x - 2y + 3z$ could be?

> **HINTS AND TIPS**
>
> You could just try all combinations, but if you think for a moment you will find that the $6y$ term must give the largest number. This will give yo[u] a clue to the other terms[.]

9 The formula for the area, A, of a rectangle with length l and width w is $A = lw$.

The formula for the area, T, of a triangle with base b and height h is $T = \frac{1}{2}bh$.

Find values of l, w, b and h so that $A = T$.

10 Find the value of $\dfrac{8 \times 4h}{5}$ when: **a** $h = 5$ **b** $h = 10$ **c** $h = 2\frac{1}{2}$

11 Find the value of $\dfrac{25 - 3p}{2}$ when: **a** $p = 4$ **b** $p = -4$ **c** $p = 1[0]$

12 Evaluate $\frac{x}{3}$ when: **a** $x = 6$ **b** $x = 24$ **c** $x = -30$

13 Evaluate $\frac{A}{4}$ when: **a** $A = 12$ **b** $A = 10$ **c** $A = -20$

14 Evaluate $\frac{12}{y}$ when: **a** $y = 2$ **b** $y = 0.5$ **c** $y = -6$

15 Evaluate $\frac{24}{x}$ when: **a** $x = 2$ **b** $x = 3$ **c** $x = 16$

16 A holiday cottage costs 150 dollars per day to rent.

A group of friends decide to rent the cottage for seven days.

a Which formula represents the cost of the rental for each person if there are n people in the group? Assume that they share the cost equally.

$$\frac{150}{n} \qquad \frac{150}{7n} \qquad \frac{1050}{n} \qquad \frac{150n}{n}$$

> **HINTS AND TIPS**
>
> To check your choice in part **a**, make up some numbers and try them in the formula. For example, take $n = 5$.

b Eventually 10 people go on the holiday. When they get the bill, they find that there is a discount for a seven-day rental.

After the discount, they each find it cost them 12.50 dollars less than they expected.

How much does a 7-day rental cost?

17 **a** p is an odd number and q is an even number.

Say if each of these expressions is odd or even.

i $p + q$ **ii** $p^2 + q$ **iii** $2p + q$ **iv** $p^2 + q^2$

> **HINTS AND TIPS**
>
> There are many answers for **b**, and **a** should give you a clue.

b x, y and z are all odd numbers.

Write an expression, using x, y and z, so that the value of the expression is always even.

18 A formula for the cost of delivery, in dollars, of orders from a warehouse is:

$$D = 2M - \frac{C}{5}$$

where D is the cost of the delivery, M is the distance in kilometres from the store and C is the cost of the goods to be delivered.

> **HINTS AND TIPS**
>
> Note: a rebate is a refund of some of the money that someone has already paid for goods or services.

a How much does the delivery cost when $M = 30$ and $C = 200$?

b Bob buys goods worth $300 and lives 10 kilometres from the store.

i The formula gives the cost of delivery as a negative value. What is this value?

ii Explain why Bob will not get a rebate from the store.

c Maya buys goods worth $400. She calculates that her cost of delivery will be zero. What is the greatest distance that Maya could live from the store?

The **subject** of a formula is the **variable** (letter) in the formula which stands on its own, usually the left-hand side of the equals sign. For example, x is the subject of each of the following formul

$$x = 5t + 4 \qquad x = 4(2y - 7) \qquad x = \frac{1}{t}$$

To change the existing subject to a different variable, you have to **rearrange** the formula to ge that variable on its own on the left-hand side.

EXAMPLE 3

Make m the subject of this formula. $\qquad T = m - 3$

Add 3 to both sides. $\qquad T + 3 = m$

Reverse the formula. $\qquad m = T + 3$

EXAMPLE 4

From the formula $P = 4t$, express t in terms of P.

(This is another common way of asking you to make t the subject.)

Divide both sides by 4: $\qquad \dfrac{P}{4} = \dfrac{4t}{4}$

Reverse the formula: $\qquad t = \dfrac{P}{4}$

EXAMPLE 5

From the formula $C = 2m^2 + 3$, make m the subject.

Subtract 3 from both sides so that the $2m^2$ is on its own $\qquad C - 3 = 2m^2$

Divide both sides by 2: $\qquad \dfrac{C - 3}{2} = \dfrac{2m^2}{2}$

Reverse the formula: $\qquad m^2 = \dfrac{C - 3}{2}$

Take the square root on both sides: $\qquad m = \sqrt{\dfrac{C - 3}{2}}$

1 $T = 3k$ Make k the subject.

2 $X = y - 1$ Express y in terms of X.

3 $Q = \dfrac{p}{3}$ Express p in terms of Q.

4 $A = 4r + 9$ Make r the subject.

5 $W = 3n - 1$ Make n the subject.

6 $p = m + t$ **a** Make m the subject.   **b** Make t the subject.

7 $g = \dfrac{m}{v}$ Make m the subject.

8 $t = m^2$ Make m the subject.

9 $C = 2\pi r$ Make r the subject.

10 $A = bh$ Make b the subject.

11 $P = 2l + 2w$ Make l the subject.

12 $m = p^2 + 2$ Make p the subject.

13 The formula for converting degrees Fahrenheit to degrees Celsius is $C = \frac{5}{9}(F - 32)$.

      a Show that when $F = -40$, C is also equal to -40.

      b Find the value of C when $F = 68$.

      c Show that the formula can be rearranged as $F = \dfrac{9C}{5} + 32$

14 $v = u + at$ **a** Make a the subject.   **b** Make t the subject.

15 $A = \dfrac{1}{4}\pi d^2$ Make d the subject.

16 $W = 3n + t$ **a** Make n the subject.   **b** Express t in terms of n and W.

17 $x = 5y - w$ **a** Make y the subject.   **b** Express w in terms of x and y.

18 $k = 2p^2$ Make p the subject.

19 $v = u^2 - t$ **a** Make t the subject.   **b** Make u the subject.

20 $k = m + n^2$ **a** Make m the subject.   **b** Make n the subject.

21 $T = 5r^2$ Make r the subject.

22 $K = 5n^2 + w$ **a** Make w the subject.   **b** Make n the subject.

More complicated formulae

To find the value of a variable you need to make it the subject of the formula. You often need to rearrange the formula to do this. Some formulae will need a number of separate steps to rearrange them.

EXAMPLE 6

Make y the subject of $a = c + \dfrac{d}{y^2}$

Subtract c from both sides: $\qquad a - c = \dfrac{d}{y^2}$

Multiply both sides by y^2: $\qquad y^2(a - c) = d$

Divide both sides by $a - c$: $\qquad y^2 = \dfrac{d}{a - c}$

Square root of y^2: $\qquad y = \sqrt{\dfrac{d}{a - c}}$

EXERCISE 11D

HIGHER

1 $a^2 + b^2 = c^2$

    a Find a if $b = 6.0$ and $c = 6.5$

    b Make a the subject.

2 $s = ut + \frac{1}{2}at^2$ is a formula used in mechanics.

    a Find s if $u = -5$, $t = 4$ and $a = 10$

    b Make a the subject.

3 $a = \dfrac{b + 2}{c}$

    a Make b the subject.

    b Make c the subject.

4 $p = \dfrac{r}{t - 3}$

    Make t the subject

5 $a = \dfrac{12}{1 + \sqrt{e}}$

    Make e the subject

6 $v^2 = u^2 + 2as$

   a Find v if $u = 3$, $a = 2$ and $s = 4$.

   b Make u the subject.

   c Make s the subject.

7 $T = 2\pi\sqrt{\dfrac{L}{G}}$

   a Make L the subject.

   b Show that $G = L\left(\dfrac{2\pi}{T}\right)^2$

8 $D = \pi R^2 - \pi r^2$

   a Make R the subject.

   b Make r the subject.

   c Make π the subject.

9 $3x^2 - 4y^2 = 11$

   a Find x if $y = 4$.

   b Make x the subject.

   c Make y the subject.

10 $T = 2\sqrt{\left(\dfrac{a}{c+3}\right)}$

   a Make a the subject.

   b Make c the subject.

11 $a^2 = b^2 + c^2 - 2bcT$

  Make T the subject.

12 $uv = fu + fv$ is a formula used in optics.

   a Find f if $u = 20$ and $v = 30$.

   b Make f the subject.

   c Make u the subject.

   d Make v the subject.

XAMPLE 7

Make x the subject of the formula	$3(x - y) = 2(x + y)$
Multiply out the brackets:	$3x - 3y = 2x + 2y$
Subtract $2x$ from both sides:	$x - 3y = 2y$
Add $3y$ to both sides:	$x = 5y$

Why this chapter matters

How can algebra save your life? Read on...

When you drive a car you must leave a safe distance between you and the car in front. What is a safe distance? How long will it take you to stop?

The **stopping distance** has two parts:

● **Thinking distance** is the distance travelled before your brain reacts and you apply the brakes

● **Braking distance** is the time it takes the car to come to a complete stop once the brakes have been applied

Stopping distances vary according to the weather conditions, the road conditions and vehicle. This table shows stopping distances for a car with good brakes on a dry, level road with good visibility.

Typical stopping distances

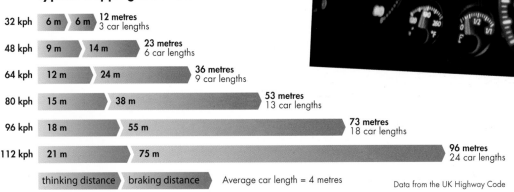

32 kph	6 m	6 m	**12 metres** 3 car lengths
48 kph	9 m	14 m	**23 metres** 6 car lengths
64 kph	12 m	24 m	**36 metres** 9 car lengths
80 kph	15 m	38 m	**53 metres** 13 car lengths
96 kph	18 m	55 m	**73 metres** 18 car lengths
112 kph	21 m	75 m	**96 metres** 24 car lengths

thinking distance ▷ braking distance   Average car length = 4 metres

Data from the UK Highway Code

You will notice that the thinking distance increases steadily with speed. It increases by 3 metres for every 16 km/h increase in speed. If you double the speed you double the thinking distance. You can show this in a simple algebraic expression:

thinking distance in metres = $0.1875x$
(where x is the speed in kilometres/hour).

The relationship between speed and braking distance is more complicated. It can be shown as:

braking distance = $0.006x^2$

So:

total stopping distance = $0.1875x + 0.006x^2$

This is a quadratic expression and you can use it to work out stopping distance at any speed (value of x). You will learn about both simple and quadratic expressions in this chapter.

12 Algebraic manipulation

ics	Level	Key words
Simplifying expressions	**FOUNDATION**	simplifying, coefficient, like terms
Expanding brackets	**FOUNDATION**	expand
Factorisation	**FOUNDATION**	common factors, factorise, factorisation
Expanding two brackets	**FOUNDATION**	quadratic expression, quadratic expansion
Multiplying more complex expressions	**HIGHER**	linear
Quadratic factorisation	**FOUNDATION**	brackets
Factorising $ax^2 + bx + c$	**HIGHER**	
More than two brackets	**HIGHER**	
Algebraic fractions	**HIGHER**	cancel, single fractions

hat you need to be able to do in the examinations:

FOUNDATION	HIGHER
Evaluate expressions by substituting numerical values for letters.	• Expand the product of two or more linear expressions.
Collect like terms.	• Understand the concept of a quadratic expression and be able to factorise such expressions.
Multiply a single term over a bracket.	• Manipulate algebraic fractions where the numerator and/or the denominator can be numeric, linear or quadratic.
Take out single common factors.	
Expand the product of two simple linear expressions.	
Understand the concept of a quadratic expression and factorise expressions of the form $x^2 + bx + c$	

Simplifying an algebraic expression means making it neater and, usually, shorter by combining its terms where possible.

Multiplying expressions

When you multiply algebraic expressions, first you multiply the numbers, then the letters.

EXAMPLE 1

Simplify:

a $2 \times t$
b $m \times t$
c $2t \times 5$
d $3y \times 2m$

The convention is to write the number first then the letters. The number in front of the letter is called the **coefficient**.

a $2 \times t = 2t$
b $m \times t = mt$
c $2t \times 5 = 10t$
d $3y \times 2m = 6my$

In an examination you will not be penalised for writing $2ba$ instead of $2ab$, but you will be penalised if you write $ab2$ as this can be confused with powers, so *always* write the number f

EXAMPLE 2

Simplify:

a $t \times t$
b $3t \times 4t$
c $3t^2 \times 4t$
d $2t^3 \times 4t^2$

a $t \times t = t^2$

b $3t \times 4t = 3 \times t \times 4 \times t = 12t^2$ (multiply 3 and 4)

c $3t^2 \times 4t = 3 \times t \times t \times 4 \times t = 12t^3$ $(t \times t \times t = t^3)$

d $2t^2 \times 4t^3 = 2 \times t \times t \times 4 \times t \times t \times t = 8t^5$ $(t \times t \times t \times t \times t = t^5)$

EXERCISE 12A

1 Simplify the following expressions.

a $2 \times 3t$
b $5y \times 3$
c $2w \times 4$

d $5b \times b$
e $2w \times w$
f $4p \times 2p$

g $3t \times 2t$
h $5t \times 3t$
i $m \times 2t$

j $5t \times q$
k $n \times 6m$
l $3t \times 2q$
m $5h \times 2k$
n $3p \times 7r$

> **HINTS AND TIPS**
>
> Remember to multiply numbers and add indices

2 a Which of the following expressions are equivalent?

$2m \times 6n$ $4m \times 3n$ $2m \times 6m$ $3m \times 4n$

b The expressions $2x$ and x^2 are the same for only two values of x. What are these values?

3 A square and a rectangle have the same area.

The rectangle has sides $2x$ cm and $8x$ cm.

What is the length of a side of the square?

4 Simplify the following expressions.

a $y^2 \times y$ **b** $3m \times m^2$ **c** $4t^2 \times t$

d $3n \times 2n^2$ **e** $t^2 \times t^2$ **f** $h^3 \times h^2$

g $3n^2 \times 4n^3$ **h** $3a^4 \times 2a^3$ **i** $k^5 \times 4k^2$

j $-t^2 \times -t$ **k** $-4d^2 \times -3d$ **l** $-3p^4 \times -5p^2$

m $3mp \times p$ **n** $3mn \times 2m$ **o** $4mp \times 2mp$

Collecting like terms

Like terms are those that are multiples of the same variable or of the same combination of variables. For example, a, $3a$, $9a$, $\frac{1}{4}a$ and $-5a$ are all like terms.

So are $2xy$, $7xy$ and $-5xy$, and so are $6x^2$, x^2 and $-3x^2$.

Collecting like terms generally involves two steps.

- Collect like terms into groups.
- Then simplify the like terms in each group.

Only like terms can be added or subtracted to simplify an expression. For example:

$a + 3a + 9a - 5a$ simplifies to $8a$

$2xy + 7xy - 5xy$ simplifies to $4xy$

Note that the variable does not change. All you have to do is find the coefficients.

For example

$6x^2 + x^2 - 3x^2 = (6 + 1 - 3)x^2 = 4x^2$

But an expression such as $4p + 8t + 5x - 9$ *cannot* be simplified, because $4p$, $8t$, $5x$ and 9 are *not like terms*.

EXAMPLE 3

Simplify:

$7x^2 + 3y - 6z + 2x^2 + 3z - y + w + 9$

Write out the expression: $7x^2 + 3y - 6z + 2x^2 + 3z - y + w + 9$

Then collect like terms: $\boxed{7x^2 + 2x^2}$ $\boxed{+3y - y}$ $\boxed{-6z + 3z}$ $\boxed{+w}$ $\boxed{+9}$

Then simplify them: $9x^2$ $+$ $2y$ $-$ $3z$ $+ w + 9$

So, the expression in its simplest form is:

$9x^2 + 2y - 3z + w + 9$

EXAMPLE 4

The length of this rectangle is twice the width.
Find expressions for the perimeter and the area.

Perimeter = $2x + x + 2x + x = 6x$

Area = $2x \times x = 2x^2$

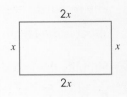

EXERCISE 12B

1 Joseph is given \$$t$, John has \$3 more than Joseph, and Joy has \$$2t$.

   a How much more money has Joy than Joseph?

   b How much do the three of them have altogether?

2 Write down an expression for the perimeter of each of these shapes.

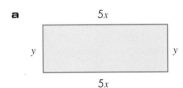

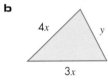

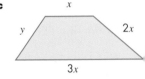

3 Simplify:

   a $a + a + a + a + a$ **b** $c + c + c + c + c + c$

   c $4e + 5e$ **d** $f + 2f + 3f$

   e $5j + j - 2j$ **f** $9q - 3q - 3q$

   g $3r - 3r$ **h** $2w + 4w - 7w$

   i $5x^2 + 6x^2 - 7x^2 + 2x^2$ **j** $8y^2 + 5y^2 - 7y^2 - y^2$

   k $2z^2 - 2z^2 + 3z^2 - 3z^2$

> **HINTS AND TIPS**
>
> The term a has a coefficient of 1, i.e. $a = 1$ but you do not need to write the 1.

4 Simplify:

   a $3x + 4x$ **b** $5t - 2t$

   c $-2x - 3x$ **d** $-k - 4k$

   e $m^2 + 2m^2 - m^2$ **f** $2y^2 + 3y^2 - 5y^2$

> **HINTS AND TIPS**
>
> Remember that only **like terms** can be added or subtracted.
> If all the terms cancel ou[t] just write 0 rather than $0x^2$, for example.

5 Simplify:

   a $5x + 8 + 2x - 3$ **b** $7 - 2x - 1 + 7x$ **c** $4p + 2t + p - 2t$

   d $8 + x + 4x - 2$ **e** $3 + 2t + p - t + 2 + 4p$ **f** $5w - 2k - 2w - 3k +$

   g $a + b + c + d - a - b - d$ **h** $9k - y - 5y - k + 10$

6 Simplify these expressions if possible.

    a $c + d + d + d + c$ **b** $2d + 2e + 3d$ **c** $f + 3g + 4h$

    d $5u - 4v + u + v$ **e** $4m - 5n + 3m - 2n$ **f** $3k + 2m + 5p$

    g $2v - 5w + 5w$ **h** $2w + 4y - 7y$ **i** $5x^2 + 6x^2 - 7y + 2y$

    j $8y^2 + 5z - 7z - 9y^2$ **k** $2z^2 - 2x^2 + 3x^2 - 3z^2$

7 Find the perimeter of each of these shapes, giving your answer in its simplest form.

a 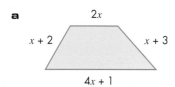   **b** 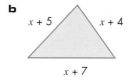   **c**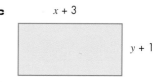

8 $3x + 5y + 2x - y = 5x + 4y$

Write down two other *different* expressions which are equal to $5x + 4y$.

9 Find the missing terms to make these equations true.

    a $4x + 5y + ... - .... = 6x + 3y$

    b $3a - 6b - .... + ... = 2a + b$

10 ABCDEF is an L-shape.

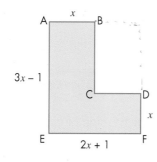

$AB = DF = x$

$AE = 3x - 1$ and $EF = 2x + 1$

> **HINTS AND TIPS**
>
> Make sure your explanation uses expressions. Do not try to explain in words alone.

    a Explain why the length $BC = 2x - 1$.

    b Find the perimeter of the shape in terms of x.

    c If $x = 2.5$ cm, what is the perimeter of the shape?

11 A teacher asks her class to work out the perimeter of this L-shape.

Tia says: 'There is information missing so you cannot work out the perimeter.'

Maria says: 'The perimeter is $4x - 1 + 4x - 1 + 3x + 2 + 3x + 2$.'

Who is correct?

Explain your answer.

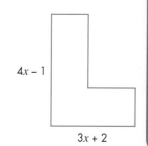

Expanding brackets

In mathematics, to '**expand**' usually means 'multiply out'. For example, expressions such as $3(y + 2)$ and $4y^2(2y + 3)$ can be expanded by multiplying them out.

Remember that there is an invisible multiplication sign between the outside number and the opening bracket. So $3(y + 2)$ is really $3 \times (y + 2)$ and $4y^2(2y + 3)$ is really $4y^2 \times (2y + 3)$.

You expand by multiplying *everything inside* the brackets by what is outside the brackets.

So in the case of the two examples above,

$$3(y + 2) = 3 \times (y + 2) = 3 \times y + 3 \times 2 = 3y + 6$$

$$4y^2(2y + 3) = 4y^2 \times (2y + 3) = 4y^2 \times 2y + 4y^2 \times 3 = 8y^3 + 12y^2$$

Look at these next examples of expansion, which show clearly how the term outside the brac has been multiplied with the terms inside them.

$2(m + 3) = 2m + 6$ $y(y^2 - 4x) = y^3 - 4xy$

$3(2t + 5) = 6t + 15$ $3x^2(4x + 5) = 12x^3 + 15x^2$

$m(p + 7) = mp + 7m$ $-3(2 + 3x) = -6 - 9x$

$x(x - 6) = x^2 - 6x$ $-2x(3 - 4x) = -6x + 8x^2$

$4t(t^3 + 2) = 4t^4 + 8t$ $3t(2 + 5t - p) = 6t + 15t^2 - 3pt$

Note: The signs change when a negative quantity is outside the brackets. For example,

$a(b + c) = ab + ac$ $a(b - c) = ab - ac$

$-a(b + c) = -ab - ac$ $-a(b - c) = -ab + ac$

$-(a - b) = -a + b$ $-(a + b - c) = -a - b + c$

Note: A minus sign on its own in front of the brackets is actually -1, so:

$$-(x + 2y - 3) = -1 \times (x + 2y - 3) = -1 \times x + -1 \times 2y + -1 \times -3 = -x - 2y + 3$$

The effect of a minus sign outside the brackets is to change the sign of everything inside the brackets.

EXERCISE 12C

1   Expand these expressions.

a $2(3 + m)$ **b** $5(2 + l)$ **c** $3(4 - y)$ **d** $4(5 + 2k)$

e $3(2 - 4f)$ **f** $2(5 - 3w)$ **g** $5(2k + 3m)$ **h** $4(3d - 2n)$

i $t(t + 3)$ **j** $k(k - 3)$ **k** $4t(t - 1)$ **l** $2k(4 - k)$

m $4g(2g + 5)$ **n** $5h(3h - 2)$ **o** $y(y^2 + 5)$ **p** $h(h^3 + 7)$

q $k(k^2 - 5)$ **r** $3t(t^2 + 4)$ **s** $3d(5d^2 - d^3)$ **t** $3w(2w^2 + t)$

u $5a(3a^2 - 2b)$ **v** $3p(4p^3 - 5m)$ **w** $4h^2(3h + 2g)$ **x** $2m^2(4m + m$

2 The local shop is offering $1 off a large tin of biscuits. Morris wants five tins.

    a If the price of one tin is t, which of the expressions below represents how much it will cost Morris to buy five tins?

$$5(t-1) \qquad 5t-1 \qquad t-5 \qquad 5t-5$$

    b Morris has $20 to spend. Will he have enough money for five tins?
Let $t = \$4.50$. Show working to justify your answer.

3 Dylan wrote the following:

$$3(5x-4) = 8x-4$$

Dylan has made two mistakes.

Explain the mistakes that Dylan has made.

> **HINTS AND TIPS**
>
> It is not enough to give the right answer. You must try to explain why Dylan wrote 8 for 3 × 5 instead of 15.

4 The expansion $2(x + 3) = 2x + 6$ can be shown by this diagram.

	x	3
2	$2x$	6

    a What expansion is shown in this diagram?

3	$6y$	9

    b Write down an expansion that is shown on this diagram.

	$12z$	8

Expand and simplify

When two brackets are expanded there are often like terms that can be collected together. Algebraic expressions should always be **simplified** as much as possible.

EXAMPLE 5

$$3(4 + m) + 2(5 + 2m) = 12 + 3m + 10 + 4m = 22 + 7m$$

EXAMPLE 6

$$3t(5t + 4) - 2t(3t - 5) = 15t^2 + 12t - 6t^2 + 10t = 9t^2 + 22t$$

EXERCISE 12D

1 Simplify these expressions.

  a $4t + 3t$

  b $3d + 2d + 4d$

  c $5e - 2e$

  d $3t - t$

  e $2t^2 + 3t^2$

  f $6y^2 - 2y^2$

  g $3ab + 2ab$

  h $7a^2d - 4a^2d$

2 Find the missing terms to make these equations true.

  a $4x + 5y + \ldots - \ldots = 6(x - y)$ **b** $3a - 6b - \ldots + \ldots = 2(a + b)$

3 ABCDEF is an 'L' shape.
AB = DE = x
AF = $3x - 1$ and EF = $2(x + 1)$

  a Explain why the length
  BC = $2x - 1$.

  b Find the perimeter of the
  shape in terms of x.

  c If $x = 2.5$ cm what is the perimeter of the shape?

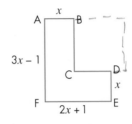

4 Expand and simplify.

  a $3(4 + t) + 2(5 + t)$

  b $5(3 + 2k) + 3(2 + 3k)$

  c $4(3 + 2f) + 2(5 - 3f)$

  d $5(1 + 3g) + 3(3 - 4g)$

5 Expand and simplify.

  a $4(3 + 2h) - 2(5 + 3h)$

  b $5(3g + 4) - 3(2g + 5)$

  c $5(5k + 2) - 2(4k - 3)$

  d $4(4e + 3) - 2(5e - 4)$

6 Expand and simplify.

  a $m(4 + p) + p(3 + m)$

  b $k(3 + 2h) + h(4 + 3k)$

  c $4r(3 + 4p) + 3p(8 - r)$

  d $5k(3m + 4) - 2m(3 - 2k)$

7 Expand and simplify.

  a $t(3t + 4) + 3t(3 + 2t)$

  b $2y(3 + 4y) + y(5y - 1)$

  c $4e(3e - 5) - 2e(e - 7)$

  d $3k(2k + p) - 2k(3p - 4k)$

8 Expand and simplify.

  a $4a(2b + 3c) + 3b(3a + 2c)$

  b $3y(4w + 2t) + 2w(3y - 4t)$

  c $5m(2n - 3p) - 2n(3p - 2m)$

  d $2r(3r + r^2) - 3r^2(4 - 2r)$

9 Fill in whole-number values so that the following expansion is true.

$$3(\ldots x + \ldots y) + 2(\ldots x + \ldots y) = 11x + 17y$$

> **HINTS AND TIPS**
>
> There is more than one answer. You don't have to give them all.

10 A rectangle with sides 5 and $3x + 2$ has a smaller rectangle with sides 3 and $2x - 1$ cut from it.

Work out the remaining area.

> **HINTS AND TIPS**
>
> Write out the expression for the difference between the two rectangles and then work it out.

2.3 ## Factorisation

Factorisation is the opposite of expansion. It puts an expression back into the brackets it may have come from.

In factorisation, you have to look for the **common factors** in *every* term of the expression.

EXAMPLE 7

Factorise *each expression.*

a $6t + 9m$ **b** $6my + 4py$ **c** $5k^2 - 25k$ **d** $10a^2b - 15ab^2$

a First look for common numerical factors.
3 is a common factor of $6t$ and $9m$.
$6t + 9m = 3(2t + 3m)$

b Look for any numerical common factors.
$6my + 4py = 2(3my + 2py)$.
Then look for any common variables.
y is a common variable giving $2y(3m + 2p)$

c 5 is a common factor of 5 and 25 and k is a common factor of k^2 and k.
$5k^2 - 25k = 5k(k - 5)$

d 5 is a common factor of 10 and 15, a is a common factor of a^2 and a, b is a common factor of b and b^2.
$10a^2b - 15ab^2 = 5ab(2a - 3b)$

Note: If you multiply out each answer, you will get the expressions you started with.

1 Factorise the following expressions.

a $6m + 12t$

b $9t + 3p$

c $8m + 12k$

d $4r + 8t$

e $mn + 3m$

f $5g^2 + 3g$

g $4w - 6t$

h $3y^2 + 2y$

i $4t^2 - 3t$

j $3m^2 - 3mp$

k $6p^2 + 9pt$

l $8pt + 6mp$

m $8ab - 4bc$

n $5b^2c - 10bc$

o $8abc + 6bed$

p $4a^2 + 6a + 8$

q $6ab + 9bc + 3bd$

r $5t^2 + 4t + at$

s $6mt^2 - 3mt + 9m^2t$

t $8ab^2 + 2ab - 4a^2b$

u $10pt^2 + 15pt + 5p^2t$

2 Three friends have a meal together. They each have a main meal costing $6.75 and a dessert costing $3.25.

Chris says that the bill will be $3 \times 6.75 + 3 \times 3.25$.

Suni says that she has an easier way to work out the bill as $3 \times (6.75 + 3.25)$.

a Explain why Chris' and Suni's methods both give the correct answer.

b Explain why Suni's method is better.

c What is the total bill?

3 Factorise the following expressions where possible. List those that do not factorise.

a $7m - 6t$

b $5m + 2mp$

c $t^2 - 7t$

d $8pt + 5ab$

e $4m^2 - 6mp$

f $a^2 + b$

g $4a^2 - 5ab$

h $3ab + 4cd$

i $5ab - 3b^2c$

4 Three students are asked to factorise the expression $12m - 8$ completely. These are their answers.

Ahmed	Bernice	Craig
$2(6m - 4)$	$4(3m - 2)$	$4m(3 - \frac{2}{m})$

All the answers are accurately factorised, but only one is the simplest one.

a Which student gave the simplest factorisation?

b Explain why the other two students' answers are not acceptable as correct answers.

5 Explain why $5m + 6p$ cannot be factorised.

Quadratic expansion

A **quadratic expression** is one in which the highest power of the variables is 2.
For example,

y^2 $3t^2 + 5t$ $5m^2 + 3m + 8$

An expression such as $(3y + 2)(4y - 5)$ can be expanded to give a quadratic expression.

Multiplying out such pairs of brackets is usually called **quadratic expansion**.

The rule for expanding expressions such as $(t + 5)(3t - 4)$ is similar to that for expanding single brackets: multiply everything in one set of brackets by everything in the other set of brackets.

There are several methods for doing this. Examples 8 to 10 show the three main methods: expansion, FOIL and the box method.

EXAMPLE 8

In the expansion method, split the terms in the first set of brackets, and multiply each of them by both terms in the second set of brackets, then simplify the outcome.

Expand $(x + 3)(x + 4)$

$$(x + 3)(x + 4) = x(x + 4) + 3(x + 4)$$
$$= x^2 + 4x + 3x + 12$$
$$= x^2 + 7x + 12$$

EXAMPLE 9

FOIL stands for First, Outer, Inner and Last. This is the order of multiplying the terms from each set of brackets.

Expand $(t + 5)(t - 2)$

First terms give: $t \times t = t^2$

Outer terms give: $t \times -2 = -2t$

Inner terms give: $5 \times t = 5t$

Last terms give: $+5 \times -2 = -10$

$$(t + 5)(t - 2) = t^2 - 2t + 5t - 10$$
$$= t^2 + 3t - 10$$

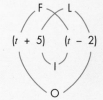

EXAMPLE 10

The "box method" can be used to lay out the multiplication.

Expand $(k - 3)(k - 2)$

$$(k - 3)(k - 2) = k^2 - 2k - 3k + 6$$
$$= k^2 - 5k + 6$$

	k	-3
k	k^2	$-3k$
-2	$-2k$	$+6$

Warning: Be careful with the signs. This is where mistakes are frequently made in questions involving the expansion of brackets.

EXERCISE 12F

Expand the expressions in questions **1–17**.

1 $(x + 3)(x + 2)$

2 $(t + 4)(t + 3)$

3 $(w + 1)(w + 3)$

4 $(m + 5)(m + 1)$

5 $(k + 3)(k + 5)$

6 $(a + 4)(a + 1)$

7 $(x + 4)(x - 2)$

8 $(t + 5)(t - 3)$

9 $(w + 3)(w - 1)$

10 $(f + 2)(f - 3)$

11 $(g + 1)(g - 4)$

12 $(y + 4)(y - 3)$

13 $(x - 3)(x + 4)$

14 $(p - 2)(p + 1)$

15 $(k - 4)(k + 2)$

16 $(y - 2)(y + 5)$

17 $(a - 1)(a + 3)$

> **HINTS AND TIPS**
>
> A common error is to get minus signs wrong.
> $-2x - 3x = -5x$ but
> $-2x \times -3 = +6x$

The expansions of the expressions in questions **18–26** follow a pattern. Work out the first few and try to spot the pattern that will allow you immediately to write down the answers to the rest.

18 $(x + 3)(x - 3)$

19 $(t + 5)(t - 5)$

20 $(m + 4)(m - 4)$

21 $(t + 2)(t - 2)$

22 $(y + 8)(y - 8)$

23 $(p + 1)(p - 1)$

24 $(5 + x)(5 - x)$

25 $(7 + g)(7 - g)$

26 $(x - 6)(x + 6)$

27 This rectangle is made up of four parts with areas of x^2, $2x$, $3x$ and 6 square units.

x^2	$2x$
$3x$	6

Work out expressions for the sides of the rectangle, in terms of x.

FOUNDATION

28 This square has an area of x^2 square units.
It is split into four rectangles.

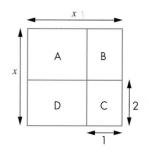

a Copy and complete the table below to show the dimensions and area of each rectangle.

Rectangle	Length	Width	Area
A	$x - 1$	$x - 2$	$(x - 1)(x - 2)$
B			
C			
D			

b Add together the areas of rectangles B, C and D.

Expand any brackets and collect terms together.

c Use the results to explain why $(x - 1)(x - 2) = x^2 - 3x + 2$.

29 **a** Expand $(x - 3)(x + 3)$

b Use the result in **a** to write down the answers to these. (Do not use a calculator or do a long multiplication.)

i 97×103 **ii** 197×203

12.5   Multiplying more complex expresions

In a pair of brackets like $(x + 4)(x - 3)$ the coefficient of x in each case is 1. We will now look at examples like $(3x + 4)(2x - 3)$ where the coefficients of x are not 1.

EXAMPLE 11

Expand $(2t + 3)(3t + 1)$

$(2t + 3)(3t + 1) = 6t^2 + 2t + 9t + 3$
$\qquad\qquad\qquad = 6t^2 + 11t + 3$

	$2t$	$+3$
$3t$	$6t^2$	$+9t$
$+1$	$+2t$	$+3$

EXAMPLE 12

Expand $(4x - 1)(3x - 5)$

$(4x - 1)(3x - 5) = 4x(3x - 5) - (3x - 5)$ [**Note:** $(3x - 5)$ is the same as $1(3x - 5)$.]
$\qquad\qquad\qquad = 12x^2 - 20x - 3x + 5$
$\qquad\qquad\qquad = 12x^2 - 23x + 5$

EXERCISE 12G

Expand the expressions in questions **1–21**.

1 $(2x + 3)(3x + 1)$

2 $(3y + 2)(4y + 3)$

3 $(3t + 1)(2t + 5)$

4 $(4t + 3)(2t - 1)$

5 $(5m + 2)(2m - 3)$

6 $(4k + 3)(3k - 5)$

7 $(3p - 2)(2p + 5)$

8 $(5w + 2)(2w + 3)$

9 $(2a - 3)(3a + 1)$

10 $(4r - 3)(2r - 1)$

11 $(3g - 2)(5g - 2)$

12 $(4d - 1)(3d + 2)$

13 $(5 + 2p)(3 + 4p)$

14 $(2 + 3t)(1 + 2t)$

15 $(4 + 3p)(2p + 1)$

16 $(6 + 5t)(1 - 2t)$

17 $(4 + 3n)(3 - 2n)$

18 $(2 + 3f)(2f - 3)$

19 $(3 - 2q)(4 + 5q)$

20 $(1 - 3p)(3 + 2p)$

21 $(4 - 2t)(3t + 1)$

> ### HINTS AND TIPS
>
> Always give answers in the form $\pm ax^2 \pm bx \pm c$ even if the quadratic coefficient is negative.

22 Expand:

a $(x + 1)(x - 1)$

b $(2x + 1)(2x - 1)$

c $(2x + 3)(2x - 3)$

d Use the results in parts **a**, **b** and **c** to write down the expansion of $(3x + 5)(3x - 5)$.

23 a Without expanding the brackets, match each expression on the left with an expression on the right. One is done for you.

$(3x - 2)(2x + 1)$ $4x^2 - 4x + 1$

$(2x - 1)(2x - 1)$ $6x^2 - x - 2$

$(6x - 3)(x + 1)$ $6x^2 + 7x + 2$

$(4x + 1)(x - 1)$ $6x^2 + 3x - 3$

$(3x + 2)(2x + 1)$ $4x^2 - 3x - 1$

b Taking any expression on the left, explain how you can match it with an expression on the right without expanding the brackets.

EXERCISE 12H

Try to spot the pattern in each of the expressions in questions **1–15** so that you can immediately write down the expansion.

1 $(2x + 1)(2x - 1)$

2 $(3t + 2)(3t - 2)$

3 $(5y + 3)(5y - 3)$

4 $(4m + 3)(4m - 3)$

5 $(2k - 3)(2k + 3)$

6 $(4h - 1)(4h + 1)$

7 $(2 + 3x)(2 - 3x)$

8 $(5 + 2t)(5 - 2t)$

9 $(6 - 5y)(6 + 5y)$

10 $(a + b)(a - b)$

11 $(3t + k)(3t - k)$

12 $(2m - 3p)(2m + 3p)$

13 $(5k + g)(5k - g)$

14 $(ab + cd)(ab - cd)$

15 $(a^2 + b^2)(a^2 - b^2)$

16 Imagine a square of side a units with a square of side b units cut from one corner.

   a What is the area remaining after the small square is cut away?

   b The remaining area is cut into rectangles, A, B and C, and rearranged as shown.

     Write down the dimensions and area of the rectangle formed by A, B and C.

   c Explain why $a^2 - b^2 = (a + b)(a - b)$.

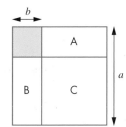
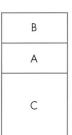

17 Explain why the areas of the shaded regions are the same.

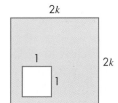
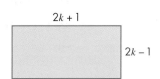

Expanding squares

Whenever you see a **linear** bracketed term squared you must write the brackets down twice and then use whichever method you prefer to expand.

EXAMPLE 13

Expand $(x + 3)^2$

$(x + 3)^2 = (x + 3)(x + 3)$
$= x(x + 3) + 3(x + 3)$
$= x^2 + 3x + 3x + 9$
$= x^2 + 6x + 9$

EXAMPLE 14

Expand $(3x - 2)^2$

$$(3x - 2)^2 = (3x - 2)(3x - 2)$$
$$= 9x^2 - 6x - 6x + 4$$
$$= 9x^2 - 12x + 4$$

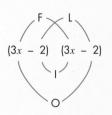

EXERCISE 12I

Expand the squares in questions **1–24** and simplify.

HINTS AND TIPS

Remember *always* write down the brackets twice. Do not try to take any short cuts.

1 $(x + 5)^2$ **2** $(m + 4)^2$ **3** $(6 + t)^2$

4 $(3 + p)^2$ **5** $(m - 3)^2$ **6** $(t - 5)^2$

7 $(4 - m)^2$ **8** $(7 - k)^2$

9 $(3x + 1)^2$ **10** $(4t + 3)^2$ **11** $(2 + 5y)^2$ **12** $(3 + 2m)^2$

13 $(4t - 3)^2$ **14** $(3x - 2)^2$ **15** $(2 - 5t)^2$ **16** $(6 - 5r)^2$

17 $(x + y)^2$ **18** $(m - n)^2$ **19** $(2t + y)^2$ **20** $(m - 3n)^2$

21 $(x + 2)^2 - 4$ **22** $(x - 5)^2 - 25$ **23** $(x + 6)^2 - 36$ **24** $(x - 2)^2 - 4$

25 A teacher asks her class to expand $(3x + 1)^2$.

Marcela's answer is $9x^2 + 1$.

Paulo's answer is $3x^2 + 6x + 1$.

a Explain the mistakes that Marcela has made.

b Explain the mistakes that Paulo has made.

c Work out the correct answer.

26 Use the diagram to show algebraically and diagrammatically that:

$$(2x - 1)^2 = 4x^2 - 4x + 1$$

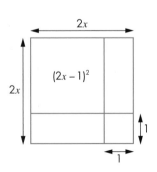

Quadratic factorisation

Factorisation involves putting a quadratic expression back into its **brackets** (if possible). We start with the factorisation of quadratic expressions of the type:

$$x^2 + ax + b$$

where a and b are integers.

There are some simple rules that will help you to factorise.

- The expression inside each set of brackets will start with an x, and the signs in the quadratic expression show which signs to put after the xs.

- When the second sign in the expression is a plus, the signs in both sets of brackets are the same as the first sign.

  $x^2 + ax + b = (x + ?)(x + ?)$ Since everything is positive.

  $x^2 - ax + b = (x - ?)(x - ?)$ Since negative × negative = positive

- When the *second* sign is a *minus*, the signs in the brackets are *different*.

  $x^2 + ax - b = (x + ?)(x - ?)$ Since positive × negative = negative

  $x^2 - ax - b = (x + ?)(x - ?)$

- Next, look at the *last* number, b, in the expression. When multiplied together, the two numbers in the brackets must give b.

- Finally, look at the coefficient *of x*, which is a. The *sum* of the two *numbers* in the brackets will give a.

EXAMPLE 15

Factorise $x^2 - x - 6$

Because of the signs we know the brackets must be $(x + ?)(x - ?)$.

Two numbers that have a product of -6 and a sum of -1 are -3 and $+2$.

So, $x^2 - x - 6 = (x + 2)(x - 3)$

EXAMPLE 16

Factorise $x^2 - 9x + 20$

Because of the signs we know the brackets must be $(x - ?)(x - ?)$.

Two numbers that have a product of $+20$ and a sum of -9 are -4 and -5.

So, $x^2 - 9x + 20 = (x - 4)(x - 5)$

EXERCISE 12J

FOUNDATION

Factorise the expressions in questions **1–40**.

1 $x^2 + 5x + 6$

2 $t^2 + 5t + 4$

3 $m^2 + 7m + 10$

4 $k^2 + 10k + 24$

5 $p^2 + 14p + 24$

6 $r^2 + 9r + 18$

7 $w^2 + 11w + 18$

8 $x^2 + 7x + 12$

9 $a^2 + 8a + 12$

10 $k^2 + 10k + 21$

11 $f^2 + 22f + 21$

12 $b^2 + 20b + 96$

13 $t^2 - 5t + 6$

14 $d^2 - 5d + 4$

15 $g^2 - 7g + 10$

16 $x^2 - 15x + 36$

17 $c^2 - 18c + 32$

18 $t^2 - 13t + 36$

19 $y^2 - 16y + 48$

20 $j^2 - 14j + 48$

21 $p^2 - 8p + 15$

22 $y^2 + 5y - 6$

23 $t^2 + 2t - 8$

24 $x^2 + 3x - 10$

25 $m^2 - 4m - 12$

26 $r^2 - 6r - 7$

27 $n^2 - 3n - 18$

28 $m^2 - 7m - 44$

29 $w^2 - 2w - 24$

30 $t^2 - t - 90$

31 $h^2 - h - 72$

32 $t^2 - 2t - 63$

33 $d^2 + 2d + 1$

34 $y^2 + 20y + 100$

35 $t^2 - 8t + 16$

36 $m^2 - 18m + 81$

37 $x^2 - 24x + 144$

38 $d^2 - d - 12$

39 $t^2 - t - 20$

40 $q^2 - q - 56$

HINTS AND TIPS

First decide on the signs in the brackets, then look at the numbers.

41 This rectangle is made up of four parts. Two of the parts have areas of x^2 and 6 square units.

The sides of the rectangle are of the form $x + a$ and $x + b$.

There are two possible answers for a and b.

Work out both answers and copy and complete the areas in the other parts of the rectangle.

In a quadratic expression like $x^2 - 6x + 8$ the coefficient of x^2 is 1. This was the case for all the questions in the last section. We can extend the method shown to factorise quadratic expressions such as $3x^2 + x - 2$ or $4x^2 + 8x - 5$ where the coefficient of x^2 is not 1.

EXAMPLE 17

Factorise $3x^2 + 8x + 4$

- First, note that both signs are positive. So the signs in the brackets must be $(?x + ?)(?x + ?)$.
- As 3 has only 3×1 as factors, the brackets must be $(3x + ?)(x + ?)$.
- Next, note that the factors of 4 are 4×1 and 2×2.
- That gives three possible pair of brackets:

  $(3x + 4)(x + 1) = 3x^2 + 4x + 3x + 4$ $(3x + 1)(x + 4) = 3x^2 + 12x + x + 4$
  $(3x + 2)(x + 2) = 3x^2 + 6x + 2x + 4$

- Only the last one gives the correct expansion, $3x^2 + 8x + 4 = (3x + 2)(x + 2)$

EXAMPLE 18

Factorise $6x^2 - 7x - 10$

- First, note that both signs are negative. So the signs in the brackets must be $(?x + ?)(?x - ?)$.
- As 6 has 6×1 and 3×2 as factors, the brackets could be $(6x \pm ?)(x \pm ?)$ or $(3x \pm ?)(2x \pm ?)$.
- Two numbers that multiply to make -10 could be: 2 and -5; -2 and 5; 1 and -10; or -1 and 10
- There are 16 different ways in which these numbers could be put in the two brackets.

  $(3x + 2)(2x - 5)$ $(3x - 5)(2x + 2)$ $\mathbf{(3x - 2)(2x + 5)}$ $(3x + 5)(2x - 2)$
  $(3x + 1)(2x - 10)$ $(3x - 1)(2x + 10)$ $\mathbf{(3x + 10)(2x - 1)}$ $\mathbf{(3x - 10)(2x + 1)}$
  $(6x + 2)(x - 5)$ $\mathbf{(6x - 5)(x + 2)}$ $(6x - 2)(x + 5)$ $\mathbf{(6x + 5)(x - 2)}$
  $\mathbf{(6x + 1)(x - 10)}$ $(6x - 1)(x + 10)$ $(6x + 10)(x - 1)$ $(6x - 10)(x + 1)$

  However, we can immediately reject any bracket which can be factorised by taking out a constant, such as $(2x + 2)$ or $(6x - 10)$, because this cannot be done with the original expression.

  We only need to look at the expressions in **bold**. The list reduces to

  $(3x + 2)(2x - 5)$ $(3x - 2)(2x + 5)$ $(3x + 10)(2x - 1)$ $(3x - 10)(2x + 1)$
  $(6x - 5)(x + 2)$ $(6x + 5)(x - 2)$ $(6x + 1)(x - 10)$ $(6x - 1)(x + 10)$

  You need to try each of these until you get the correct one.

  You should not need to try them all. With experience you will get better at deciding which one to try first.

- Check that the correct answer is $6x^2 - 7x - 10 = (6x + 5)(x - 2)$

Although this seems to be very complicated, it becomes quite easy with practice and experience.

EXERCISE 12K

Factorise the expressions in questions **1–12**.

1 $2x^2 + 5x + 2$

2 $7x^2 + 8x + 1$

3 $4x^2 + 3x - 7$

4 $24t^2 + 19t + 2$

5 $15t^2 + 2t - 1$

6 $16x^2 - 8x + 1$

7 $6y^2 + 33y - 63$

8 $4y^2 + 8y - 96$

9 $8x^2 + 10x - 3$

10 $6t^2 + 13t + 5$

11 $3x^2 - 16x - 12$

12 $7x^2 - 37x + 10$

13 This rectangle is made up of four parts, with areas of $12x^2$, $3x$, $8x$ and 2 square units.

Work out expressions for the sides of the rectangle, in terms of x.

$12x^2$	$3x$
$8x$	2

14 Three students are asked to factorise the expression $6x^2 + 30x + 36$ completely.

These are their answers.   **Adam**   **Bella**   **Cara**

$(6x + 12)(x + 3)$ $(3x + 6)(2x + 6)$ $(2x + 4)(3x + 9)$

All the answers are correctly factorised.

a Explain why one quadratic expression can have three different factorisations.

b Which of the following is the most complete factorisation?

$2(3x + 6)(x + 3)$ $6(x + 2)(x + 3)$ $3(x + 2)(2x + 6)$

Explain your choice.

c What is the geometrical significance of the answers to parts **a** and **b**?

12.8 More than two brackets

Sometimes you will have expressions with more than two brackets multiplied together. You can expand these by multiplying out two brackets at a time.

EXAMPLE 19

Expand $(x + 2)(x - 1)(x - 2)$ and simplify the result as much as possible.

Start with the first two brackets.

$(x + 2)(x - 1) \qquad = x^2 + 2x - x - 2$

$\qquad\qquad\qquad\quad = x^2 + x - 2$

Multiply this by the third bracket

$(x^2 + x - 2)(x - 2) \; = x^3 + x^2 - 2x^2 - 2x - 2x + 4$

$\qquad\qquad\qquad\quad = x^3 - x^2 - 4x + 4$

	x^2	$+x$	-2
x	x^3	$+x^2$	$-2x$
-2	$-2x^2$	$-2x$	$+4$

You can multiply the brackets in any order.

In Example 19 you could start with the first and third brackets:

$(x + 2)(x - 2) = x^2 + 2x - 2x - 4$

$\qquad\qquad\quad = x^2 - 4$

And then multiply by the second bracket: $(x^2 - 4)(x - 1) = x^3 - x^2 - 4x + 4$

	x^2	-4
x	x^3	$-4x$
-1	$-x^2$	$+4$

EXERCISE 12L

1 Expand and simplify

   a $(x + 1)(x + 2)$ **b** $(x + 1)(x + 2)(x + 3)$

2 Expand and simplify

   a $(x + 2)(x - 1)(x + 2)$ **b** $(x + 5)(x - 3)(x + 1)$

3 Expand and simplify

   a $(x + 3)^2$ **b** $x(x + 3)^2$ **c** $(x - 1)(x + 3)^2$

4 Expand and simplify

   a $(x + 10)(x - 2)(x + 2)$ **b** $(x - 1)(x + 4)(x - 1)$

5 Expand and simplify

   a $(x + 1)^2(x + 3)$ **b** $(x - 1)^2(x + 3)$

6 Expand and simplify

   a $(x + 1)^3$ **b** $(x - 1)^3$ **c** $(x + 2)^3$ **d** $(x - 2)^3$

7 **a** Explain how the area of this square   **b** Explain how the volume of this cube
       shows that $(x + 1)^2 = x^2 + 2x + 1$ shows that $(x + 1)^3 = x^3 + 3x^2 + 3x + 1$

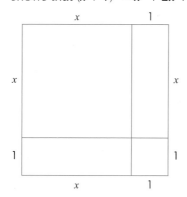

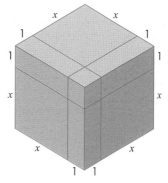

8 $(x + 1)(x - 2)(x + c)$ expanded and simplified is $x^3 + 2x^2 - 5x - 6$

   Work out the value of c.

HIGHER

9 Expand and simplify

   a $(2x + 1)(x - 2)$ **b** $(2x + 1)(x - 2)(x + 3)$

10 Expand and simplify

   a $(3x - 1)(4x + 1)$ **b** $(3x - 1)(4x + 1)(x - 2)$

11 Expand and simplify

   a $(2x - 5)(3x - 2)$ **b** $(2x - 5)(3x - 2)(2x - 1)$

12.9 Algebraic fractions

The following five rules are used to work out the value of fractions.

Cancelling: $\dfrac{ac}{ad} = \dfrac{c}{d}$

                  a is a common factor of the numerator and the denominator and so can be removed from both.

Addition: $\dfrac{a}{b} + \dfrac{c}{d} = \dfrac{ad + bc}{bd}$

Subtraction: $\dfrac{a}{b} - \dfrac{c}{d} = \dfrac{ad - bc}{bd}$

Multiplication: $\dfrac{a}{b} \times \dfrac{c}{d} = \dfrac{ac}{bd}$

Division: $\dfrac{a}{b} \div \dfrac{c}{d} = \dfrac{a}{b} \times \dfrac{d}{c} = \dfrac{ad}{bc}$

To divide by a fraction, multiply by the reciprocal (see chapter 3).

Note that a, b, c and d can be numbers, other letters or algebraic expressions. Remember:

- use brackets, if necessary
- factorise if you can
- **cancel** if you can.

EXAMPLE 20

Simplify **a** $\dfrac{1}{x} + \dfrac{x}{2y}$ **b** $\dfrac{2}{b} - \dfrac{a}{2b}$

a Using the addition rule: $\dfrac{1}{x} + \dfrac{x}{2y} = \dfrac{(1)(2y) + (x)(x)}{(x)(2y)} = \dfrac{2y + x^2}{2xy}$

b Using the subtraction rule: $\dfrac{2}{b} - \dfrac{a}{2b} = \dfrac{(2)(2b) - (a)(b)}{(b)(2b)} = \dfrac{4b - ab}{2b^2}$

$$= \dfrac{\cancel{b}(4 - a)}{2b\cancel{^2}} = \dfrac{4 - a}{2b}$$

Note: There are different ways of working out fraction calculations. Part **b** could have been done by making the denominator of each fraction the same.

$$\dfrac{(2)2}{(2)b} - \dfrac{a}{2b} = \dfrac{4 - a}{2b}$$

EXAMPLE 21

Simplify **a** $\dfrac{x}{3} \times \dfrac{x + 2}{x - 2}$ **b** $\dfrac{x}{3} \div \dfrac{2x}{7}$

a Using the multiplication rule: $\dfrac{x}{3} \times \dfrac{x + 2}{x - 2} = \dfrac{(x)(x + 2)}{(3)(x - 2)} = \dfrac{x^2 + 2x}{3x - 6}$

You can cancel common factors to simplify. Note that it is sometimes preferable to leave an algebraic fraction in a factorised form.

b Using the division rule: $\dfrac{x}{3} \div \dfrac{2x}{7} = \dfrac{(x)(7)}{(3)(2x)} = \dfrac{7}{6}$

EXAMPLE 22

Write $\dfrac{3}{x - 1} - \dfrac{2}{x + 1}$ as a **single fraction** as simply as possible.

Using the subtraction rule:

$$\dfrac{3}{x - 1} - \dfrac{2}{x + 1} = \dfrac{3(x + 1) - 2(x - 1)}{(x - 1)(x + 1)}$$

$$= \dfrac{3x + 3 - 2x + 2}{(x - 1)(x + 1)}$$

$$= \dfrac{x + 5}{(x - 1)(x + 1)}$$

EXAMPLE 23

Simplify this expression. $\dfrac{2x^2 + x - 3}{4x^2 - 9}$

Factorise the numerator and denominator: $\dfrac{(2x + 3)(x - 1)}{(2x + 3)(2x - 3)}$

Denominator is the difference of two squares.

Cancel any common factors: $\dfrac{\cancel{(2x + 3)}(x - 1)}{\cancel{(2x + 3)}(2x - 3)}$

The remaining fraction is the answer: $\dfrac{(x - 1)}{(2x - 3)}$

EXERCISE 12M

HIGHER

1 Simplify each of these.

a $\dfrac{x}{2} + \dfrac{x}{3}$ b $\dfrac{3x}{4} + \dfrac{x}{5}$ c $\dfrac{3x}{4} + \dfrac{2x}{5}$ d $\dfrac{x}{2} + \dfrac{y}{3}$

e $\dfrac{xy}{4} + \dfrac{2}{x}$ f $\dfrac{x + 1}{2} + \dfrac{x + 2}{3}$ g $\dfrac{2x + 1}{2} + \dfrac{3x + 1}{4}$ h $\dfrac{x}{5} + \dfrac{2x + 1}{3}$

i $\dfrac{x - 2}{2} + \dfrac{x - 3}{4}$ j $\dfrac{x - 4}{4} + \dfrac{2x - 3}{2}$

2 Simplify each of these.

a $\dfrac{x}{2} - \dfrac{x}{3}$ b $\dfrac{3x}{4} - \dfrac{x}{5}$ c $\dfrac{3x}{4} - \dfrac{2x}{5}$ d $\dfrac{x}{2} - \dfrac{y}{3}$

e $\dfrac{xy}{4} - \dfrac{2}{y}$ f $\dfrac{x + 1}{2} - \dfrac{x + 2}{3}$ g $\dfrac{2x + 1}{2} - \dfrac{3x + 1}{4}$ h $\dfrac{x}{5} - \dfrac{2x + 1}{3}$

i $\dfrac{x - 2}{2} - \dfrac{x - 3}{4}$ j $\dfrac{x - 4}{4} - \dfrac{2x - 3}{2}$

3 Simplify each of these.

a $\dfrac{x}{2} \times \dfrac{x}{3}$ b $\dfrac{2x}{7} \times \dfrac{3y}{4}$ c $\dfrac{4x}{3y} \times \dfrac{2y}{x}$ d $\dfrac{4y^2}{9x} \times \dfrac{3x^2}{2y}$

e $\dfrac{x}{2} \times \dfrac{x - 2}{5}$ f $\dfrac{x - 3}{15} \times \dfrac{5}{2x - 6}$ g $\dfrac{2x + 1}{2} \times \dfrac{3x + 1}{4}$ h $\dfrac{x}{5} \times \dfrac{2x + 1}{3}$

i $\dfrac{x - 2}{2} \times \dfrac{4}{x - 3}$ j $\dfrac{x - 5}{10} \times \dfrac{5}{x^2 - 5x}$

4 Simplify each of these. Factorise and cancel where appropriate.

a $\dfrac{3x}{4} + \dfrac{x}{4}$ b $\dfrac{3x}{4} - \dfrac{x}{4}$ c $\dfrac{3x}{4} \times \dfrac{x}{4}$

d $\dfrac{3x}{4} \div \dfrac{x}{4}$ e $\dfrac{3x + 1}{2} + \dfrac{x - 2}{5}$ f $\dfrac{3x + 1}{2} - \dfrac{x - 2}{5}$

g $\dfrac{3x + 1}{2} \times \dfrac{x - 2}{5}$ h $\dfrac{x^2 - 9}{10} \times \dfrac{5}{x - 3}$ i $\dfrac{2x^2}{9} - \dfrac{2y^2}{3}$

HIGHER

5 Write these expressions as single fractions as simply as possible.

a $\dfrac{2}{x+1} + \dfrac{5}{x+2}$

b $\dfrac{4}{x-2} + \dfrac{7}{x+1}$

c $\dfrac{3}{4x+1} - \dfrac{4}{x+2}$

d $\dfrac{2}{2x-1} - \dfrac{6}{x+1}$

e $\dfrac{3}{2x-1} - \dfrac{4}{3x-1}$

6 For homework a teacher asks her class to simplify the expression $\dfrac{x^2-x-2}{x^2+x-6}$.

This is Tom's answer:

$$\frac{\cancel{x^2-x-2}^{\,-1}}{\cancel{x^2+x-6}_{\,+3}}$$

$$= \frac{-x-1}{x+3} = \frac{x+1}{x+3}$$

When she marked the homework, the teacher was in a hurry and only checked the answer, which was correct.

Tom made several mistakes. What are they?

7 An expression of the form $\dfrac{ax^2+bx-c}{dx^2-e}$ simplifies to $\dfrac{x-1}{2x-3}$.

What was the original expression?

8 Write these expressions as single fractions.

a $\dfrac{4}{x+1} + \dfrac{5}{x+2}$

b $\dfrac{18}{4x-1} - \dfrac{1}{x+1}$

c $\dfrac{2x-1}{2} - \dfrac{6}{x+1}$

d $\dfrac{3}{2x-1} - \dfrac{4}{3x-1}$

9 Simplify the following expressions.

a $\dfrac{x^2+2x-3}{2x^2+7x+3}$

b $\dfrac{4x^2-1}{2x^2+5x-3}$

c $\dfrac{6x^2+x-2}{9x^2-4}$

d $\dfrac{4x^2+x-3}{4x^2-7x+3}$

e $\dfrac{4x^2-25}{8x^2-22x+5}$

Why this chapter matters

We use equations to explain some of the most important things in the world.

Three of the most important are shown on this page.

Why does the Moon keep orbiting the Earth and not fly off into space?

This is explained by Newton's law of universal gravitation, which describes the gravitational attraction between two bodies:

$$F = G \times \frac{m_1 \times m_2}{r^2}$$

where F is the force between the bodies, G is the gravitational constant, m_1 and m_2 are the masses of the two bodies and r is the distance between them.

Why don't planes fall out of the sky?

This is explained by Bernoulli's principle, which states that as the speed of a fluid increases, its pressure decreases. This is what causes the difference in air pressure between the top and bottom of an aircraft wing, as shown in the diagram on the left.

In its simplest form, the equation can be written as:

$$p + q = p_0$$

where p = static pressure, q = dynamic pressure and p_0 is the total pressure.

How can a small amount of plutonium have enough energy to wipe out a city?

This is explained by Einstein's theory of special relativity, which connects mass and energy in the equation:

$$E = mc^2$$

where E is the energy, m is the mass and c is the speed of light. As the speed of light is nearly 300 000 kilometres per second, the amount of energy in a small mass is huge. If this can be released, it can be used for good (as in nuclear power stations) or harm (as in nuclear bombs).

13

Solutions of equations

What you need to be able to do in the examinations:

FOUNDATION	HIGHER
• Solve linear equations, with integer or fractional coefficients, in one unknown in which the unknown appears on either side or both sides of the equation.	• Solve quadratic equations by factorisation, by completing the square or by using the quadratic formula.
• Set up simple linear equations from given data.	• Form and solve quadratic equations from data given in a context.
• Solve quadratic equations by factorisation.	• Solve simultaneous equations in two unknowns, one equation being linear and the other being quadratic.
• Calculate the exact solution of two simultaneous equations in two unknowns.	

A teacher gave these instructions to her class.

What algebraic expression represents the teacher's statement?

- Think of a number.
- Double it.
- Add 3.

This is what two of her students said.

Can you work out Kim's answer and the number that Freda started with?

Kim's answer will be $2 \times 5 + 3 = 13$.

I chose the number 5

Freda's answer can be set up as an **equation**.

An equation is formed when an expression is put equal to a number or another expression.

You are expected to deal with equations that have only one **variable** or letter.

The **solution** to an equation is the value of the variable that makes the equation true.
For example, the equation for Freda's answer is

$$2x + 3 = 10$$

where x represents Freda's number.

The value of x that makes this true is $x = 3\frac{1}{2}$.

My final answer was 10.

Kim

Freda

To **solve** an equation like $2x + 3 = 10$, do the same thing to each side of the equation until we have x on its own.

$$2x + 3 = 10$$

Subtract 3 from both sides: $\quad 2x + 3 - 3 = 10 - 3$

$$2x = 7$$

Divide both sides by 2: $\quad \dfrac{2x}{2} = \dfrac{7}{2}$

$$x = 3\frac{1}{2}$$

Here is another example:

Mary had two bags of marbles, each of which contained the same number of marbles, and five spare marbles.

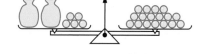

She put them on scales and balanced them with 17 single marbles.

How many marbles were there in each bag?

If x is the number of marbles in each bag, then the equation representing Mary's balanced scales is:

$$2x + 5 = 17$$

Take five marbles from each pan:

$$2x + 5 - 5 = 17 - 5$$
$$2x = 12$$

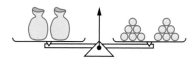

Now halve the number of marbles on each pan.

That is, divide both sides by 2:

$$\frac{2x}{2} = \frac{12}{2}$$
$$x = 6$$

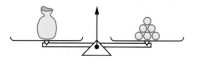

Checking the answer gives $2 \times 6 + 5 = 17$, which is correct.

EXAMPLE 1

Solve each of these equations by 'doing the same to both sides'.

a $3x - 5 = 16$

Add 5 to both sides.

$$3x - 5 + 5 = 16 + 5$$
$$3x = 21$$

Divide both sides by 3.

$$\frac{3x}{3} = \frac{21}{3}$$
$$x = 7$$

Checking the answer gives:

$$3 \times 7 - 5 = 16$$

which is correct.

b $\frac{x}{2} + 2 = 10$

Subtract 2 from both sides.

$$\frac{x}{2} + 2 - 2 = 10 - 2$$
$$\frac{x}{2} = 8$$

Multiply both sides by 2.

$$\frac{x}{2} \times 2 = 8 \times 2$$
$$x = 16$$

Checking the answer gives:

$$16 \div 2 + 2 = 10$$

which is correct.

EXERCISE 13A

1 Solve each of the following equations by 'doing the same to both sides'. Remember to check that each answer works for its original equation.

a $x + 4 = 60$

b $3y - 2 = 4$

c $3x - 7 = 11$

d $5y + 3 = 18$

HINTS AND TIPS

Be careful with negative numbers.

e $7 + 3t = 19$

f $5 + 4f = 15$

g $3 + 6k = 24$

h $4x + 7 = 17$

i $5m - 3 = 17$

j $\dfrac{w}{3} - 5 = 2$

k $\dfrac{x}{2} + 3 = 12$

l $\dfrac{m}{7} - 3 = 5$

m $\dfrac{x}{5} + 3 = 3$

n $\dfrac{h}{7} + 2 = 1$

o $\dfrac{w}{3} + 10 = 4$

p $\dfrac{x}{3} - 5 = 7$

q $\dfrac{y}{2} - 13 = 5$

r $\dfrac{f}{6} - 2 = 8$

s $\dfrac{x + 3}{2} = 5$

t $\dfrac{t - 5}{2} = 3$

u $\dfrac{3x + 10}{2} = 8$

v $\dfrac{2x + 1}{3} = 5$

w $\dfrac{5y - 2}{4} = 3$

x $\dfrac{6y + 3}{9} = 1$

y $\dfrac{2x - 3}{5} = 4$

z $\dfrac{5t - 3}{4} = 1$

2

Think of a number. Divide it by 3 and subtract 6.

Teacher

My answer is −1.

Mandy

My starting number is 6.

Andy

a What answer did Andy get?

b What number did Mandy start with?

3 A teacher asked her class to solve the equation $2x - 1 = 7$.

Mustafa wrote:

$$2x - 1 = 7$$

$$2x - 1 - 1 = 7 - 1$$

$$2x = 6$$

$$2x - 2 = 6 - 2$$

$$x = 4$$

Elif wrote:

$$2x - 1 = 7$$

$$2x - 1 + 1 = 7 + 1$$

$$2x = 8$$

$$2x \div 2 = 8 \div 2$$

$$x = 4$$

When the teacher read out the correct answer of 4, both students ticked their work as correct.

a Which student used the correct method?

b Explain the mistakes the other student made.

Brackets

When you have an equation that contains **brackets**, you first must multiply out the brackets and then solve the resulting equation.

EXAMPLE 2

Solve $5(x + 3) = 25$

First multiply out the brackets to get:

$$5x + 15 = 25$$

Subtract 15: $5x = 25 - 15 = 10$

Divide by 5: $\dfrac{5x}{5} = \dfrac{10}{5}$

$$x = 2$$

An alternative method is to divide by the number outside the bracket.

EXAMPLE 3

Solve $3(2x - 7) = 15$

Divide both sides by 3: $2x - 7 = 5$

Add 7: $2x = 12$

Divide by 2: $x = 6$

Make sure you can use both methods.

EXERCISE 13B

FOUNDATION

1 Solve each of the following equations. Some of the answers may be decimals or negative numbers. Remember to check that each answer works for its original equation.

a $2(x + 5) = 16$

b $5(x - 3) = 20$

c $3(t + 1) = 18$

d $4(2x + 5) = 44$

e $2(3y - 5) = 14$

f $5(4x + 3) = 135$

g $4(3t - 2) = 88$

h $6(2t + 5) = 42$

i $2(3x + 1) = 11$

j $4(5y - 2) = 42$

k $6(3k + 5) = 39$

l $5(2x + 3) = 27$

m $9(3x - 5) = 9$

n $2(x + 5) = 6$

o $5(x - 4) = -25$

p $3(t + 7) = 15$

q $2(3x + 11) = 10$

r $4(5t + 8) = 12$

> **HINTS AND TIPS**
>
> Once the brackets have been expanded the equations become straightforward. Remember to multiply everything inside the brackets with what is outside.

2 Fill in values for a, b and c so that the answer to this equation is $x = 4$.

$a(bx + 3) = c$

Equations with the variable on both sides

When a variable appears on both sides of an equation, collect all the terms containing the lett on the left-hand side of the equation. But when there are more of the letters on the right-hand side, it is easier to turn the equation round. When an equation contains brackets, they must be multiplied out first.

EXAMPLE 4

Solve this equation. $5x + 4 = 3x + 10$

There are more xs on the left-hand side, so leave the equation as it is.

Subtract $3x$ from both sides: $2x + 4 = 10$

Subtract 4 from both sides: $2x = 6$

Divide both sides by 2: $x = 3$

EXAMPLE 5

Solve this equation. $\qquad$ $2x + 3 = 6x - 5$

There are more xs on the right-hand side, so turn the equation round.

$$6x - 5 = 2x + 3$$

Subtract $2x$ from both sides: $\qquad$ $4x - 5 = 3$

Add 5 to both sides: $\qquad$ $4x = 8$

Divide both sides by 4: $\qquad$ $x = 2$

EXAMPLE 6

Solve this equation. $\qquad$ $3(2x + 5) + x = 2(2 - x) + 2$

Multiply out both brackets: $\qquad$ $6x + 15 + x = 4 - 2x + 2$

Simplify both sides: $\qquad$ $7x + 15 = 6 - 2x$

There are more xs on the left-hand side, so leave the equation as it is.

Add $2x$ to both sides: $\qquad$ $9x + 15 = 6$

Subtract 15 from both sides: $\qquad$ $9x = -9$

Divide both sides by 9: $\qquad$ $x = -1$

XERCISE 13C

1 Solve each of the following equations.

a $2x + 3 = x + 5$

b $5y + 4 = 3y + 6$

c $4a - 3 = 3a + 4$

d $5t + 3 = 2t + 15$

e $7p - 5 = 3p + 3$

f $6k + 5 = 2k + 1$

g $4m + 1 = m + 10$

h $8s - 1 = 6s - 5$

> **HINTS AND TIPS**
>
> **Remember:** 'Change sides, change signs'. Show all your working. Rearrange before you simplify. If you try to do these at the same time you could get it wrong.

2 Hasan says:

I am thinking of a number. I multiply it by 3 and subtract 2.

Miriam says:

I am thinking of a number. I multiply it by 2 and add 5.

Hasan and Miriam find that they both thought of the same number and both got the same final answer.

What number did they think of?

> **HINTS AND TIPS**
>
> Set up expressions; make them equal and solve.

FOUNDATION

3 Solve each of the following equations.

a $2(d + 3) = d + 12$ **b** $5(x - 2) = 3(x + 4)$

c $3(2y + 3) = 5(2y + 1)$ **d** $3(h - 6) = 2(5 - 2h)$

e $4(3b - 1) + 6 = 5(2b + 4)$ **f** $2(5c + 2) - 2c = 3(2c + 3) + 7$

4 Explain why the equation $3(2x + 1) = 2(3x + 5)$ cannot be solved.

> **HINTS AND TIPS**
>
> Expand the brackets and collect terms on one side as usual. What happens?

5 Explain why there are an infinite number of solutions to the equation:

$2(6x + 9) = 3(4x + 6)$

13.2 Setting up equations

Equations are used to represent situations, so that you can solve real-life problems. Many real-life problems can be solved by setting them up as linear equations and then solving the equation.

EXAMPLE 7

The rectangle shown has a perimeter of 40 cm.

Find the value of x.

The perimeter of the rectangle is:

$$3x + 1 + x + 3 + 3x + 1 + x + 3 = 40$$

This simplifies to: $8x + 8 = 40$

Subtract 8 from both sides: $8x = 32$

Divide both sides by 8: $x = 4$

$3x + 1$

$x + 3$

EXERCISE 13D

Set up an equation to represent each situation described below. Then solve the equation. Remember to check each answer.

1 A man buys a daily paper from Monday to Saturday for d cents. He buys a Sunday paper for $1.80. His weekly paper bill is $7.20.

What is the price of his daily paper?

2 The diagram shows a rectangle.

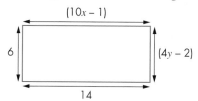

(10*x* – 1)

6

(4*y* – 2)

14

a What is the value of *x*? **b** What is the value of *y*?

3 In this rectangle, the length is 3 cm more than the width. The perimeter is 12 cm.

(*x* + 3)

x

a What is the value of *x*?

b What is the area of the rectangle?

4 Masha has two bags, each of which contains the same number of sweets. She eats four sweets. She then finds that she has 30 sweets left. How many sweets were there in each bag to start with?

5 Flooring costs $12.75 per square metre.

The shop charges $35 for fitting. The final bill was $137.

How many square metres of flooring were fitted?

6 Moshin bought eight garden chairs. When he got to the till he used a $10 voucher as part payment. His final bill was $56.

a Set this problem up as an equation, using *c* as the cost of one chair.

b Solve the equation to find the cost of one chair.

7 This diagram shows the traffic flow through a traffic system in a town centre.

Cars enter at A and at each junction the fractions show the proportion of cars that take each route.

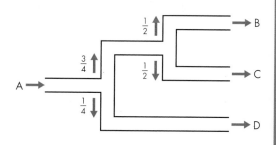

$\frac{1}{2}$ → B

$\frac{3}{4}$

$\frac{1}{2}$ → C

A →

$\frac{1}{4}$

→ D

a 1200 cars enter at A. How many come out of each of the exits, B, C and D?

b If 300 cars exit at B, how many cars entered at A?

c If 500 cars exit at D, how many exit at B?

8 A rectangular room is 3 m longer than it is wide. The perimeter is 16 m.

Floor tiles cost 9 dollars per square metre. How much will it cost to cover the floor?

9 A boy is Y years old. His father is 25 years older than he is.
The sum of their ages is 31. How old is the boy?

10 Another boy is X years old. His sister is twice as old as he is. The sum of their ages is 27.
How old is the boy?

11 The diagram shows a square.

Find x if the perimeter is 44 cm.

$(4x - 1)$

12 Max thought of a number. He then multiplied his number by 3. He added 4 to the answer.
He then doubled that answer to get a final value of 38. What number did he start with?

13 The angles of a triangle, in degrees, are $2x$, $x + 5$
and $x + 35$.

a Write down an equation to show that the angles add
up to 180 degrees.

b Solve your equation to find the value of x.

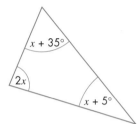

14 Five friends went for a meal in a restaurant. The bill was $\$x$.
They decided to add a \$10 tip and split the bill between them.

Each person paid \$9.50.

a Set up this problem as an equation.

b Solve the equation to work out the bill before the tip was added.

15 A teacher asked her class to find three angles of a triangle
that were consecutive even numbers.

Tammy wrote: $x + x + 2 + x + 4 = 180$
$$3x + 6 = 180$$
$$3x = 174$$
$$x = 58$$

So the angles are 58°, 60° and 62°.

The teacher then asked the class to find four angles of a quadrilateral that are consecutive
even numbers.

Can this be done? Explain your answer.

16 Maria has a large and a small bottle of cola. The large
bottle holds 50 cl more than the small bottle.

From the large bottle she fills four cups and has 18 cl
left over.

From the small bottle she fills three cups and has 1 cl
left over.

How much does each bottle hold?

More complex equations

More complicated equations will require a number of steps to reach a solution. Often they can be solved in a number of different ways.

EXAMPLE 8

Solve the equation $\dfrac{x+1}{3} - \dfrac{x-3}{2} = x - 4$

Method 1

Write the expression on the left as a single fraction:

$$\frac{2(x+1) - 3(x-3)}{3 \times 2} = x - 4$$

Multiply by 6: $\qquad 2(x+1) - 3(x-3) = 6(x-4)$

Remove the brackets: $\qquad 2x + 2 - 3x + 9 = 6x - 24$

$$11 - x = 6x - 24$$

Rearrange: $\qquad\qquad\qquad 35 = 7x$

$$\Rightarrow \quad x = 5$$

Method 2

The LCM of 2 and 3 is 6, so multiply by 6:

$$6 \times \frac{x+1}{3} - 6 \times \frac{x-3}{2} = 6(x-4)$$

$$\Rightarrow \quad 2(x+1) - 3(x-3) = 6(x-4)$$

This gets rid of the fractions. Now continue as before.

EXERCISE 13E

1 Solve the following equations.

a $\dfrac{3x+5}{2} = x + 6$

b $\dfrac{12-x}{3} = x - 8$

c $\dfrac{2x-3}{2} = \dfrac{3x+8}{4}$

d $\dfrac{3x+1}{2} = \dfrac{9x-5}{5}$

e $\dfrac{6x+5}{x} = 8$

f $10 - x = \dfrac{18-3x}{2}$

2 Solve the following equations.

a $\dfrac{x+1}{2} + \dfrac{x+2}{4} = 3$

b $\dfrac{x+2}{4} + \dfrac{x+1}{7} = 3$

c $\dfrac{4x+1}{3} - \dfrac{x+2}{4} = 2$

d $\dfrac{2x-1}{3} + \dfrac{3x+1}{4} = 7$

e $\dfrac{2x+1}{2} + \dfrac{x+1}{7} = 1$

f $\dfrac{3x+1}{5} - \dfrac{5x-1}{7} = 0$

3 Solve the following equations.

a $\dfrac{x}{3} + \dfrac{x}{4} = \dfrac{x+1}{2}$

b $\dfrac{12-x}{2} = \dfrac{11-x}{3}$

c $\dfrac{x+1}{4} + \dfrac{x+2}{3} = 12 - x$

4 The perimeter of this triangle is 18. Calculate the length of each side.

5 The angles of a triangle are x, $\dfrac{2x + 10}{2}$, and $\dfrac{3x}{2}$ degrees.

Use the fact that the angles of a triangle add up to 180 degrees to calculate the size of each angle.

13.4 Solving quadratic equations by factorisation

All the equations in this chapter so far have been **linear equations**. We shall now look at equations which involve **quadratic expressions** such as $x^2 - 2x - 3$ which contain the square of the variable.

Solving the quadratic equation $x^2 + ax + b = 0$

To solve a **quadratic equation** such as $x^2 - 2x - 3 = 0$, you first have to be able to factorise it. Work through Examples 9 to 11 below to see how this is done.

EXAMPLE 9

Solve $x^2 + 6x + 5 = 0$.

This factorises into $(x + 5)(x + 1) = 0$.

The only way this expression can ever equal 0 is if the value of one of the brackets is 0.
Hence either $(x + 5) = 0$ or $(x + 1) = 0$
$\Rightarrow$ $\qquad$ $x + 5 = 0$ or $x + 1 = 0$
$\Rightarrow$ $\qquad$ $x = -5$ or $x = -1$

So the solution is $x = -5$ or $x = -1$.

There are two possible values for x.

EXAMPLE 10

Solve $x^2 + 3x - 10 = 0$.

This factorises into $(x + 5)(x - 2) = 0$.

Hence either $(x + 5) = 0$ or $(x - 2) = 0$
$\Rightarrow$ $\qquad$ $x + 5 = 0$ or $x - 2 = 0$
$\Rightarrow$ $\qquad$ $x = -5$ or $x = 2$

So the solution is $x = -5$ or $x = 2$.

EXAMPLE 11

Solve $x^2 - 6x + 9 = 0$.

This factorises into $(x - 3)(x - 3) = 0$.

That is: $(x - 3)^2 = 0$

Hence, there is only one solution, $x = 3$.

EXERCISE 13F

Solve the equations in questions **1–12**.

1 $(x + 2)(x + 5) = 0$ **2** $(t + 3)(t + 1) = 0$ **3** $(a + 6)(a + 4) = 0$

4 $(x + 3)(x - 2) = 0$ **5** $(x + 1)(x - 3) = 0$ **6** $(t + 4)(t - 5) = 0$

7 $(x - 1)(x + 2) = 0$ **8** $(x - 2)(x + 5) = 0$ **9** $(a - 7)(a + 4) = 0$

10 $(x - 3)(x - 2) = 0$ **11** $(x - 1)(x - 5) = 0$ **12** $(a - 4)(a - 3) = 0$

First factorise, then solve the equations in questions **13–26**.

13 $x^2 + 5x + 4 = 0$ **14** $x^2 + 11x + 18 = 0$ **15** $x^2 - 6x + 8 = 0$

16 $x^2 - 8x + 15 = 0$ **17** $x^2 - 3x - 10 = 0$ **18** $x^2 - 2x - 15 = 0$

19 $t^2 + 4t - 12 = 0$ **20** $t^2 + 3t - 18 = 0$ **21** $x^2 - x - 2 = 0$

22 $x^2 + 4x + 4 = 0$ **23** $m^2 + 10m + 25 = 0$ **24** $t^2 - 8t + 16 = 0$

25 $t^2 + 8t + 12 = 0$ **26** $a^2 - 14a + 49 = 0$

27 A woman is x years old. Her husband is three years younger.

The product of their ages is 550.

a Set up a quadratic equation to represent this situation.

b How old is the woman?

> **HINTS AND TIPS**
>
> If one solution to a real-life problem is negative, reject it and only give the positive answer.

28 A rectangular field is 40 m longer than it is wide. The area is 48 000 square metres.

The farmer wants to place a fence all around the field.

How long will the fence be?

> **HINTS AND TIPS**
>
> Let the width be x, set up a quadratic equation and solve it to get x.

First rearrange the equations in questions **29–37**, then solve them.

29 $x^2 + 10x = -24$ **30** $x^2 - 18x = -32$

31 $x^2 + 2x = 24$ **32** $x^2 + 3x = 54$

33 $t^2 + 7t = 30$ **34** $x^2 - 7x = 44$

35 $t^2 - t = 72$ **36** $x^2 = 17x - 72$

37 $x^2 + 1 = 2x$

> **HINTS AND TIPS**
>
> Put your equation into the form:
> $x^2 + ax + b = 0$

38 A teacher asks her class to solve $x^2 - 3x = 4$.

This is Mario's answer.

$x^2 - 3x - 4 = 0$

$(x - 4)(x + 1) = 0$

Hence $x - 4 = 0$ or $x + 1 = 0$

$x = 4$ or -1

This is Sylvan's answer.

$x(x - 3) = 4$

Hence $x = 4$ or $x - 3 = 4 \Rightarrow x = -3 + 4 = -1$

When the teacher reads out the answer of $x = 4$ or -1, both students mark their work as correct.

Who used the correct method and what mistakes did the other student make?

13.5 More factorisation in quadratic equations

The general quadratic equation is of the form $ax^2 + bx + c = 0$ where a, b and c are positive or negative whole numbers. (It is easier to make sure that a is always positive.) Before any quadratic equation can be solved by factorisation, it must be rearranged to this form.

The method is similar to that used to solve equations of the form $x^2 + ax + b = 0$.
That is, you have to find two factors of $ax^2 + bx + c$ with a product of 0.

EXAMPLE 12

Solve these quadratic equations.   **a** $12x^2 - 28x = -15$ **b** $30x^2 - 5x - 5 = 0$

a First, rearrange the equation to the general form.

$12x^2 - 28x + 15 = 0$

This factorises into $(2x - 3)(6x - 5) = 0$.

The only way this product can equal 0 is if the value of one of the brackets is 0.

Hence:

$$\text{either } 2x - 3 = 0 \quad \text{or} \quad 6x - 5 = 0$$
$$\Rightarrow 2x = 3 \quad \text{or} \quad 6x = 5$$
$$\Rightarrow x = \tfrac{3}{2} \quad \text{or} \quad x = \tfrac{5}{6}$$

So the solution is $x = 1\tfrac{1}{2}$ or $x = \tfrac{5}{6}$

Note: It is more accurate to give your answer as a fraction when it is appropriate.

If you have to round a solution it becomes less accurate.

---

b This equation is already in the general form and it will factorise to $(15x + 5)(2x - 1) = 0$ or $(3x + 1)(10x - 5) = 0$.

Look again at the equation. There is a common factor of 5 which can be taken out to give:

$$5(6x^2 - x - 1) = 0$$

This is much easier to factorise to $5(3x + 1)(2x - 1) = 0$, which can be solved to give $x = -\frac{1}{3}$ or $x = \frac{1}{2}$

Special cases

Sometimes the values of b and c are zero. (Note that if a is zero the equation is no longer a quadratic equation but a linear equation.)

EXAMPLE 13

Solve these quadratic equations.   **a** $3x^2 - 4 = 0$ **b** $6x^2 - x = 0$

a Rearrange to get $3x^2 = 4$.

Divide both sides by 3: $x^2 = \frac{4}{3}$

Take the square root on both sides: $x = \pm\sqrt{\frac{4}{3}} = \pm\frac{2}{\sqrt{3}} = \pm\frac{2\sqrt{3}}{3}$

Note: A square root can be positive or negative. The symbol $\pm$ states that the square root has a positive and a negative value, *both* of which must be used in solving for x.

b There is a common factor of x, so factorise as $x(6x - 1) = 0$.

There is only one set of brackets this time but each factor can be equal to zero, so $x = 0$ or $6x - 1 = 0$.

Hence, $x = 0$ or $\frac{1}{6}$

EXERCISE 13G

Give your answers either in rational form or as mixed numbers.

1 Solve these equations.

a $3x^2 + 8x - 3 = 0$

b $6x^2 - 5x - 4 = 0$

c $5x^2 - 9x - 2 = 0$

d $4t^2 - 4t - 35 = 0$

e $18t^2 + 9t + 1 = 0$

f $3t^2 - 14t + 8 = 0$

g $6x^2 + 15x - 9 = 0$

h $12x^2 - 16x - 35 = 0$

i $15t^2 + 4t - 35 = 0$

j $28x^2 - 85x + 63 = 0$

k $24x^2 - 19x + 2 = 0$

l $16t^2 - 1 = 0$

m $4x^2 + 9x = 0$

n $25t^2 - 49 = 0$

o $9m^2 - 24m - 9 = 0$

> **HINTS AND TIPS**
>
> Look out for the special cases where b or c is zero.

HIGHER

2 Rearrange these equations into the general form and then solve them.

a $x^2 - x = 42$

b $8x(x + 1) = 30$

c $(x + 1)(x - 2) = 40$

d $13x^2 = 11 - 2x$

e $(x + 1)(x - 2) = 4$

f $10x^2 - x = 2$

g $8x^2 + 6x + 3 = 2x^2 + x + 2$

h $25x^2 = 10 - 45x$

i $8x - 16 - x^2 = 0$

j $(2x + 1)(5x + 2) = (2x - 2)(x - 2)$

k $5x + 5 = 30x^2 + 15x + 5$

l $2m^2 = 50$

m $6x^2 + 30 = 5 - 3x^2 - 30x$

n $4x^2 + 4x - 49 = 4x$

o $2t^2 - t = 15$

3 Here are three equations.

A: $(x - 1)^2 = 0$ B: $3x + 2 = 5$ C: $x^2 - 4x = 5$

a Give some mathematical fact that equations A and B have in common.

b Give a mathematical reason why equation B is different from equations A and C.

4 Pythagoras' theorem states that the sum of the squares of the two short sides of a right-angled triangle equals the square of the long side (hypotenuse).

A right-angled triangle has sides $5x - 1$, $2x + 3$ and $x + 1$ cm.

a Show that $20x^2 - 24x - 9 = 0$

b Find the area of the triangle.

$2x + 3$

$5x$

$x + 1$

13.6 Solving quadratic equations by completing the square **H**

Another method for solving quadratic equations is **completing the square**.

You will remember that:

$$(x + a)^2 = x^2 + 2ax + a^2$$

which can be rearranged to give:

$$x^2 + 2ax = (x + a)^2 - a^2$$

This is the basic principle behind completing the square.

There are three basic steps in rewriting $x^2 + px + q$ in the form $(x + a)^2 + b$.

Step 1: Ignore q and just look at the first two terms, $x^2 + px$.

Step 2: Rewrite $x^2 + px$ as $\left(x + \dfrac{p}{2}\right)^2 - \left(\dfrac{p}{2}\right)^2$.

Step 3: Bring q back to get $x^2 + px + q = \left(x + \dfrac{p}{2}\right)^2 - \left(\dfrac{p}{2}\right)^2 + q$.

EXAMPLE 14

Rewrite each expression in the form $(x \pm a)^2 \pm b$.

a $x^2 + 6x - 7$

b $x^2 - 8x + 3$

a Ignore -7 for the moment.

Rewrite $x^2 + 6x$ as $(x + 3)^2 - 9$.

(Expand $(x + 3)^2 - 9 = x^2 + 6x + 9 - 9 = x^2 + 6x$. The 9 is subtracted to get rid of the constant term when the brackets are expanded.)

Now bring the -7 back, so $x^2 + 6x - 7 = (x + 3)^2 - 9 - 7$

Combine the constant terms to get the final answer:

$$x^2 + 6x - 7 = (x + 3)^2 - 16$$

b Ignore $+3$ for the moment.

Rewrite $x^2 - 8x$ as $(x - 4)^2 - 16$.

(Note that you still subtract $(-4)^2$, as $(-4)^2 = +16$.)

Now bring the $+3$ back, so $x^2 - 8x + 3 = (x - 4)^2 - 16 + 3$.

Combine the constant terms to get the final answer:

$$x^2 - 8x + 3 = (x - 4)^2 - 13$$

EXAMPLE 15

Rewrite $x^2 + 4x - 7$ in the form $(x + a)^2 - b$. Hence solve the equation $x^2 + 4x - 7 = 0$, giving your answers to 2 decimal places.

Note that:
$$x^2 + 4x = (x + 2)^2 - 4$$

So:
$$x^2 + 4x - 7 = (x + 2)^2 - 4 - 7 = (x + 2)^2 - 11$$

When $x^2 + 4x - 7 = 0$, you can rewrite the equations completing the square as:
$$(x + 2)^2 - 11 = 0$$

Rearranging gives $(x + 2)^2 = 11$.

Taking the square root of both sides gives:
$$x + 2 = \pm \sqrt{11}$$
$$\Rightarrow x = -2 \pm \sqrt{11}$$

This answer could be left like this, but you are asked to calculate it to 2 decimal places.

$$\Rightarrow x = 1.32 \text{ or } -5.32 \text{ (to 2 decimal places)}$$

To solve $ax^2 + bx + c = 0$ when a is not 1, start by dividing through by a.

EXAMPLE 16

Solve by completing the square.

$2x^2 - 6x - 7 = 0$

Divide by 2: $x^2 - 3x - 3.5 = 0$

$$x^2 - 3x = (x - 1.5)^2 - 2.25$$

So: $x^2 - 3x - 3.5 = (x - 1.5)^2 - 5.75$

When: $x^2 - 3x - 3.5 = 0$

then: $(x - 1.5)^2 = 5.75$

$$x - 1.5 = \pm \sqrt{5.75}$$

$$x = 1.5 \pm \sqrt{5.75}$$

$$x = 3.90 \text{ or } x = -0.90$$

EXERCISE 13H

1 Write an equivalent expression in the form $(x \pm a)^2 - b$.

a $x^2 + 4x$ **b** $x^2 + 14x$ **c** $x^2 - 6x$ **d** $x^2 + 6x$

e $x^2 - 3x$ **f** $x^2 - 9x$ **g** $x^2 + 13x$ **h** $x^2 + 10x$

i $x^2 + 8x$ **j** $x^2 - 2x$ **k** $x^2 + 2x$

2 Write an equivalent expression in the form $(x \pm a)^2 - b$.
Question **1** will help with **a** to **h**.

a $x^2 + 4x - 1$ **b** $x^2 + 14x - 5$ **c** $x^2 - 6x + 3$ **d** $x^2 + 6x + 7$

e $x^2 - 3x - 1$ **f** $x^2 + 6x + 3$ **g** $x^2 - 9x + 10$ **h** $x^2 + 13x + 35$

i $x^2 + 8x - 6$ **j** $x^2 + 2x - 1$ **k** $x^2 - 2x - 7$ **l** $x^2 + 2x - 9$

3 Solve each equation by completing the square. Leave a square root sign in your answer
where appropriate. The answers to question **2** will help.

a $x^2 + 4x - 1 = 0$ **b** $x^2 + 14x - 5 = 0$ **c** $x^2 - 6x + 3 = 0$

d $x^2 + 6x + 7 = 0$ **e** $x^2 - 3x - 1 = 0$ **f** $x^2 + 6x + 3 = 0$

g $x^2 - 9x + 10 = 0$ **h** $x^2 + 13x + 35 = 0$ **i** $x^2 + 8x - 6 = 0$

j $x^2 + 2x - 1 = 0$ **k** $x^2 - 2x - 7 = 0$ **l** $x^2 + 2x - 9 = 0$

4 Solve by completing the square. Give your answers to 2 decimal places.

a $x^2 + 2x - 5 = 0$ **b** $x^2 - 4x - 7 = 0$ **c** $x^2 + 2x - 9 = 0$

5 Solve these equations by completing the square. Leave your answers in square root form.

a $2x^2 - 6x - 3 = 0$ **b** $4x^2 - 8x + 1 = 0$ **c** $2x^2 + 5x - 10 = 0$

d $0.5x^2 - 7.5x + 8 = 0$

HIGHER

6 Ahmed rewrites the expression $x^2 + px + q$ by completing the square. He does this correctly and gets $(x - 7)^2 - 52$.

What are the values of p and q?

7 Jorge writes the steps to solve $x^2 + 6x + 7 = 0$ by completing the square. He writes them on sticky notes. Unfortunately he drops the sticky notes and they get out of order. Try to put the notes in the correct order.

Add 2 to both sides	Subtract 3 from both sides	Write $x^2 + 6x + 7 = 0$ as $(x + 3)^2 - 2 = 0$	Take the square root of both sides

---

3.7 Solving quadratic equations by the quadratic formula H

It is possible to use completing the square to produce a **quadratic formula**. This formula can be used to solve any quadratic equation that can be solved (is **soluble**).

The solution of the equation $ax^2 + bx + c = 0$ is given by:

$$x = \frac{-b \pm \sqrt{b^2 - 4ac}}{2a}$$

where a and b are the **coefficients** of x^2 and x respectively and c is the **constant** term.

EXAMPLE 17

Solve $5x^2 - 11x - 4 = 0$, giving solutions correct to 2 decimal places.

Take the quadratic formula:

$$x = \frac{-b \pm \sqrt{b^2 - 4ac}}{2a}$$

and put $a = 5$, $b = -11$ and $c = -4$, which gives:

$$x = \frac{-(-11) \pm \sqrt{(-11)^2 - 4(5)(-4)}}{2(5)}$$

Note that the values for a, b and c have been put into the formula in brackets. This is to avoid mistakes in calculation. It is a very common mistake to get the sign of b wrong or to think that -11^2 is -121. Using brackets will help you do the calculation correctly.

$$x = \frac{11 \pm \sqrt{121 + 80}}{10} = \frac{11 \pm \sqrt{201}}{10}$$

$$\Rightarrow x = 2.52 \text{ or } -0.32$$

Note: The calculation has been done in stages. With a calculator it is possible just to work out the answer, but make sure you can use your calculator properly. If not, break the calculation down. Remember the rule 'if you try to do two things at once, you will probably get one of them wrong'.

Examination tip: If you are asked to solve a quadratic equation to one or two decimal places, you can be sure that it can be solved only by the quadratic formula.

EXERCISE 13I

Use the quadratic formula to solve the equations in questions **1–15**. Give your answers to 2 decimal places.

1 $2x^2 + x - 8 = 0$

2 $3x^2 + 5x + 1 = 0$

HINTS AND TIPS

Use brackets when substituting and do not try to work two things out at the same time.

3 $x^2 - x - 10 = 0$

4 $5x^2 + 2x - 1 = 0$

5 $7x^2 + 12x + 2 = 0$

6 $3x^2 + 11x + 9 = 0$

7 $4x^2 + 9x + 3 = 0$

8 $6x^2 + 22x + 19 = 0$

9 $x^2 + 3x - 6 = 0$

10 $3x^2 - 7x + 1 = 0$

11 $2x^2 + 11x + 4 = 0$

12 $4x^2 + 5x - 3 = 0$

13 $4x^2 - 9x + 4 = 0$

14 $7x^2 + 3x - 2 = 0$

15 $5x^2 - 10x + 1 = 0$

16 A rectangular lawn is 2 m longer than it is wide.

The area of the lawn is 21 m^2. The gardener wants to edge the lawn with edging strips, which are sold in lengths of 1 m. How many will she need to buy?

17 Shaun is solving a quadratic equation, using the formula.

He correctly substitutes values for a, b and c to get:

$$x = \frac{3 \pm \sqrt{37}}{2}$$

What is the equation Shaun is trying to solve?

18 Hasan uses the quadratic formula to solve $4x^2 - 4x + 1 = 0$.

Miriam uses factorisation to solve $4x^2 - 4x + 1 = 0$.

They both find something unusual in their solutions.

Explain what this is, and why.

13.8 Simple simultaneous equations

All the equations we have looked at so far have just one unknown.

Sometimes there is more than one unknown variable in a problem. In that case we will have several simultaneous equations to solve.

EXAMPLE 18

Tariq is twice as old as Meera. Their total age is 39 years. How old are they?

Suppose Tariq is x years old and Meera is y years old.

Tariq is twice as old as Meera: $\quad x = 2y$ (equation 1)

Their total age is 39 years: $\quad x + y = 39$ (equation 2)

We have *two* unknowns and *two* equations to use to find them.

Substitute $2y$ for x in equation (2):

$$2y + y = 39$$
$$\Rightarrow \quad 3y = 39$$
$$\Rightarrow \quad y = 13$$

Now use equation (1) to find x: $\quad x = 2 \times 13 = 26$

Tariq is 26 and Meera is 13.

EXAMPLE 19

Ari has some tomatoes and onions. The total number of items is 20.

There are four more tomatoes than onions.

How many of each does he have?

Suppose there are x tomatoes and y onions.

The total is 20: $\quad x + y = 20 \quad$ (equation 1)

Four more tomatoes: $\quad x - y = 4 \quad$ (equation 2)

The easiest way to solve these two equations is to add the two together.

On the left hand side: $\quad x + y + x - y = 2x$

On the right hand side: $\quad 20 + 4 = 24$

So: $\quad 2x = 24$

$$x = 12$$

Substitute this value in equation 1:

$$12 + y = 20$$
$$\Rightarrow \quad y = 8$$

There are 12 tomatoes and 8 onions.

EXERCISE 13J

1 Solve each of these pairs of simultaneous equations.

a $x + y = 15$
$\quad y = 2x$

b $x = 3y$
$\quad x + y = 24$

c $x + y = 60$
$\quad y = 4x$

2 Solve each of these pairs of simultaneous equations.

a $y = x + 12$
$y = 3x$

b $y = x - 10$
$x = 5y$

c $x + 4 = y$
$y = 9x$

3 Solve each of these pairs of simultaneous equations.

a $x + y = 20$
$x - y = 6$

b $y + x = 23$
$y - x = 5$

c $x + y = 6$
$x - y = 14$

4 Solve each of these pairs of simultaneous equations.

a $y = 2x + 3$
$y = 8x$

b $x + y = 20$
$y = 3x - 2$

c $y = 2x + 4$
$y = 10 - x$

5 Carmen and Anish are carrying some books. There are 40 books altogether.

Carmen has 4 times as many as Anish.

How many does each one have?

6 Ari writes down two numbers. The total is 37. The difference between them is 14.

What are the numbers?

7 Luis records the temperature at midday and again at midnight.

He notices that the temperatures add up to 5 and the difference between them is 11.

What are the temperatures?

We will now look in detail at several ways of solving simultaneous equations.

Elimination method

Here, you solve simultaneous equations by the *elimination method*. There are six steps in this method.

Step 1 is to make the coefficients of one of the variables the same.

Step 2 is to **eliminate** this variable by adding or subtracting the equations.

Step 3 is to solve the resulting linear equation in the other variable.

Step 4 is to substitute the value found back into one of the original equations.

Step 5 is to solve the resulting equation.

Step 6 is to check that the two values found satisfy the original equations.

EXAMPLE 20

Solve the equations: $6x + y = 15$ and $4x + y = 11$

Label the equations so that the method can be clearly explained.

$6x + y = 15$ (1)

$4x + y = 11$ (2)

Step 1: Since the y-term in both equations has the same coefficient there is no need to balance them.

Step 2: Subtract one equation from the other. (Equation (1) minus equation (2) will give positive values.)

$$(1) - (2) \qquad 2x = 4$$

Step 3: Solve this equation: $\qquad x = 2$

Step 4: Substitute $x = 2$ into one of the original equations. (Usually the one with smallest numbers involved.)

So substitute into: $4x + y = 11$

which gives: $\qquad 8 + y = 11$

Step 5: Solve this equation: $\qquad y = 3$

Step 6: Test the solution in the original equations. So substitute $x = 2$ and $y = 3$ into $6x + y$, which gives $12 + 3 = 15$ and into $4x + y$, which gives $8 + 3 = 11$. These are correct, so you can confidently say the solution is $x = 2$ and $y = 3$.

EXAMPLE 21

Solve these equations.
$$5x + y = 22 \quad (1)$$
$$2x - y = 6 \quad (2)$$

Step 1: Both equations have the same y-coefficient but with different signs so there is no need to balance them.

Step 2: As the signs are different, add the two equations, to eliminate the y-terms.

$$(1) + (2) \qquad 7x = 28$$

Step 3: Solve this equation: $\qquad x = 4$

Step 4: Substitute $x = 4$ into one of the original equations, $5x + y = 22$, which gives: $\qquad 20 + y = 22$

Step 5: Solve this equation: $\qquad y = 2$

Step 6: Test the solution by putting $x = 4$ and $y = 2$ into the original equations, $2x - y = 6$, which gives $8 - 2 = 6$ and $5x + y = 22$ which gives $20 + 2 = 22$. These are correct, so the solution is $x = 4$ and $y = 2$.

Substitution method

This is an alternative method. Which method you use depends very much on the coefficients of the variables and the way that the equations are written in the first place. There are five steps in the substitution method.

Step 1 is to rearrange one of the equations into the form $y = \ldots$ or $x = \ldots$.

Step 2 is to substitute the right-hand side of this equation into the other equation in place of the variable on the left-hand side.

Step 3 is to expand and solve this equation.

Step 4 is to substitute the value into the $y = \ldots$ or $x = \ldots$ equation.

Step 5 is to check that the values work in both original equations.

EXAMPLE 22

Solve the simultaneous equations: $y = 2x + 3$, $3x + 4y = 1$

Because the first equation is in the form $y = \ldots$ it suggests that the substitution method should be used.

Again label the equations to help with explaining the method.

$$y = 2x + 3 \quad (1)$$
$$3x + 4y = 1 \quad (2)$$

Step 1: As equation (1) is in the form $y = \ldots$ there is no need to rearrange an equation.

Step 2: Substitute the right-hand side of equation (1) into equation (2) for the variable y.

$$3x + 4(2x + 3) = 1$$

Step 3: Expand and solve the equation. $\quad 3x + 8x + 12 = 1, 11x = -11, x = -1$

Step 4: Substitute $x = -1$ into $y = 2x + 3$: $\quad\quad y = -2 + 3 = 1$

Step 5: Test the values in $y = 2x + 3$ which gives $1 = -2 + 3$ and $3x + 4y = 1$, which gives $-3 + 4 = 1$. These are correct so the solution is $x = -1$ and $y = 1$.

EXERCISE 13K

1 Solve these simultaneous equations.

Use elimination for **a** to **i**. Use substitution for **j** to **l**.

a $4x + y = 17$
$2x + y = 9$

b $5x + 2y = 13$
$x + 2y = 9$

c $2x + y = 7$
$5x - y = 14$

d $3x + 2y = 11$
$2x - 2y = 14$

e $3x - 4y = 17$
$x - 4y = 3$

f $3x + 2y = 16$
$x - 2y = 4$

g $x + 3y = 9$
$x + y = 6$

h $2x + 5y = 16$
$2x + 3y = 8$

i $3x - y = 9$
$5x + y = 11$

j $2x + 5y = 37$
$y = 11 - 2x$

k $4x - 3y = 7$
$x = 13 - 3y$

l $4x - y = 17$
$x = 2 + y$

2 In this sequence, the next term is found by multiplying the previous term by a and then adding b. a and b are positive whole numbers.

3 14 47 … …

a Explain why $3a + b = 14$

b Set up another equation in a and b.

c Solve the equations to solve for a and b.

d Work out the next two terms in the sequence.

FOUNDATION

Balancing coefficients in one equation only

You could solve all the examples in Exercise 13I, question **1** by adding or subtracting the equations in each pair, or by substituting without rearranging. This does not always happen. The next examples show what to do when there are no identical terms, or when you need to rearrange.

EXAMPLE 23

Solve these equations.
$$3x + 2y = 18 \qquad (1)$$
$$2x - y = 5 \qquad (2)$$

Step 1: Multiply equation (2) by 2. There are other ways to balance the coefficients but this is the easiest and leads to less work later. With practice, you will get used to which will be the best way to balance the coefficients.

$$2 \times (2) \qquad 4x - 2y = 10 \qquad (3)$$

Label this equation as number (3).

Be careful to multiply every term and not just the y-term. You could write:

$$2 \times (2x - y = 5) \Rightarrow 4x - 2y = 10 \qquad (3)$$

Step 2: As the signs of the y-terms are opposite, add the equations.

$$(1) + (3) \qquad 7x = 28$$

Be careful to add the correct equations. This is why labelling them is useful.

Step 3: Solve this equation: $\qquad x = 4$

Step 4: Substitute $x = 4$ into any equation, say $2x - y = 5 \Rightarrow 8 - y = 5$

Step 5: Solve this equation: $\qquad y = 3$

Step 6: Check: (1), $3 \times 4 + 2 \times 3 = 18$ and (2), $2 \times 4 - 3 = 5$, which are correct so the solution is $x = 4$ and $y = 3$.

EXAMPLE 24

Solve the simultaneous equations:
$$3x + y = 5 \qquad (1)$$
$$5x - 2y = 12 \qquad (2)$$

Step 1: Multiply the first equation by 2: $6x + 2y = 10 \qquad (3)$

Step 2: Add (2) + (3): $11x = 22$

Step 3: Solve: $x = 2$

Step 4: Substitute back: $3 \times 2 + y = 5$

Step 5: Solve: $y = -1$

Step 6: Check: (1) $3 \times 2 - 1 = 5$ and (2) $5 \times 2 - 2 \times (-1) = 10 + 2 = 12$, which are correct.

EXERCISE 13L

1 Solve parts **a** to **c** by the substitution method and the rest by first changing one of the equations in each pair to obtain identical terms, and then adding or subtracting the equations to eliminate those terms.

a $5x + 2y = 4$
$4x - y = 11$

b $4x + 3y = 37$
$2x + y = 17$

c $x + 3y = 7$
$2x - y = 7$

d $2x + 3y = 19$
$6x + 2y = 22$

e $5x - 2y = 26$
$3x - y = 15$

f $10x - y = 3$
$3x + 2y = 17$

g $3x + 5y = 15$
$x + 3y = 7$

h $3x + 4y = 7$
$4x + 2y = 1$

i $5x - 2y = 24$
$3x + y = 21$

j $5x - 2y = 4$
$3x - 6y = 6$

k $2x + 3y = 13$
$4x + 7y = 31$

l $3x - 2y = 3$
$5x + 6y = 12$

2 a Francesca is solving the simultaneous equations $4x - 2y = 8$ and $2x - y = 4$.

She finds a solution of $x = 5$, $y = 6$ which works for both equations.

Explain why this is not a unique solution.

b Dimitri is solving the simultaneous equations $6x + 2y = 9$ and $3x + y = 7$.

Why is it impossible to find a solution that works for both equations?

13.9 More complex simultaneous equations

There are also cases where *both* equations have to be changed to obtain identical terms. The next example shows you how this is done.

Note: The substitution method is not suitable for these types of equations as you end up with fractional terms.

EXAMPLE 25

Solve these equations.

$4x + 3y = 27$ (1)

$5x - 2y = 5$ (2)

Both equations have to be changed to obtain identical terms in either x or y. However, you can see that if you make the y-coefficients the same, you will add the equations. This is always safer than subtraction. We do this by multiplying the first equation by 2 (the y-coefficient of equation 2) and the second equation by 3 (the y-coefficient of equation 1).

Step 1: (1) $\times$ 2 or 2 $\times$ $(4x + 3y = 27)$ $\Rightarrow$ $8x + 6y = 54$ (3)

(2) $\times$ 3 or 3 $\times$ $(5x - 2y = 5)$ $\Rightarrow$ $15x - 6y = 15$ (4)

Label the new equations (3) and (4).

Step 2: Eliminate one of the variables: (3) + (4) $23x = 69$

Step 3: Solve the equation: $x = 3$

Step 4: Substitute into equation (1): $12 + 3y = 27$

Step 5: Solve the equation: $y = 5$

Step 6: Check: (1), $4 \times 3 + 3 \times 5 = 12 + 15 = 27$, and (2), $5 \times 3 - 2 \times 5 = 15 - 10 = 5$, which are correct so the solution is $x = 3$ and $y = 5$.

EXERCISE 13M

1 Solve the following simultaneous equations.

a $2x + 5y = 15$
$3x - 2y = 13$

b $2x + 3y = 30$
$5x + 7y = 71$

c $2x - 3y = 15$
$5x + 7y = 52$

d $3x - 2y = 15$
$2x - 3y = 5$

e $5x - 3y = 14$
$4x - 5y = 6$

f $3x + 2y = 28$
$2x + 7y = 47$

g $2x + y = 4$
$x - y = 5$

h $5x + 2y = 11$
$3x + 4y = 8$

i $x - 2y = 4$
$3x - y = -3$

j $3x + 2y = 2$
$2x + 6y = 13$

k $6x + 2y = 14$
$3x - 5y = 10$

l $2x + 4y = 15$
$x + 5y = 21$

m $3x - y = 5$
$x + 3y = -20$

n $3x - 4y = 4.5$
$2x + 2y = 10$

o $x - 5y = 15$
$3x - 7y = 17$

2 Here are four equations.

A: $5x + 2y = 1$

B: $4x + y = 9$

C: $3x - y = 5$

D: $3x + 2y = 3$

Here are four sets of (x, y) values.

$(1, -2), (-1, 3), (2, 1), (3, -3)$

Match each pair of (x, y) values to a pair of equations.

> **HINTS AND TIPS**
>
> You could solve each possible set of pairs but there are six to work out. Alternatively you can substitute values into the equations to see which work.

3 Find the area of the triangle enclosed by these three equations.

$y - x = 2$ $x + y = 6$ $3x + y = 6$

4 Find the area of the triangle enclosed by these three equations.

$x - 2y = 6$ $x + 2y = 6$ $x + y = 3$

> **HINTS AND TIPS**
>
> Find the point of intersection of each pair of equations, plot the points on a grid and use any method to work out the area of the resulting triangle.

HIGHER

You have already seen the method of substitution for solving **linear** simultaneous equations.

You can use a similar method when you need to solve a pair of equations, one of which is linear and the other of which is **non-linear**. But you must always substitute from the linear into the non-linear.

EXAMPLE 26

Solve these simultaneous equations.

$$x^2 + y^2 = 5$$
$$x + y = 3$$

Call the equations (1) and (2):

$$x^2 + y^2 = 5 \quad (1)$$
$$x + y = 3 \quad (2)$$

Rearrange equation (2) to obtain:

$$x = 3 - y$$

Substitute this into equation (1), which gives:

$$(3 - y)^2 + y^2 = 5$$

Expand and rearrange into the general form of the quadratic equation:

$$9 - 6y + y^2 + y^2 = 5$$
$$2y^2 - 6y + 4 = 0$$

Divide by 2:

$$y^2 - 3y + 2 = 0$$

Factorise:

$$(y - 1)(y - 2) = 0$$
$$\Rightarrow y = 1 \text{ or } 2$$

Substitute for y in equation (2):

When $y = 1$, $x = 2$ and when $y = 2$, $x = 1$

Note that you should always give answers as a pair of values in x and y.

EXAMPLE 27

Find the solutions of the pair of simultaneous equations: $y = x^2 + x - 2$ and $y = 2x + 4$

This example is slightly different, as both equations are given in terms of y,

so substituting for y gives:

$$2x + 4 = x^2 + x - 2$$

Rearranging into the general quadratic:

$$x^2 - x - 6 = 0$$

Factorising and solving gives:

$$(x + 2)(x - 3) = 0$$

$$x = -2 \text{ or } 3$$

Substituting back to find y:

When $x = -2$, $y = 0$

When $x = 3$, $y = 10$

So the solutions are $(-2, 0)$ and $(3, 10)$.

EXERCISE 13N

1 Solve these pairs of linear simultaneous equations using the substitution method.

a $2x + y = 9$
$x - 2y = 7$

b $3x - 2y = 10$
$4x + y = 17$

c $x - 2y = 10$
$2x + 3y = 13$

2 Solve these pairs of simultaneous equations.

a $xy = 2$
$y = x + 1$

b $xy = -4$
$2y = x + 6$

3 Solve these pairs of simultaneous equations.

a $x^2 + y^2 = 25$
$x + y = 7$

b $x^2 + y^2 = 9$
$y = x + 3$

c $x^2 + y^2 = 13$
$5y + x = 13$

4 Solve these pairs of simultaneous equations.

a $y = x^2 + 2x - 3$
$y = 2x + 1$

b $y = x^2 - 2x - 5$
$y = x - 1$

c $y = x^2 - 2x$
$y = 2x - 3$

5 Solve these pairs of simultaneous equations.

a $y = x^2 + 3x - 3$ and $y = x$

b $x^2 + y^2 = 13$ and $x + y = 1$

c $x^2 + y^2 = 5$ and $y = x + 1$

d $y = x^2 - 3x + 1$ and $y = 2x - 5$

e $y = x^2 - 3$ and $y = x + 3$

f $y = x^2 - 3x - 2$ and $y = 2x - 6$

g $x^2 + y^2 = 41$ and $y = x + 1$

HIGHER

Why this chapter matters

Line graphs are used in many media, including newspapers and the textbooks of most of the subjects that you learn in school.

Graphs show the relationship between variables. Often one of these variables is time and the graph shows how the other variable changes over time.

For example, this graph on the right shows how the exchange rate between the dollar and the pound changed over five months in 2009.

American Dollars to 1 GBP

USD
1.71841
1.64663
1.57485
1.50306
1.43128
1.3595

Apr 8 May 8 Jun 8 Jul 7 Aug 5

Graphs like this make it easy to see what is happening to a variable – much easier than looking at lists of data. Here you can see instantly that the pound was worth a growing number of dollars over the five months.

A graph can show several variables to make it easier to compare them. The graph below shows data about a racing car going round a circuit. It compares the driver's acceleration and deceleration (speeding up and slowing down) with his steering. The green line is acceleration and the pink line is steering.

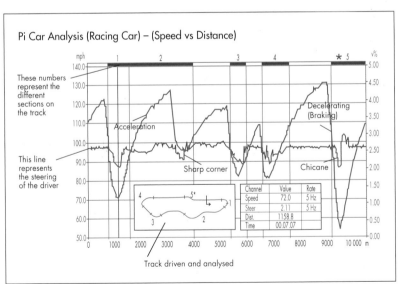

The graph gives the team engineers and trainers an instant picture of the way the driver goes round the course. It would be difficult to compare all this data in any other way.

Graphs in practical situations

ics	Level	Key words
Conversion graphs	**FOUNDATION**	scale, estimate, conversion graph
Travel graphs	**FOUNDATION**	average speed, distance–time graph
Speed—time graphs	**FOUNDATION**	speed—time graph, constant speed

What you need to be able to do in the examinations:

FOUNDATION

Interpret information presented in a range of linear and non-linear graphs.
Draw and interpret straight line conversion graphs.

Conversion graphs

Look at Examples 1 and 2, and make sure that you can understand the conversions. You need to be able to read these types of graph by finding a value on one axis and following it through to the other axis. Make sure you understand the **scales** on the axes to help you **estimate** the answers.

EXAMPLE 1

This is a **conversion graph** between litres and gallons.

a How many litres are there in 5 gallons?

b How many gallons are there in 15 litres?

From the graph you can see that:

a 5 gallons are approximately equivalent to 23 litres.

b 15 litres are approximately equivalent to $3\frac{1}{4}$ gallons.

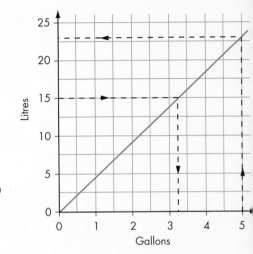

EXAMPLE 2

This is a graph of the charges made for units of electricity used in the home.

a How much will a customer who uses 500 units of electricity be charged?

b How many units of electricity will a customer who is charged $20 have used?

From the graph you can see that:

a A customer who uses 500 units of electricity will be charged $45.

b A customer who is charged $20 will have used about 150 units.

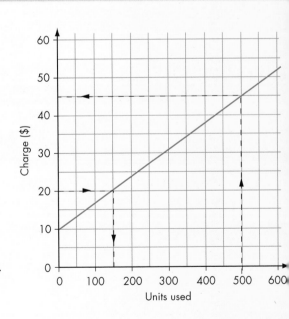

EXERCISE 14A

1 Mass can be measured in kilograms or pounds. This is a conversion graph between kilograms (kg) and pounds (lb).

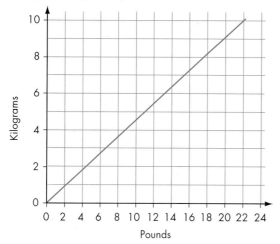

Pounds

a Use the graph to make an approximate conversion of:

  i 18 lb to kilograms

  ii 5 lb to kilograms

  iii 4 kg to pounds

  iv 10 kg to pounds.

b Approximately how many pounds are equivalent to 1 kg?

c Explain how you could use the graph to convert 48 lb to kilograms.

2 Distances can be measured in centimetres or inches. This is a conversion graph between inches (in) and centimetres (cm).

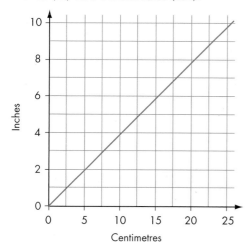

Centimetres

a Use the graph to make an approximate conversion of:

  i 4 in to centimetres

  ii 9 in to centimetres

  iii 5 cm to inches

  iv 22 cm to inches.

b Approximately how many centimetres are equivalent to 1 inch?

c Explain how you could use the graph to convert 18 inches to centimetres.

3 This graph was produced to show approximately how much the British pound (£) is worth in Singapore dollars ($).

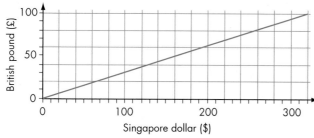

Singapore dollar ($)

a Use the graph to make an approximate conversion of:

  i £100 to Singapore dollars

  ii £30 to Singapore dollars

  iii $150 to British pounds

  iv $250 to British pounds.

b Approximately how many Singapore dollars are equivalent to £1?

c What would happen to the conversion line on the graph if the pound is worth fewer Singapore dollars?

4 A company hired out heaters. They used the following graph to estimate what the charge would be.

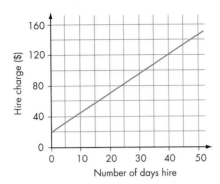

a Use the graph to find the approximate charge for hiring a heater for:

   i 40 days

   ii 25 days.

b Use the graph to find out how many days' hire you would get for a cost of:

   i $100

   ii $140.

5 A conference centre had the following chart on the office wall so that the staff could see the approximate cost of a conference, based on the number of people attending it.

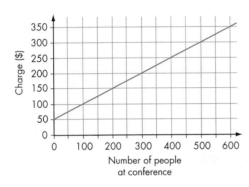

a Use the graph to estimate the charge for:

   i 100 people

   ii 550 people.

b Use the graph to estimate how many people can attend a conference at the centre for a cost of:

   i $300

   ii $175.

6 The graph shows the original prices of items and their selling prices after sales tax is added.

a Use the chart to find the selling price of goods if their original prices are:

   i $60

   ii $25.

b What were the original prices of goods selling at:

   i $100

   ii $45?

7 In Europe temperatures are measured in Celsius and in the USA they are measured in Fahrenheit. Here is a conversion graph for the two scales.

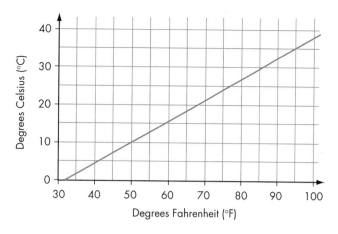

a Use the graph to estimate a conversion of:

   i 35°C to Fahrenheit

   ii 20°C to Fahrenheit

   iii 50°F to Celsius

   iv 90°F to Celsius.

b Water freezes at 0°C. What temperature is this in Fahrenheit?

8 I lost my fuel bill, but while talking to my friends I found out that:

Bill, who had used 850 units, was charged $57.50

Wendy, who had used 320 units, was charged $31

Rhanni, who had used 540 units, was charged $42.

a Plot the given information and draw a straight-line graph. Use a scale from 0 to 900 on the horizontal units axis, and from $0 to $60 on the vertical cost axis.

b Use your graph to find what I will be charged for 700 units.

9 Distances can be measured in kilometres or miles.

80 kilometres is approximately the same as 50 miles.

a Use this information to draw a conversion graph between kilometres and miles.

b Use the graph to convert 30 miles into kilometres.

c Use the graph to convert 25 kilometres into miles.

10 A candle gets shorter as it burns.

After 10 hours it has burnt down by 13 centimetres.

a Show this information on a graph.

b How much did the candle burn down in 7 hours?

c How long did it take to burn down by 5 centimetres?

11 This table shows how far a snail has moved after different periods of time.

Time in minutes	5	15	30
Distance in centimetres	13	39	78

a Draw a graph to show this information.

b How long did the snail take to move 60 centimetres?

As the name suggests, a travel graph gives information about how far someone or something h
travelled over a given time period.

Travel graphs are sometimes called **distance–time graphs**.

A travel graph is read in a similar way to the conversion graphs you have just done. But you ca
also find the **average speed** from a distance–time graph, using the formula:

$$\text{average speed} = \frac{\text{total distance travelled}}{\text{total time taken}}$$

EXAMPLE 3

The distance–time graph
shown right represents a car
journey from Murcia to
Cartagena, a distance of
50 km, and back again.

a What can you say about
points B, C and D?

b What can you say about
the journey from D to F?

c Work out the average
speed for each of the five
stages of the journey.

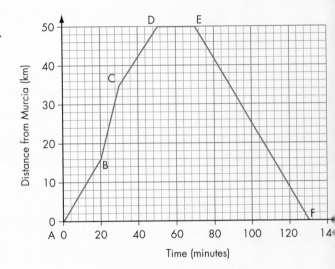

From the graph:

a B: After 20 minutes the car was 16 km away from Murcia.

C: After 30 minutes the car was 35 km away from Murcia.

D: After 50 minutes the car was 50 km away from Murcia, so at Cartagena.

b D–F: The car stayed at Cartagena for 20 minutes, and then took 60 minutes for the
return journey.

c The average speeds over the five stages of the journey are worked out as follows.

    A to B represents 16 km in 20 minutes.

        20 minutes is $\frac{1}{3}$ of an hour, so multiply by 3 to give distance/hour.

        Multiplying both numbers by 3 gives 48 km in 60 minutes, which is 48 km/h.

    B to C represents 19 km in 10 minutes.

        Multiplying both numbers by 6 gives 114 km in 60 minutes, which is 114 km/h.

    C to D represents 15 km in 20 minutes.

        Multiplying both numbers by 3 gives 45 km in 60 minutes, which is 45 km/h.

    D to E represents a stop: no further distance travelled.

    E to F represents the return journey of 50 km in 60 minutes, which is 50 km/h.

  So, the return journey was at an average speed of 50 km/h.

EXERCISE 14B

1 Paulo was travelling in his car to a meeting 240 km away. This distance–time graph illustrates his journey.

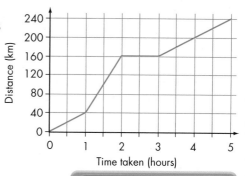

a How long after he set off did he:

   i stop for his break

   ii set off after his break

   iii get to his meeting place?

b At what average speed was he travelling:

   i over the first hour

   ii over the second hour

   iii for the last part of his journey?

c The meeting was scheduled to start at 10.30 am. What is the latest time he should have left home?

> **HINTS AND TIPS**
>
> If a part of a journey takes 30 minutes, just double the distance to get the average speed.

2 Farid was travelling by car in Europe on his holiday. This distance–time graph illustrates his journey.

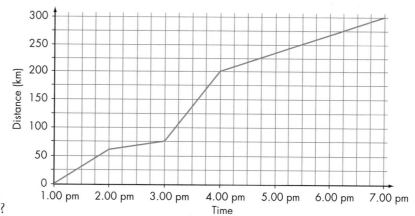

a His greatest speed was on the motorway.

   i How far did he travel on the motorway?

   ii What was his average speed on the motorway?

b **i** When did he travel the most slowly?   **ii** What was his lowest average speed?

3 A small bus set off from Auzio at 12 noon to pick up Mike and his family. It then went on to pick up Mike's parents and grandparents. It then travelled further, dropping them all off at a hotel. The bus then went on a further 10 km to pick up another party and it took them back to Auzio. This distance–time graph illustrates the journey.

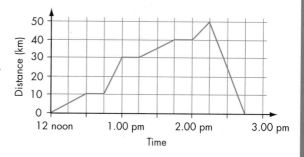

a How far from Auzio did Mike's parents and grandparents live?

b How far from Auzio is the hotel at which they all stayed?

c What was the average speed of the bus on its way back to Auzio?

4 Reu and Yuto took part in a 5000 m race. It is illustrated in this graph.

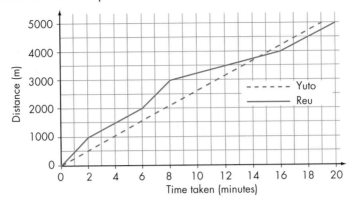

a Yuto ran a steady race. What is his average speed in:

i metres per minute

ii kilometres per hour?

b Reu ran in spurts. What was his highest average speed?

c Who won the race and by how much?

5 Three friends, Patrick, Araf and Sean, ran a 1000 m race. The race is illustrated on the distance–time graph shown here.

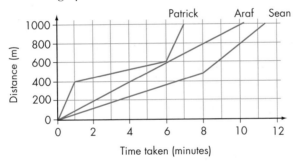

a Describe how each of them completed the race.

b i What is Araf's average speed in m/s?

ii What is this speed in km/h?

6 A walker sets off at 0900 from point P to walk along a trail at a steady pace of 6 km per hour.

90 minutes later, a cyclist sets off from P on the same trail at a steady pace of 15 km per hour.

a Draw a graph to illustrate the journeys of the walker and cyclist.

b At what time did the cyclist overtake the walker?

HINTS AND TIPS

Mark a grid with a horizontal axis as time from 0900–1300, and the vertical axis as distance from 0 to 24. Draw lines for both walker and cyclist. Remember that the cyclist doesn't start until 1030.

7 Three school friends set off from school at the same time, 1545. They all lived 12 km away from the school. The distance–time graph illustrates their journeys.

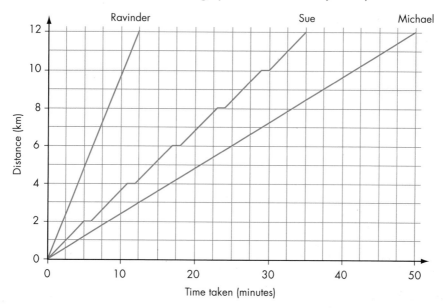

One of them went by bus, one cycled and one was taken by car.

a i Explain how you know that Sue used the bus.

ii Who went by car?

b At what times did each of them get home?

c i When the bus was moving, it covered 2 km in 5 minutes. What is this speed in kilometres per hour?

ii Overall, the bus covered 12 km in 35 minutes. What is this speed, in kilometres per hour?

iii How many stops did the bus make before Sue got off?

8 A girl walks at a steady speed of 1.5 metres/second for 20 seconds.

She stops for 20 seconds.

Then she walks at 2 metres/second for 10 seconds.

a How far has she walked altogether?

b Show her journey on a distance–time graph.

c What is her average speed for the whole journey?

9 A car drives along a motorway for 2 hours at a steady speed of 100 km/hour.

The driver has a break of 30 minutes.

She then drives back to where she started.

The whole trip takes 5 hours.

a Draw a distance–time graph to show the journey.

b What is the average speed for the second half of the journey?

FOUNDATION

10 A man walks to the top of a high hill.

He starts at 0900 and by 1000 he has travelled 5 km.

From 1000 to 1200 he covers another 8 km.

He takes an hour to complete the final 2 km to the top of the hill.

a What time did he reach the top?

b How far did he walk?

c Illustrate the journey with a distance–time graph.

d Find the average speed for each stage of the walk and for the whole journey.

14.3 Speed–time graphs

This is the distance–time graph for a cyclist.

The graph is a straight line. This shows that she is moving at **constant speed**.

$$\text{Speed} = \frac{\text{distance}}{\text{time}} = \frac{80}{20} = 4 \text{ m/s}$$

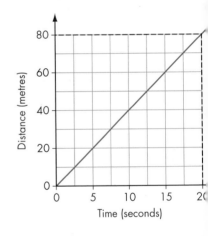

This is a **speed–time graph** of the same situation.

The horizontal line shows that she is moving at a constant speed.

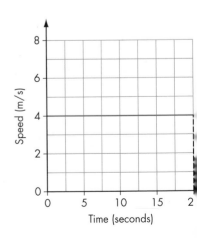

EXERCISE 14C

1 The speed–time graph shows a car accelerating for 40 seconds and then decelerating.

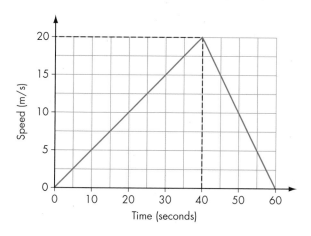

  a What is the maximum speed of the car?

  b What was the speed after 10 seconds?

  c When was the speed greater than 15 m/s?

2 The speed–time graph shows a train slowing down as it approaches a station.

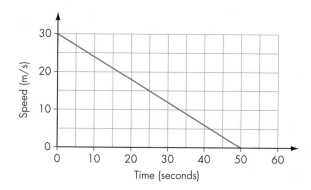

  a How long did it take the train to stop?

  b For how long was the train travelling faster than 10 m/s?

3 The speed-time graph shows the speed of a boat over thirty seconds.

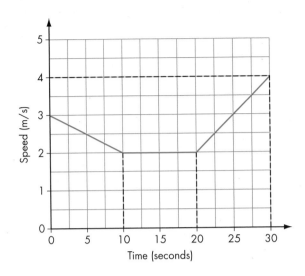

a When was the boat slowing down?

b For how long did the boat travel at a constant speed?

c What was the highest speed achieved?

4 The graph illustrates a journey for a car.

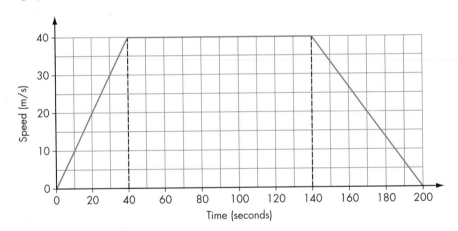

a What was the highest speed?

b How long was the car moving at more than 20 m/s?

5 The red car in this graph is travelling at a constant speed.

As the red car passes it, the blue car starts from rest and accelerates for 15 seconds in the same direction as the red car.

The blue car then continues at a constant speed.

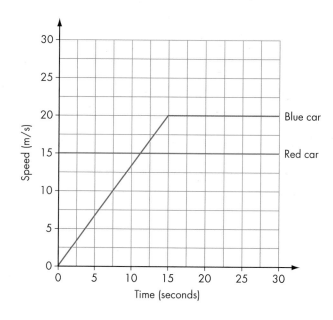

a What was the speed of the red car?

b When were the cars travelling at the same speed?

c How far did the red car travel in 30 seconds?

d How far did the blue car travel between 15 and 30 seconds?

The simplest type of graph is a straight line graph. This shows that the value of one variable on the graph is always affected in a certain way by changes in the value of the other.

A graph gives us a good visual impression of the way two variables are related to each other.

This graph shows the results of an experiment which measured the voltage in an electrical circuit when different currents were flowing. The points are approximately in a straight line.

We can use this graph to find the voltage for any current we choose. For example, if we read up from 0.6 amps on the x-axis to the graph line we find that the voltage on the y-axis is 0.66 volts and if the current is 0.8 amps the voltage is 0.88 volts.

We can see that there is a constant relationship between current and voltage: the voltage in volts is 1.1 × the current in amps. We can show this as an equation:

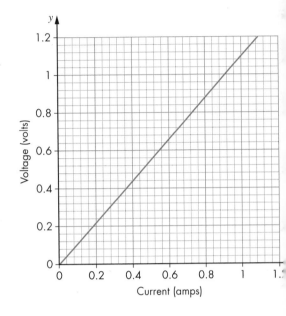

$y = 1.1x$ (voltage is on the y-axis and current is on the x-axis)

We can use this equation to find the voltage for any current we choose. We can also use it to find the current for any voltage. Once we know the relationship between variables, using the equation is quicker and easier than using the graph.

This connection between the geometry of graphs and the algebra of equations was made by a French mathematician, René Descartes, in the 17th century CE. This picture shows him at work.

The coordinates in such graphs are called *Cartesian coordinates* (in Latin his name was Cartesius) and the *xy* grid they appear on is called the *Cartesian plane*.

The grid can be extended to include negative values as well. The graphs and equations shown on it are not always straight line ones as we shall see in later chapters.

15

Straight line graphs

Topics	Level	Key words
1 Using coordinates	FOUNDATION	quadrant, coordinates, negative coordinates, Cartesian coordinates, equation of line, Cartesian plane, midpoint, line segment
2 Drawing straight line graphs	FOUNDATION	straight line graph
3 More straight lines	FOUNDATION	slope, gradient
4 The equation $y = mx + c$	FOUNDATION	intercept, coefficient
5 Finding equations	HIGHER	
6 Parallel and perpendicular lines	HIGHER	parallel, perpendicular, negative reciprocal
7 Graphs and simultaneous equations	HIGHER	

What you need to be able to do in the examinations:

FOUNDATION	HIGHER
<ul><li>Understand and use conventions for rectangular Cartesian coordinates.</li><li>Plot points (x, y) in any of the four quadrants.</li><li>Locate points with given coordinates.</li><li>Determine the coordinates:<ul><li>of points identified by geometrical information</li><li>of the midpoint of a line segment, given the coordinates of the two end points.</li></ul></li><li>Find the gradient of a straight line.</li><li>Recognise that equations of the form $y = mx + c$ are straight line graphs with gradient m and intercept on the y-axis at the point $(0, c)$.</li><li>Generate points and plot graphs of linear functions.</li></ul>	<ul><li>Calculate the gradient of a straight line given the coordinates of two points.</li><li>Find the equation of a straight line parallel to a given line.</li><li>Find the equation of a straight line perpendicular to a given line.</li><li>Interpret simultaneous equations as lines and the common solution as the point of intersection.</li></ul>

Using coordinates

A set of axes can form four sectors called **quadrants**, but so far, all the points you have read or plotted on graphs have been **coordinates** in the first quadrant (the top right section of a grid). The grid below shows you how to read and plot coordinates in all four quadrants and how to find the equations of vertical and horizontal lines. This involves using **negative coordinates**.

The coordinates of a point are given in the form (x, y), where x is the number along the x-axis and y is the number up the y-axis. These are sometimes called **Cartesian coordinates** after their inventor, René Descartes.

The coordinates of the four points on the grid are:

$A(2, 3)$ $\qquad$ $B(-1, 2)$ $\qquad$ $C(-3, -4)$ $\qquad$ $D(1, -3)$

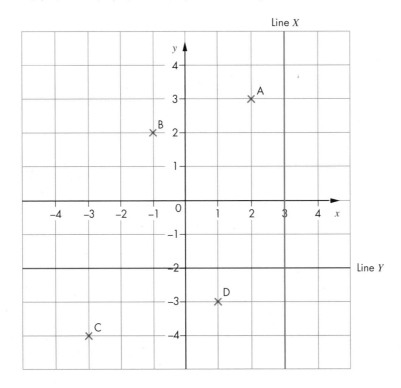

The x-coordinate of all the points on line X is 3.
So you can say the **equation of line** X is $x = 3$.

The y-coordinate of all the points on line Y is -2.
So you can say the equation of line Y is $y = -2$.

The grid is sometimes called the **Cartesian plane**.

Note: The equation of the x-axis is $y = 0$ and the equation of the y-axis is $x = 0$.

Midpoints

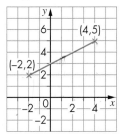

The **midpoint** of a **line segment** is the same distance from each end.

To find the coordinates of the midpoint, add the coordinates of the end points and divide by 2.

In the example shown:

$$\text{Midpoint is } \left(\frac{-2 + 4}{2}, \frac{2 + 5}{2} \right) = (1, 3.5)$$

EXERCISE 15A

1 **a** Write down the coordinates of A, B, C, D and E.

   b Write down the equation of the straight line through:

     i A and C

     ii B and D

     iii D and E

   c Which point is on the line with equation $y = 5$?

   d Which point is on the line with equation $x = -3$?

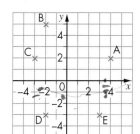

2 **a** Write down the coordinates of P and Q.

   b Find the coordinates of the midpoint of PQ.

   c Find the coordinates of the midpoint of:

     i QR

     ii RS

     iii SP

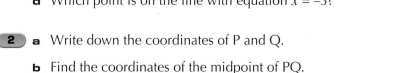

   d The line $y = -2$ has three points on one side and one point on the other.

     Which is the point on its own? *R.*

3 **a** On the grid in question **1**, find the midpoints of AE and CD.

   b Find the equation of the line through both midpoints.

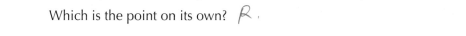

An equation of the form $y = mx + c$ where m and c are numbers will give a **straight line graph**.

EXAMPLE 1

Draw the graph of $y = 4x - 5$ for values of x from 0 to 5. This is usually written as $0 \leqslant x \leqslant 5$.

Choose three values for x: these should be the highest and lowest x-values and one in between.

Work out the y-values by substituting the x-values into the equation.

Keep a record of your calculations in a table, as shown below.

x	0	3	5
y			

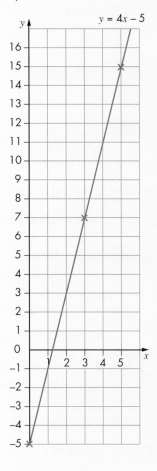

$y = 4x - 5$

When $x = 0$, $y = 4(0) - 5 = -5$
This gives the point $(0, -5)$.

When $x = 3$, $y = 4(3) - 5 = 7$
This gives the point $(3, 7)$.

When $x = 5$, $y = 4(5) - 5 = 15$
This gives the point $(5, 15)$.

Hence your table is:

x	0	3	5
y	−5	7	15

You now have to decide the extent (range) of the axes. You can find this out by looking at the coordinates that you have so far.

The smallest x-value is 0, the largest is 5.
The smallest y-value is −5, the largest is 15.

Now draw the axes, plot the points and complete the graph.

It is nearly always a good idea to choose 0 as one of the x-values. In an examination, the range for the x-values will usually be given and the axes will already be drawn.

EXERCISE 15B

Read through these hints before drawing the following straight line graphs.

- Use the highest and lowest values of x given in the range.

- Do not pick x-values that are too close together, such as 1 and 2. Try to space them out so that you can draw a more accurate graph.

- Always label your graph with its equation. This is particularly important when you are drawing two graphs on the same set of axes.

- Create a table of values. You will often have to complete these in your examinations.

1 Draw the graph of $y = 3x + 4$ for x-values from 0 to 5 ($0 \leqslant x \leqslant 5$).

2 Draw the graph of $y = 2x - 5$ for $0 \leqslant x \leqslant 5$.

3 Draw the graph of $y = \frac{x}{2} - 3$ for $0 \leqslant x \leqslant 10$.

4 Draw the graph of $y = 3x + 5$ for $-3 \leqslant x \leqslant 3$.

5 Draw the graph of $y = \frac{x}{3} + 4$ for $-6 \leqslant x \leqslant 6$.

> **HINTS AND TIPS**
>
> Complete the table of values first, then you will know the extent of the y-axis.

6 **a** On the same set of axes, draw the graphs of $y = 3x - 2$ and $y = 2x + 1$ for $0 \leqslant x \leqslant 5$.

  b At which point do the two lines intersect?

7 **a** On the same axes, draw the graphs of $y = 4x - 5$ and $y = 2x + 3$ for $0 \leqslant x \leqslant 5$.

  b At which point do the two lines intersect?

8 **a** On the same axes, draw the graphs of $y = \frac{x}{3} - 1$ and $y = \frac{x}{2} - 2$ for $0 \leqslant x \leqslant 12$.

  b At which point do the two lines intersect?

9 **a** On the same axes, draw the graphs of $y = 3x + 1$ and $y = 3x - 2$ for $0 \leqslant x \leqslant 4$.

  b Do the two lines intersect? If not, why not?

10 **a** Copy and complete the table to draw the graph of $x + y = 5$ for $0 \leqslant x \leqslant 5$.

x	0	1	2	3	4	5
y	5		3		1	

  b Now draw the graph of $x + y = 7$ for $0 \leqslant x \leqslant 7$.

11 A line has the equation $y = 1.5x + 3$.

  Decide whether each of these points is on the line or not:

  a (6, 12)   **b** (0, 4.5)

  c (2, 6)   **d** (10, 13)

  e (−2, 0)   **f** (−4, −6)

The equation of a straight line can take a number of different forms.

EXAMPLE 2

Draw the straight line with the equation $x - y = 2$.

We need to find pairs of numbers with a difference of 2.

5 and 3 is one pair, so $(5, 3)$ is on the line.

Other possibilities are 2 and 0; 6 and 4; –2 and –4.

These give the points $(2, 0)$, $(6, 4)$ and $(–2, –4)$.

Plotting these on a grid we can join them to give the line.

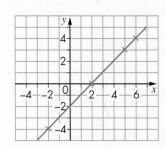

EXAMPLE 3

Draw the straight line with the equation $2x + 3y = 12$.

We need to find pairs of numbers which satisfy this equation.

It is always useful to find where the line crosses the axes.

If $x = 0$, then $0 + 3y = 12 \Rightarrow y = 4$ so $(0, 4)$ is on the line.

If $y = 0$, then $2x + 0 = 12 \Rightarrow x = 6$ so $(6, 0)$ is on the line.

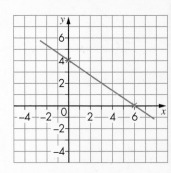

EXERCISE 15C

1 A straight line has the equation $x + y = 2$.

   a Which of the following points are on the line?

      i $(1, 1)$ **ii** $(3, –1)$

      iii $(0, 2)$ **iv** $(–3, 5)$

      v $(4, –2)$ **vi** $(–2, 4)$

   b Draw the graph of the line.

2 A line has the equation $x - y = 0$.

    a Show that (5, 5) is on the line.

    b Write down the coordinates of three more points on the line.

    c Draw a graph of the line.

3 A line has the equation $y - x = 3$.

    a Show that (3, 6) and (–3, 0) are on the line.

    b Find three more points on the line.

    c Draw a graph of the line.

4 Draw a graph of the line with equation $x + y = 0$.

5 A line has the equation $x + 2y = 6$.

    a Where does it cross the x-axis?

    b Where does it cross the y-axis?

    c Draw a graph of $x + 2y = 6$.

6 Draw a graph of the line $2x + y = 8$.

7 Draw a graph of the line $x + 3y = 9$.

8 Draw a graph of the line $5x + 2y = 10$.

9 Draw a graph of the line $2y - x = 4$.

Gradient

A ramp rises by 15cm over a horizontal distance of 1.20 metres.

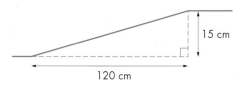

The **gradient** of the ramp $= \dfrac{\text{vertical distance}}{\text{horizontal distance}} = \dfrac{15}{120} = \dfrac{1}{8}$

The larger the gradient, the steeper the **slope**.

We can define the gradient of a line on a graph in the same way.

EXAMPLE 4

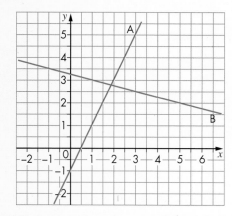

Find the gradients of lines A and B.

Draw a triangle under each line. Choose any convenient points on the line. Here is line A.

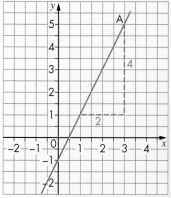

The gradient of line A is $\dfrac{4}{2} = 2$

It does not matter which points you choose, the gradient will always be 2.

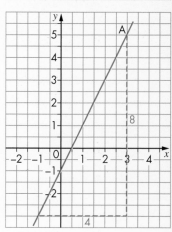

A different triangle but gradient $= \dfrac{8}{4} = 2$ as before.

Here is line B.

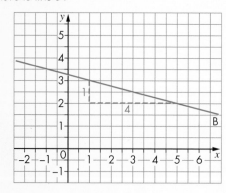

When a line on a graph slopes down from left to right, the gradient is negative.

Gradient $= -\dfrac{1}{4}$

EXERCISE 15D

1 **a** Find the gradient of each of the following ramps:

A: A rise of 10 cm over a horizontal distance of 2 metres.

B: A rise of 8 cm over 0.5 metres.

C: A rise of 50cm over 4 metres.

b Which slope is the steepest?

c Which is the least steep?

2 A road rises steadily, a height of 500 metres over a horizontal distance of 2.5 kilometres.

What is the gradient of the road?

3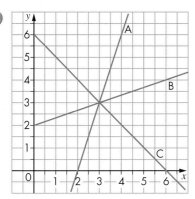

Find the gradients of lines A, B and C.

4 Find the gradients of the straight lines through each of these pairs of points.

a (0, 0) and (2, 8)

b (0, 0) and (8, 2)

c (3, 0) and (5, 5)

d (4, 0) and (5, 10)

e (0, 8) and (4, 0)

f (0, 0) and (10, –2)

g (0, 5) and (7, 5)

h (0, 9) and (6, 0)

HINTS AND TIPS

Plot the points on a graph.

The equation $y = mx + c$

If the equation of the line is $y = mx + c$, then m is the gradient of the line.

EXAMPLE 5

Show that the line with the equation $y = 0.5x - 1$ has a gradient of 0.5.

x	-2	0	2	4	6
y	-2	-1	0	1	2

This is a graph of the line.

Using the shaded triangle, we can see that the gradient is

$\frac{3}{6} = 0.5$

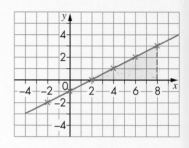

What does the 'c' in $y = mx + c$ represent?

From the graph above you can see it is where the line crosses the y-axis.

$y = 0.5x - 1$ passes through $(0, -1)$.

Note that a line that slopes downwards from left to right has a negative gradient.

This line has a gradient of -1.

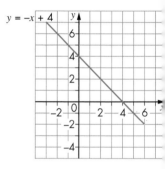

$y = -x + 4$

Summary

When a graph can be expressed in the form $y = mx + c$, the **coefficient** of x, m, is the gradient, and the constant term, c, is the **intercept** on the y-axis.

This means that if you know the gradient, m, of a line and its intercept, c, on the y-axis, you can write down the equation of the line immediately.

For example, if $m = 3$ and $c = -5$, the equation of the line is $y = 3x - 5$.

This gives a method of finding the equation of any line drawn on a pair of coordinate axes.

EXAMPLE 6

Find the equation of the line shown in diagram **A**.

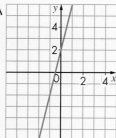 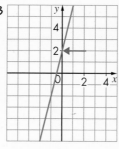 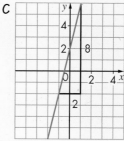

First, find where the graph crosses the y-axis (diagram **B**).

So $c = 2$

Next, measure the gradient of the line (diagram **C**).

y-step = 8
x-step = 2
gradient = $8 \div 2 = 4$

So $m = 4$

Finally, write down the equation of the line: $y = 4x + 2$

EXERCISE 15E

1 You drew the graphs of the lines with these equations in Exercise 15B, questions **1–5**.

In each case state the gradient and the intercept on the y-axis.

Then check from your drawing that you are correct.

a $y = 3x + 4$

b $y = 2x - 5$

c $y = \dfrac{x}{2} - 3$

d $y = 3x + 5$

e $y = \dfrac{x}{3} + 4$

2 Give the equation of each of these lines, all of which have positive gradients. (Each square represents one unit.)

a
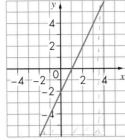

b
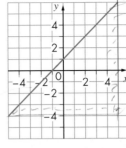

c
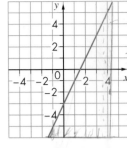

d
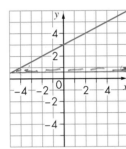

e
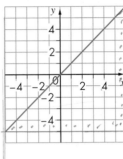

f
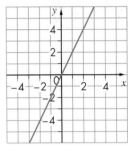

3 In each of these grids, there are two lines. (Each square represents one unit.)

a
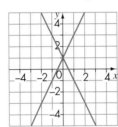

b
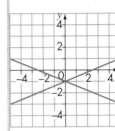

c
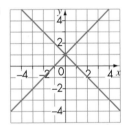

For each grid find the equation of each of the lines.

4 Give the equation of each of these lines, all of which have negative gradients. (Each square represents one unit.)

a
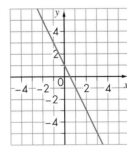

b
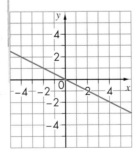

c
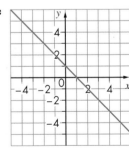

d
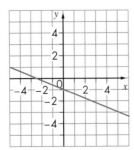

e
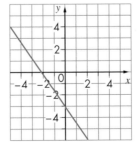

5 The line $y = 4x + c$ passes through $(1, 7)$.

    a Find the value of c.
        b Where does the line cross the y–axis?

6 The line $y = mx - 6$ passes through $(3, 6)$.

    a Find the value of m.
        b What is the gradient of the line?

15.5 Finding equations

The equation of a line

If we know two points on a straight line we have enough information to find the equation of the line between them.

Consider the line through $(-2, 2)$ and $(4, 5)$.

Using the triangle shown,

$$\text{gradient} = \frac{\text{difference between } y\text{-coordinates}}{\text{difference between } x\text{-coordinates}}$$

$$= \frac{5 - 2}{4 - -2}$$

$$= \frac{3}{6}$$

$$= 0.5$$

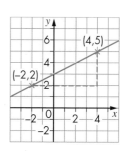

So the equation must be $y = 0.5x + c$ for some value of c.

Use the coordinates of either of the points to find c.

Using $x = 4$ and $y = 5$:

$$5 = 0.5 \times 4 + c$$
$$5 = 2 + c$$
$$c = 3$$

So the equation is $y = 0.5x + 3$.

EXERCISE 15F

1 Find the gradient of the lines through these pairs of points.

    a $(4, 0)$ and $(6, 6)$
    b $(0, 3)$ and $(8, 7)$

    c $(2, -2)$ and $(4, 6)$
    d $(1, 5)$ and $(5, 1)$

    e $(-4, 6)$ and $(6, 1)$
    f $(-5, -3)$ and $(4, 3)$

HINTS AND TIPS

It will help to plot them on a grid.

2 Find the equations of the lines joining these pairs of points.

    a (0, – 3) and (4, 5)   **b** (–4, 2) and (2, 5)

    c (–1, –6) and (2, 6)   **d** (1, 5) and (4, –4)

3 Find the midpoints of the line segments joining the points in question 1.

4 A is (–3, 5), B is (1, 1) and C is (5, 9).

    a Draw the triangle ABC on a coordinate grid.

    b Find the equation of the straight line through A and C.

    c Find the midpoint of AB.

    d Find the equation of the straight line through the midpoints of AC and BC.

5 Find the equations of the lines joining these pairs of points:

    a (2, 2) and (6, 5)   **b** (–3, 2) and (9, 8)

    c (1, 5) and (5, –3)   **d** (–6, –4) and (2, 4)

15.6   Parallel and perpendicular lines   H

If two lines are **parallel**, then their gradients are equal.

If two lines are **perpendicular**, their gradients are **negative reciprocals** of each other.

Consider the line AB. Point A is at (2, –1) and point B is at (4, 5).

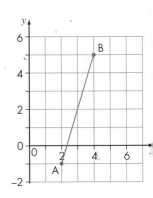

Finding the equation of a parallel line

The gradient of AB is 3, so any parallel line can be written in the form $y = 3x + c$.

To find the equation of the parallel line that passes through the point C at (2, 8), substitute $x = 2$ and $y = 8$ into the equation $y = 3x + c$:

$$8 = 3 \times 2 + c$$

$$\Rightarrow c = 2$$

So the parallel line that passes through (2, 8) is $y = 3x + 2$.

Finding the equation of a perpendicular line

The gradient of the perpendicular line is the negative reciprocal of 3, which is $-\frac{1}{3}$.

To find the equation of the perpendicular line that passes through the midpoint of AB, find the midpoint and substitute for x and y into the equation $y = -\frac{1}{3}x + c$, or sketch the perpendicular line on the grid.

To find the x-coordinate of the midpoint of a line between two points, you add the x-coordinates of the points and divide by 2. To find the y-coordinate, you add the y-coordinates of the points and divide by 2. This gives you a pair of coordinates.

The midpoint of AB is (3, 2).

The perpendicular line passes through (0, 3).

So the equation of the perpendicular line through the midpoint of AB is $y = -\frac{1}{3}x + 3$.

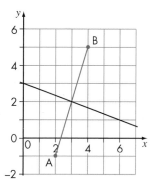

EXAMPLE 7

Two points A and B are A (0, 1) and B (2, 4).

a Work out the equation of the line AB.

b Write down the equation of the line parallel to AB and passing through the point (0, 5).

c Write down the gradient of a line perpendicular to AB.

d Write down the equation of a line perpendicular to AB and passing through the point (0, 2).

a The gradient of AB is $\frac{3}{2}$ and the intercept is (0, 1), so the equation is $y = \frac{3}{2}x + 1$.

b The gradient is the same $\left(\frac{3}{2}\right)$ and the intercept is (0, 5), so the equation is $y = \frac{3}{2}x + 5$.

c The perpendicular gradient is the negative reciprocal $-\frac{2}{3}$.

d The gradient is $-\frac{2}{3}$ and the intercept is (0, 2), so the equation is $y = -\frac{2}{3}x + 2$.

HIGHER

EXAMPLE 8

Find the equation of line that is perpendicular to the line $y = \frac{1}{2}x - 3$ and passes through $(0, 5)$.

The gradient of the new line will be the negative reciprocal of $\frac{1}{2}$ which is -2.

The point $(0, 5)$ is the intercept on the y-axis so the equation of the line is $y = -2x + 5$.

EXERCISE 15G

1 Here are the equations of three lines.

A: $y = 3x - 2$ B: $y = 3x + 1$ C: $y = -\frac{1}{3}x + 1$

a Give a reason why line A is the odd one out of the three.

b Give a reason why line C is the odd one out of the three.

c Which of the following would be a reason why line B is the odd one out of the three?

  i Line B is the only one that intersects the negative x-axis.

  ii Line B is not parallel to either of the other two lines.

  iii Line B does not pass through $(0, -2)$.

2 Write down the negative reciprocals of the following numbers.

  a 2   **b** -3 **c** $\frac{1}{2}$

  d $-\frac{2}{3}$ **e** 1.5   **f** $\frac{4}{3}$

3 Four of these lines make a rectangle. Which four?

  $y = 3x + 5$ $y = 5x + 3$

  $3x + y = 6$ $x + 3y = 10$

  $y = 8 - \frac{1}{3}x$ $y = 3(x + 2)$

4 Match the pairs of perpendicular lines.

  $x = 6$ $x + y = 5$

  $y = 8x - 9$ $2y = x + 4$

  $2x + y = 9$ $y = -\frac{1}{8}x + 6$

  $5y = 2x + 15$ $y = 0.1x + 2$

  $y = 33 - 10x$ $y = -2$

  $2y + 5x = 2$ $y = x + 4$

HIGHER

5 Write down the equations of these lines.

 a Parallel to $y = \frac{1}{2}x + 3$ and passes through $(0, -2)$

 b Parallel to $y = -x + 2$ and passes through $(0, 3)$

 c Perpendicular to $y = 3x + 2$ and passes through $(0, -1)$

 d Perpendicular to $y = -\frac{1}{3}x - 2$ and passes through $(0, 5)$

6 The line segment AB joins A(10, 11) and B(12, 3).

 a Find the gradient of AB.

 b State the gradient of the line perpendicular to AB.

 c Find the midpoint of AB.

 d The line L is perpendicular to AB and passes through the midpoint of AB.
 Show that the equation of L is given by $4y - x = 17$.

7 Find the equation of the line perpendicular to $y = 4x - 3$, passing though $(-4, 3)$.

8 Here are the coordinates of the vertices of three quadrilaterals

 i A(−2, 3) B(3, 2) C(4, 3) D(−1, 4)

 ii A(3, 5) B(9, 9) C(7, 12) D(1, 8)

 iii A(0, 3) B(10, 7) C(6, 6) D(1, 4)

 a For each one, work out the gradients of AB, BC, CD and DA.

 b Determine which quadrilateral is a rectangle, which is a parallelogram and which is a trapezium, explaining your answers.

9 Find the equation of the perpendicular bisector of the line segment AB where A is (1, 2) and B is (3, 6).

10 A is the point (0, 4), B is the point (4, 6) and C is the point (2, 0).

 a Find the equation of the line BC.

 b Show that the point of intersection of the perpendicular bisectors of AB and AC is (3, 3).

 c Show algebraically that this point lies on the line BC.

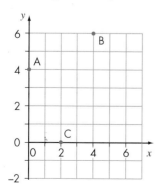

11 The points A(1, 34), B(27, 12) and C(21, –6) lie on the circumference of a circle. Given that a radius always bisects a chord, find the coordinates of the centre of the circle

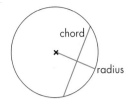

15.7 Graphs and simultaneous equations

Consider the equations $x + 2y = 10$ and $x + y = 8$.

The simultaneous solution is $x = 6$ and $y = 2$.

Here is a graph of the lines with equations $x + 2y = 10$ and $x + y = 8$.

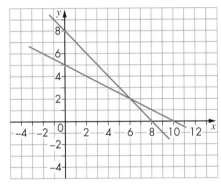

They cross at (6, 2).

The solution of a pair of simultaneous equations can always be interpreted as the coordinates of the point where the corresponding lines cross.

EXERCISE 15H

HIGHER

1 *Use the graph* to solve these simultaneous equations:

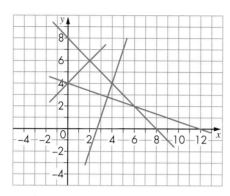

a $x + y = 8$

$y - x = 4$

b $x + 3y = 12$

$x + y = 8$

c $x + y = 8$

$3x - y = 8$

d $y - x = 4$

$x + 3y = 12$

2 Draw graphs to solve these pairs of simultaneous equations:

a $x + 2y = 20$

$2x + y = 16$

b $x - y = 2$

$2x + 3y = 24$

c $2x - y = 12$

$3x + 2y = 18$

There are many curves that can be seen in everyday life. Did you know that all these curves can be represented mathematically?

Below are a few examples of simple curves that you may have come across. Can you think of others?

In mathematics, curves can take many shapes. These can be demonstrated using a cone, as shown on the right and below. If you make a cone out of plasticine or modelling clay, you can see this for yourself. As you look at these curves, try to think of where you have seen them in your own life.

If you slice the cone parallel to the base, the shape you are left with is a circle.

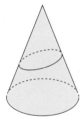

If you slice the cone at an angle to the base, the shape you are left with is an ellipse

If you slice the cone vertically, the shape you are left with is a hyperbola.

The curve that will be particularly important in this chapter is the parabola. Car headlights are shaped like parabolas.

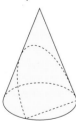

If you slice the cone parallel to its side, the shape you are left with is a parabola.

All parabolas can be represented by *quadratic graphs*. During the course of this chapter you will be looking at how to use *quadratic equations* to draw graphs that have this kind of curve.

The suspension cables on this bridge are also parabolas.

hapter

Graphs of functions

ɔics		Level	Key words
1	Quadratic graphs	FOUNDATION	quadratic graph, quadratic equation, parabola
2	Solving equations with quadratic graphs	HIGHER	
3	Other graphs	HIGHER	cubic
4	Estimating gradients	HIGHER	gradient, tangent
5	Graphs of sin x, cos x and tan x	HIGHER	sin x, cos x, tan x, periodic, period
6	Transformations of graphs	HIGHER	translation, stretch

Vhat you need to be able to do in the examinations:

FOUNDATION	HIGHER

● Generate points and plot graphs of quadratic functions.	● Plot and draw graphs with equation: $y = Ax^3 + Bx^2 + Cx + D$ in which:
	   ● the constants are integers and some could be zero
	   ● the letters x and y can be replaced with any other two letters
	 or:
	   $y = Ax^3 + Bx^2 + Cx + D + \dfrac{E}{x} + \dfrac{E}{x^2}$ in which:
	   ● the constants are numerical and at least three of them are zero
	   ● the letters x and y can be replaced with any other two letters
	 or:
	   ● $y = \sin x$, $y = \cos x$, $y = \tan x$ for angles of any size
	 ● Find the gradients of non-linear graphs.
	 ● Find the intersection points of two graphs, one linear (y_1) and one non-linear (y_2), and recognise that the solutions correspond to the solutions of $y_2 - y_1 = 0$.
	 ● Apply to the graph of $y = f(x)$ the transformations $y = f(x) + a$, $y = f(ax)$, $y = f(x + a)$, $y = af(x)$ for linear, quadratic, sine and cosine functions.

A graph with a ∪ or ∩ shape is a quadratic graph.

A **quadratic graph** has an equation that involves x^2.

All of the following are **quadratic equations** and each would produce a quadratic graph.

$$y = x^2$$
$$y = x^2 + 5$$
$$y = x^2 - 3x$$
$$y = x^2 + 5x + 6$$
$$y = x^2 + 2x - 5$$

EXAMPLE 1

Draw the graph of $y = x^2$ for $-3 \leqslant x \leqslant 3$.

First make a table, as shown below.

x	−3	−2	−1	0	1	2	3
$y = x^2$	9	4	1	0	1	4	9

Now draw axes, with $-3 \leqslant x \leqslant 3$ and $0 \leqslant y \leqslant 9$, plot the points and join them to make a smooth curve.

This is the graph of $y = x^2$.
This type of graph is often referred to as a **parabola**.

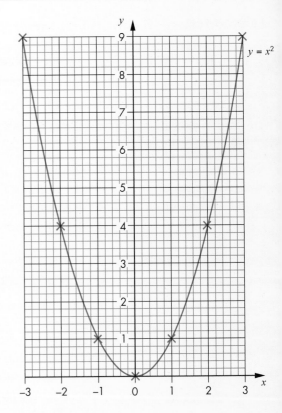

Here are some of the more common ways in which quadratic graphs are drawn inaccurately.

- When the points are too far apart, a curve tends to 'wobble'.

  Wobbly curve

- Drawing curves in small sections leads to 'feathering'.

  Feathering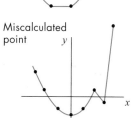

- The place where a curve should turn smoothly is drawn 'flat'.

  Flat bottom

- A curve is drawn through a point that, clearly, has been incorrectly plotted.

  Miscalculated point

A quadratic curve drawn correctly will *always* be a smooth curve.

Here are some tips that will make it easier for you to draw smooth, curved graphs.

- If you are *right-handed*, you might like to turn your piece of paper or your exercise book round so that you draw from left to right. Your hand may be steadier this way than if you try to draw from right to left or away from your body. If you are *left-handed*, you may find drawing from right to left the more accurate way.

- Move your pencil over the points as a practice run without drawing the curve.

- Do one continuous curve and only stop at a plotted point.

- Use a *sharp* pencil and do not press too heavily, so that you may easily rub out mistakes.

EXERCISE 16A

1 **a** Copy and complete the table for $y = x^2 + 2$.

x	−3	−2	−1	0	1	2	3
$y = x^2 + 2$	11			2		6	

  b Draw a graph of $y = x^2 + 2$ for $-3 \leqslant x \leqslant 3$.

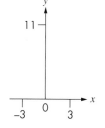

FOUNDATION

2 **a** Copy and complete the table for $y = x^2 - 3x$ for $-3 \leqslant x \leqslant 5$.
Use your table to plot the graph.

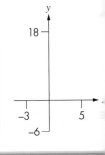

x	−3	−2	−1	0	1	2	3	4	5
x^2						4			
$-3x$						−6			
y						−2			

b Use your graph to find the value of y when $x = 3.5$.

c What are the coordinates of the lowest point on the graph?

d What is the equation of the line of symmetry?

e Use your graph to solve the equation $x^2 - 3x = 5$.

> **HINTS AND TIPS**
>
> You may find you do not need the second and third rows.

3 **a** Copy and complete the table for the graph of $x^2 - 2x - 8$ for $-3 \leqslant x \leqslant 5$. Use your table to plot the graph.

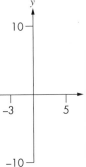

x	−3	−2	−1	0	1	2	3	4	5
y						−8			

b Use your graph to find the value of y when $x = 0.5$.

c Use your graph to solve the equation $y = x^2 - 2x - 8 = 3$.

4 **a** Copy and complete the table for $h = t^2 - 5t + 4$ for $-1 \leqslant t \leqslant 6$.
Use your table to plot the graph.

t	−1	0	1	2	3	4	5	6
h		4			−2			

b What are the coordinates of the lowest point on the graph?

c What is the equation of the line of symmetry?

d Use your graph to find the value of h when $t = -0.5$.

e Use your graph to solve the equation $t^2 - 5t + 4 = 3$.

5 **a** Draw a graph of $y = 8 - x^2$ for $-3 \leqslant x \leqslant 3$.

b Use the graph to solve the equation $8 - x^2 = 6$.

6 **a** Draw a graph of $y = 5x - x^2$ for $0 \leqslant x \leqslant 5$.

b What is the equation for the line of symmetry?

Solving equations with quadratic graphs

We can use graphs to solve quadratic equations.

EXAMPLE 2

a Draw the graph of $y = x^2 + 2x - 3$ for $-4 \leqslant x \leqslant 2$.

b Use your graph to find the value of y when $x = 1.6$.

c Draw the line $y = x - 1$ on the graph.

d Use the graph to solve the equation $x^2 + 2x = x + 2$.

a Draw a table as follows to help work out each step of the calculation.

x	−4	−3	−2	−1	0	1	2
$y = x^2 + 2x - 3$	5	0	−3	−4	−3	0	5

b To find the corresponding y-value for any value of x, you start on the x-axis at that x-value, go up to the curve, across to the y-axis and read off the y-value. This procedure is marked on the graph with arrows.

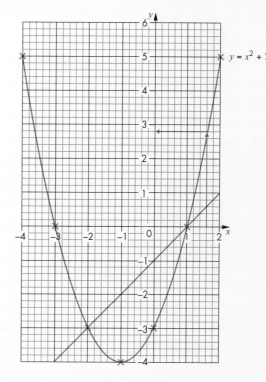

Always show these arrows because even if you make a mistake and misread the scales, you may still get a mark.

So when $x = 1.6$, $y = 2.8$.

c $y = x - 1$ is a straight line with gradient 1 and intercept −1 on the y-axis.

d $x^2 + 2x = x + 2$

Subtract 3 from both sides:

$x^2 + 2x - 3 = x - 1$

The solution of this is given by the intersections of $y = x^2 + 2x - 3$ and $y = x - 1$.

From the graph, $x = -2$ or 1. There are two possible values of x.

Notice that the graph is symmetrical. The equation of the line of symmetry is $x = -1$, and the lowest point is $(-1, -4)$.

EXERCISE 16B

1 **a** Copy and complete the table for $y = x^2 + 2x - 1$ for $-4 \leqslant x \leqslant 2$.
Use your table to plot the graph.

x	−4	−3	−2	−1	0	1	2
y	7						

b Use your graph to find the y-value when $x = -2.5$.

c Draw the line $y = x + 1$.

d Use the graph to solve the equation $x^2 + 2x - 1 = x + 1$.

e Show that the equation in part **d** can be rearranged as $x^2 + x - 2 = 0$.
Solve this by factorisation and check that the answer is the same as part **d**.

2 **a** Copy and complete the table to draw the graph of $y = 12 - x^2$ for $-4 \leqslant x \leqslant 4$.

x	−4	−3	−2	−1	0	1	2	3	4
y				11				3	

b Use your graph to find the y-value when $x = 1.5$.

c Use your graph to solve the equation $12 - x^2 = 0$.

d Draw the line with the equation $y = 6 - x$.

e Use your graph to solve the equation $12 - x^2 = 6 - x$.

3 **a** Copy and complete the table to draw the graph of $y = x^2 + 4x$ for $-5 \leqslant x \leqslant 2$.

x	−5	−4	−3	−2	−1	0	1	2
x^2	25			4			1	
+4x	−20			−8			4	
y	5			−4			5	

b Where does the graph cross the x-axis?

c Use your graph to find the y-value when $x = -2.5$.

d Use your graph to solve the equation $x^2 + 4x = 3$.

e Use the graph to solve the equation $x^2 + 4x + 3 = 0$.

f Use the graph to solve the equation $x^2 + 3x = 0$.

g Use the graph to solve the equation $x^2 + 3x = 4$.

> **HINTS AND TIPS**
>
> What is $x^2 + 4x$?

4 **a** Copy and complete the table to draw the graph of $s = t^2 - 6t + 3$ for $-1 \leqslant t \leqslant 7$.

t	−1	0	1	2	3	4	5	6	7
s	10			−5			−2		

b Where does the graph cross the t-axis?

c Use your graph to find the s-value when $t = 3.5$.

d Use your graph to solve the equation $t^2 - 6t + 3 = 5$.

5 $y = 5x - x^2$

a Copy and complete this table of values.

x	−1	0	1	2	3	4	5	6
y								

b Draw a graph of $y = 5x - x^2$ for $-1 \leqslant x \leqslant 6$.

c What is the highest point on the graph?

d What is the equation of the line of symmetry on the graph?

e Solve the equation $5x - x^2 - 2 = 0$.

16.3 Other graphs H

Cubic graphs

A **cubic** function or graph is one that contains a term in x^3. The following are examples of cubic graphs.

$$y = x^3 \qquad y = x^3 + 3x \qquad y = x^3 + x^2 + x + 1$$

The techniques used to draw them are exactly the same as those for quadratic graphs.

For example, here is a table of values and graph of $y = x^3 - x^2 - 4x + 4$.

x	−3	−2.5	−2	−1.5	−1	−0.5	0	0.5	1	1.5	2	2.5	3
y	−20.00	−7.88	0.00	4.38	6.00	5.63	4.00	1.88	0.00	−0.88	0.00	3.38	10.00

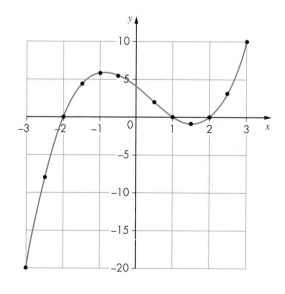

Reciprocal graphs

Graphs with equations involving $\frac{a}{x}$ or $\frac{a}{x^2}$ are called reciprocal graphs.

EXAMPLE 3

Complete this table of values for $y = x + \frac{12}{x}$

x	1	2	3	4	5	6	7
y			7				8.7

Draw the graph of $y = x + \frac{12}{x}$ for $1 \leqslant x \leqslant 7$.

Here is the completed table and graph:

x	1	2	3	4	5	6	7
y	13	8	7	7	7.4	8	8.7

The graph is very steep if x is close to 1.

$x = 0$ must be excluded for the possible values of x

because $\frac{12}{x}$ cannot be calculated if $x = 0$.

If the graph is extended to include negative values of x, it looks like this.

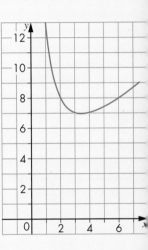

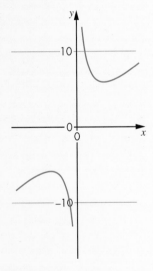

There are two separate parts to the graph. Reciprocal graphs often have separate parts like this.

EXERCISE 16C

1 Draw the graph of $y = x^3$ for $-2 \leqslant x \leqslant 2$.

2   **a** Complete the table to draw the graph of $y = 0.5x^3$ for $-2.5 \leqslant x \leqslant 2.5$.

x	−2.5	−2	−1.5	−1	−0.5	0	0.5	1	1.5	2	2.5
y			−1.69			0.00					7.81

    b Use your graph to solve the equation $0.5x^3 = 6$.

3   **a** Complete the table to draw the graph of $v = r^3 + 3$ for $-2.5 \leqslant r \leqslant 2.5$.

r	−2.5	−2	−1.5	−1	−0.5	0	0.5	1	1.5	2	2.5
v	−12.63			2.00		3.00	3.13			11.00	

    b Use your graph to solve the equation $r^3 + 3 = 0$.

4   **a** Complete this table of values for $y = 4 + \dfrac{8}{x}$

x	1	2	3	4	5	6
y			6.67			5.33

    b Draw a graph of $y = 4 + \dfrac{8}{x}$ for $1 \leqslant x \leqslant 6$.

5   **a** Complete the table to draw the graph of $y = x^3 - 2x + 5$ for $-2 \leqslant x \leqslant 2$.

x	−2	−1.5	−1	−0.5	0	0.5	1	1.5	2
y	1.00	4.63			5.00	4.13			

    b Use your graph to solve the equation $x^3 - 2x + 5 = 3$.

6   **a** Complete this table of values for $y = \dfrac{20}{x^2}$. Give the values to 2 decimal places.

x	1	2	3	4	5
y			2.22		

    b Draw a graph of $y = \dfrac{20}{x^2}$ for $2 \leqslant x \leqslant 5$.

    c Draw a graph of $y = \dfrac{20}{x^2}$ for $-5 \leqslant x \leqslant -2$.

7   **a** Complete this table of values for $y = \dfrac{100}{x^2} - 0.5x$.

x	4	5	6	7	8
y	4.25			−1.46	

    b Draw a graph of $y = \dfrac{100}{x^2} - 0.5x$ for $4 \leqslant x \leqslant 8$.

    c Use the graph to solve the equation $\dfrac{100}{x^2} - 0.5x = 0$.

The **gradient** of a curve varies from point to point. At points on this curve to the left of A or to th[e] right of B the gradient is positive.

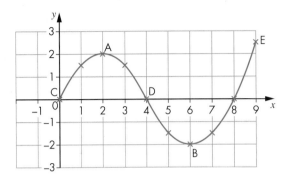

Between A and B the gradient is negative.

The curve is steepest near C, D or E and the gradient there will have the greatest magnitude.

We estimate the gradient at any point on a curve by drawing a **tangent** at that point. This is a straight line which touches the curve at that point and has the same gradient. We can find the gradient of the tangent, which is a straight line, by drawing a triangle.

EXAMPLE 4

Find the gradient at the point P with the coordinates (7, −1.5).

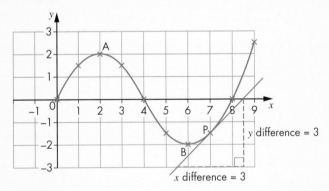

Draw a tangent at point P.

Draw a triangle and find the differences between the x-coordinates and y-coordinates.

$$\text{gradient} = \frac{\text{difference in } y}{\text{difference in } x} = \frac{3}{3} = 1$$

The gradient of the curve at P is 1.

Note that at A and B the gradient is 0.

EXERCISE 16D

1 The straight line is a tangent to the curve at the point P with coordinates (2.5, 1).

Find the gradient of the curve at P.

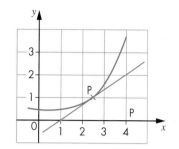

2 A and B have coordinates (–1, 1.5) and (3, 2). Tangents to the curve have been drawn at A and B.

Calculate the gradient of the curve at A and at B.

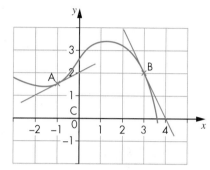

3 **a** Use this table of values to draw a graph of y for $0 \leqslant x \leqslant 4$.

x	0	1	2	3	4
y	2	1.5	2	3.5	6

b Draw a tangent to the curve at the point (3, 3.5).

c Estimate the gradient of the curve at the point (3, 3.5).

d At which point is the gradient of the curve 0?

4 **a** Copy and complete this table of values for $0.1x^3$.

x	0	0.5	1	1.5	2	2.5	3
$0.1x^3$		0.01	0.1		0.8		

b Draw a graph of $y = 0.1x^3$ for $0 \leqslant x \leqslant 3$.

c Draw a tangent to the curve at (2, 0.8).

d Estimate the gradient of the curve at (2, 0.8).

5 **a** Copy and complete this table of values for $\frac{5}{x}$.

x	2	3	4	5	6
$\frac{5}{x}$			1.25		0.83

b Draw a graph of $y = \frac{5}{x}$ for $2 \leqslant x \leqslant 6$.

c Estimate the gradient of the curve at (4, 1.25).

A calculator will give the values of **sin x**, **cos x** and **tan x** for any angle x.

A graph plotter will draw graphs of these trigonometric functions.

Here is a graph of $y = \sin x$

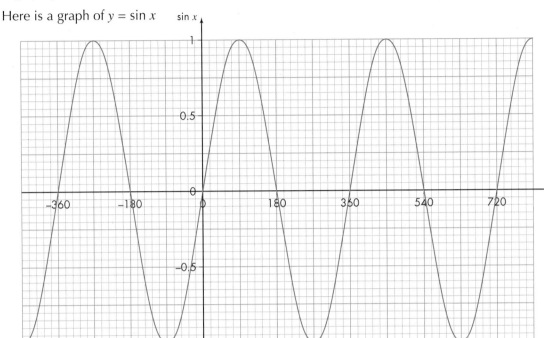

The graph is **periodic**. It repeats every 360°. The **period** is 360°.

You can imagine the graph continuing in the same way in both directions.

The graph is symmetric. It has rotational symmetry of order 2 about (0, 0)

Here is a graph of $y = \cos x$

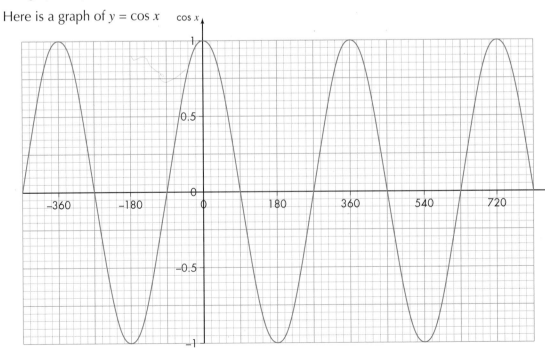

This is also periodic with a period of 360°. It is a translation of the sine graph.

Here is the graph of $y = \tan x$

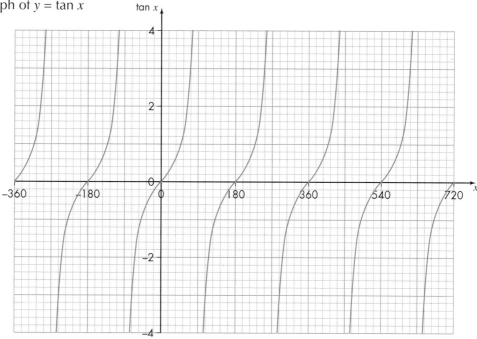

This graph looks very different to sin x and cos x

EXERCISE 16E

1 From the graph you can see that sin $x = 0$ if $x = 0°$ or $180°$

Find three more possible values of x if sin $x = 0$

2 Find three possible values of x if cos $x = 0$

3 **a** What is the largest possible value of sin x?

b What is the smallest possible value of sin x?

4 **a** What is the largest possible value of cos x?

b What is the smallest possible value of cos x?

5 Find three possible values of x if sin $x = 1$

6 Find three possible values of x if sin $x = -1$

7 Find three possible values of x if cos $x = 1$

8 Find three possible values of x if cos $x = -1$

9 The graph of $y = \tan x$ is periodic. What is the period?

10 Find four possible values of x if $\tan x = 0$

11 Look at the graph of $y = \sin x$

Two lines of symmetry are $x = 90°$ and $x = -90°$

Write down the equations of two more lines of symmetry.

12 Write down the equations of three lines of symmetry for the graph of $y = \cos x$

13 The graph of $y = \sin x$ has rotational symmetry of order two about $(0, 0)$.

Write down the coordinates of two more centres of rotational symmetry for this graph.

14 Describe the rotational symmetry of the graph of $y = \cos x$

15 $\sin 30° = 0.5$

Use the symmetry of the graph to find two more possible values of x if $\sin x = 0.5$

16 $\cos 60° = 0.5$

Use the symmetry of the graph to find two more possible values of x if $\cos x = 0.5$

16.6 Transformations of graphs

Here are three graphs.

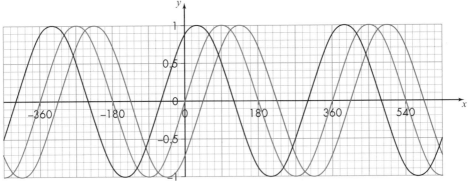

The equations are $y = \sin x$ and $y = \sin(x + 60)$ and $y = \sin(x - 45)$

You can see that $y = \sin(x + 60)$ is a **translation** of $y = \sin x$ by $\begin{pmatrix} -60 \\ 0 \end{pmatrix}$ and $y = \sin(x - 45)$ is a translation of $y = \sin x$ by $\begin{pmatrix} 45 \\ 0 \end{pmatrix}$.

This is an example of a result that is true for any graph.

If you replace x by $x + a$ you translate the graph by $\begin{pmatrix} -a \\ 0 \end{pmatrix}$

You can write it like this :

> The graph of $y = f(x + a)$ is a translation of $y = f(x)$ by $\begin{pmatrix} -a \\ 0 \end{pmatrix}$

EXAMPLE 5

Sketch the graphs of $y = x^2$ and $y = (x - 3)^2$

The graph of $y = x^2$ is a quadratic curve with the lowest point at $(0, 0)$.

The graph of $y = (x - 3)^2$ is a translation by $\binom{3}{0}$ or 3 units to the right.

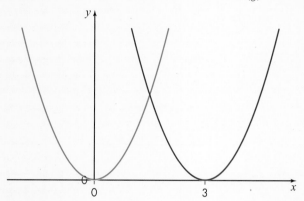

Every point on the graph of $y = x^2$ moves 3 units to the right. $(0, 0)$ goes to $(0, 3)$ and $(-1, 1)$ goes to $(2, 1)$.

Here are some other transformations of graphs.

> The graph of $y = f(x) + a$ is a translation of $y = f(x)$ by $\binom{0}{a}$

For example, here are the graphs of $y = \sin x$ and $y = \sin x + 1$

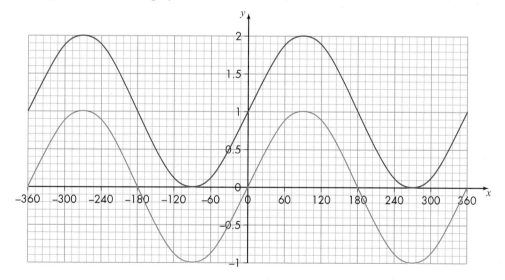

Every point on the graph of $y = \sin x$ moves up 1 unit. $(0, 0)$ goes to $(0, 1)$ and $(90, 1)$ goes to $(90, 2)$.

> The graph of $y = af(x)$ is a **stretch** of $y = f(x)$ by a factor of a from the x-axis.

For example, here are the graphs of $y = \cos x$ and $y = 2\cos x$

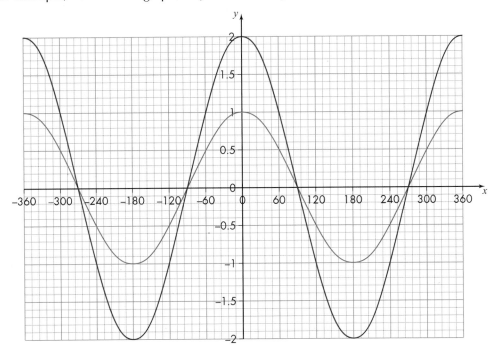

The graph of $y = \cos x$ has been stretched up and down and every point is now twice as far from the x-axis. (0, 1) goes to (0, 2) and (180, –1) goes to (180, –2).

> The graph of $y = f(ax)$ is a stretch of $y = f(x)$ by a factor of $\frac{1}{a}$ from the y-axis

For example, here are the graphs of $y = \sin x$ and $y = \sin 2x$

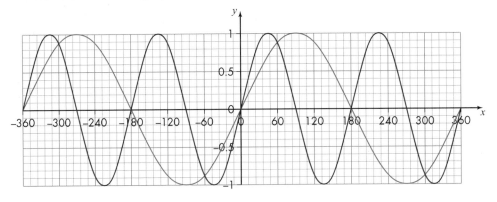

The period of the graph of $y = \sin 2x$ is 180°.

The graph of $y = \sin x$ has been squashed horizontally and every point is now only half the distance from the y-axis. (90, 1) goes to (45, 1) and (360, 0) goes to (180, 0).

EXAMPLE 6

a Sketch the graph of $y = x^2$

b On the same axes sketch the graph of $y = (\frac{1}{3}x)^2$

The second graph is a stretch of the first from the y-axis with a stretch factor of 3.

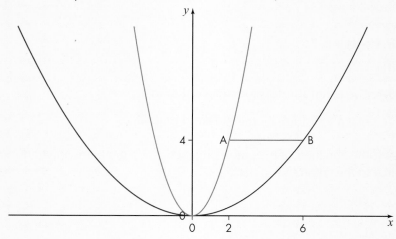

Your sketch should look like this.

For example, A(2, 4) moves to B(6, 4).

EXERCISE 16F

1 On the same axes, sketch the following graphs. Describe the transformation(s) that take(s) the graph in part **a** to each of the other graphs. If your graph gets too 'crowded', draw a new set of axes and re-draw part **a**.

a $y = x^2$

b $y = x^2 + 3$

c $y = x^2 - 1$

d $y = (x + 3)^2$

2 On the same axes, sketch the following graphs. Describe the transformation(s) that take(s) the graph in part **a** to each of the other graphs. If your graph gets too 'crowded', draw a new set of axes and re-draw part **a**.

a $y = \sin x$

b $y = \sin (x + 90°)$

c $y = \sin (x - 45°)$

d $y = \sin x + 2$

HIGHER

HIGHER

3 Sketch each of these pairs of graphs on the same axes for $0° \leqslant x \leqslant 360°$.

a $y = \cos x$ and $y = 2\cos x$ **b** $y = \cos x$ and $y = \cos 2x$

c $y = \cos x$ and $y = \cos x + 2$

4 **a** On the same axes sketch these straight lines:

    i $y = x + 3$

    ii $y = 2x + 3$

    iii $y = 2(x + 3)$

b Describe a transformation that takes **i** onto **ii**.

c Describe a transformation that takes **i** onto **iii**.

5 What is the equation of the graph obtained when the following transformations are applied to the graph of $y = \cos x$?

a Translation of $\begin{pmatrix} 0 \\ 3 \end{pmatrix}$

b Translation of $\begin{pmatrix} -30 \\ 0 \end{pmatrix}$

6 This graph shows three curves.

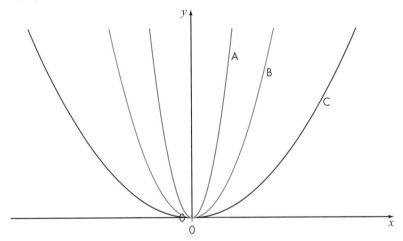

The equations of the curves are $y = x^2$ and $y = (2x)^2$ and $y = (\tfrac{1}{2}x)^2$

Match each equation to the correct letter.

7 Explain why the graphs of $y = \sin x$ and $y = \cos (x - 90)$ are identical.

8 The graphs below are both transformations of $y = x^2$. The coordinates of two points are marked on each graph. Use this information to work out the equation of each graph.

a

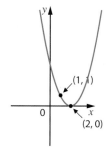

(1, 1)

0

(2, 0)

b

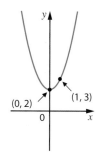

(0, 2)

(1, 3)

0

Patterns often appear in numbers. Prime numbers, square numbers and multiples all form patterns. Mathematical patterns also appear in nature.

There are many mathematical patterns that appear in nature. The most famous of these is probably the **Fibonacci** series.

1   1   2   3   5   8   13   21   ...

This is formed by adding the two previous terms to get the next term.

The sequence was discovered by the Italian, Leonardo Fibonacci, in 1202, when he was investigating the breeding patterns of rabbits!

Since then, the pattern has been found in many other places in nature. The spirals found in a nautilus shell and in the seed heads of a sunflower plant also follow the Fibonacci series.

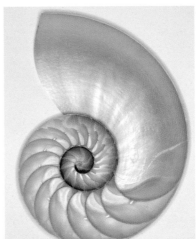
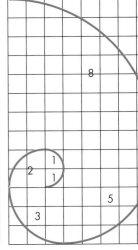

Fractals form another kind of pattern.

Fractals are geometric patterns that are continuously repeated on a smaller and smaller scale.

A good example of a fractal is this: start with an equilateral triangle and draw an equilateral triangle, a third of the size of the original, on the middle of each side. Keep on repeating this and you will get an increasingly complex-looking shape.

The pattern shown here is called the Koch snowflake. It is named after the Swedish mathematician, Helge von Koch (1870–1924).

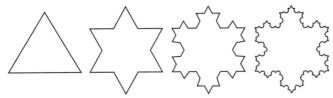

Fractals are commonly found in nature, a good example being the complex patterns found in plants, such as the leaves of a fern.

Integer sequences

ics	Level	Key words
Number sequences	**FOUNDATION**	sequence, term, difference, consecutive
The nth term of a sequence	**FOUNDATION**	nth term, coefficient
Finding the nth term of an arithmetic sequence	**FOUNDATION**	arithmetic sequence
The sum of an arithmetic sequence	**HIGHER**	first term, common difference

What you need to be able to do in the examinations:

FOUNDATION	HIGHER
Generate terms of a sequence using term-to-term and position-to-term definitions of the sequence. Find subsequent terms of an integer sequence and the rule for generating it. Use linear expressions to describe the nth term of an arithmetic sequence.	• Understand and use common difference (d) and first term (a) in an arithmetic sequence. • Know and use nth term = $a + (n - 1)d$ • Find the sum of the first n terms of an arithmetic series (S_n).

A number **sequence** is an ordered set of numbers with a rule for finding every number in the sequence. The rule that takes you from one number to the next could be a simple addition or multiplication, but often it is more tricky than that. So you need to look very carefully at the pattern of a sequence.

Each number in a sequence is called a **term** and is in a certain position in the sequence.

Look at these sequences and their rules.

3, 6, 12, 24, ... doubling the previous term each time ... 48, 96, ...

2, 5, 8, 11, ... adding 3 to the previous term each time ... 14, 17, ...

1, 10, 100, 1000, ... multiplying the previous term by 10 each time ... 10 000, 100 000

1, 8, 15, 22, ... adding 7 to the previous term each time ... 29, 36, ...

These are all quite straightforward once you have looked for the link from one term to the next (**consecutive** terms).

Differences

For some sequences you need to look at the **differences** between consecutive terms to determine the pattern.

EXAMPLE 1

Find the next two terms of the sequence 1, 3, 6, 10, 15, ...

Looking at the differences between consecutive terms:

```
1   3    6    10    15
  ↑   ↑    ↑    ↑
  2   3    4    5
```

So the sequence continues as follows.

```
1   3    6    10    15    21    28
  ↑   ↑    ↑    ↑   ⌐  ⌐ ⌐  ⌐
  2   3    4    5    +6    +7
```

So the next two terms are 21 and 28.

The differences usually form a number sequence of their own, so you need to find the *sequence of the differences* before you can expand the original sequence.

EXERCISE 17A

1 Look at the following number sequences. Write down the next three terms in each and explain how each sequence is formed.

   a 1, 3, 5, 7, …   **b** 2, 4, 6, 8, …

   c 5, 10, 20, 40, …   **d** 1, 3, 9, 27, …

   e 4, 10, 16, 22, …   **f** 3, 8, 13, 18, …

   g 2, 20, 200, 2000, …   **h** 7, 10, 13, 16, …

   i 10, 19, 28, 37, …   **j** 5, 15, 45, 135, …

   k 2, 6, 10, 14, …   **l** 1, 5, 25, 125, …

2 By considering the differences in the following sequences, write down the next two terms in each case.

   a 1, 2, 4, 7, 11, …   **b** 1, 2, 5, 10, 17, …

   c 1, 3, 7, 13, 21, …   **d** 1, 4, 10, 19, 31, …

   e 1, 9, 25, 49, 81, …   **f** 1, 2, 7, 32, 157, …

   g 1, 3, 23, 223, 2223, … **h** 1, 2, 4, 5, 7, 8, 10, …

   i 2, 3, 5, 9, 17, …   **j** 3, 8, 18, 33, 53, …

3 Look at the sequences below. Find the rule for each sequence and write down its next three terms.

   a 3, 6, 12, 24, …   **b** 3, 9, 15, 21, 27, …

   c 128, 64, 32, 16, 8, …   **d** 50, 47, 44, 41, …

   e 2, 5, 10, 17, 26, …   **f** 5, 6, 8, 11, 15, 20, …

   g 5, 7, 8, 10, 11, 13, … **h** 4, 7, 10, 13, 16, …

   i 1, 3, 6, 10, 15, 21, … **j** 1, 2, 3, 4, …

   k 100, 20, 4, 0.8, …   **l** 1, 0.5, 0.25, 0.125, …

4 Look carefully at each number sequence below. Find the next two numbers in the sequence and try to explain the pattern.

   a 1, 1, 2, 3, 5, 8, 13, …

   b 1, 4, 9, 16, 25, 36, …

   c 3, 4, 7, 11, 18, 29, …

   d 1, 8, 27, 64, 125, …

> **HINTS AND TIPS**
>
> These patterns do not go up by the same value each time so you will need to find another connection between the terms.

5 Triangular numbers are found as follows.

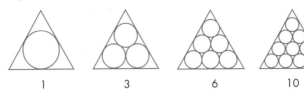

1   3   6   10

Find the next four triangular numbers.

6 Hexagonal numbers are found as follows.

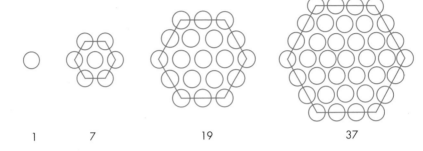

1   7   19   37

Find the next three hexagonal numbers.

7 The first term that these two sequences have in common is 17:

8, 11, 14, 17, 20, ...

1, 5, 9, 13, 17, ...

What are the next two terms that the two sequences have in common?

8 Two sequences are:

2, 5, 8, 11, 14, ...

3, 6, 9, 12, 15, ...

Will the two sequences ever have a term in common? Yes or no?
Justify your answer.

17.2 The nth term of a sequence

Finding the rule

When using a number sequence, you sometimes need to know, say, its 50th term, or even a
higher term in the sequence. To do so, you need to find the rule that produces the sequence i
its general form.

It may be helpful to look at the problem backwards. That is, take a rule and see how it produc
a sequence. The rule is given for the general term, which is called the **nth term**.

FOUNDATION

EXAMPLE 2

The nth term of a sequence is $3n + 1$, where $n = 1, 2, 3, 4, 5, 6, \ldots$. Write down the first five terms of the sequence.

Substituting $n = 1, 2, 3, 4, 5$ in turn:

$(3 \times 1 + 1), (3 \times 2 + 1), (3 \times 3 + 1), (3 \times 4 + 1), (3 \times 5 + 1), \ldots$
    4   7   10   13   16

So the sequence is 4, 7, 10, 13, 16, ...

Notice that in Example 2 the difference between each term and the next is always 3, which is the **coefficient** of n (the number attached to n). Also, the constant term is the difference between the first term and the coefficient, that is, $4 - 3 = 1$.

Here are two useful sequences.

The even numbers are 2, 4, 6, 8, 10, 12, …
The nth even number is $2n$

The odd numbers are 1, 3, 5, 7, 9, 11, …
The nth odd number is $2n - 1$

EXERCISE 17B

1 Here are the nth terms of some sequences. Write down the first five terms of each sequence.

  a $2n + 1$ for $n = 1, 2, 3, 4, 5$ **b** $3n - 2$ for $n = 1, 2, 3, 4, 5$

  c $5n + 2$ for $n = 1, 2, 3, 4, 5$

2 Write down the first five terms of the sequence that has as its nth term:

  a $n + 3$ **b** $3n - 1$ **c** $5n - 2$ **d** $4n + 5$

> **HINTS AND TIPS**
>
> Substitute numbers into the expressions until you can see how the sequence works.

3 A haulage company uses this formula to calculate the cost of transporting n pallets.

For $n \leqslant 5$, the cost will be $\$(40n + 50)$
For $6 \leqslant n \leqslant 10$, the cost will be $\$(40n + 25)$
For $n \geqslant 11$, the cost will be $\$40n$

  a How much will the company charge to transport 7 pallets?

  b How much will the company charge to transport 15 pallets?

  c A company is charged $170 for transporting pallets. How many pallets did they transport?

  d Another haulage company uses the formula $\$50n$ to calculate the cost for transporting n pallets.

    At what value of n do the two companies charge the same amount?

FOUNDATION

4 The nth term of a sequence is $3n + 7$.
The nth term of another sequence is $4n - 2$.

These two series have several terms in common but only one term that is common and ▌
the same position in the sequence.

Without writing out the sequences, show how you can tell, using the expressions for the ▌
nth term, that this is the 9th term.

5 Here is a sequence. 1, 2, 3, 4, 5, 6, …

The nth term is n.

Use this fact to write down the nth term of each of these sequences.

a 2, 3, 4, 5, 6, …   **b** 4, 5, 6, 7, 8, ……   **c** 11, 12, 13, 14, 15, ..▌

d 0, 1, 2, 3, 4, …..   **e** 2, 4, 6, 8, 10, ….   **f** 3, 6, 9, 12, 15, 18, ..▌

6 Match the sequence to the nth term. One is done for you.

a 2, 4, 6, 8, 10, … $2n - 1$

b 1, 3, 5, 7, 9, … $2n$

c 4, 6, 8, 10, 12, … $2n + 1$

d 3, 5, 7, 9, 11, … $2n + 2$

e 5, 7, 9, 11, 13, …. $2n + 3$

7 Match the sequence to the nth term. One is done for you.

a 4, 8, 12, 16, 20, … $4n - 1$

b 5, 9, 13, 17, 21, … $4n$

c 7, 11, 15, 19, 23, … $4n + 1$

d 3, 7, 11, 15, 19, … $4n + 2$

e 6, 10, 14, 18, 22, …. $4n + 3$

8 Find the nth term.

a 5, 10, 15, 20, 25, …   **b** 6, 12, 18, 24, 30, …   **c** 8, 16, 24, 32, 40, …

Finding the nth term of an arithmetic sequence

In an **arithmetic sequence** the difference between one term and the next is always the same. All the sequences in the last exercise were arithmetic sequences.

HINTS AND TIPS

The stress when you say the word 'arithmetic' here is in the letter e: arithmétic.

For example:

2, 5, 8, 11, 14, … difference of 3

The nth term of this sequence is given by $3n - 1$.

Here is another linear sequence.

5, 7, 9, 11, 13, … difference of 2

The nth term of this sequence is given by $2n + 3$.

So, you can see that the nth term of a linear sequence is *always* of the form $An + b$, where:

- A, the coefficient of n, is the difference between each term and the next term (consecutive term)
- b is the difference between the first term and A.

EXAMPLE 3

From the arithmetic sequence 5, 12, 19, 26, 33, … , find:

a the nth term   **b** the 50th term.

a The difference between consecutive terms is 7. So the first part of the nth term is $7n$.
Subtract the difference, 7, from the first term, 5, which gives $5 - 7 = -2$.
So the nth term is given by $7n - 2$.

b The 50th term is found by substituting $n = 50$ into the rule, $7n - 2$.
$$50\text{th term} = 7 \times 50 - 2 = 350 - 2$$
$$= 348$$

EXERCISE 17C

1 Find the next two terms and the nth term in each of these arithmetic sequences.

HINTS AND TIPS

Remember to look at the differences and the first term.

a 3, 5, 7, 9, 11, …

b 5, 9, 13, 17, 21, …

c 8, 13, 18, 23, 28, …

d 2, 8, 14, 20, 26, …

e 5, 8, 11, 14, 17, …

f 2, 9, 16, 23, 30, …

g 1, 5, 9, 13, 17, …

h 3, 7, 11, 15, 19, …

i 2, 5, 8, 11, 14, …

j 2, 12, 22, 32, …

k 8, 12, 16, 20, …

l 4, 9, 14, 19, 24, …

FOUNDATION

2 Find the *n*th term and the 50th term in each of these arithmetic sequences.

a 4, 7, 10, 13, 16, …

b 7, 9, 11, 13, 15, …

c 3, 8, 13, 18, 23, …

d 1, 5, 9, 13, 17, …

e 2, 10, 18, 26, …

f 5, 6, 7, 8, 9, …

g 6, 11, 16, 21, 26, …

h 3, 11, 19, 27, 35, …

i 1, 4, 7, 10, 13, …

j 21, 24, 27, 30, 33, …

k 12, 19, 26, 33, 40, …

l 1, 9, 17, 25, 33, …

3 For each sequence **a** to **j**, find:

   i the *n*th term

   ii the 100th term.

a 5, 9, 13, 17, 21, …

b 3, 5, 7, 9, 11, 13, …

c 4, 7, 10, 13, 16, …

d 8, 10, 12, 14, 16, …

e 9, 13, 17, 21, …

f 6, 11, 16, 21, …

g 0, 3, 6, 9, 12, …

h 2, 8, 14, 20, 26, …

i 7, 15, 23, 31, …

j 25, 27, 29, 31, …

17.4 The sum of an arithmetic sequence

Here are the first six terms of an arithmetic sequence 7, 11, 15, 19, 23, 27

> In an arithmetic sequence we use the letter *a* for the **first term** and *d* for the **common difference**.

In this case, $a = 7$ and $d = 4$

A formula for the *n*th term of an arithmetic sequence is $a + (n - 1)d$

We write S_n for the sum of the first *n* terms of a sequence.

In this case, $S_2 = 7 + 11 = 18$ and $S_4 = 7 + 11 + 15 + 19 = 52$

> A formula for the sum of the first *n* terms of an arithmetic sequence is
> $S_n = \frac{n}{2}[2a + (n - 1)d]$

In the example above the formula gives the 6th term as $7 + 5 \times 4 = 27$ which is correct.

The sum of the first 6 terms, $S_6 = \frac{6}{2}[2 \times 7 + 5 \times 4] = 3[14 + 20] = 102$

Check that this is correct.

EXAMPLE 4

Work out: **a** the 20th odd number   **b** the sum of the first 20 odd numbers

The odd numbers are an arithmetic sequence with $a = 1$ and $d = 2$

Use the formulae above with $n = 20$

The 20th odd number is $1 + (20 - 1) \times 2 = 1 + 19 \times 2 = 39$

The sum of the first 20 odd numbers, $S_{20} = \frac{20}{2}[2 \times 1 + (20 - 1) \times 2]$

$$= 10\,[2 + 38]$$
$$= 10 \times 40 = 400$$

Use the formulae for the nth term and for S_n in the following exercise.

EXERCISE 17D

1. Here are the first five terms of an arithmetic sequence: 10, 13, 16, 19, 22

   a Write down the values of a and d.   **b** Work out the 12th term.

   c Work out the sum of the first 12 terms.

2. Here is the start of an arithmetic sequence: 20, 22, 24, 26, 28, …

   Work out:

   a the 8th term   **b** the sum of the first 5 terms

   c the sum of the first 25 terms.

3. Here are the first four terms of an arithmetic sequence: 3, 9, 15, 21

   a Work out the 10th term.   **b** Work out the sum of the first 10 terms.

   c Work out the sum of the first 15 terms.

4. Work out:

   a the sum of the first 20 even numbers   **b** the sum of the first 50 even numbers

5. **a** Work out the sum of:

   i the first 8 odd numbers   **ii** the first 12 odd numbers

   iii the first 15 odd numbers.

   b Can you see a pattern in your answers to part **a**?

6. Delia's uncle gives her $10 on her first birthday, $20 on her second birthday, $30 on her third birthday and so on.

   How much has he given her altogether after her 18th birthday?

HIGHER

Indices are a useful way to write numbers. They show how different numbers are related to one another and they can make it easier to multiply or divide, or to compare the sizes of different numbers.

You probably already know about powers of numbers from chapters 5 and 9.

We find powers of 10 when numbers are written in standard form such as 3.7×10^6 or 8.92×10^{-5}

The first is $3\,700\,000$ and the second is $0.000\,089\,2$

Using powers is a useful 'short cut'.
For example, the centre of the galaxy Andromeda is $24\,000\,000\,000\,000\,000\,000\,000$ km from our Sun.
It is much easier to write 24×10^{18}!

We also find powers in squares and cubes such as 7^2 and 14^3.

Remember that $7^2 = 7 \times 7$ and $14^3 = 14 \times 14 \times 14$. This also saves us time in writing out the numbers.

The small number which shows the power is called an index:

- In 7^2 the index is 2 and in 14^3 the index is 3.
- In 3.7×10^6 or 8.92×10^{-5} the indices are 6 and -5.

The plural of index is indices. In the examples above 2, 3, 6 and -5 are indices.

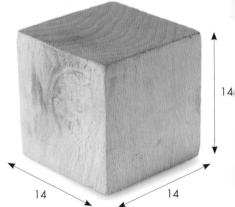

In this chapter you will discover more useful ways to use indices. You might be surprised to find you can write fractions using indices. You can probably see how 8 can be written as a power of two (2^3) but it is not so obvious how $\frac{1}{8}$ can also be written as a power of two.

Warning

The word 'index' has a number of other meanings in English. For example, you will find an index at the back of this book. In this chapter the word is always used to mean a power.

18

Indices

Topics	Level	Key words
1 Using indices	**FOUNDATION**	index, indices, power, power 1, power 0
2 Negative indices	**HIGHER**	negative index, reciprocal
3 Multiplying and dividing with indices	**FOUNDATION**	
4 Fractional indices	**HIGHER**	

What you need to be able to do in the examinations:

FOUNDATION	HIGHER
Use index notation and index laws for multiplication and division of positive integer powers. Use index notation for positive integer powers. Use index laws in simple cases.	• Use index notation involving fractional, negative and zero powers. • Use index laws to simplify and evaluate numerical expressions involving integer, fractional and negative powers.

An **index** is a convenient way of writing repetitive multiplications. The plural of index is **indice**

The index tells you the number of times a number is multiplied by itself. For example:

$4^6 = 4 \times 4 \times 4 \times 4 \times 4 \times 4$ six lots of 4 multiplied together (we call this "4 to the **power** 6

$6^4 = 6 \times 6 \times 6 \times 6$ four lots of 6 multiplied together (we call this "6 to the power 4

$7^3 = 7 \times 7 \times 7$

$12^2 = 12 \times 12$

EXAMPLE 1

   a   Write each of these numbers out in full.

      i 4^3 **ii** 6^2 **iii** 7^5 **iv** 12^4

   b   Write the following multiplications using powers.

      i $3 \times 3 \times 3 \times 3 \times 3 \times 3 \times 3 \times 3$

      ii $13 \times 13 \times 13 \times 13 \times 13$

      iii $7 \times 7 \times 7 \times 7$

      iv $5 \times 5 \times 5 \times 5 \times 5 \times 5 \times 5$

   a   **i** $4^3 = 4 \times 4 \times 4$ **ii** $6^2 = 6 \times 6$

      iii $7^5 = 7 \times 7 \times 7 \times 7 \times 7$ **iv** $12^4 = 12 \times 12 \times 12 \times 12$

   b   **i** $3 \times 3 \times 3 \times 3 \times 3 \times 3 \times 3 \times 3 = 3^8$

      ii $13 \times 13 \times 13 \times 13 \times 13 = 13^5$

      iii $7 \times 7 \times 7 \times 7 = 7^4$

      iv $5 \times 5 \times 5 \times 5 \times 5 \times 5 \times 5 = 5^7$

Using indices on your calculator

The power button on your calculator will probably look like this $\boxed{x^\blacksquare}$.

To work out 5^7 on your calculator use the power key.

$5^7 = \boxed{5} \ \boxed{x^\blacksquare} \ \boxed{7} = 78\,125$

Two special powers

Power 1

Any number to the power 1 is the same as the number itself. This is always true so normally you do not write the power 1.

For example: $5^1 = 5$ $32^1 = 32$ $(-8)^1 = -8$

Power 0

Any number to the power 0 is equal to 1.

For example: $5^0 = 1$　　　　$32^0 = 1$　　　　$(-8)^0 = 1$

You can check these results on your calculator.

EXERCISE 18A

1　Write these expressions using index notation. Do not work them out yet.

　　a　$2 \times 2 \times 2 \times 2$ 　　　　　　　**b**　$3 \times 3 \times 3 \times 3 \times 3$

　　c　7×7 　　　　　　　　　　　**d**　$5 \times 5 \times 5$

　　e　$10 \times 10 \times 10 \times 10 \times 10 \times 10 \times 10$ 　　**f**　$6 \times 6 \times 6 \times 6$

　　g　4 　　　　　　　　　　　　　**h**　$1 \times 1 \times 1 \times 1 \times 1 \times 1 \times 1$

　　i　$0.5 \times 0.5 \times 0.5 \times 0.5$ 　　　　**j**　$100 \times 100 \times 100$

2　Write these power terms out in full. Do not work them out yet.

　　a　3^4　　　**b**　9^3　　　**c**　6^2　　　**d**　10^5　　　**e**　2^{10}

　　f　8^1　　　**g**　0.1^3　　　**h**　2.5^2　　　**i**　0.7^3　　　**j**　1000^2

3　Using the power key on your calculator (or another method), work out the values of the power terms in question **1**.

4　Using the power key on your calculator (or another method), work out the values of the power terms in question **2**.

5　A storage container is in the shape of a cube. The length of the container is 5 m.

　　To work out the volume of a cube, use the formula:

　　　volume = (length of edge)^3

　　Work out the total storage space in the container.

6　Write each number as a power of a different number.

　　The first one has been done for you.

　　a　$32 = 2^5$　　　**b**　100　　　**c**　8　　　**d**　25

7　Without using a calculator, work out the values of these power terms.

　　a　2^0　　　**b**　4^1　　　**c**　5^0　　　**d**　1^9　　　**e**　1^{235}

8　The answers to question **7**, parts **d** and **e**, should tell you something special about powers of 1. What is it?

9　Write the answer to question **1**, part **j** as a power of 10.

Negative indices

A **negative index** is a convenient way of writing the **reciprocal** of a number or term. (That is, o divided by that number or term.) For example:

$$x^{-a} = \frac{1}{x^a}$$

Here are some other examples.

$$5^{-2} = \frac{1}{5^2} \qquad 3^{-1} = \frac{1}{3} \qquad 5x^{-2} = \frac{5}{x^2}$$

EXAMPLE 2

Rewrite the following in the form 2^n.

a 8 b $\frac{1}{4}$ c -32 d $-\frac{1}{64}$

a $8 = 2 \times 2 \times 2 = 2^3$ b $\frac{1}{4} = \frac{1}{2^2} = 2^{-2}$

c $-32 = -2^5$ d $-\frac{1}{64} = -\frac{1}{2^6} = -2^{-6}$

What about x^0?

In fact $x^0 = 1$ for any value of x. This can seem surprising but it fits the pattern in this table:

n	3	2	1	0	−1	−2
2^n	8	4	2	1	$\frac{1}{2}$	$\frac{1}{4}$
3^n	27	9	3	1	$\frac{1}{3}$	$\frac{1}{9}$

The numbers in the second row divide by 2 as you move from left to right.

The numbers in the third row divide by 3 as you move from left to right.

2^0 and 3^0 are both equal to 1.

The rules for multiplying and dividing with indices still apply, whether the indices are positive negative or zero.

EXERCISE 18B

1 Write down each of these in fraction form.

a 5^{-3} b 6^{-1} c 10^{-5} d 3^{-2} e 8^{-2}

f 9^{-1} g w^{-2} h t^{-1} i x^{-m} j $4m^{-3}$

2 Write down each of these in negative index form.

a $\frac{1}{3^2}$ b $\frac{1}{5}$ c $\frac{1}{10^3}$ d $\frac{1}{m}$ e $\frac{1}{t^n}$

HINTS AND TIPS

If you move a power from top to bottom, or vice versa, the sign changes. Negative power means the reciprocal: it does not mean the answer is negative.

3 Change each of the following expressions into an index form of the type shown.

    a All of the form 2^n

        i 16   **ii** $\frac{1}{2}$ **iii** $\frac{1}{16}$ **iv** −8

    b All of the form 10^n

        i 1000   **ii** $\frac{1}{10}$ **iii** $\frac{1}{100}$ **iv** 1 million

    c All of the form 5^n

        i 125   **ii** $\frac{1}{5}$ **iii** $\frac{1}{25}$ **iv** 1

    d All of the form 3^n

        i 9   **ii** $\frac{1}{27}$ **iii** 1   **iv** −243

4 Rewrite each of the following expressions in fraction form.

    a $5x^{-3}$ **b** $6t^{-1}$ **c** $7m^{-2}$ **d** $4q^{-4}$ **e** $10y^{-5}$

    f $\frac{1}{2}x^{-3}$ **g** $\frac{1}{2}m^{-1}$ **h** $\frac{3}{4}t^{-4}$ **i** $\frac{4}{5}y^{-3}$ **j** $\frac{7}{8}x^{-5}$

5 Write each fraction in index form.

    a $\frac{7}{x^3}$ **b** $\frac{10}{p}$ **c** $\frac{5}{t^2}$ **d** $\frac{8}{m^5}$ **e** $\frac{1}{64}$

6 Find the value of each of the following.

    a $x = 5$

        i x^2 **ii** x^{-3} **iii** $4x^{-1}$

    b $t = 4$

        i t^3 **ii** t^{-2} **iii** $5t^{-4}$

    c $m = 2$

        i m^3 **ii** m^{-5} **iii** $9m^{-1}$

    d $w = 10$

        i w^6 **ii** w^{-3} **iii** $25w^{-2}$

7 Write in index form:

    a $a^{-3} \times a^{-4}$ **b** $a^{-2} \times a^4$ **c** $a^2 \div a^{-2}$ **d** $a^{-3} \div a^2$ **e** $(a^{-3})^2$ **f** $(a^{-2})^{-3}$

When you *multiply* powers of the same number or variable, you add the indices. For example

$3^4 \times 3^{-2} = 3^2 \qquad 4 + -2 = 2$

$2^{-3} \times 2^{-4} = 2^{-7} \qquad -3 + -4 = -7$

$a^x \times a^y = a^{x+y}$

When you *divide* powers of the same number or variable, you subtract the indices. For examp

$5^2 \div 5^3 = 5^{-1} \qquad 2 - 3 = -1$

$a^{-2} \div a^{-3} = a^1 = a \qquad -2 - -3 = 1$

$a^x \times a^y = a^{x-y}$

When you raise a power to a further power, you multiply the indices. For example:

$(4^2)^3 = 4^6 \qquad 2 \times 3 = 6$

$(a^{-3})^2 = a^{-6} \qquad -3 \times 2 = -6$

$(a^x)^y = a^{xy}$

EXERCISE 18C

1 Write these as single powers of 5.

   a $5^2 \times 5^2$ **b** 5×5^2 **c** $5^2 \times 5^{-4}$ **d** $5^{-1} \times 5^{-3}$ **e** $5^{-3} \times 5^4$

2 Write these as single powers of 6.

   a $6^5 \div 6^2$ **b** $6^4 \div 6$ **c** $6^2 \div 6^4$ **d** $6 \div 6^{-4}$ **e** $6^{-1} \div 6^{-3}$

3 Simplify these and write them as single powers of *a*.

   a $a^2 \times a$ **b** $a^3 \times a^2$ **c** $a^{-2} \times a^{-2}$

   d $a^6 \div a^2$ **e** $b^{-1} \div b^2$ **f** $b \div b^{-2}$

4 **a** $a^x \times a^y = a^{10}$

      Write down a possible pair of values of *x* and *y*.

   b $a^x \div a^y = a^{10}$

      Write down a possible pair of values of *x* and *y*.

5 Write these as single powers of 4.

   a $(4^2)^3$ **b** $(4^3)^5$ **c** $(4^2)^{-1}$

   d $(4^{-2})^{-2}$ **e** $(4^{-3})^{-1}$ **f** $(4^{-3})^3$

6 Simplify the following:

   a $\dfrac{a^2}{a^3}$ **b** $\dfrac{a^4}{a^6}$ **c** $\dfrac{a^4}{a}$ **d** $\dfrac{a}{a^4}$ **e** $\dfrac{a^5}{a^4}$

7 If $a^5 \times a^n = a^2$ find the value of n.

8 If $b^{-1} \div b^n = b$ find the value of n.

9 Simplify:

   a $(a^2)^3$ **b** $(t^{-1})^3$ **c** $(k^{-2})^{-1}$ **d** $\sqrt{x^6}$

10 Simplify these expressions.

   a $6a^3 \div 2a^2$ **b** $12a^5 \div 3a^2$

   c $15a^5 \div 5a$ **d** $\dfrac{18a}{3a^2}$

   e $\dfrac{24a^5}{6a^2}$ **f** $\dfrac{30a}{6a^5}$

> **HINTS AND TIPS**
>
> Deal with numbers and indices separately and do not confuse the rules.
> For example: $12a^5 \div 4a^2$
> $= (12 \div 4)\ (a^5 \div a^2)$

11 Write down **two** possible:

   a multiplication questions with an answer of $12x^2y^5$

   b division questions with an answer of $12x^2y^5$.

12 a, b and c are three different positive integers.

   What is the smallest possible value of a^2b^3c?

8.4   Fractional indices

Indices of the form $\dfrac{1}{n}$

Consider the problem $7^x \times 7^x = 7$. This can be written as:

$$7^{(x + x)} = 7$$
$$7^{2x} = 7^1 \;\Rightarrow\; 2x = 1 \;\Rightarrow\; x = \tfrac{1}{2}$$

If you now substitute $x = \tfrac{1}{2}$ back into the original equation, you see that:

$$7^{\frac{1}{2}} \times 7^{\frac{1}{2}} = 7$$

So $7^{\frac{1}{2}}$ is the same as $\sqrt{7}$.

You can similarly show that $7^{\frac{1}{3}}$ is the same as $\sqrt[3]{7}$. And that, generally:

$$x^{\frac{1}{n}} = \sqrt[n]{x} \;(n\text{th root of } x)$$

So in summary:

Power $\frac{1}{2}$ is the same as positive square root.

Power $\frac{1}{3}$ is the same as cube root.

Power $\frac{1}{n}$ is the same as nth root.

For example:

$$49^{\frac{1}{2}} = \sqrt{49} = 7 \qquad 8^{\frac{1}{3}} = \sqrt[3]{8} = 2 \qquad 10\,000^{\frac{1}{4}} = \sqrt[4]{10\,000} = 10 \qquad 36^{-\frac{1}{2}} = \frac{1}{\sqrt{36}} = \frac{1}{6}$$

If you have an expression in the form $\left(\dfrac{a}{b}\right)^n$ it can be calculated as $\dfrac{a^n}{b^n}$ and then written as a fraction.

EXAMPLE 3

Write $\left(\dfrac{16}{25}\right)^{\frac{1}{2}}$ as a fraction.

We can find the power of the numerator and denominator separately.

$$\left(\frac{16}{25}\right)^{\frac{1}{2}} = \frac{16^{\frac{1}{2}}}{25^{\frac{1}{2}}} = \frac{4}{5}$$

EXERCISE 18D

1 Evaluate the following.

a $25^{\frac{1}{2}}$	**b** $100^{\frac{1}{2}}$	**c** $64^{\frac{1}{2}}$	**d** $81^{\frac{1}{2}}$	**e** $625^{\frac{1}{2}}$
f $27^{\frac{1}{3}}$	**g** $64^{\frac{1}{3}}$	**h** $1000^{\frac{1}{3}}$	**i** $125^{\frac{1}{3}}$	**j** $512^{\frac{1}{3}}$
k $144^{\frac{1}{2}}$	**l** $400^{\frac{1}{2}}$	**m** $625^{\frac{1}{4}}$	**n** $81^{\frac{1}{4}}$	**o** $100\,000^{\frac{1}{5}}$
p $729^{\frac{1}{6}}$	**q** $32^{\frac{1}{5}}$	**r** $1024^{\frac{1}{10}}$	**s** $1296^{\frac{1}{4}}$	**t** $216^{\frac{1}{3}}$
u $16^{-\frac{1}{2}}$	**v** $8^{-\frac{1}{3}}$	**w** $81^{-\frac{1}{4}}$	**x** $3125^{-\frac{1}{5}}$	**y** $1\,000\,000$

2 Evaluate the following.

a $\left(\dfrac{25}{36}\right)^{\frac{1}{2}}$	**b** $\left(\dfrac{100}{36}\right)^{\frac{1}{2}}$	**c** $\left(\dfrac{64}{81}\right)^{\frac{1}{2}}$	**d** $\left(\dfrac{81}{25}\right)^{\frac{1}{2}}$	**e** $\left(\dfrac{25}{64}\right)^{\frac{1}{2}}$
f $\left(\dfrac{27}{125}\right)^{\frac{1}{3}}$	**g** $\left(\dfrac{8}{512}\right)^{\frac{1}{3}}$	**h** $\left(\dfrac{1000}{64}\right)^{\frac{1}{3}}$	**i** $\left(\dfrac{64}{125}\right)^{\frac{1}{3}}$	**j** $\left(\dfrac{512}{343}\right)^{\frac{1}{3}}$

3 Use the general rule for raising a power to another power to prove that $x^{\frac{1}{n}}$ is equivalent to

4 Which of these is the odd one out?

$$16^{-\frac{1}{4}} \qquad 64^{-\frac{1}{2}} \qquad 8^{-\frac{1}{3}}$$

Show how you decided.

5 Imagine that you are the teacher.

Write down how you would teach the class that $27^{-\frac{1}{3}}$ is equal to $\frac{1}{3}$.

6 $x^{-\frac{2}{3}} = y^{\frac{1}{3}}$

Find values for x and y that make this equation work.

7 Solve these equations:

a $2^x = 8$ **b** $8^x = 2$ **c** $4^x = 1$ **d** $16^x = 4$ **e** $100^x = 10$

f $81^x = 3$ **g** $16^x = 2$ **h** $125^x = 5$ **i** $1000^x = 10$ **j** $400^x = 20$

k $512^x = 8$ **l** $128^x = 2$

Indices of the form $\frac{a}{b}$

Here are two examples of this form.

$$t^{\frac{2}{3}} = t^{\frac{1}{3}} \times t^{\frac{1}{3}} = (\sqrt[3]{t})^2 \qquad\qquad 81^{\frac{3}{4}} = (\sqrt[4]{81})^3 = 3^3 = 27$$

If you have an expression of the form $\left(\dfrac{a}{b}\right)^{-n}$ we can invert it to calculate it as a fraction:

$$\left(\frac{a}{b}\right)^{-n} = \left(\frac{b}{a}\right)^n$$

XAMPLE 4

Evaluate the following. **a** $16^{-\frac{1}{4}}$ **b** $32^{-\frac{4}{5}}$

When dealing with the negative index remember that it means reciprocal.

Do problems like these one step at a time.

Step 1: Rewrite the calculation as a fraction by dealing with the negative power.

Step 2: Take the root of the base number given by the denominator of the fraction.

Step 3: Raise the result to the power given by the numerator of the fraction.

Step 4: Write out the answer as a fraction.

a Step 1: $16^{-\frac{1}{4}} = \left(\dfrac{1}{16}\right)^{\frac{1}{4}}$ **Step 2:** $16^{\frac{1}{4}} = \sqrt[4]{16} = 2$ **Step 3:** $2^1 = 2$ **Step 4:** $16^{-\frac{1}{4}} = \dfrac{1}{2}$

b Step 1: $32^{-\frac{4}{5}} = \left(\dfrac{1}{32}\right)^{\frac{4}{5}}$ **Step 2:** $32^{\frac{1}{5}} = \sqrt[5]{32} = 2$ **Step 3:** $2^4 = 16$ **Step 4:** $32^{-\frac{4}{5}} = \dfrac{1}{16}$

XAMPLE 5

Write $\left(\dfrac{8}{27}\right)^{-\frac{2}{3}}$ as a fraction.

$$\left(\frac{8}{27}\right)^{-\frac{2}{3}} = \left(\frac{27}{8}\right)^{\frac{2}{3}}$$

$$= \frac{27^{\frac{2}{3}}}{8^{\frac{2}{3}}}$$

$$= \frac{(\sqrt[3]{27})^2}{(\sqrt[3]{8})^2} = \frac{3^2}{2^2} = \frac{9}{4}$$

The rules for multiplying and dividing with indices still apply for fractional indices. For example:

$$a^{\frac{1}{2}} \times a^{-2} = a^{-\frac{3}{2}}$$

$$a^{\frac{1}{2}} \div a^{-2} = a^{\frac{5}{2}}$$

EXERCISE 18E

1 Evaluate the following.

   a $32^{\frac{4}{5}}$ **b** $125^{\frac{2}{3}}$

   c $1296^{\frac{3}{4}}$ **d** $243^{\frac{4}{5}}$

2 Rewrite the following in index form.

   a $\sqrt[3]{t^2}$ **b** $\sqrt[4]{m^3}$

   c $\sqrt[5]{k^2}$ **d** $\sqrt{x^3}$

3 Evaluate the following.

   a $8^{\frac{2}{3}}$ **b** $27^{\frac{2}{3}}$

   c $16^{\frac{3}{2}}$ **d** $625^{\frac{5}{4}}$

4 Evaluate the following.

   a $25^{-\frac{1}{2}}$ **b** $36^{-\frac{1}{2}}$

   c $16^{-\frac{1}{4}}$ **d** $81^{-\frac{1}{4}}$

   e $16^{-\frac{1}{2}}$ **f** $8^{-\frac{1}{3}}$

   g $32^{-\frac{1}{5}}$ **h** $27^{-\frac{1}{3}}$

5 Evaluate the following.

   a $25^{-\frac{3}{2}}$ **b** $36^{-\frac{3}{2}}$

   c $16^{-\frac{3}{4}}$ **d** $81^{-\frac{3}{4}}$

   e $64^{-\frac{4}{3}}$ **f** $8^{-\frac{2}{3}}$

   g $32^{-\frac{2}{5}}$ **h** $27^{-\frac{2}{3}}$

6 Evaluate the following.

   a $100^{-\frac{5}{2}}$ **b** $144^{-\frac{1}{2}}$

   c $125^{-\frac{2}{3}}$ **d** $9^{-\frac{3}{2}}$

   e $4^{-\frac{5}{2}}$ **f** $64^{-\frac{5}{6}}$

   g $27^{-\frac{4}{3}}$ **h** $169^{-\frac{1}{2}}$

7 Which of these is the odd one out?

   $16^{-\frac{3}{4}}$ $64^{-\frac{1}{2}}$ $8^{-\frac{2}{3}}$

Show how you decided.

8 Imagine that you are the teacher.

Write down how you would teach the class that $27^{-\frac{2}{3}}$ is equal to $\frac{1}{9}$.

9 Write as fractions:

a $\left(\dfrac{9}{4}\right)^{\frac{3}{2}}$

b $\left(\dfrac{27}{125}\right)^{\frac{2}{3}}$

c $\left(\dfrac{16}{9}\right)^{\frac{5}{2}}$

d $\left(\dfrac{4}{49}\right)^{\frac{3}{2}}$

e $\left(\dfrac{64}{27}\right)^{\frac{2}{3}}$

f $\left(\dfrac{16}{81}\right)^{\frac{3}{4}}$

g $\left(\dfrac{125}{64}\right)^{\frac{4}{3}}$

h $\left(\dfrac{64}{729}\right)$

10 Write as fractions:

a $\left(\dfrac{3}{5}\right)^{-2}$

b $\left(\dfrac{4}{3}\right)^{-3}$

c $\left(\dfrac{9}{5}\right)^{-3}$

d $\left(\dfrac{2}{3}\right)^{-5}$

e $\left(\dfrac{9}{4}\right)^{-\frac{3}{2}}$

f $\left(\dfrac{4}{9}\right)^{-\frac{5}{2}}$

g $\left(\dfrac{8}{27}\right)^{-\frac{2}{3}}$

h $\left(\dfrac{49}{25}\right)^{-\frac{3}{2}}$

i $\left(\dfrac{125}{64}\right)^{-\frac{2}{3}}$

j $\left(\dfrac{25}{64}\right)^{-\frac{3}{2}}$

k $\left(\dfrac{16}{81}\right)^{-\frac{5}{4}}$

l $\left(\dfrac{2187}{128}\right)^{-\frac{5}{7}}$

11 Simplify the following:

a $x^{\frac{3}{2}} \times x^{\frac{5}{2}}$

b $x^{\frac{1}{2}} \times x^{-\frac{3}{2}}$

c $(8y^3)^{\frac{2}{3}}$

d $5x^{\frac{3}{2}} \div \frac{1}{2}x^{-\frac{1}{2}}$

e $4x^{\frac{1}{2}} \times 5x^{-\frac{3}{2}}$

f $\left(\dfrac{27}{y^3}\right)^{-\frac{1}{3}}$

12 Simplify the following:

a $x^{\frac{1}{2}} \times x^{\frac{1}{2}}$

b $d^{-\frac{1}{2}} \times d^{-\frac{1}{2}}$

c $t^{\frac{1}{2}} \times t$

d $(x^{\frac{1}{2}})^4$

e $(y^2)^{\frac{1}{4}}$

f $a^{\frac{1}{2}} \times a^{\frac{3}{2}} \times a^2$

13 Simplify the following:

a $x \div x^{\frac{1}{2}}$

b $y^{\frac{1}{2}} \div y^{1\frac{1}{2}}$

c $a^{\frac{1}{3}} \times a^{\frac{4}{3}}$

d $t^{\frac{1}{2}} \times t^{\frac{3}{2}}$

e $\dfrac{1}{d^{-2}}$

f $\dfrac{k^{\frac{1}{2}} \times k^{\frac{3}{2}}}{k^2}$

In many real-life situations, variables are connected by a rule or relationship. It may be that as one variable increases the other increases. Alternatively, it may be that as one variable increases the other decreases.

This chapter looks at how quantities vary when they are related in some way.

As this plant gets older it becomes taller.

As the storm increases the number of sunbathers decreases.

As this car gets older it is worth less (and eventually it is worthless!).

As more songs are downloaded, there is less money left on the voucher.

Try to think of other variables that are connected in this way.

Chapter

19

Direct and inverse proportion

Topics	Level	Key words
1 Direct proportion	HIGHER	direct proportion, constant of proportionality
2 Inverse proportion	HIGHER	inverse proportion

What you need to be able to do in the examinations:

HIGHER

- Set up problems involving direct or inverse proportion and relate algebraic solutions to graphical representation of the equations.

There is **direct proportion** between two variables when one variable is a multiple of the other. That is, their ratio is a constant.

For example:

1 kilogram = 2.2 pounds   There is a multiplying factor of 2.2 between kilograms and pounds.

Area of a circle = πr^2 There is a multiplying factor of π between the area of a circle and the square of its radius.

A problem involving direct proportion usually requires you first to find this multiplying factor (called the **constant of proportionality**), then to use it to solve a problem.

The symbol for proportion is $\propto$.

So the statement 'Pay is directly proportional to time' can be mathematically written as:

pay $\propto$ time

which implies that:

pay $= k \times$ time

where k is the constant of proportionality.

There are four steps to be followed when you are using proportionality to solve problems.

Step 1: Set up the statement, using the proportionality symbol (you may use symbols to represent the variables).

Step 2: Set up the equation, using a constant of proportionality.

Step 3: Use given information to work out the value of the constant of proportionality.

Step 4: Substitute the value of the constant of proportionality into the equation and use this equation to find unknown values.

EXAMPLE 1

The cost of an item is directly proportional to the time spent making it. An item taking 6 hours to make costs $30. Find:

a the cost of an item that takes 5 hours to make

b the length of time it takes to make an item costing $40.

Step 1: Let C dollars be the cost of making an item and t hours the time it takes.

$$C \propto t$$

Step 2: Setting up the equation gives:

$$C = kt$$

where k is the constant of proportionality.

Note that you can 'replace' the proportionality sign $\propto$ with $= k$ to obtain the proportionality equation.

Step 3: Since $C = 30$ when $t = 6$, then $30 = 6k$

$$\Rightarrow \frac{30}{6} = k$$

$$\Rightarrow k = 5$$

Step 4: So the formula is $C = 5t$.

a When $t = 5$ $C = 5 \times 5 = 25$

So the cost is $25.

b When $C = 40$ $40 = 5 \times t$

$$\Rightarrow \quad \frac{40}{5} = t$$

$$\Rightarrow \quad t = 8$$

So the time spent making the item is 8 hours.

As $C = 5t$, a graph of C against t will be a straight line through the origin of gradient 5.

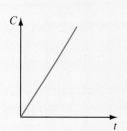

EXERCISE 19A

For questions **1** to **4**, first find k, the constant of proportionality, and then the formula connecting the variables.

1 T is directly proportional to M. If $T = 20$ when $M = 4$, find:

     a T when $M = 3$ **b** M when $T = 10$.

HIGHER

2 W is directly proportional to F. If $W = 45$ when $F = 3$, find:

   a W when $F = 5$ **b** F when $W = 90$.

3 Q varies directly with P. If $Q = 100$ when $P = 2$, find:

   a Q when $P = 3$ **b** P when $Q = 300$.

4 X varies directly with Y. If $X = 17.5$ when $Y = 7$, find:

   a X when $Y = 9$ **b** Y when $X = 30$.

5 The distance covered by a train is directly proportional to the time taken for the journey. The train travels 105 kilometres in 3 hours.

   a What distance will the train cover in 5 hours?

   b How much time will it take for the train to cover 280 kilometres?

6 The cost of fuel delivered to your door is directly proportional to the mass received. When 250 kg is delivered, it costs 47.50 dollars.

   a How much will it cost to have 350 kg delivered?

   b How much would be delivered if the cost were 33.25 dollars?

7 The number of children who can play safely in a playground is directly proportional to the area of the playground. A playground with an area of 210 m^2 is safe for 60 children.

   a How many children can safely play in a playground of area 154 m^2?

   b A playgroup has 24 children. What is the smallest playground area in which they could safely play?

8 The number of spaces in a car park is directly proportional to the area of the car park.

   a A car park has 300 parking spaces in an area of 4500 m².

      It is decided to increase the area of the car park by 500 m² to make extra spaces.

      How many extra spaces will be made?

   b The old part of the car park is redesigned so that the original area has 10% more parking spaces.

      How many more spaces than in the original car park will there be altogether if the number of spaces in the new area is directly proportional to the number in the redesigned car park?

9 The number of passengers in a bus queue is directly proportional to the time that the person at the front of the queue has spent waiting.

Karen is the first to arrive at a bus stop. When she has been waiting 5 minutes the queue has 20 passengers.

A bus has room for 70 passengers.

How long had Karen been in the queue if the bus fills up from empty when it arrives and all passengers get on?

Direct proportions involving squares, cubes and square roots

The process is the same as for a linear direct proportion, as the next example shows.

EXAMPLE 2

The cost of a circular badge is directly proportional to the square of its radius. The cost of a badge with a radius of 2 cm is $0.68. Find:

a the cost of a badge of radius 2.4 cm   **b** the radius of a badge costing $1.53.

Step 1: Let C be the cost in dollars and r the radius of a badge in centimetres.
$$C \propto r^2$$

Step 2: Setting up the equation gives:
$$C = kr^2$$
where k is the constant of proportionality.

Step 3: $C = 0.68$ when $r = 2$. So:
$$0.68 = 4k$$
$$\Rightarrow \frac{0.68}{4} = k \Rightarrow k = 0.17$$

Step 4: So the formula is $C = 0.17r^2$.

a When $r = 2.4$ $C = 0.17 \times 2.4^2 = 0.98$ to 2 decimal places.
Rounding gives the cost as $0.98.

b When $C = 1.53$ $1.53 = 0.17r^2$
$$\Rightarrow \frac{1.53}{0.17} = 9 = r^2$$
$$\Rightarrow r = \sqrt{9} = 3$$

Hence, the radius is 3 cm.

As $C = 0.17r^2$, a graph of C against r will be a quadratic graph passing through the origin:

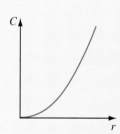

EXERCISE 19B

For questions **1** to **6**, first find k, the constant of proportionality, and then the formula connecting the variables.

1 T is directly proportional to x^2. If $T = 36$ when $x = 3$, find:

a T when $x = 5$ **b** x when $T = 400$.

HIGHER

2 W is directly proportional to M^2. If $W = 12$ when $M = 2$, find:

    a W when $M = 3$ **b** M when $W = 75$.

3 E varies directly with $\sqrt{C}$. If $E = 40$ when $C = 25$, find:

    a E when $C = 49$ **b** C when $E = 10.4$.

4 X is directly proportional to $\sqrt{Y}$. If $X = 128$ when $Y = 16$, find:

    a X when $Y = 36$ **b** Y when $X = 48$.

5 P is directly proportional to f^3. If $P = 400$ when $f = 10$, find:

    a P when $f = 4$ **b** f when $P = 50$.

6 The cost of serving tea and biscuits varies directly with the square root of the number of people at the buffet. It costs \$25 to serve tea and biscuits to 100 people.

    a How much will it cost to serve tea and biscuits to 400 people?

    b For a cost of \$37.50, how many people could be served tea and biscuits?

7 In an experiment, the temperature, in °C, varied directly with the square of the pressure, in atmospheres (atm). The temperature was 20 °C when the pressure was 5 atm.

    a What will the temperature be at 2 atm?

    b What will the pressure be at 80 °C?

8 The mass, in grams, of ball bearings varies directly with the cube of the radius, measured in millimetres. A ball bearing of radius 4 mm has a mass of 115.2 g.

    a What will be the mass of a ball bearing of radius 6 mm?

    b A ball bearing has a mass of 48.6 g. What is its radius?

9 The energy, in J, of a particle varies directly with the square of its speed, in m/s. A particle moving at 20 m/s has 50 J of energy.

    a How much energy has a particle moving at 4 m/s?

    b At what speed is a particle moving if it has 200 J of energy?

10 The cost, in dollars, of a trip varies directly with the square root of the number of miles travelled. The cost of a 100-mile trip is 35 dollars.

    a What is the cost of a 500-mile trip (to the nearest dollar)?

    b What is the distance of a trip costing 70 dollars?

11 A sculptor is making statues.

The amount of clay used is directly proportional to the cube of the height of the statue.

A statue is 10 cm tall and uses 500 cm³ of clay.

How much clay will a similar statue use if it is twice as tall?

12 The cost of making different-sized machines is proportional to the time taken.

A small machine costs $100 and takes two hours to make.

How much will a large machine cost that takes 5 hours to build?

13 The sketch graphs show each of these proportion statements.

a $y \propto x^2$

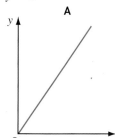

b $y \propto x$

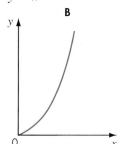

c $y \propto \sqrt{x}$

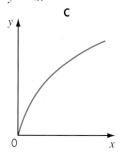

Match each statement to the correct sketch.

14 Here are two tables.

Match each table to a graph in question **13**.

a

x	1	2	3
y	3	12	27

b

x	1	2	3
y	3	6	9

19.2 Inverse proportion

There is **inverse proportion** between two variables when one variable is directly proportional to the *reciprocal* of the other. That is, the product of the two variables is constant. So, as one variable increases, the other decreases.

For example, the faster you travel over a given distance, the less time it takes. So there is an inverse variation between speed and time. Speed is inversely proportional to time:

$$S \propto \frac{1}{T} \text{ and so } S = \frac{k}{T}$$

which can be written as $ST = k$.

EXAMPLE 3

M is inversely proportional to R. If $M = 9$ when $R = 4$, find the value of:

a M when $R = 2$ **b** R when $M = 3$.

Step 1: $M \propto \dfrac{1}{R}$

Step 2: Setting up the equation gives:

$$M = \frac{k}{R}$$

where k is the constant of proportionality.

Step 3: $M = 9$ when $R = 4$. So $9 = \dfrac{k}{4}$

$\Rightarrow 9 \times 4 = k \Rightarrow k = 36$

Step 4: The formula is $M = \dfrac{36}{R}$

a When $R = 2$, then $M = \dfrac{36}{2} = 18$

b When $M = 3$, then $3 = \dfrac{36}{R} \Rightarrow 3R = 36 \Rightarrow R = 12$

As $M = \dfrac{36}{R}$, a graph of M against R will look like this:

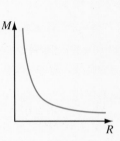

EXAMPLE 4

r is inversely proportional to the square root of a.

Find the missing value in this table.

a	25	400
r	20	

Step 1: $r \propto \dfrac{1}{\sqrt{a}}$

Step 2: $r = \dfrac{k}{\sqrt{a}}$

Step 3: $20 = \dfrac{k}{\sqrt{25}} \Rightarrow k = 20 \times 5 = 100$

Step 4: The formula is $r = \dfrac{100}{\sqrt{a}}$

When $a = 400$, $r = \dfrac{100}{\sqrt{400}} = \dfrac{100}{20} = 5$

EXERCISE 19C

For questions **1** to **6**, first find the formula connecting the variables.

1 T is inversely proportional to m.
If $T = 6$ when $m = 2$, find:

  a T when $m = 4$

  b m when $T = 4.8$.

2 W is inversely proportional to x.
If $W = 5$ when $x = 12$, find:

  a W when $x = 3$

  b x when $W = 10$.

HIGHER

3 Q varies inversely with $(5 - t)$.
If $Q = 8$ when $t = 3$, find:

   a Q when $t = 10$

   b t when $Q = 16$.

4 M varies inversely with t^2.
If $M = 9$ when $t = 2$, find:

   a M when $t = 3$

   b t when $M = 1.44$.

5 y is inversely proportional to the square of x. If $y = 4$ when $x = 2$, find:

   a y when $x = 1$

   b x when $y = \frac{1}{4}$.

6 V is inversely proportional to the cube of R.

R	6	10	
V	12.5		0.8

Find the missing values in the table.

7 The grant available to a group of students was inversely proportional to the number of students. When 30 students needed a grant, they received $60 each.

   a What would the grant have been if 120 students had needed one?

   b If the grant had been $50 each, how many students would have received it?

8 While doing underwater tests in an ocean, scientists noticed that the temperature, in °C, was inversely proportional to the depth, in kilometres. When the temperature was 6 °C, the scientists were at a depth of 4 km.

   a What would the temperature have been at a depth of 8 km?

   b At what depth would they find the temperature at 2 °C?

9 A new engine had serious problems. The distance it went, in kilometres, without breaking down was inversely proportional to the square of its speed in metres per second (m/s). When the speed was 12 m/s, the engine lasted 3 km.

   a Find the distance covered before a breakdown, when the speed is 15 m/s.

   b On one test, the engine broke down after 6.75 km. What was the speed?

10 Which statement is represented by the graph?

Give a reason for your answer.

   A $y \propto x$ B $y \propto \dfrac{1}{x}$ C $y \propto \sqrt{x}$

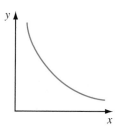

11 A student is investigating the dimensions of cylinders with a fixed mass.

The diameter d cm is inversely proportional to the square root of the height h cm.

When the height is 25 cm the diameter is 12 cm.

Work out the diameter when the height is 36 cm.

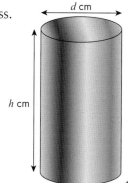

Why this chapter matters

The use of graphs to represent and analyse problems has been used by many companies to reduce their costs and increase productivity.

The use of linear graphs to solve planning problems was developed at the start of the Second World War in 1939.

It was used to work out ways to supply armaments as efficiently as possible. It was such a powerful tool that the British and Americans did not want the Germans to know about it, so it was not made public until 1947.

George Dantzig was one of the inventors of the methods used. He came late to a lecture at University one day and saw two problems written on the blackboard. He copied them, thinking they were the homework assignment.

He solved both problems, but apologised to the lecturer later as he found them a little harder than the usual homework, so he took a few days to solve them and was late handing them in.

The lecturer was astonished. The problems he had written on the board were not homework but examples of 'impossible problems'. Not any more!

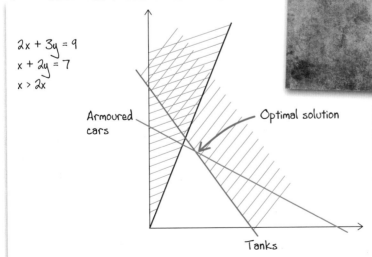

$$2x + 3y = 9$$
$$x + 2y = 7$$
$$x > 2x$$

Armoured cars

Optimal solution

Tanks

20 Inequalities and regions

Topics	Level	Key words
1 Linear inequalities	**FOUNDATION**	inequality, number line
2 Quadratic inequalities	**HIGHER**	solution set
3 Graphical inequalities	**FOUNDATION**	boundary, region, required
4 More than one inequality	**FOUNDATION**	
5 More complex inequalities	**HIGHER**	

What you need to be able to do in the examinations:

FOUNDATION	HIGHER
• Understand and use the convention for open and closed intervals on a number line.	• Solve quadratic inequalities in one unknown and represent the solution set on a number line.
• Solve simple linear inequalities in one variable and represent the solution set on a number line.	• Identify harder examples of regions defined by linear inequalities.
• Represent simple linear inequalities on rectangular Cartesian graphs.	
• Identify regions on rectangular Cartesian graphs defined by simple linear inequalities.	

Inequalities behave similarly to equations. You use the same rules to solve linear inequalities as you use for linear equations.

There are four inequality signs:

$<$ means 'less than'

$>$ means 'greater than'

$\leqslant$ means 'less than or equal to'

$\geqslant$ means 'greater than or equal to'.

EXAMPLE 1

Solve $2x + 3 < 14$

Subtract 3 from both sides: $\quad 2x < 14 - 3$

$2x < 11$

Divide both sides by 2: $\quad \dfrac{2x}{2} < \dfrac{11}{2}$

$\Rightarrow x < 5.5$

This means that x can take any value below 5.5 but *not* the value 5.5.

If you divide by a negative number when you are solving an inequality you must change the sign.

'less than' becomes 'more than'

'more than' becomes 'less than'

EXAMPLE 2

Solve the inequality $10 - 2x \geqslant 3$

Subtract 10 from both sides: $\quad -2x \geqslant -7$

Divide both sides by -2: $\quad x \leqslant 3.5$

We reverse inequality signs when multiplying or dividing both sides by a negative number. So the inequality has changed in the last line.

We could do example 2 in a different way:

$$10 - 2x \geqslant 3$$

Add $2x$ to both sides: $\quad 10 \geqslant 2x + 3$

Subtract 3 from both sides: $\quad 7 \geqslant 2x$

Divide both sides by 2: $\quad 3.5 \geqslant x$

The sign does not change this time.

$$3.5 \geqslant x \text{ is equivalent to } x \leqslant 3.5$$

Sometimes there is more than one inequality.

EXAMPLE 3

a Solve the inequality $5 \leqslant 4n + 13 < 29$

b If n is an integer, write down the possible values of n.

a In this case do the same thing to all three terms.

Subtract 13 from all three expressions: $\quad -8 \leqslant 4n < 16$

Divide each part by 4: $\quad -2 \leqslant n < 4$

This is the solution.

b The solution shows that n is between -2 and 4.

If n is an integer, the possible values of n are $-2, -1, 0, 1, 2$ and 3.

The inequalities mean that -2 is included but 4 is not.

EXERCISE 20A

1 Solve the following linear inequalities.

a $x + 4 < 7$	**b** $t - 3 > 5$	**c** $p + 2 \geqslant 12$
d $2x - 3 < 7$	**e** $4y + 5 \leqslant 17$	**f** $3t - 4 > 11$
g $\frac{x}{2} + 4 < 7$	**h** $\frac{y}{5} + 3 \leqslant 6$	**i** $\frac{t}{3} - 2 \geqslant 4$
j $3(x - 2) < 15$	**k** $5(2x + 1) \leqslant 35$	**l** $2(4t - 3) \geqslant 34$

2 Write down the largest integer value of x that satisfies each of the following.

a $x - 3 \leqslant 5$, where x is positive

b $x + 2 < 9$, where x is positive and even

c $3x - 11 < 40$, where x is a square number

d $5x - 8 \leqslant 15$, where x is positive and odd

e $2x + 1 < 19$, where x is positive and prime

3 Write down the smallest integer value of x that satisfies each of the following.

a $x - 2 \geqslant 9$, where x is positive

b $x - 2 > 13$, where x is positive and even

c $2x - 11 \geqslant 19$, where x is a square number

FOUNDATION

4 Ahmed went to town with $20 to buy two CDs. His bus fare was $3. The CDs were both the same price. When he reached home he still had some money in his pocket. What was the most each CD could cost?

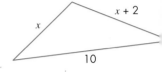

HINTS AND TIPS

Set up an inequality and solve it.

5 **a** Explain why you cannot make a triangle with three sticks of length 3 cm, 4 cm and 8 cm.

b Three sides of a triangle are x, $x + 2$ and 10 cm.

x is a whole number.

What is the smallest value x can take?

6 Five cards have inequalities and equations marked on them.

| $x > 0$ | $x < 3$ | $x \geqslant 4$ | $x = 2$ | $x = 6$ |

The cards are shuffled and then turned over, one at a time.

If two consecutive cards have any numbers in common, then a point is scored.

If they do not have any numbers in common, then a point is deducted.

a The first two cards below score −1 because $x = 6$ and $x < 3$ have no numbers in common. Explain why the total for this combination scores 0.

| $x = 6$ | $x < 3$ | $x > 0$ | $x = 2$ | $x \geqslant 4$ |

b What does this combination score?

| $x > 0$ | $x = 6$ | $x \geqslant 4$ | $x = 2$ | $x < 3$ |

c Arrange the cards to give a maximum score of 4.

7 Solve the following linear inequalities.

a $4x + 1 \geqslant 3x - 5$ **b** $5t - 3 \leqslant 2t + 5$

c $3y - 12 \leqslant y - 4$ **d** $2x + 3 \geqslant x + 1$

e $5w - 7 \leqslant 3w + 4$ **f** $2(4x - 1) \leqslant 3(x + 4)$

8 Solve the following linear inequalities.

a $\dfrac{x + 4}{2} \leqslant 3$ **b** $\dfrac{x - 3}{5} > 7$

c $\dfrac{2x + 5}{3} < 6$ **d** $\dfrac{4x - 3}{5} \geqslant 5$

e $\dfrac{2t - 2}{7} > 4$ **f** $\dfrac{5y + 3}{5} \leqslant 2$

9 In this question n is always an integer.

    a Find the largest possible value of n if $2n + 3 < 12$

    b Find the largest possible value of n if $\frac{n}{5} < 20$

    c Find the smallest possible value of n if $3(n - 7) \geqslant 10$

    d Find the smallest possible value of n if $\frac{6n - 2}{7} \geqslant 9$

    e Find the smallest possible value of n if $3n + 14 \leqslant 8n - 13$

10 **a** If $20 - x > 4$, which of the following are possible values of x?

        $-10 \quad 0 \quad 10 \quad 20 \quad 30$

    b Solve the inequality $20 - x > 4$

11 Solve the following inequalities:

    a $15 - x > 6$ **b** $18 - x \leqslant 7$ **c** $6 \geqslant 9 - x$

12 Solve these inequalities:

    a $20 - 2x \leqslant 5$ **b** $3 - 4x \geqslant 11$ **c** $25 - 3x > 7$

    d $2(6 - x) < 9$ **e** $\frac{10 - 2x}{5} \leqslant 4$ **f** $\frac{8 - 4x}{3} > 2$

13 Solve these inequalities.

    a $12 < 3x \leqslant 24$ **b** $10 < x - 5 < 18$ **c** $0 \leqslant 5x - 10 \leqslant 100$

    d $4 < k + 10 < 8$ **e** $10 \leqslant 20y \leqslant 35$ **f** $13 < 4t + 1 < 53$

14 n is an integer and $20 < 2n < 30$. List the possible value of n.

15 The sides of a triangle are n cm, $n + 1$ cm and $n + 2$ cm.

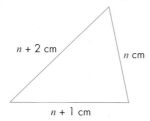

$n + 2$ cm n cm

$n + 1$ cm

    a Find an expression for the perimeter of the triangle, in centimetres.

    b The perimeter of the triangle is more than 90 cm but less than 100 cm. Write down an inequality to show this.

    c If n is an integer, find the possible values of n.

The number line

The solution to a linear inequality can be shown on the **number line** by using the following conventions.

 $x \leqslant$  $x \geqslant$  $x <$  x

The inequalities $\leqslant$ and $\geqslant$ include the boundary point. The inequalities $<$ and $>$ do not. The shading is used to show this.

Below are five examples.

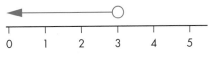

represents $x < 3$

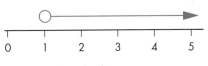

represents $x > 1$

represents $x \leqslant -2$

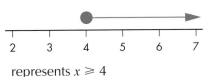

represents $x \geqslant 4$

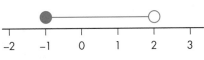

represents $-1 \leqslant x < 2$

EXAMPLE 4

    a Write down the inequality shown by this diagram.

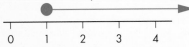

    b **i** Solve the inequality $2x + 3 < 11$.

       ii Mark the solution on a number line.

    c Write down the integers that satisfy both the inequalities in **a** and **b**.

    a The inequality shown is $x \geqslant 1$.

    b **i** $2x + 3 < 11$

           $\Rightarrow 2x < 8$

           $\Rightarrow x < 4$

       ii

    c The integers that satisfy both inequalities are 1, 2 and 3.

EXERCISE 20B

1 Write down the inequality that is represented by each diagram below.

a

b

c

d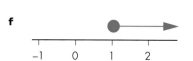

e

f

2 Draw diagrams to illustrate these inequalities.

   a $x \leqslant 3$ **b** $x > -2$ **c** $x \geqslant 0$ **d** $x < 5$

   e $x \geqslant -1$ **f** $2 < x \leqslant 5$ **g** $-1 \leqslant x \leqslant 3$ **h** $-3 < x < 4$

3 Solve the following inequalities and illustrate their solutions on number lines.

   a $x + 4 \geqslant 8$ **b** $x + 5 < 3$ **c** $4x - 2 \geqslant 12$ **d** $2x + 5 < 3$

   e $2(4x + 3) < 18$ **f** $\dfrac{x}{2} + 3 \leqslant 2$ **g** $\dfrac{x}{5} - 2 > 8$ **h** $\dfrac{x}{3} + 5 \geqslant 3$

4 Max went to the supermarket with \$1.20. He bought three apples costing x cents each and a chocolate bar costing 54 cents. When he got to the till, he found he didn't have enough money.

Max took one of the apples back and paid for two apples and the chocolate bar. He counted his change and found he had enough money to buy a 16 cent candy.

   a Explain why $3x + 54 > 120$ and solve the inequality.

   b Explain why $2x + 54 \leqslant 104$ and solve the inequality.

   c Show the solution to both of these inequalities on a number line.

   d What is the possible price of an apple?

5 On copies of the number lines below, draw two inequalities so that only the integers $\{-1, 0, 1, 2\}$ are common to both inequalities.

FOUNDATION

6 Which number is being described?

x is a square number

$2x + 3 > 5$

7 Solve the following inequalities and illustrate their solutions on number lines.

a $\dfrac{2x + 5}{3} > 3$ **b** $\dfrac{3x + 4}{2} \geqslant 11$ **c** $\dfrac{2x + 8}{3} \leqslant 2$ **d** $\dfrac{2x - 1}{3} \geqslant -3$

8 Solve these inequalities and show the solutions on number lines.

a $12 < 6x < 48$ **b** $9 \leqslant 2x + 13 \leqslant 19$

20.2 Quadratic inequalities H

Suppose $x^2 > 16$

What can we say about x?

First, change the inequality to an equals sign.

$x^2 = 16$

There are two solutions, $x = 4$ or -4.

These divide the number line into three sections.

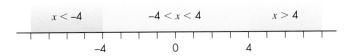

By choosing a value in each section in turn and squaring it we can see that:

If $x > 4$ then $x^2 > 16$ (e.g. $5^2 = 25 > 16$)
If $-4 < x < 4$ then $x^2 < 16$ (e.g. $2^2 = 4 < 16$)
If $x < -4$ then $x^2 > 16$ (e.g. $(-5)^2 = 25 > 16$)

So the **solution set** for $x^2 > 16$ is in two parts:

$x < -4$ or $x > 4$

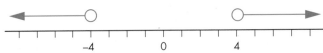

Notice that the boundary values are not included in this case because $4^2 = (-4)^2 = 16$.

Neither of them are less than 16.

The solution to $x^2 \geqslant 16$ is similar but in this case the boundary values are included because the inequality includes 16 as a possible value.

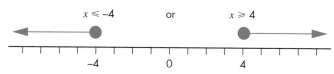

EXERCISE 20C

1 Solve these inequalities and show the solution sets on a number line.

    a $x^2 \leqslant 16$ **b** $x^2 < 4$ **c** $x^2 > 6.25$ **d** $x^2 \geqslant 1$

2 Solve these inequalities.

    a $2x^2 < 18$ **b** $3x^2 > 75$ **c** $4x^2 \geqslant 9$ **d** $4x^2 \leqslant 1$

3 Solve these inequalities.

    a $x^2 - 4 \leqslant 0$ **b** $x^2 - 12.25 > 0$ **c** $8x^2 - 50 < 0$ **d** $9 - x^2 \geqslant 0$

4 This diagram shows the solution set for a quadratic inequality.

Write down the inequality as simply as possible.

5 Show the solution set on a number line if both $x^2 \leqslant 36$ and $x^2 \geqslant 9$.

20.3 Graphical inequalities

A linear inequality can be plotted on a graph. The result is a **region** that lies on one side or the other of a straight line. You will recognise an inequality by the fact that it looks like an equation but instead of the equals sign it has an inequality sign: $<$, $>$, $\leqslant$ or $\geqslant$.

The following are examples of linear inequalities that can be represented on a graph.

    $y < 3$ $x > 7$ $-3 \leqslant y < 5$ $y \geqslant 2x + 3$ $2x + 3y < 6$ $y \leqslant x$

The method for graphing an inequality is to draw the **boundary** line that defines the inequality. This is found by replacing the inequality sign with an equals sign.

A common convention is to use a solid line when the boundary is included and a broken line when the boundary is not included. This means a solid line for $\leqslant$ and $\geqslant$ and a broken one for $<$ and $>$. This convention is not required in the IGCSE and we shall not use it in this book. All boundaries will be shown by a solid line.

After the boundary line has been drawn, shade the **required** region.

To confirm on which side of the line the region lies, choose any point that is not on the boundary line and test it in the inequality. If it satisfies the inequality, that is the side required. If it doesn't, the other side is required.

Work through the six inequalities in the following example to see how the procedure is applie

---

EXAMPLE 5

Show each of the following inequalities on a graph.

a $y \leqslant 3$ **b** $x > 7$ **c** $-3 \leqslant y < 5$

d $y \leqslant 2x + 3$ **e** $2x + 3y < 6$ **f** $y \leqslant x$

a Draw the line $y = 3$. Test a point that is not on the line. The **origin** is always a good choice if possible, as 0 is easy to test.

Putting 0 into the inequality gives $0 \leqslant 3$. The inequality is satisfied and so the region containing the origin is the side we want.

Shade it in.

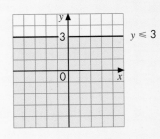

b Draw the line $x = 7$.

Test the origin $(0, 0)$, which gives $0 > 7$. This is not true, so you want the other side of the line from the origin.

Shade it in.

In this case the boundary line is not included.

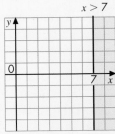

c Draw the lines $y = -3$ and $y = 5$.

Test a point that is not on either line, say $(0, 0)$. Zero is between -3 and 5, so the required region lies between the lines.

Shade it in.

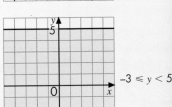

d Draw the line $y = 2x + 3$.

Test a point that is not on the line, $(0, 0)$. Putting these x- and y-values in the inequality gives $0 \leqslant 2(0) + 3$, which is true. So the region that includes the origin is what you want.

Shade it in.

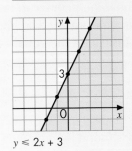

EXAMPLE 5 continued

e Draw the line $2x + 3y = 6$.

The easiest way is to find out where it crosses the axes.

If $x = 0$, $3y = 6 \Rightarrow y = 2$. Crosses y-axis at $(0, 2)$.

If $y = 0$, $2x = 6 \Rightarrow x = 3$. Crosses x-axis at $(3, 0)$.

Draw the line through these two points.

Test a point that is not on the line, say $(0, 0)$. Is it true that $2(0) + 3(0) < 6$? The answer is yes, so the origin is in the region that you want.
Shade it in.

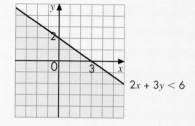

f Draw the line $y = x$.

This time the origin is on the line, so pick any other point, say $(1, 3)$. Putting $x = 1$ and $y = 3$ in the inequality gives $3 \leqslant 1$. This is not true, so the point $(1, 3)$ is not in the region you want.

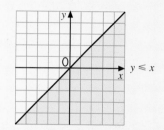

EXERCISE 20D

1 **a** Draw the line $x = 2$.
   b Shade the region defined by $x \leqslant 2$.

2 **a** Draw the line $y = -3$.
   b Shade the region defined by $y > -3$.

3 **a** Draw the line $x = -2$.
   b Draw the line $x = 1$ on the same grid.
   c Shade the region defined by $-2 \leqslant x \leqslant 1$.

4 **a** Draw the line $y = -1$.
   b Draw the line $y = 4$ on the same grid.
   c Shade the region defined by $-1 < y \leqslant 4$.

5 **a** On the same grid, draw the regions defined by these inequalities.
   i $-3 \leqslant x \leqslant 6$ **ii** $-4 < y \leqslant 5$

   b Are the following points in the region defined by both inequalities?
   i $(2, 2)$ **ii** $(1, 5)$ **iii** $(-2, -4)$

6 **a** Draw the line $y = 2x - 1$.
   b Shade the region defined by $y < 2x - 1$.

7 **a** Draw the line $x + y = 4$.
   b Shade the region defined by $x + y \leqslant 4$.

8 **a** Draw the line $y = \frac{1}{2}x + 3$.
   b Shade the region defined by $y \geqslant \frac{1}{2}x + 3$.

9 Shade the region defined by $x + y \geqslant 3$.

FOUNDATION

More than one inequality

When several inequalities are given they define a **region**.

EXAMPLE 6

Shade the region where $x \geqslant 1$, $y \geqslant 2$ and $x + y \leqslant 8$

Start by drawing the lines $x = 1$, $y = 2$ and $x + y = 8$

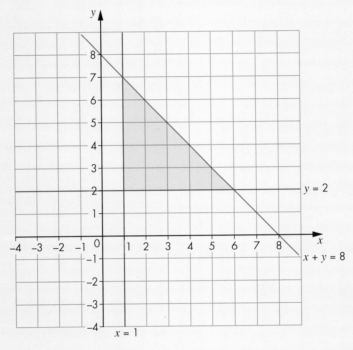

Check which side of each line is required. Points inside the shaded triangle satisfy all three inequalities.

EXERCISE 20E

1 Shade the region where $y > 4$ and $x > 5$.

2 Shade the region where $x > 0$, $y \geqslant 3$ and $x + y \leqslant 7$.

3 Shade the region where $x \leqslant 7$, $y \leqslant 6$ and $x + y \geqslant 3$.

4 Shade the region where $x \leqslant 10$, $y \geqslant 0$ and $y \leqslant x$.

5 Shade the region where $x + y > 4$, $x + y < 8$.

6 Shade the region where $x + y \geqslant 0$, $y \geqslant 3$.

7 Write down two inequalities to describe this region:

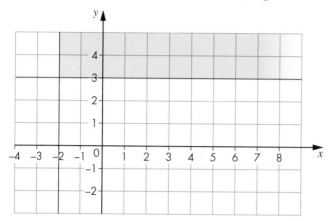

8 Write down three inequalities to define this region:

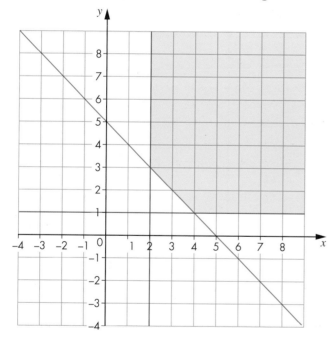

9 Show by shading the region where $y \geqslant x$, $x + y \leqslant 10$ and $x \geqslant 2$.

10 Show by shading the region where $y < x$ and $x < 0$.

When more complex inequalities are involved the same procedure can be used. Draw the boundary lines first then decide on the required region.

EXAMPLE 7

Shade the region defined by $y \geq 2x - 4$, $x + 2y \leq 12$ and $3x + y \geq 6$

The line $y = 2x - 4$ has gradient 2 and intercept -4 on the y-axis.

The inequality indicates the region above this line.

To draw the line $x + 2y = 12$, find where it crosses the axes.

If $x = 0$, $2y = 12 \Rightarrow y = 6$ so it goes through $(0, 6)$.

If $y = 0$, $x = 12$ so it goes through $(12, 0)$.

We want the region below this line.

The line $3x + y = 6$ passes through $(0, 6)$ and $(2, 0)$.

We want the region above this line.

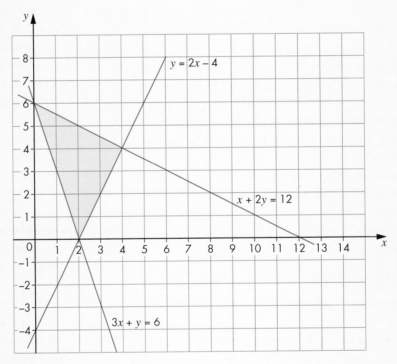

The required region is shaded.

EXERCISE 20F

HIGHER

1 **a** Draw the line $y = x$.

b Draw the line $2x + 5y = 10$ on the same diagram.

c Draw the line $2x + y = 6$ on the same diagram.

d Shade the region defined by $y \geqslant x$, $2x + 5y \geqslant 10$ and $2x + y \leqslant 6$.

e Are the following points in the region defined by these inequalities?

i (1, 1)　　**ii** (2, 2)　　**iii** (1, 3)

> **HINTS AND TIPS**
>
> Find the points where the line crosses each axis.

2 **a** On the same grid, shade the regions defined by the following inequalities.

i $y > x - 3$　　**ii** $3y + 4x \leqslant 24$　　**iii** $x \geqslant 2$

b Are the following points in the region defined by all three inequalities?

i (1, 1)　　**ii** (2, 2)　　**iii** (3, 3)　　**iv** (4, 4)

3 **a** On a graph draw the lines $y = x$, $x + y = 8$ and $y = 2$.

b Label the region R where $y \leqslant x$, $x + y \leqslant 8$ and $y \geqslant 2$. Shade the region that is required.

4 **a** On a graph draw the lines $y = x - 4$, $y = 0.5x$ and $y = -x$.

b Show the region S where $y \geqslant x - 4$, $y \leqslant 0.5x$ and $y \geqslant -x$. Shade the region that is required.

c What is the largest y-coordinate of a point in S?

d What is the smallest y-coordinate of a point in S?

e What is the largest value of $x + y$ for a point in S?

5 Explain how you would find which side of the line represents the inequality $y < x + 2$.

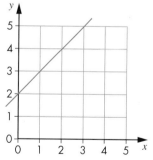

6 The region marked R is bounded by the lines $x + y = 3$, $y = \frac{1}{2}x + 3$ and $y = 5x - 15$.

a What three inequalities are satisfied in region R?

b What is the greatest value of $x + y$ in region R?

c What is the greatest value of $x - y$ in region R?

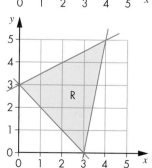

Why this chapter matters

Notation is important in mathematics. Try writing an equation or formula in words instead of symbols and you will see why notation makes things easier to understand.

If you drop a coin, how long does it take to reach the ground? That depends on the height you drop it from. We say that the time taken is a *function* of the height.

By dropping the coin from different heights and measuring the time it takes to fall, it would be possible to find a formula for the time in terms of the height.

Here are some other examples where one variable is a *function* of another:

- The cost of posting a parcel is a function of its mass.
- The time taken for a journey is a function of the distance travelled.
- The stopping distance of a car is a function of its speed.
- The cost of a second hand car is a function of its age.
- The time taken to download a computer file is a function of the size of the file.

The idea of an *inverse* occurs frequently in mathematics. It is not a difficult idea and it is a useful one. Putting a hat on and taking it off are inverse operations. Switching a light on and switching it off are also inverse operations.

Here are some examples of inverse operations in mathematics:

- Add 3 and subtract 3
- Multiply by 5 and divide by 5
- Rotate 90° clockwise and rotate 90° anticlockwise
- Square a number and find the square root of the square number

It is not always possible to find an inverse. Sadly the inverse of breaking a glass does not exist.

21

Functions

Topics	Level	Key words
1 Function notation	HIGHER	function
2 Domain and range	HIGHER	domain, range, mapping
3 Inverse functions	HIGHER	inverse
4 Composite functions	HIGHER	composite
5 More about composite functions	HIGHER	

What you need to be able to do in the examinations:

HIGHER

- Understand the concept that a function is a mapping between elements of two sets.
- Use function notation of the form $f(x) = \ldots$ and $f: x \rightarrow \ldots$
- Understand the terms domain and range and which values may need to be excluded from the domain.
- Understand and find the composite function fg and the inverse function f^{-1}.

You are familiar with equations written using x and y such as $y = 3x - 4$ or $y = 2x^2 + 5x - 3$.

These equations are showing that y is a **function** of x. This means that the value of y depends on the value of x so that y changes when x changes.

Sometimes it is useful to use a different notation to show this. We could write the first equation above as $f(x) = 3x - 4$ and refer to it as 'function f'.

It is then easy to show the result of using different values for x. For example:

- "the value of $f(x)$ when x is 5" can be written as $f(5)$.

  So $f(5) = 3 \times 5 - 4 = 11$

- $f(1)$ means "the value of $f(x)$ when x is 1".

  So $f(1) = 3 \times 1 - 4 = -1$

  and $f(-1) = -7$

If there are different functions in the same problem different letters can be used, for example:

$g(x) = 2x^2 + 5x - 3$ or 'function g'.

Sometimes instead of writing $f(x) = 3x - 4$ we write $f: x \rightarrow 3x - 4$. These two forms mean exactly the same thing.

EXERCISE 21A

1 $f(x) = 2x + 6$. Find:

   a $f(3)$ **b** $f(10)$ **c** $f(\frac{1}{2})$ **d** $f(-4)$ **e** $f(-1.5)$

2 $g(x) = \dfrac{x^2 + 1}{2}$. Find:

   a $g(0)$ **b** $g(3)$ **c** $g(10)$ **d** $g(-2)$ **e** $g(-\frac{1}{2})$

3 $f: x \rightarrow x^3 - 2x + 1$. Find:

   a $f(2)$ **b** $f(-2)$ **c** $f(100)$ **d** $f(0)$ **e** $f(\frac{1}{2})$

4 $g: x \rightarrow 2^x$. Find:

   a $g(2)$ **b** $g(5)$ **c** $g(0)$ **d** $g(-1)$ **e** $g(-3)$

5 $h(x) = \dfrac{x + 1}{x - 1}$. Find:

   a $h(2)$ **b** $h(3)$ **c** $h(-1)$ **d** $h(0)$ **e** $h(1\frac{1}{2})$

6 $f: x \rightarrow 2x + 5$

   a If $f(a) = 20$, what is the value of a?   **b** If $f(b) = 0$, what is the value of b?

   c If $f(c) = c$, what is the value of c?

HIGHER

7 $g(x) = \sqrt{x + 3}$

    a Find $g(33)$.
    b If $g(a) = 10$, find a.
    c If $g(b) = 2.5$, find b.

8 $f(x) = 2x - 8$ and $g(x) = 10 - x$.

    a If $f(x) = g(x)$, what is the value of x?

    b Sketch the graphs of $y = f(x)$ and $y = g(x)$. At what point do they cross?

9 $h: x \rightarrow \dfrac{12}{x} + 1$ $k: x \rightarrow 2x - 1$

    a Find $h(6)$.
    b Find $k(-1)$.

    c Solve the equation $h(x) = k(3)$.
    d Solve the equation $k(x) = h(-12)$.

21.2 Domain and range   H

Consider the function $f: x \rightarrow \sqrt{x - 3}$

    $f(3) = 0$

    $f(4) = 1$

    $f(19) = 4$

    $f(103) = 10$

We can think of this as giving us a connection between two sets of numbers.

The starting set is called the **domain.** The resulting set of numbers is called the **range**.

We call this connection between the two sets a **mapping**.

We cannot find the square root of a negative number so numbers less than 3 must be excluded from the domain of f in this case.

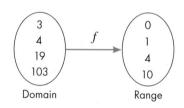

EXAMPLE 1

$g: x \rightarrow \dfrac{1}{x}$

    a What number must be excluded from the domain of g?

    b If the domain is $\{x: 1 \leqslant x \leqslant 2\}$, what is the range?

    a We cannot evaluate $\dfrac{1}{0}$ so 0 must be excluded from the domain of g.

    b $g(1) = \dfrac{1}{1} = 1$ $g(2) = \dfrac{1}{2}$

    The range will be all the numbers between $\dfrac{1}{2}$ and 1.

    We could write this as $\{y: \dfrac{1}{2} \leqslant y \leqslant 1\}$

HIGHER

1 What values of x must be excluded from the domains of the following functions?

    a $f: x \to \sqrt{x}$ **b** $g: x \to \dfrac{1}{x + 1}$ **c** $h: x \to \sqrt{x + 1}$

    d $j: x \to \dfrac{1}{2x + 1}$

2 $f(x) = x^2 + 1$

    Find the ranges for each of these domains:

    a $\{3, 4, 5\}$ **b** $\{-2, -1, 0, 1, 2\}$ **c** $\{x: 1 \leqslant x \leqslant 2\}$

    d $\{x: x \geqslant 10\}$ **e** $\{x: x \leqslant -10\}$

3 Suppose the domain is $\{1, 2, 3, 4\}$. Find the range for these functions:

    a $f(x) = (x - 2)^2$ **b** $g(x) = \dfrac{1}{x}$ **c** $h(x) = 2x + 3$

    d $f(x) = 6 - x$ **e** $g(x) = (x - 1)(x - 4)$

4 $f(x) = x^2$

    Explain why -2 could be in the domain but cannot be in the range.

5 The domain of a function f is $\{1, 2, 3, 4\}$ and the range is $\{2, 3, 4, 5\}$.

    Say whether each of these is a possible description of :

    a $f: x \to x + 1$ **b** $f: x \to 2x$ **c** $f: x \to 6 - x$

21.3 Inverse functions

Suppose $f(x) = 2x + 6$.

Then $f(1) = 8$, $f(3) = 12$ and $f(-4) = -2$.

The **inverse** of f is the function which has the opposite effect and 'undoes' f. We write the inverse of f as f^{-1}.

Since f above means 'multiply by 2 and then add 6', the inverse will be 'subtract 6 and then divide by 2':

$$f^{-1}(x) = \frac{x - 6}{2}$$

So $f^{-1}(8) = 1$, $f^{-1}(12) = 3$ and $f^{-1}(-2) = -4$.

You can find the inverse in the following way:

Step 1: Write $y = f(x)$

$$y = 2x + 6$$

Step 2: Rearrange to make x the subject.

$$y - 6 = 2x$$

$$\frac{y - 6}{2} = x$$

$$x = \frac{y - 6}{2}$$

Step 3: Replace y by x in the result.

$$f^{-1}(x) = \frac{x - 6}{2}$$

EXERCISE 21C

1 Find $f^{-1}(x)$ for the following functions.

  a $f(x) = x + 7$
  b $f(x) = 8x$
  c $f(x) = \frac{x}{5}$
  d $f(x) = x - 3$

2 $f: x \rightarrow \frac{x}{2} + 6$. Find the following:

  a $f(4)$
  b $f^{-1}(8)$
  c $f(-2)$
  d $f^{-1}(5)$

3 Find $f^{-1}(x)$ for the following functions:

  a $f(x) = \frac{x}{3} - 2$
  b $f(x) = 4(x - 5)$
  c $f(x) = \frac{x + 4}{5}$
  d $f(x) = \frac{(3x - 6)}{2}$
  e $f(x) = 3(\frac{x}{2} + 4)$
  f $f(x) = 4x^3$

4 $g(x) = \frac{2x + 5}{3}$. Find:

  a $g^{-1}(3)$
  b $g^{-1}(2)$
  c $g^{-1}(0)$

5 $f(x) = 10 - x$

  a Find an expression for $f^{-1}(x)$.
  b What do you notice about $f(x)$ and $f^{-1}(x)$?

6 Find $f^{-1}(x)$ in the following cases.

  a $f(x) = \frac{8}{x}$
  b $f(x) = \frac{20}{x} - 1$
  c $f(x) = \frac{2}{x + 1}$

7 $f(x) = 2x - 4$

  a Find $f^{-1}(x)$.
  b On the same axes draw graphs of $y = f(x)$ and $y = f^{-1}(x)$.
  c Where do the lines cross?

8 $f(x) = \frac{x + 5}{2}$

  Solve the equation $f(x) = f^{-1}(x)$.

9 $f^{-1}(x) = 3x - 2$

  Find $f(x)$.

HIGHER

Suppose we have a function given by $f(x) = 2x$.

So f means 'double it'.

$$f(2) = 4$$
$$f(3) = 6$$
$$f(5) = 10$$
and $f(-3) = -6$

Now suppose we have another function, g, and $g(x) = x - 3$.

So g means 'take away 3'.

$$g(4) = 1$$
$$g(6) = 3$$
$$g(10) = 7$$
and $g(-6) = -9$

Let's put those side by side to make a **composite** function:

$f(2) = 4$	and	$g(4) = 1$
$f(3) = 6$	and	$g(6) = 3$
$f(5) = 10$	and	$g(10) = 7$
$f(-3) = -6$	and	$g(-6) = -9$

In words they say:

Start with 2, double it, subtract 3, the answer is 1.
Start with 3, double it, subtract 3, the answer is 3.
Start with 5, double it, subtract 3, the answer is 7.
Start with -3, double it, subtract 3, the answer is -9.

We write that in symbols as:

$$gf(2) = 1$$
$$gf(3) = 3$$
$$gf(5) = 7$$
$$gf(-3) = -9$$

> **HINTS AND TIPS**
>
> gf means 'first f, then g'.

EXAMPLE 2

$h(x) = x^2$ and $k(x) = x + 4$. Find:

a $kh(3)$ **b** $kh(-2)$ **c** $hk(5)$

a h means 'square it' and k means 'add 4'.

    $kh(3)$ means 'start with 3, square it, then add 4'.

    So $kh(3) = 3^2 + 4 = 9 + 4 = 13$

b $kh(-2) = (-2)^2 + 4 = 4 + 4 = 8$

c $hk(5)$ is the other way round.

    It means 'start with 5, add 4, then square it'.

    $hk(5) = (5 + 4)^2 = 9^2 = 81$

Going back to f and g we had before:

$f(x) = 2x$ and $g(x) = x - 3$

$gf(x)$ means 'first double x, then subtract 3'. So we write:

$gf(x) = 2x - 3$.

What about $fg(x)$?

That means 'first subtract 3, then double it'. So we write:

$fg(x) = 2(x - 3)$

Looking at example 1, $h(x) = x^2$ and $k(x) = x + 4$.

$kh(x) = x^2 + 4$

and $hk(x) = (x + 4)^2$

EXERCISE 21D

1 $s(x) = x + 4$ and $t(x) = \frac{x}{2}$

   a   Find $s(2)$ and $ts(2)$.   **b**   Find $s(3)$ and $ts(3)$.   **c**   Find $s(6)$ and $ts(6)$.

   d   Find an expression for $ts(x)$.   **e**   Find $t(2)$ and $st(2)$.   **f**   Find $t(3)$ and $st(3)$.

   g   Find $t(-10)$ and $st(-10)$.   **h**   Find an expression for $st(x)$.

2 $c(x) = x^3$ and $d(x) = 2x$

   a   Find $d(3)$ and $cd(3)$.   **b**   Find $d(5)$ and $cd(5)$.

   c   Find an expression for $cd(x)$.   **d**   Find $c(4)$ and $dc(4)$.

   e   Find an expression for $dc(x)$.

3 $r(x) = \sqrt{x}$ and $a(x) = 2x + 1$

   a   Find $a(0)$, $a(4)$ and $a(12)$.   **b**   Find $ra(0)$, $ra(4)$ and $ra(12)$.   **c**   Find an expression for $ra(x)$.

4 $m(x) = 3x$

   a   Find $m(2)$ and $mm(2)$.   **b**   Find $m(4)$ and $mm(4)$.   **c**   Find an expression for $mm(x)$.

5 $f(x) = 3x$ and $g(x) = x - 6$

   a   Find an expression for $fg(x)$.   **b**   Find an expression for $gf(x)$.

6 $a(x) = x + 4$ and $b(x) = x - 7$

   Show that $ab(x)$ and $ba(x)$ are identical.

Suppose $f(x) = x^2 + 2$ and $g(x) = 2x - 3$.

Can we find an expression for $gf(x)$?

First let's start with a value of x, say $x = 3$.

$$f(3) = 3^2 + 2 = 11$$

Now apply g to that answer.

$$g(11) = 2 \times 11 - 3 = 19$$

So $gf(3) = 19$

$gf(x)$ means 'start with x, apply f, and then apply g to the answer.'

If we start with x and 'apply f' we get $x^2 + 2$.

Now take that answer and 'apply g', double it and subtract 3:

$$gf(x) = 2(x^2 + 2) - 3$$

We could simplify that:

$$gf(x) = 2(x^2 + 2) - 3.$$
$$= 2x^2 + 4 - 3$$
$$= 2x^2 + 1$$

So $gf(x) = 2x^2 + 1$

Test your answer with a number. For example, if x is 3:

$$gf(3) = 2 \times 3^2 + 1$$
$$= 18 + 1$$
$$= 19 \text{ as before.}$$

EXERCISE 21E

1 $f(x) = \dfrac{x - 3}{2}$ and $g(x) = 3x + 1$. Find:

    a $fg(3)$ **b** $gf(3)$ **c** $fg(6)$ **d** $gf(6)$

2 $f(x) = x^2 - x$ and $g(x) = \dfrac{x}{2} + 3$. Find:

    a $fg(4)$ **b** $gf(4)$ **c** $fg(1)$ **d** $gf(1)$

3 $f(x) = 2^x$ and $g(x) = 2x - 1$. Find:

    a $gf(2)$ **b** $fg(2)$ **c** $ff(3)$ **d** $gg(6)$

4 $f(x) = 3x + 1$ and $g(x) = 2x - 2$

    a Find an expression for $gf(x)$. Write your answer as simply as possible.

    b Find $fg(x)$. Write your answer as simply as possible.

5 In the following cases find $fg(x)$. Write your answer as simply as possible.

    a $f(x) = x^2$ and $g(x) = 3x + 4$ **b** $f(x) = 2x + 3$ and $g(x) = 3x - 4$

    c $f(x) = \dfrac{x}{2} + 4$ and $g(x) = 4x - 2$ **d** $f(x) = 12 - x$ and $g(x) = 2x + 8$

6 $h(x) = 10 - x$ and $k(x) = 20 - x$. Find:

    a $hk(x)$ **b** $kh(x)$ **c** $kk(x)$

7 $h(x) = x^2$ and $k(x) = \dfrac{12}{x}$. Find:

    a $hh(x)$ **b** $hk(x)$ **c** $kh(x)$ **d** $kk(x)$

8 $m(x) = x^2 + 2x$ and $n(x) = 2x - 1$

    a Find $mm(2)$ **b** Show that $nn(x) = 4x - 3$

    c Show that $mn(x) = 4x^2 - 2x + 1$

9 $f(x) = \dfrac{1}{x - 4}$ and $g(x) = 3x + 1$

    a Find $fg(x)$ **b** Find $gf(x)$, writing your answer as a single fraction.

10 $h(x) = \dfrac{x + 4}{2}$ and $k(x) = 3x - 5$. Find:

    a $h^{-1}k(x)$ **b** $hh(x)$

11 Suppose $f(x) = 0.5(x + 9)$.

    a Show that $f(1) = 5$.

    b Find $f(5)$.

    c Find $f(b)$ where b is the answer to part **b**.

    d Continue in this way, using the last answer as the next value of x, to find the next six values.

    e What is happening to the answers?

12 What happens in question **11** if you start with $f(25)$ instead of $f(1)$?

13 $f(x) = 2x + 3$

    a Find $f^{-1}(x)$ **b** Find $f^{-1}f(100)$

    c Find $f^{-1}f(x)$ **d** Find $ff^{-1}(x)$

    e If $g(x)$ is a function with inverse $g^{-1}(x)$, what is $g^{-1}g(100)$?

When you look at a graph, it shows you how changes in one variable lead to changes in another. Calculus helps us to find the rate at which this change happens.

When you look at a straight line graph you can easily calculate the gradient which tells you how quickly one variable changes compared to the other. But when the line on a graph is curved the gradient is changing all the time. Using calculus you can calculate the gradient and so work out how quickly change is happening at any point.

For example:

- When you walk up a hill your height compared to ground level changes with respect to your position on the slope

- When you drive a car your position and your velocity vary with respect to time

- The brightness of a light bulb varies depending on the electric current flowing through it

- Hot drinks cool down and ice melts at different rates as time passes

We can use calculus to understand how changes like this happen.

A form of calculus was used in Ancient Greece and China. In the seventeenth century CE a German mathematician called Leibniz and an English scientist, Isaac Newton, started to develop the form of calculus we use today.

Calculus is now an essential tool in almost all areas of science, including engineering, physics, chemistry, biology, economics and computer science.

22

Calculus

Topics	Level	Key words
1 The gradient of a curve	HIGHER	gradient
2 More complex curves	HIGHER	rate of change, differentiate, differentiation
3 Turning points	HIGHER	turning point, maximum, minimum
4 Motion of a particle	HIGHER	particle, displacement, velocity, acceleration

What you need to be able to do in the examinations:

HIGHER

- Understand the concept of a variable rate of change.
- Differentiate integer powers of x.
- Determine gradients, rates of change and turning points (maxima and minima) by differentiation and relate these to graphs.
- Distinguish between maxima and minima by considering the general shape of the graph.
- Apply calculus to linear kinetics (motion in a straight line) and to other simple practical problems.

This is a graph of $y = x^2 - 3x + 4$.

What is the **gradient** at the point P(3, 4)?

In an earlier chapter you learnt how to find the gradient by drawing a tangent to the curve at P. It is hard to do this accurately and we need to draw the graph first.

We will now look at another method.

We choose two points on the curve that are close to P, one each side.

Let's choose A (2.9, 3.71) and B (3.1, 4.31).

2.9 and 3.1 are both close to P.

We substitute these values for x in the equation $y = x^2 - 3x + 4$ to get the corresponding y values of 3.71 and 4.31.

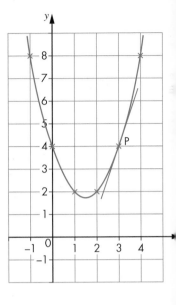

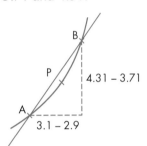

Imagine we draw a straight line through A and B. We can assume this will have *a* similar gradient to the tangent at P.

Gradient of AB = $\dfrac{4.31 - 3.71}{3.1 - 2.9} = 3$

If we choose any two points close to P we will get a similar result.

In fact the gradient of the curve at P (3, 4) is 3.

This method could be used at any point on the curve and it will show the following:

The gradient at any point on the curve $y = x^2 - 3x + 4$ is $2x - 3$.

For example, at (3, 4) the gradient is $2 \times 3 - 3 = 3$
at (4, 8) the gradient is $2 \times 4 - 3 = 5$
at (1, 2) the gradient is $2 \times 1 - 3 = -1$

Looking at the graph should convince you that these seem to be reasonable values.

The notation we use to represent the gradient of a curve is $\dfrac{dy}{dx}$ (read it as "dee y by dee x").

So the tangent to the curve on the graph shows that:

If $y = x^2 - 3x + 4$

$\dfrac{dy}{dx} = 2x - 3$

This is the case for all equations of this form. So the general result is:

If $\quad y = ax^2 + bx + c$

then $\quad \dfrac{dy}{dx} = 2ax + b$

We can use this to calculate the gradient at any point on a quadratic curve. We no longer need to draw the graph and this method can be used for any point on the curve.

EXAMPLE 1

A curve has the equation $y = 0.5x^2 + 4x - 3$.

a Find the gradient at $(0, -3)$ and at $(2, 7)$.

b Find the coordinates of the point where the gradient is 0.

a Using the general result above, if $y = 0.5x^2 + 4x - 3$,

$\qquad$ then $\quad \dfrac{dy}{dx} = 2 \times 0.5x + 4$

$\qquad \Rightarrow \quad \dfrac{dy}{dx} = x + 4$

If $x = 0, \dfrac{dy}{dx} = 0 + 4 = 4$. The gradient at $(0, -3)$ is 4.

If $x = 2, \dfrac{dy}{dx} = 2 + 4 = 6$. The gradient at $(2, 7)$ is 6.

b If the gradient is 0, then $\dfrac{dy}{dx} = 0$

$\qquad \Rightarrow \quad x + 4 = 0$

$\qquad \Rightarrow \qquad x = -4$

If $x = -4, y = 0.5 \times (-4)^2 + 4 \times (-4) - 3 = -11$

$\qquad \Rightarrow$ The gradient is 0 at $(-4, -11)$.

EXERCISE 22A

1 A curve has the equation $y = x^2 - 2x$.

a Copy and complete this table of values:

x	-2	-1	0	1	2	3	4
$x^2 - 2x$	8	3				3	8

b Sketch the graph of $y = x^2 - 2x$.　　c Find $\dfrac{dy}{dx}$.

d Find the gradient of the curve at $(3, 3)$.　　e Find the gradient of the curve at $(4, 8)$.

f Find the gradient at two more points on the curve.

g At what point on the graph is the gradient 0?

h By looking at your graph, check that your answers to parts **d**, **e**, **f** and **g** seem sensible.

HIGHER

2 $y = x^2 - 6x + 15$

    a Find $\dfrac{dy}{dx}$.
                 b Find the gradient at (0, 15).

    c Find the gradient at (5, 10).

    d Find the coordinates of the point where the gradient is 2.

3 $y = 2x^2 - 10$

    a Find $\dfrac{dy}{dx}$.
                 b Find the gradient at (2, –2).

    c Find the gradient at (–1, –9).
           d Find the point where the gradient is 12

4 This is the graph of $y = 4x - x^2$

    a Find $\dfrac{dy}{dx}$.

    b Find the gradient at each point where the curve crosses the x-axis.

    c Where is the gradient equal to 2?

    d Where is the gradient equal to 1?

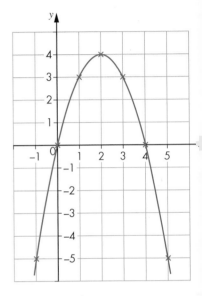

5 Find $\dfrac{dy}{dx}$ for each of the following:

    a $y = x^2 + x + 1$
                 b $y = x^2 - 7x + 3$

    c $y = 4x^2 - x + 6$
              d $y = 0.3x^2 - 1.5x + 7.2$

    e $y = 6 - 2x + x^2$
               f $y = 10 + 3x - x^2$

    g $y = 2x + 5$
                   h $y = 4$

6 If $y = (x + 4)(x - 2)$ what is $\dfrac{dy}{dx}$?

7 Find $\dfrac{dy}{dx}$ for each of the following:

    a $y = 2x(x + 1)$

    b $y = (x + 2)(x + 5)$

    c $y = (x + 3)(x - 3)$

> **HINTS AND TIPS**
>
> First multiply out the brackets.

8 A curve has the equation $y = x^2 + 2x - 5$

    a Where does the curve cross the y-axis?

    b What is the gradient of the curve at that point?

HIGHER

You have seen that if $y = x^2$ then $\dfrac{dy}{dx} = 2x$.

This table shows the value of $\dfrac{dy}{dx}$ for some other curves.

y	$\dfrac{dy}{dx}$	
1	0	$\rightarrow$ The line $y = 1$ has gradient 0
x	1	$\rightarrow$ The line $y = x$ has gradient 1
x^2	$2x$	
x^3	$3x^2$	
x^4	$4x^3$	

There is a general pattern here:

If $y = x^n$ then $\dfrac{dy}{dx} = nx^{n-1}$

We say that $\dfrac{dy}{dx}$ is the **rate of change** of y with respect to x.

If a is a constant and $y = ax^n$ then $\dfrac{dy}{dx} = anx^{n-1}$

For example, if $y = 5x^2$ then $\dfrac{dy}{dx} = 5 \times 2x = 10x$.

If $y = 4x^3$ then $\dfrac{dy}{dx} = 12x^2$.

If $y = 6$ then $\dfrac{dy}{dx} = 0$. So the line with equation $y = 6$ is horizontal and has gradient 0.

EXAMPLE 2

What is the gradient of the curve with equation $y = x^3 - 3x^2 + 4x + 7$ at the point $(2, 11)$?

$$\frac{dy}{dx} = 3x^2 - 6x + 4$$

If $x = 2$, $\dfrac{dy}{dx} = 3 \times 2^2 - 6 \times 2 + 4 = 4$

The gradient at $(2, 11)$ is 4.

The process of finding $\dfrac{dy}{dx}$ is called **differentiation**.

In example 2 we **differentiated** each term in turn:

Differentiate x^3 to get $3x^2$

Differentiate $-3x^2$ to get $-6x$

Differentiate $4x$ to get 4 $\qquad \dfrac{dy}{dx} = 3x^2 - 6x + 4$

Differentiate 7 to get 0

Negative powers

Can we differentiate $y = \dfrac{1}{x}$ or $y = \dfrac{1}{x^2}$?

We can write these as $y = x^{-1}$ and $y = x^{-2}$.

The rule that if $y = x^n$ then $\dfrac{dy}{dx} = nx^{n-1}$ applies in these cases too.

If $y = x^{-1}$, $\dfrac{dy}{dx} = -1 \times x^{-2} = -x^{-2}$ or $-\dfrac{1}{x^2}$.

If $y = x^{-2}$, $\dfrac{dy}{dx} = -2 \times x^{-3} = -2x^{-3}$ or $-\dfrac{2}{x^3}$.

EXAMPLE 3

If $y = x + \dfrac{2}{x}$, find the gradient at $(2, 3)$.

$$y = x + 2x^{-1} \Rightarrow \dfrac{dy}{dx} = 1 + 2 \times (-1)x^{-2}$$

$$= 1 - 2x^{-2}$$

$$= 1 - \dfrac{2}{x^2}$$

If $x = 2$, $\dfrac{dy}{dx} = 1 - \dfrac{2}{4} = \dfrac{1}{2}$

So the gradient at $(2, 3)$ is $\dfrac{1}{2}$

EXERCISE 22B

HIGHER

1 The equation of a curve is $y = 2x^3$.

   a Find $\dfrac{dy}{dx}$.

   b Find the gradient of the curve at $(1, 2)$ and $(2, 16)$.

2 The equation of a curve is $y = x^3 - 6x^2 + 8x$.

   a Find $\dfrac{dy}{dx}$.

   b Show that $(0, 0)$, $(2, 0)$ and $(4, 0)$ are all on this curve.

   c Find the rate of change of y with respect to x at each point in part **b**.

3 The graph shows the curve $y = \dfrac{12}{x}$ for $0 < x \leqslant 12$

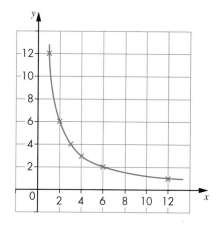

   a Find $\dfrac{dy}{dx}$.

   b Find the gradient at $(2, 6)$ and at $(4, 3)$.

   c Where is the gradient -1?

4 Find $\dfrac{dy}{dx}$ for each of the following:

   a $y = 2x^4$

   b $y = x + \dfrac{3}{x}$

   c $y = 5x^3 - 2x + 4$

   d $y = \dfrac{4}{x^2}$

   e $y = 3x^3 + 5x - 7$

   f $y = 10 - x^3$

   g $y = x(x^3 - 1)$

   h $y = \dfrac{x^2 + 1}{x}$

5 This is a sketch of the curve $y = x^4 - 4x^2$.

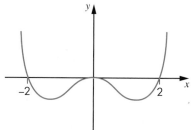

Not to scale

Find the gradient at the points where the curve meets the x-axis.

6 A curve has the equation $y = \dfrac{20}{x^2}$.

   a Find the two points where the curve meets the line $y = 5$.

   b Find the gradient of the curve at each of the meeting points.

7 A curve has the equation $y = \dfrac{1}{3}x^3 - 5x + 4$.

Show that there are two points on the curve where the rate of change of y with respect to x is 4. Find the coordinates of the two points.

A point where the gradient is zero is called a **turning point**.

A and B are turning points.

At any turning point $\frac{dy}{dx} = 0$

A is called a **maximum** point because it is higher than the points near it.

B is called a **minimum** point.

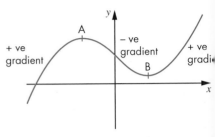

EXAMPLE 4

Find the turning points of $y = x^3 - 12x + 4$ and state whether each is a maximum or a minimum point.

If $y = x^3 - 12x + 4$:

$$\frac{dy}{dx} = 3x^2 - 12$$

At a turning point $\frac{dy}{dx} = 0$:

$$\Rightarrow \quad 3x^2 - 12 = 0$$

$$\Rightarrow \qquad 3x^2 = 12$$

$$\Rightarrow \qquad\quad x^2 = 4$$

$$\Rightarrow \qquad\qquad x = 2 \text{ or } -2$$

If $x = 2$, $y = 8 - 24 + 4 = -12$ $\qquad \Rightarrow \qquad$ $(2, -12)$ is a turning point

If $x = -2$, $y = -8 + 24 + 4 = 20$ $\qquad \Rightarrow \qquad$ $(-2, 20)$ is a turning point

A rough sketch makes it look likely that $(-2, 20)$ is a maximum point and $(2, -12)$ is a minimum point.

If you are not sure, check the gradient on each side of the point.

For $(2, -12)$:

x	1.9	2	2.1
$\frac{dy}{dx}$	−1.17	0	1.23
gradient	negative	0	positive

This is $3 \times 2.1^2 - 12$

$(2, -12)$ is a minimum point.

Because the gradient changes from negative to zero to positive as we move from left to right, $(2, -12)$ must be a minimum point.

For $(-2, 20)$:

x	-2.1	-2	-1.9
$\dfrac{dy}{dx}$	1.23	0	-1.17
$gradient$	positive	0	negative

$(-2, 20)$ is a minimum point.

Because the gradient changes from positive to zero to negative as

we move from left to right, $(-2, 20)$ must be a maximum point.

A sketch of the curve.

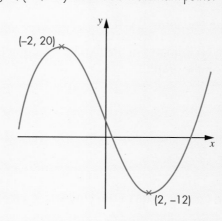

EXERCISE 22C

1 $y = x^2 - 4x + 3$

    a   Find $\dfrac{dy}{dx}$.

    b   Show that the curve has one turning point and find its coordinates.

    c   State whether it is a maximum or minimum point.

2   **a**   Find the turning point of the curve $y = x^2 + 6x - 3$.

    b   Is it a maximum or a minimum point?

3 $y = 1 + 5x - x^2$

    a   Find $\dfrac{dy}{dx}$.

    b   Find the turning point of the curve.

    c   Is it a maximum or a minimum point?

HIGHER

HIGHER

4 This is the sketch of $y = x + \dfrac{1}{x}$ for $x > 0$

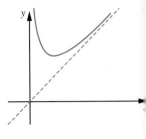

  a Find $\dfrac{dy}{dx}$.

  b Find the coordinates of the minimum point shown on the graph.

5 This is a sketch of $y = x^3 - 3x^2$

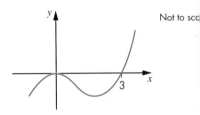

Not to sc[a]

  a Find $\dfrac{dy}{dx}$.

  b Solve the equation $\dfrac{dy}{dx} = 0$.

  c Find the coordinates of the two turning points shown on the graph.

6 $y = x^2 - 3x - 10$

  a Show that the graph crosses the x-axis at $(-2, 0)$ and $(5, 0)$.

  b Find $\dfrac{dy}{dx}$.

  c Find the turning point of the curve. Is it a maximum or a minimum?

  d The curve has a line of symmetry. What is the equation of this line?

7 A curve has the equation $y = 3x + \dfrac{12}{x^2}$

  a Find $\dfrac{dy}{dx}$.

  b The curve has one turning point. Find its coordinates.

  c Is it a maximum or a minimum point?

8 The equation of a curve is $y = 2x^3 - 6x + 4$.

  a Find $\dfrac{dy}{dx}$.

  b Find the turning points of the curve and state whether each one is a maximum or a minimum point.

9 A rectangle has a perimeter of 30 cm.

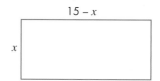

15 − x

x

a Explain why, if one side of the rectangle is x cm, the other will be $(15 − x)$ cm.

Suppose the area of the rectangle is A cm^2 then $A = x(15 − x)$.

b Find $\dfrac{dA}{dx}$.

c Find the turning point of the graph $A = x(15 − x)$.

d Is the turning point a maximum or a minimum?

e What does the result of **c** and **d** tell you about the area of the rectangle?

> **HINTS AND TIPS**
>
> A is used instead of y here. Differentiate in the usual way.

22.4 Motion of a particle **H**

A **particle** is a stone or any small object.

Imagine a particle moving backwards and forwards in a straight line.

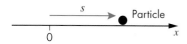

The **displacement**, s, is given by the position of the particle relative to a fixed point, O.

On one side of O the displacement is positive. On the other side it is negative.

The particle will have **velocity**, v.

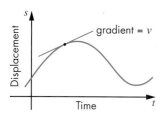

The velocity is the gradient of the displacement–time graph.

We can write $v = \dfrac{ds}{dt}$.

The velocity is the rate of change of displacement with respect to time.

> **HINTS AND TIPS**
>
> We usually use t instead of x when time is the variable.

EXAMPLE 5

The displacement of an object after t seconds is given by $s = 3 + 4t - t^2$ metres.

Find the displacement and the velocity at 1 second intervals from 0 to 5.

Differentiate $s = 3 + 4t - t^2$

$$v = \frac{ds}{dt} = 4 - 2t$$

We can show the values of s and v at different times in a table:

Time t	0	1	2	3	4	5
Displacement $s = 3 + 4t - t^2$	3	6	7	6	3	−2
Velocity $v = 4 - 2t$	4	2	0	−2	−4	−6

We could use these values to draw graphs of s and v.

The particle is initially 3 m away from O.

At first it moves away from O, until it is 7 m away.

Then it moves back towards O and passes it after about 4.6 seconds.

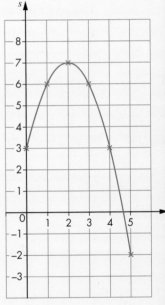

The velocity is initially 4 m/s and then decreases.

The velocity is positive when the displacement is increasing and negative when the displacement is decreasing.

When $t = 2$, the gradient of the displacement graph is zero and the velocity is zero.

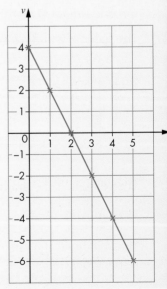

The gradient of the velocity–time graph is the **acceleration**.

Acceleration is the rate of change of velocity with respect to time.

In Example 5: $v = 4 - 2t$

So acceleration: $a = \dfrac{dv}{dt} = -2$ m/s^2

The velocity–time graph is a straight line and the gradient is –2.

The velocity decreases by 2 m/s every second.

EXERCISE 22D

1 The displacement, s metres, after t seconds, of a particle moving in a straight line is given by $s = t^2 - 3t + 5$.

   a Differentiate s to find v.

   b Find s and v when $t = 1$.

   c Find s and v when $t = 3$.

   d Find s and v when $t = 4$.

   e Find the acceleration of the particle.

   f When is the velocity equal to 0?

2 The displacement of a particle after t seconds is given by $s = t^3 - t^2$ metres.

   a Differentiate s to find v.

   b Differentiate v to find a.

   c Find s, v and a when $t = 0$.

   d Find s, v and a when $t = 1$.

   e Find s, v and a when $t = 2$.

3 The displacement, s metres, of a particle after t seconds is given by:

   $$s = t^2 - 4t + 6 \qquad 0 \leqslant t \leqslant 6$$

   a Find expressions for the velocity and the acceleration.

   b Find the displacement and the velocity when $t = 6$.

   c When is the velocity zero?

4 The displacement, s metres, of a particle from a fixed point O after t seconds is given by:

   $$s = 8t^3 - 12t^2 \qquad 0 \leqslant t \leqslant 3$$

   a Find expressions for the velocity and the acceleration after time t.

   b Find the acceleration when $t = 2$.

   c When is the acceleration equal to 0?

HIGHER

5 The displacement in metres of a particle after t seconds is given by:

$$s = 5t + \frac{8}{t} \qquad 1 \leqslant t \leqslant 5$$

a Find an expression for the velocity.

b Find an expression for the acceleration.

c Find the displacement, the velocity and the acceleration when $t = 2$.

6 The displacement of a particle from a fixed point after t seconds is given by:

$$s = 48t - 4t^3 \qquad 0 \leqslant t \leqslant 4$$

a Find the velocity and the acceleration after t seconds.

b When is the acceleration -60 m/s^2?

c When is the velocity 0 m/s?

d What is the displacement when the velocity is 0 m/s?

7 This is a graph of the displacement, s metres, of a particle from a fixed point after t seconds where $0 \leqslant t \leqslant 5$. The displacement is given by $s = 12t(5 - t)$.

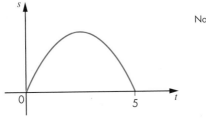

Not to scale

a Find an expression for the velocity of the particle.

b The graph has a turning point. What is the value of t at this point?

c What is the maximum displacement of the particle from the fixed point?

8 The graph shows the displacement, s metres, of a particle after t seconds.

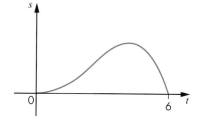

Not to scale

$$s = t^2(6 - t) \qquad 0 \leqslant t \leqslant 6$$

a Find expressions for the velocity and the acceleration.

b When is the acceleration zero?

c When is the velocity zero?

d What is the maximum displacement of the particle?

9 A particle is moving in a straight line. The displacement, s metres, after t seconds is given by:

$$s = t^4 - 10t^3 \qquad 0 \leqslant t \leqslant 12$$

a The initial displacement is 0. At what other time is the displacement 0?

b Find expressions for the velocity and the acceleration.

c When is the velocity zero?

d When is the acceleration zero?

HIGHER

PAPER 1F

1 **a** Simplify $5k + 7k - 2k$ [1]

    b Simplify $e \times 4 \times g$ [1]

    c Solve $6m + 5 = 17$ [2]

    d Factorise $15r + 10$ [1]

    e Simplify $y^7 \times y^2$ [1]

Edexcel Limited Paper 1F Q8 Jan 16

2 This rule gives the cost, in euros, of hiring a bicycle for a number of days.

> Cost in euros = 8 × (number of days) + 15

Marina hires a bicycle for 4 days.

a Work out the cost in euros.   [2]

Cyril hires a bicycle for a number of days.

The cost is 71 euros.

b Work out the number of days.   [2]

Edexcel Limited Paper 1F Q11 May 15

3

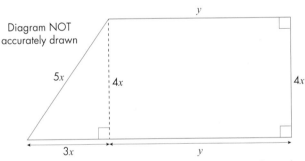

Diagram NOT accurately drawn

The shape in the diagram is made from a rectangle and a right-angled triangle.

The diagram shows, in terms of x and y, the lengths, in centimetres, of the sides of the rectangle and of the triangle.

The perimeter, P cm, of the shape is given by the formula

    $P = 12x + 2y$

a Work out the value of P when $x = 3$ and $y = 7$ [2]

b Work out the value of x when $P = 43$ and $y = 6.5$ [3]

c Find, in terms of x and y, a formula for the area, A cm^2, of the shape.

    Give your area as simply as possible.   [2]

Edexcel Limited Paper 1F Q12 May 13

4 $w = 4x - 5y$

a $x = 7, y = 4$

    Work out the value of w.   [2]

FOUNDATION

b $w = 100$, $y = 22$

Work out the value of x. [2]

c $x = 6t$, $y = 2t$

Find a formula for w in terms of t.

Give your answer in its simplest form. [2]

Edexcel Limited Paper 1F Q12 May 15

5 **a** Solve $3x + 5 = 26$

Show clear algebraic working. [2]

b Solve $4(5y - 1) = 3(6y + 7)$

Show clear algebraic working. [3]

Edexcel Limited Paper 1F Q15 Jan 15

6 **a** Write $2^3 \times 2^4$ as a single power of 2 [1]

b $280 = 2^n \times 5 \times 7$

Find the value of n. [2]

Edexcel Limited Paper 1F Q17 Jan 14

7 Solve $3(2x + 5) = 4 - x$ [3]

Edexcel Limited Paper 1F Q18 Jan 16

8 **a** Expand and simplify $3(2c - 5) - 2(c - 4)$ [2]

b Simplify $(4e^3)^2$ [2]

c Expand and simplify $(a + 5)(a - 1)$ [2]

Edexcel Limited Paper 1F Q20 Jan 15

9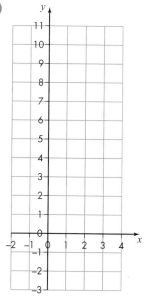

a On the grid, draw the graph of $y = 2x + 3$ for values of x from -2 to 4 [3]

b Show, by shading on the grid, the region that satisfies **all three** of the inequalities

$x \leqslant 3$ and $y \geqslant 2$ and $y \leqslant 2x + 3$

Label your region **R**. [2]

Edexcel Limited Paper 1F Q20 Jan 16

PAPER 2F

1 Here are the first five terms of a number sequence.

32   29   26   23   20

a Work out the next two terms of the sequence. [2]

b Explain how you worked out your answer. [1]

The 15th term of this sequence is −10

c Work out the 14th term of this sequence. [1]

Edexcel Limited Paper 2H Q5 Jun 15

2 Budget Taxis use this rule to work out the cost in euros (€), for taxi journeys.

> 3 euros
> plus
> 2 euros for each kilometre travelled

a Claude travelled 4 kilometres in a Budget taxi.

What was the cost of Claude's journey? [2]

b Bridgette travelled in a Budget taxi.

The cost was 35 euros.

How many kilometres did Bridgette travel? [2]

Economy taxis use this rule to work out their cost, in euros (€), for taxi journeys.

> 8 euros
> plus
> 1 euro for each kilometre travelled

c Find the distance, in kilometres, for which the cost of a journey in a Budget taxi is the same as the cost of a journey in an Economy taxi. [2]

Edexcel Limited Paper 2F Q8 Jan 14

3 Solve $6(3y + 5) = 39$

Show clear algebraic working. [3]

Edexcel Limited Paper 2F Q13 Jan 14

4 a Expand $x(x + 2)$ [1]

b Simplify $6t − 3 − 8t + 7$ [2]

c Solve the inequality $4x − 7 > 3$ [2]

Edexcel Limited Paper 2F Q22 Jan 16

5 $D = 3e^2 + 4e$

Work out the value of D when $e = −5$ [2]

Edexcel Limited Paper 2F Q18 Jan 16

6 **a** Expand and simplify fully $4(2y + 6) - 3(2y - 7)$ [2]

 b Expand and simplify fully $(x - 6)(x - 4)$ [2]

 c Simplify fully $\dfrac{v^4 \times v^7}{v^5}$ [2]

Edexcel Limited Paper 2F Q17 Jan 14

7 The diagram shows a triangle.

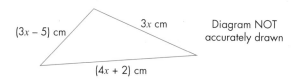

$(3x - 5)$ cm $3x$ cm   Diagram NOT accurately drawn

$(4x + 2)$ cm

The lengths of the sides of the triangle are $3x$ cm, $(3x - 5)$ cm and $(4x + 2)$ cm.
The perimeter of the triangle is 62 cm.
Work out the value of x.
Show clear algebraic working. [4]

Edexcel Limited Paper 2F Q20 Jan 15

8 Solve the simultaneous equations
$$5x + y = 17$$
$$x + y = 3$$

Show clear algebraic working. [3]

Edexcel Limited Paper 2F Q23 Jan 16

9 **a** Solve the inequality $3x + 8 < 35$ [2]

 b Write down the inequality shown on the number line. [2]

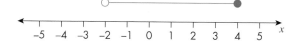

Edexcel Limited Paper 2F Q22 Jan 15

10 Solve $\dfrac{3 - 5m}{4} = 8$

Show clear algebraic working. [3]

Edexcel Limited Paper 2F Q15c Jan 16

11 **a** Solve the inequalities $-6 \leqslant 3n < 9$ [2]

 b n is an integer.
  Write down all the values of n which satisfy $-6 \leqslant 3n < 9$ [2]

Edexcel Limited Paper 2F Q23 May 13

PAPER 3H

1   **a**   Simplify $4p^3q^5 \times 6p^2q$ [2]

     b   Simplify $(5x^2y^4)^3$ [2]

     c   Factorise $9a^2 - b^2$ [2]

Edexcel Limited Paper 3H Q14 May 13

2   **a**   Factorise $a^2 - b^2$ [1]

     $N = 2^{22} - 1$

     b   Write N as the product of two integers, both of which are greater than 1000   [2]

Edexcel Limited Paper 3H Q14 May 15

3

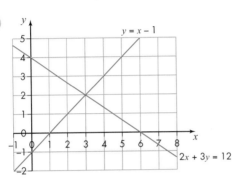

The diagram shows two straight lines.

The equations of the lines are $y = x - 1$ and $2x + 3y = 12$

     a   Write down the solution of the simultaneous equations

         $y = x - 1$

         $2x + 3y = 12$ [1]

     b   Find an equation of the line which is parallel to the line with equation $2x + 3y = 12$ and passes through the point (0, 10)   [4]

     c   On the grid, mark with a cross (×) each point which satisfies both the inequalities $y > x - 1$ and $2x + 3y < 12$ and whose coordinates are **positive integers**.   [2]

Edexcel Limited Paper 3H Q15 May 13

4   The diagram at the top of the next page shows the graph of $y = f(x)$ for $-3.5 \leqslant x \leqslant 1.5$

     a   Find $f(0)$ [1]

     b   For which values of k does the equation $f(x) = k$ have only one solution?   [2]

     c   Find an estimate for the gradient of the curve at the point where $x = -2.5$ [3]

         $g(x) = \dfrac{1}{2 + x}$

     d   State which value of x must be excluded from any domain of g [1]

     e   Find $fg(-3)$ [2]

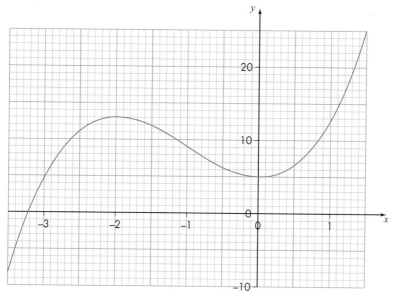

Edexcel Limited Paper 3H Q15 Jan 16

5 Solve the inequality $5x^2 - 13 < 32$

Show clear algebraic working. [3]

Edexcel Limited Paper 3H Q16 Jan 16

6 $f:x \mapsto 2x^2 + 1$ $g:x \mapsto \dfrac{2x}{x-1}$ where $x \neq 1$

a Express the composite function gf in the from $gf:x \mapsto \ldots$

Give your answer as simply as possible. [2]

b Express the inverse function g^{-1} in the form $g^{-1}:x \mapsto \ldots$ [3]

Edexcel Limited Paper 3H Q20 Jan 15

7 $f(x) = \dfrac{3}{x+1} + \dfrac{1}{x-2}$

a State one value of x which cannot be included in any domain of f. [1]

b Find the value of $f(0)$. [1]

c Find the value of x for which $f(x) = 0$.

Show clear algebraic working. [3]

Edexcel Limited Paper 3H Q17 May 15

8 Solve the equation $\dfrac{3}{(x+2)} + \dfrac{4}{(x-3)} = 2$

Show clear algebraic working. [5]

Edexcel Limited Paper 3H Q23 May 14

9 Given that $(5 - \sqrt{x})^2 = y - 20\sqrt{2}$ where x and y are positive integers, find the value of x and the value of y. [3]

Edexcel Limited Paper 3H Q17 Jan 14

PAPER 4H

1 Use algebra to show that the recurring decimal $0.3\dot{8} = \dfrac{7}{18}$ [2]

Edexcel Limited Paper 4H Q20 May 14

2 n is a positive integer.

    a Explain why $2n + 1$ is an odd number for all values of n. [1]

    b Show, using algebra, that the sum of any 4 consecutive odd numbers is always a multiple of 8 [3]

Edexcel Limited Paper 4H Q20 Jun 15

3 Find an equation of the line that is parallel to the line $y = 4 - 2x$ and passes through the point (3, 7) [3]

Edexcel Limited Paper 4H Q12 Jan 16

4 Here is the graph of $y = x^2 - 2x - 1$

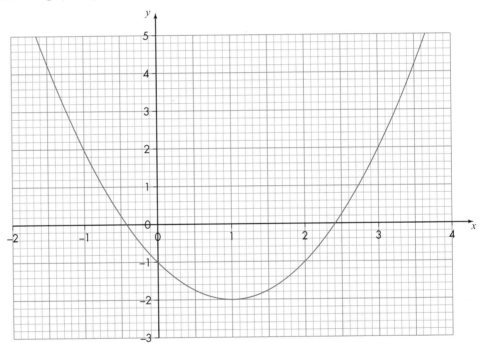

    a Use the graph to solve the equation $x^2 - 2x - 1 = 2$ [2]

The equation $x^2 + 5x - 7 = 0$ can be solved by finding the points of intersection of the line $y = ax + b$ with the graph of $y = x^2 - 2x - 1$

    b Find the value of a and the value of b. [4]

Edexcel Limited Paper 4H Q15 Jan 15

5 Rationalise the denominator of $\dfrac{3c - \sqrt{c}}{\sqrt{c}}$

Simplify your answer. [2]

Edexcel Limited Paper 4H Q19b Jun 15

6 **a** Simplify $\left(\dfrac{8e^6}{f^{12}}\right)^{\frac{1}{3}}$ [2]

b Factorise fully $2y^2 - 72$ [2]

c Simplify $\dfrac{2p^2 - p - 15}{p^2 - 3p}$ [3]

Edexcel Limited Paper 4H Q17 Jan 16

7 Write $5 - (x - 2) \div \left(\dfrac{x^2 - 4}{x - 3}\right)$ as a single fraction.

Simplify your answer fully. [4]

Edexcel Limited Paper 4H Q23 Jan 15

8 A farmer has 120 metres of fencing.

He is going to make a rectangular enclosure *PQRS* with the fencing.

He is also going to divide the enclosure into two equal parts by fencing along *MN*.

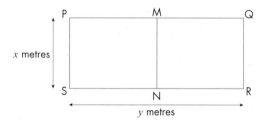

The width of the enclosure is x metres.

The length of the enclosure is y metres.

a **i** Show that $y = 60 - 1.5x$ [1]

The area of the enclosure *PQRS* is A m^2

ii Show that $A = 60x - 1.5x^2$ [2]

b Find $\dfrac{dA}{dx}$ [2]

c Find the maximum value of A. [3]

Edexcel Limited Paper 4H Q15 Jan 14

9 Solve the simultaneous equations

$$2x - y = 7$$
$$x^2 + y^2 = 34$$

Show clear algebraic working. [7]

Edexcel Limited Paper 4H Q22 Jan 14

Why this chapter matters

Angles describe an amount of turn around a point. It is important to be able to measure them and understand their properties.

Why this chapter matters

In a regular polygon all the angles are the same size and all the sides are the same length. The shape of the polygon depends on how many angles and sides it has:

- a triangle has 3 sides and angles
- a square has 4
- a pentagon has 5
- a hexagon has 6.

The patterns below are made by putting together different regular polygons. How many different polygons can you see in them? Can you see any octagons (8-sided shapes) or decagons (10-sided shapes)?

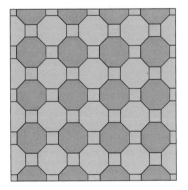 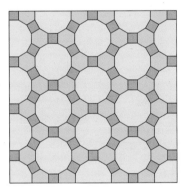 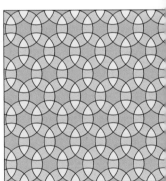

Some shapes fit together better than others because of the size of their angles.

Bees construct their hives from hexagon shapes which can join together without gaps.

Squares and rectangles also fit together easily which makes them an ideal shape to use in building.

This chapter looks at angles, the shapes they form, and their propertie

23

Angle properties

Topics	Level	Key words
1 Angle facts	FOUNDATION	angles at a point, angles on a straight line, opposite angles, vertically opposite angles
2 Parallel lines	FOUNDATION	corresponding angles, alternate angles, allied angles, interior angles
3 Angles in a triangle	FOUNDATION	equilateral triangle, isosceles triangle, right-angled triangle
4 Angles in a quadrilateral	FOUNDATION	quadrilateral, parallelogram, rhombus, kite, trapezium
5 Regular polygons	FOUNDATION	polygon, regular polygon, exterior angles, pentagon, hexagon, octagon, square
6 Irregular polygons	FOUNDATION	irregular polygon, heptagon, nonagon, decagon
7 Tangents and chords	FOUNDATION	radius, tangent, chord, circumference, arc, sector, segment
8 Angles in a circle	HIGHER	subtended, diameter, semi-circle
9 Cyclic quadrilaterals	HIGHER	cyclic quadrilateral, opposite segment, supplementary
10 Alternate segment theorem	HIGHER	alternate segment, alternate segment theorem
11 Intersecting chords	HIGHER	

What you need to be able to do in the examinations:

FOUNDATION	HIGHER
• Distinguish between acute, obtuse, reflex and right angles. • Use angle properties of lines. • Recognise types of triangles, polygons and quadrilaterals and understand their angle properties. • Identify regular polygons and calculate their interior and exterior angles. • Recognise the key terms for parts of a circle. • Understand chord and tangent properties of circles. • Give informal reasons for solutions to geometrical problems.	• Understand and use the internal and external intersecting chord properties. • Recognise cyclic quadrilaterals and their angle properties. • Understand and use the angle properties of the circle. • Provide reasons for solutions in any geometrical context involving lines, polygons and circles.

Angles on a line

The **angles on a straight line** add up to 180°.

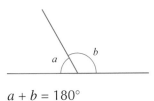

$a + b = 180°$

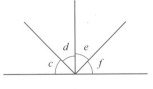

$c + d + e + f = 180°$

Draw an example for yourself (and measure a and b) to show that the statement is true.

Angles at a point

The sum of the **angles at a point** is 360°. For example:

$a + b + c + d + e = 360°$

Again, check this for yourself by drawing an example and measuring the angles.

Sometimes equations can be used to solve angle problems.

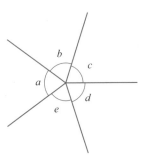

Opposite angles

Opposite angles are equal.

So $a = c$ and $b = d$.

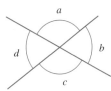

Sometimes opposite angles are called **vertically opposite angles**.

EXERCISE 23A

1 Write down the value of *x* in each of these diagrams.

a

132°   *x*

b
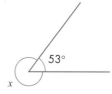
53°
x

c

x   72°

d

38°   *x*

e
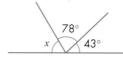
78°
x   43°

f

x
48°   51°

g
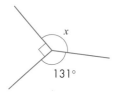
x
131°

h

x   122°

> **HINTS AND TIPS**
>
> Never measure angles in questions like these. Diagrams in examinations are not drawn accurately. Always calculate angles unless you are told to measure them.

2 Write down the value of *x* in each of these diagrams.

a
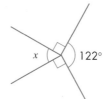
x
82°

b

105°
x

c

x
75°

3 In the diagram, angle ABD is 45° and angle CBD is 125°.

Decide whether ABC is a straight line. Write down how you decided.

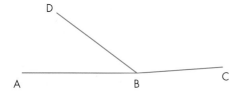
D
A   B   C

FOUNDATION

4 Calculate the value of x in each of these examples.

a

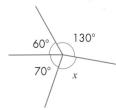

b

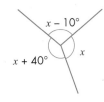

c

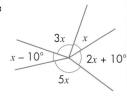

5 Calculate the value of x in each of these examples.

a

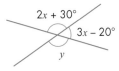

b

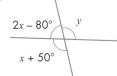

c

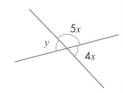

6 Calculate the value of x first and then calculate the value of y in each of these examples.

a

b

c

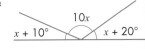

7 Shalini has a collection of tiles. They are all equilateral triangles and are all the same size.

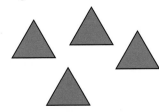

> **HINTS AND TIPS**
>
> All the angles in an equilateral triangle are 60°.

She says that six of the tiles will fit together and leave no gaps.

Explain why Shalini is correct.

8 Work out the value of y in the diagram.

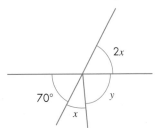

Angles in parallel lines

The arrowheads indicate that the lines are parallel and the line that crosses the parallel lines is called a transversal.

Notice that eight angles are formed.

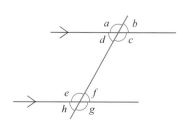

Angles like these

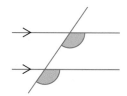

are called **corresponding angles** (Look for the letter F).

Corresponding angles are equal.

Angles like these

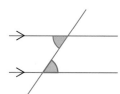

are called **alternate angles** (Look for the letter Z).

Alternate angles are equal.

Angles like these

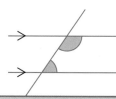

are called **allied angles** or **interior angles** (Look for the letter C).

Allied angles add to 180°.

EXERCISE 23B

1 State the sizes of the lettered angles in each diagram.

a

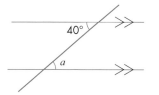

b

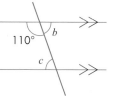

c

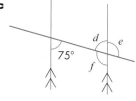

d

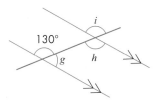

e

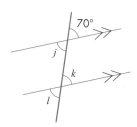

f

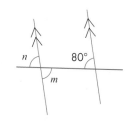

2 State the sizes of the lettered angles in each diagram.

a

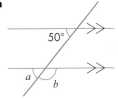

50°

a / b

b

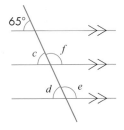

65°

c / f

d / e

c

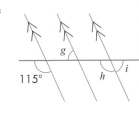

g

115° / h / i

d

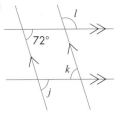

l

72°

k

j

e

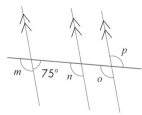

m / 75° / n / o / p

f

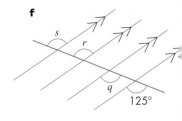

s / r

q

125°

3 State the sizes of the lettered angles in these diagrams.

a

a / 85°

b

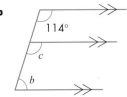

114°

c

b

4 Calculate the values of x and y in these diagrams.

a

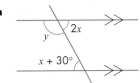

2x

y

x + 30°

b

3x

y

2x + 25°

c

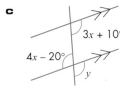

3x + 10°

4x – 20°

y

5 Calculate the values of x and y in these diagrams.

a

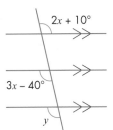

2x + 10°

3x – 40°

y

b

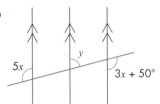

5x

y

3x + 50°

c

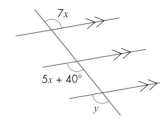

7x

5x + 40°

y

6 A company makes signs like the one below.

It has one line of symmetry.

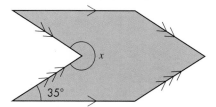

HINTS AND TIPS

Draw the line of symmetry on the shape first.

The company needs to know the size of the angle marked x on the diagram.

Work out the size of angle x.

7 In the diagram, AE is parallel to BD.

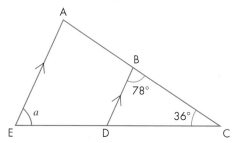

Work out the size of angle a.

8

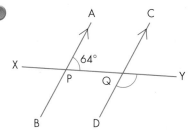

The line XY crosses the parallel lines AB and CD at P and Q.

a Work out the size of angle DQY. Give reasons for your answer.

b This is Vreni's solution.

> *Angle PQD = 64° (corresponding angles)*
> *So angle DQY = 124° (angles on a line = 190°)*

She has made a number of errors in her solution.

Write out the correct solution for the question.

9 Use the diagram to prove that the three angles in a triangle add up to 180°.

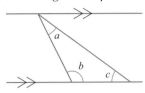

10 Prove that $p + q + r = 180°$.

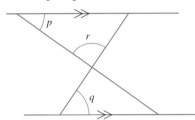

23.3 **Angles in a triangle**

The three angles in any triangle add up to 180°.

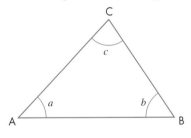

$a + b + c = 180°$

We can prove this by drawing a line through C parallel to AB.

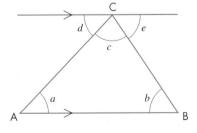

$a = d$	(alternate angles)
$b = e$	(alternate angles)
$d + e + c = 180°$	(angles on a straight line)

So $a + b + c = 180°$

Special triangles

Equilateral triangle

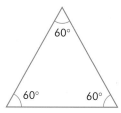

An **equilateral triangle** is a triangle with all its sides equal. Therefore, all three angles are 60°.

Isosceles triangle

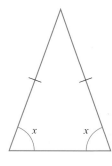

An **isosceles triangle** is a triangle with two equal sides and, therefore, with two equal interior angles (at the foot of the equal sides).

Notice how to mark the equal sides and equal angles.

Right-angled triangle

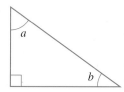

A **right-angled triangle** has an interior angle of 90°.
$a + b = 90°$

EXERCISE 23C

1 Find the size of the angle marked with a letter in each of these triangles.

a

b

c

d

e

f

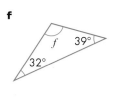

g

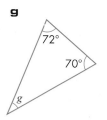

h

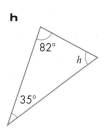

2 Do any of these sets of angles form the three angles of a triangle? Explain your answer.

    a 35°, 75°, 80°

    b 50°, 60°, 70°

    c 55°, 55°, 60°

    d 60°, 60°, 60°

    e 35°, 35°, 110°

    f 102°, 38°, 30°

FOUNDATION

3 In the triangle on the right, all the interior angles are the same.

    a What is the size of each angle?

    b What is the name of a special triangle like this?

    c What is special about the sides of this triangle?

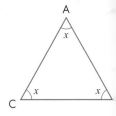

4 In the triangle on the right, two of the angles are the same.

    a Work out the size of the lettered angles.

    b What is the name of a special triangle like this?

    c What is special about the sides AC and AB of this triangle?

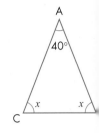

5 Find the size of the angle marked with a letter in each of these diagrams.

a 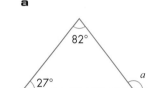   **b** 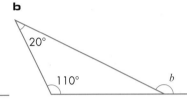   **c**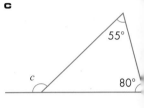

6 What is the special name for triangle DEF?

Show all your working to explain your answer.

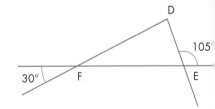

7 The diagram shows three intersecting straight lines.

Work out the values of a, b and c.

Give reasons for your answers.

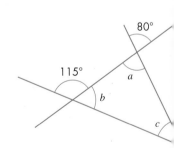

8 By using algebra, show that $x = a + b$.

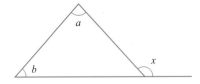

FOUNDATION

Angles in a quadrilateral

The four angles in any **quadrilateral** add up to 360°.

$$a + b + c + d = 360°$$

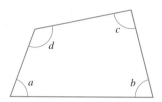

We can see this by dividing the quadrilateral into two triangles.

The six angles of the triangles have the same sum as the four angles of the quadrilateral.

Sum of angles of a quadrilateral = 180° + 180° = 360°

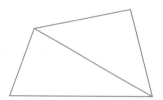

EXAMPLE 1

Three angles of a quadrilateral are 125°, 130° and 60°.

Find the fourth angle.

$$125 + 130 + 60 + x = 360$$
$$315 + x = 360$$
$$x = 360 - 315$$
$$x = 45$$

So the fourth angle is 45°.

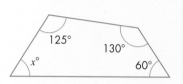

Special quadrilaterals

A **parallelogram** has opposite sides that are parallel.

Its opposite sides are equal. Its diagonals bisect each other. Its opposite angles are equal: that is, $\angle A = \angle C$ and $\angle B = \angle D$

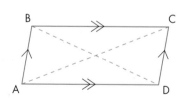

A **rhombus** is a parallelogram with all its sides equal.

Its diagonals bisect each other at right angles. Its diagonals also bisect the angles at the vertices.

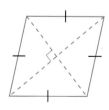

A **kite** is a quadrilateral with two pairs of equal adjacent sides.

Its longer diagonal bisects its shorter diagonal at right angles. The opposite angles between the sides of different lengths are equal.

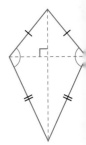

A **trapezium** has two parallel sides.

The sum of the interior angles at the ends of each non-parallel side is 180°: that is, ∠A + ∠D = 180° and ∠B + ∠C = 180°

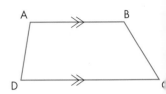

EXAMPLE 2

Find the size of the angles marked x and y in this parallelogram.

$x = 55°$ (opposite angles are equal) and $y = 125°$ ($x + y = 180°$)

EXERCISE 23D

1 Find the sizes of the lettered angles in these quadrilaterals.

a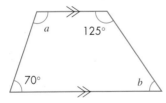

b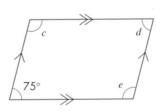

c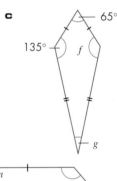

d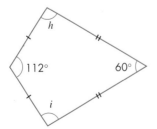

e

f

FOUNDATION

2 Calculate the values of *x* and *y* in each of these quadrilaterals.

a
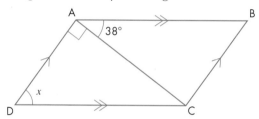

b
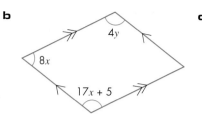

c
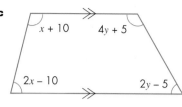

3 Find the value of *x* in each of these quadrilaterals and state what type of quadrilateral it could be.

a A quadrilateral with angles *x* + 10, *x* + 20, 2*x* + 20, 2*x* + 10

b A quadrilateral with angles *x* – 10, 2*x* + 10, *x* – 10, 2*x* + 10

c A quadrilateral with angles *x* – 10, 2*x*, 5*x* – 10, 5*x* – 10

d A quadrilateral with angles 4*x* + 10, 5*x* – 10, 3*x* + 30, 2*x* + 50

4 The diagram shows a parallelogram ABCD.

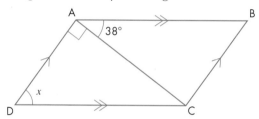

Work out the size of angle *x*, marked on the diagram.

5 Dani is making a kite and wants angle C to be half of angle A.

Work out the size of angles B and D.

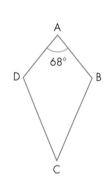

6 This quadrilateral is made from two isosceles triangles. They are both the same size.

Find the value of *y* in terms of *x*.

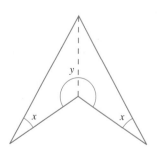

Regular polygons

Below are five **regular polygons**.

| **Square**
4 sides | **Pentagon**
5 sides | **Hexagon**
6 sides | **Octagon**
8 sides | **Decagon**
10 sides |

A **polygon** is regular if all its interior angles are equal and all its sides have the same length.

A **square** is a regular four-sided shape that has an angle sum of 360°.

So, each angle is 360° ÷ 4 = 90°.

The angles of a regular polygon

Lines from the centre of a regular pentagon divide it into five isosceles triangles.

The angle at the centre is 360° so the angle of each isosceles triangle at the centre is:

360° ÷ 5 = 72°

The other angles in each triangle are identical so each one is:

(180 − 72) ÷ 2 = 54°

So each angle of a regular pentagon is:

2 × 54° = 108°

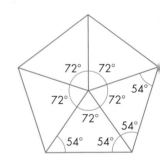

There is also an **exterior angle** at each vertex. It is:

180 − 108 = 72°

Notice that 72° = 360° ÷ 5.

General result

If a regular polygon has n sides each exterior angle is $\frac{360°}{n}$.

If you want to find the interior angle of a regular polygon it may be easier to find the exterior angle like this first. Then subtract it from 180 to find the interior angle.

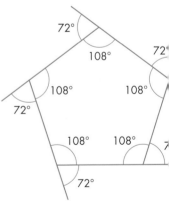

Regular polygon	Number of sides	Exterior angle	Interior angle
Equilateral triangle	3	120°	60°
Square	4	90°	90°
Pentagon	5	72°	108°
Hexagon	6	60°	120°
Octagon	8	45°	135°

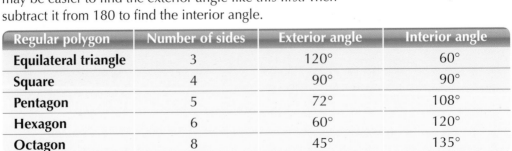

EXAMPLE 3

Calculate the size of the exterior and interior angle for a regular 12-sided polygon (a regular dodecagon).

$$\text{exterior angle} = \frac{360°}{12} = 30° \quad \text{and} \quad \text{interior angle} = 180° - 30° = 150°$$

EXERCISE 23E

1 Each diagram shows an interior angle of a regular polygon. For each polygon, answer the following.

  i   What is its exterior angle?

  ii How many sides does it have?

  iii What is the sum of its interior angles?

a

135°

b

160°

c

165°

d

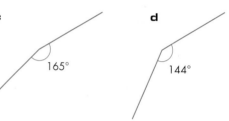

144°

2 Each diagram shows an exterior angle of a regular polygon. For each polygon, answer the following.

  i   What is its interior angle?

  ii How many sides does it have?

  iii What is the sum of its interior angles?

a

8°

b

6°

c

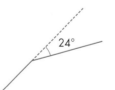

24°

d

3°

3 Each of these cannot be the interior angle of a regular polygon. Explain why.

a

173°

b

161°

c

169°

d

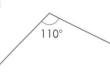

110°

FOUNDATION

4 Each of these cannot be the exterior angle of a regular polygon. Explain why.

a  7°

b  26°

c 44°

d 13°

5 Draw a sketch of a regular octagon and join each vertex to the centre.

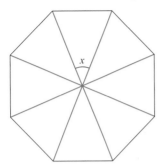

Calculate the value of the angle at the centre (marked *x*).

What connection does this have with the exterior angle?

Is this true for all regular polygons?

6 The diagram shows part of a regular polygon.

Each interior angle is 144°.

a What is the size of each exterior angle of the polygon?

b How many sides does the polygon have?

144°

23.6 Irregular polygons

A polygon is regular if all its sides are the same length and all its angles are the same size. If th is not the case it is **irregular**.

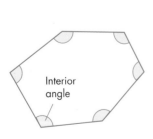

Interior angle

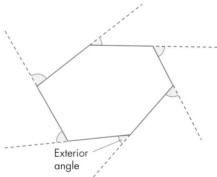

Exterior angle

The *exterior* angles of *any* polygon add up to 360°.

Interior angles

You can find the sum of the interior angles of any polygon by splitting it into triangles.

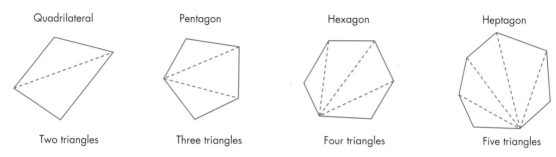

| Quadrilateral | Pentagon | Hexagon | Heptagon |
| Two triangles | Three triangles | Four triangles | Five triangles |

Since you already know that the angles in a triangle add up to 180°, you find the sum of the interior angles in a polygon by multiplying the number of triangles in the polygon by 180°, as shown in this table.

Shape	Name	Sum of interior angles
4-sided	**Quadrilateral**	2 × 180° = 360°
5-sided	**Pentagon**	3 × 180° = 540°
6-sided	**Hexagon**	4 × 180° = 720°
7-sided	**Heptagon**	5 × 180° = 900°
8-sided	**Octagon**	6 × 180° = 1080°
9-sided	**Nonagon**	7 × 180° = 1260°
10-sided	**Decagon**	8 × 180° = 1440°

As you can see from the table, for an n-sided polygon, the sum of the interior angles, S, is given by the formula:

$$S = 180(n - 2)°$$

Exterior angles

As in regular polygons the sum of all the exterior angles in an irregular polygon is 360°, but their sizes may not be the same.

The size of any specific exterior angle = 180° – the size of its adjacent interior angle.

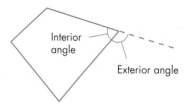

Interior angle

Exterior angle

EXAMPLE 4

Four angles of a pentagon are 100°.

How big is the fifth angle?

The interior angle sum of a pentagon is 3 × 180° = 540°.

Four angles add up to 400°.

The fifth angle must be 540 – 400 = 140°.

EXERCISE 23F

1 Calculate the sum of the interior angles of polygons with these numbers of sides.

a   10 sides
b   15 sides
c   100 sides
d   45 sides

2 Find the number of sides of polygons with these interior angle sums.

a   1260°
b   2340°
c   18 000°
d   8640°

3 Calculate the size of the lettered angles in each of these polygons.

a

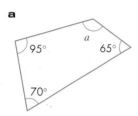

b

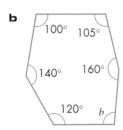

c

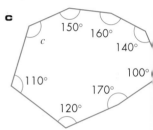

4 Find the value of x in each of these polygons.

a

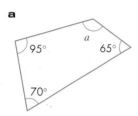

b

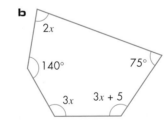

c

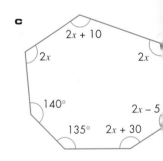

5 What is the name of the regular polygon in which the interior angles are twice its exterior angles?

6 Wesley measured all the interior angles in a polygon. He added them up to make 991°, but he had missed out one angle.

a   What type of polygon did Wesley measure?

b   What is the size of the missing angle?

7 a In the triangle ABC, angle A is 42° and angle B is 67°.

  i Calculate the value of angle C.

  ii What is the value of the exterior angle at C?

  iii What connects the exterior angle at C with the sum of the angles at A and B?

b Prove that any exterior angle of a triangle is equal to the sum of the two opposite interior angles.

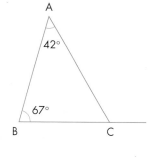

8 Two regular pentagons are placed together.

Work out the value of *a*.

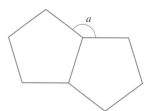

9 A joiner is making tables so that the shape of each one is half a regular octagon, as shown in the diagram.

He needs to know the size of each angle on the top.

What are the sizes of the angles?

23.7 Tangents and chords

O is the centre of the circle.

AXB is a **tangent**.

It is a straight line that touches the circle at X.

The **radius** OX is perpendicular to the tangent.

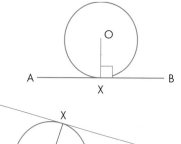

You can draw two tangents from a point outside the circle.

AX and AY are the same length.

Angles OAX and OAY are the same size.

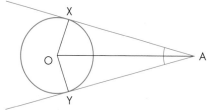

The part of the **circumference** between two points, A and B, is called an **arc**.

The arc AB and the radii OA and OB form a **sector**.

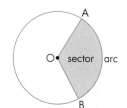

The straight line joining two points, A and B, is called a **chord**.

The part of the circle between the chord AB and the arc AB is called a **segment**.

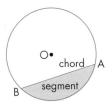

EXAMPLE 5

O is the centre of a circle. AP and BP are tangents.

Calculate the angle at P.

The angles at A and B are right angles.

OAPB is a quadrilateral so the interior angles add up to 360°.

Angle P is 360 – (90 + 90 + 154) = 26°.

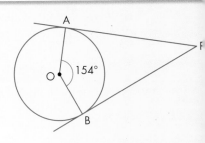

EXERCISE 23G

1 In each diagram, TP and TQ are tangents to a circle with centre O. Find each value of x.

a

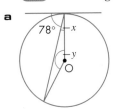

b

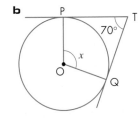

c

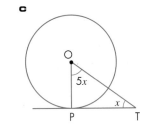

d

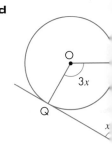

2 Each diagram shows a tangent to a circle with centre O. Find x and y in each case.

a

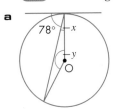

b

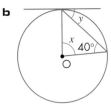

c

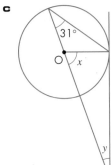

d

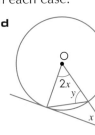

3 In the diagram, O is the centre of the circle and AB is a tangent to the circle at C.

   a Explain why triangle BCD is isosceles.

      Give reasons to justify your answer.

   b Identify a chord, a segment and a sector in the diagram.

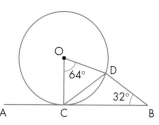

4 In each of the diagrams, TP and TQ are tangents to the circle with centre O. Find each value of x.

> **HINTS AND TIPS**
>
> Look for isosceles triangles

a

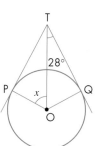

b

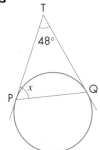

c

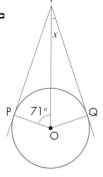

d

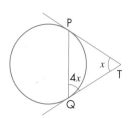

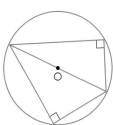

3.8   # Angles in a circle   **H**

Here are two more theorems you need to know about angles in circles.

If you draw lines from each end of an arc to the centre of a circle they form an angle at the centre. We say that the arc has **subtended** an angle at the centre.

The angle at the centre of a circle is twice the angle at the circumference that is subtended by the same arc.

This diagram shows the angles subtended by arc AB.

$$\angle AOB = 2 \times \angle ACB$$

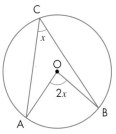

Every angle at the circumference of a **semi-circle** that is subtended by the **diameter** of the semicircle is a right angle.

Angles subtended at the circumference in the same segment of a circle are equal.

Points C_1, C_2, C_3 and C_4 on the circumference are subtended by the same arc AB.

So $\angle AC_1B = \angle AC_2B = \angle AC_3B = \angle AC_4B$

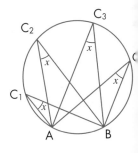

EXAMPLE 6

O is the centre of each circle. Find the angles marked a and b in each circle.

i

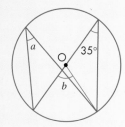

ii

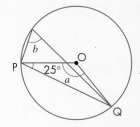

i $a = 35°$ (angles in same segment)

$b = 2 \times 35°$ (angle at centre = twice angle at circumference)

$= 70°$

ii With OP = OQ, triangle OPQ is isosceles and the sum of the angles in this triangle $= 180°$

So $a + (2 \times 25°) = 180°$

$a = 180° - (2 \times 25°)$

$= 130°$

$b = 130° \div 2$ (angle at centre = twice angle at circumference)

$= 65°$

EXAMPLE 7

O is the centre of the circle. PQR is a straight line.

Find the angle labelled a.

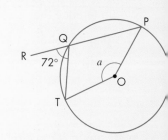

$\angle PQT = 180° - 72° = 108°$ (angles on straight line)

The reflex angle $\angle POT = 2 \times 108°$
(angle at centre = twice angle at circumference)

$= 216°$

$a + 216° = 360°$ (sum of angles around a point)

$a = 360° - 216°$

$a = 144°$

EXERCISE 23H

1 Find the angle marked x in each of these circles with centre O.

a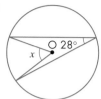

b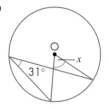

c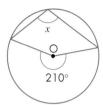

d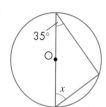

e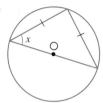

f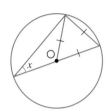

g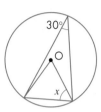

h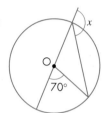

2 Find the angle marked x in each of these circles with centre O.

a

b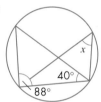

c

d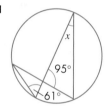

e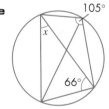

f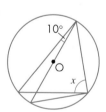

g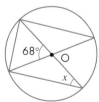

h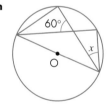

3 In the diagram, O is the centre of the circle. Find these angles.

a ∠ADB

b ∠DBA

c ∠CAD

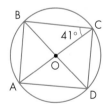

4 In the diagram, O is the centre of the circle. Find these angles.

a ∠EDF

b ∠DEG

c ∠EGF

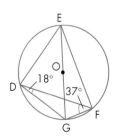

5 In the diagram XY is a diameter of the circle and ∠AZX is *a*.

Ben says that the value of *a* is 50°.

Give reasons to explain why he is wrong.

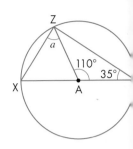

6 Find the angles marked *x* and *y* in each of these circles. O is the centre where shown.

a

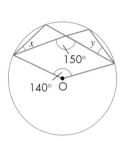

b

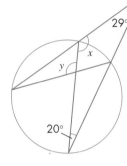

c

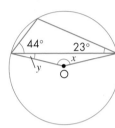

d

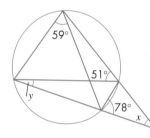

e

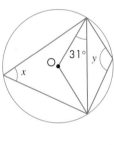

f

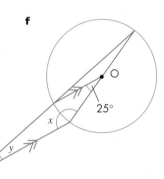

7 In the diagram, O is the centre and AD is the diameter of the circle.

Find *x*.

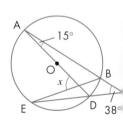

8 In the diagram, O is the centre of the circle and ∠CBD is *x*.

Show that the reflex ∠AOC is 2*x*, giving reasons to explain your answer.

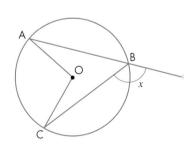

9 A, B, C and D are points on the circumference of a circle with centre O.
Angle ABO is $x°$ and angle CBO is $y°$.

a State the value of angle BAO.

b State the value of angle AOD.

c Prove that the angle subtended by the chord AC at the centre of a circle is twice the angle subtended at the circumference.

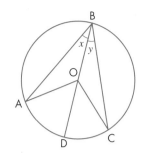

23.9 Cyclic quadrilaterals H

There are *two* segments between points P and Q.

a is the angle in one segment.

b is the angle in the **opposite segment**.

Angles in opposite segments add up to 180°.
We say they are **supplementary**.

Proof: $a + b = \frac{1}{2}c + \frac{1}{2}d = \frac{1}{2}(c + d) = \frac{1}{2} \times 360 = 180°$

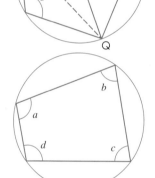

A quadrilateral whose four vertices lie on the circumference of a circle is called a **cyclic quadrilateral**.

The sum of the opposite angles of a cyclic quadrilateral is 180°.

$a + c = 180°$ and $b + d = 180°$

EXERCISE 23I

1 Find the sizes of the lettered angles in each of these circles.

a

b

c

d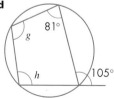

e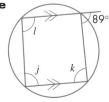

f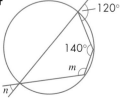

g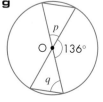

h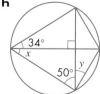

2 Find the values of *x* and *y* in each of these circles. Where shown, O marks the centre of the circle.

a

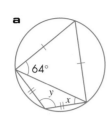

b

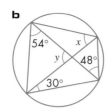

c

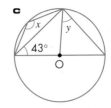

d

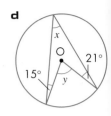

e

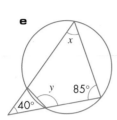

f

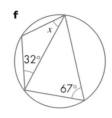

g

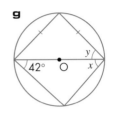

h

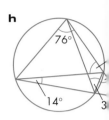

3 Find the values of *x* and *y* in each of these circles. Where shown, O marks the centre of the circle.

a

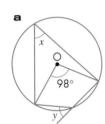

b

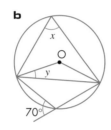

c

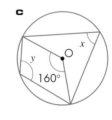

d

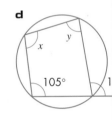

4 Find the values of *x* and *y* in each of these circles.

a

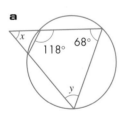

b

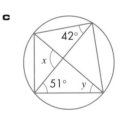

c

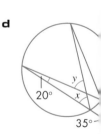

d

5 Find the values of the angles with letters in each of these circles with centre O.

a

b

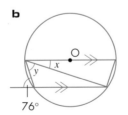

c

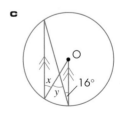

d

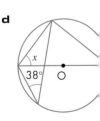

6 The cyclic quadrilateral PQRT has ∠ROQ equal to 38° where O is the centre of the circle. POT is a diameter and parallel to QR. Calculate these angles.

a ROT

b QRT

c QPT

7 In the diagram, O is the centre of the circle.

a Explain why $3x – 30° = 180°$.

b Work out the size of ∠CDO, marked y on the diagram.

Give reasons in your working.

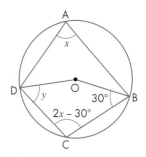

8 ABCD is a cyclic quadrilateral within a circle centre O and ∠AOC is $2x°$.

a Write down the value of ∠ABC.

b Write down the value of the reflex angle AOC.

c Prove that the sum of a pair of opposite angles of a cyclic quadrilateral is 180°.

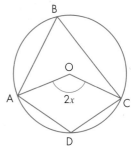

9 In the diagram, ABCE is a parallelogram.

Prove ∠AED = ∠ADE.

Give reasons in your working.

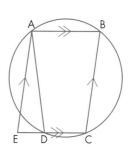

10 Two circles touch at D.

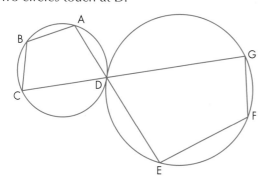

ADE and CDG are straight lines.

Explain why angles ABC and EFG must be equal in size.

PTQ is the tangent to a circle at T. The segment containing ∠TBA is known as the **alternate segment** of ∠PTA, because it is on the other side of the chord AT from ∠PTA.

Alternate segment theorem

The angle between a tangent and a chord through the point of contact is equal to the angle in the alternate segment.

∠PTA = ∠TBA

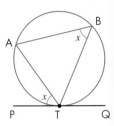

EXAMPLE 8

In the diagram, find **a** ∠ATS and **b** ∠TSR.

a ∠ATS = 80° (angle in alternate segment)

b ∠TSR = 70° (angle in alternate segment)

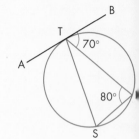

EXERCISE 23J

1 Find the size of each lettered angle.

a

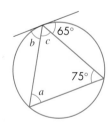

b

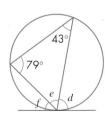

c

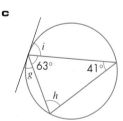

d

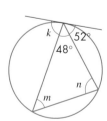

2 In each diagram, find the size of each lettered angle.

a

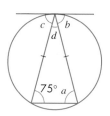

b

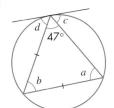

c

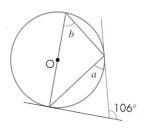

d

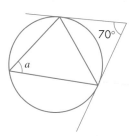

3 In each diagram, find the value of *x*.

a

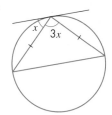

b

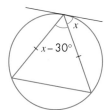

4 ATB is a tangent to each circle with centre O. Find the size of each lettered angle.

a

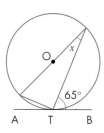

b

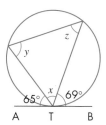

c

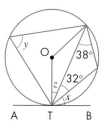

d

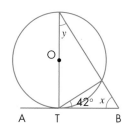

5 In the diagram, O is the centre of the circle.

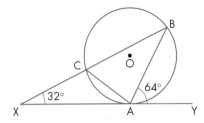

XY is a tangent to the circle at A.

BCX is a straight line.

Show that triangle ACX is isosceles.

Give reasons to justify your answer.

6 AB and AC are tangents to the circle at X and Y.

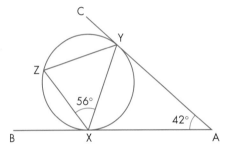

Work out the size of ∠XYZ.

Give reasons to justify your answer.

7 PT is a tangent to a circle with centre O.
AB are points on the circumference. Angle PBA is $x°$.

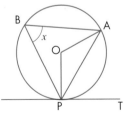

a Write down the value of angle AOP in terms of x.

b Calculate the angle OPA in terms of x.

c Prove that the angle APT is equal to the angle PBA.

Intersecting chords

The chords AB and CD intersect at P.

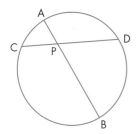

The lengths of the four lines are connected by this formula:

$$AP \times PB = CP \times PD$$

This result is true for any two chords.

EXAMPLE 9

WX and YZ are two chords of the circle.

Calculate the value of a.

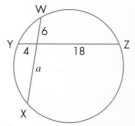

$$6 \times a = 4 \times 18$$
$$6a = 72$$
$$a = 12$$

The result is still true if the chords, when extended, meet outside the circle.

$$AP \times PB = CP \times PD$$

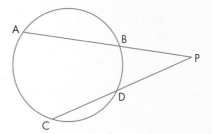

A similar result is true if one of the lines is a tangent.

AB is a chord and CP is a tangent.

In this case:

$$AP \times PB = CP^2$$

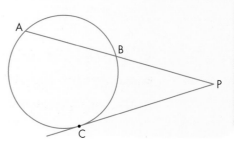

EXERCISE 23K

1 Calculate x in each of these circles.

a

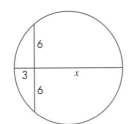

b

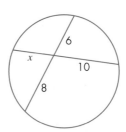

c

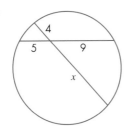

d

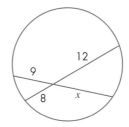

2 CP = 27 cm, DP = 10 cm and BP = 12 cm.

a Find AP.

b Find AB.

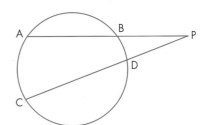

3 AB = BP = 12 cm.

CP = 20 cm.

Find DP.

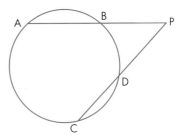

4 AB = 20 cm, BP = 50 cm, PD = 40 cm.

Find CD.

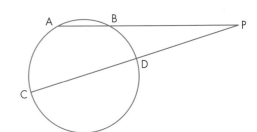

5 Find x.

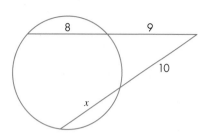

6 PT is a tangent to the circle.
PT = 18 cm, BP = 16 cm.
Calculate AP.

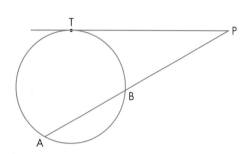

7 Calculate x.

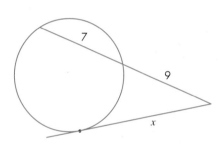

8 Calculate x.

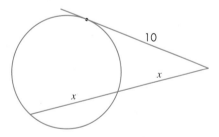

Why this chapter matters

Thales of Miletus (624–547 BCE) was a Greek philosopher. We believe he was the first person to use similar triangles to find the height of tall objects.

Thales discovered that, at a particular time of day, the height of an object and the length of its shadow were the same. He used this to calculate the height of the Egyptian pyramids.

You can apply the geometry of triangles to calculate heights using a clinometer. This means you can find the heights of trees, buildings and towers, mountains and other objects which are difficult to measure physically.

Astronomers use the geometry of triangles to measure the distance to nearby stars. They use the Earth's journey in its orbit around the Sun.

They measure the angle of the star twice, from the same point on Earth, but at opposite ends of its orbit. This gives them angle measurements at a known distance apart and from this triangle they can calculate the distance to the star.

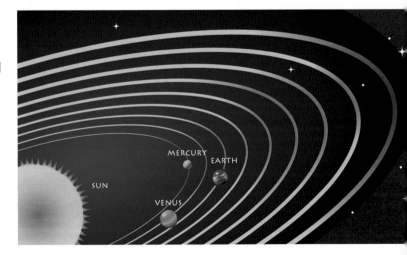

Telescopes and binoculars also use the geometry of triangles.

24 Geometrical terms and relationships

Topics	Level	Key words
1 Measuring and drawing angles	FOUNDATION	right angle, acute angle, obtuse angle, reflex angle, perpendicular, protractor
2 Bearings	FOUNDATION	bearing, three-figure bearing
3 Congruent shapes	FOUNDATION	congruent
4 Similar shapes	FOUNDATION	similar, enlargement, linear scale factor, corresponding angles, corresponding sides
5 Areas of similar triangles	HIGHER	area scale factor
6 Areas and volumes of similar shapes	HIGHER	solid shapes, volume scale factor

What you need to be able to do in the examinations:

FOUNDATION	HIGHER
• Understand congruence as meaning the same shape and size. • Understand that two or more polygons with the same shape and size are said to be congruent to each other. • Understand and use the geometrical properties that similar figures have corresponding lengths in the same ratio but corresponding angles remain unchanged. • Understand angle measure including three-figure bearings. • Measure an angle to the nearest degree.	• Understand that areas of similar figures are in the ratio of the square of corresponding sides. • Understand that volumes of similar figures are in the ratio of the cube of corresponding sides. • Use areas and volumes of similar figures in solving problems.

A whole turn is divided into 360° or four **right angles** of 90° each.

An **acute angle** is less than one right angle.

An **obtuse angle** is between one and two right angles (90° and 180°).

A **reflex angle** is between two and four right angles (180° and 360°).

Two lines are **perpendicular** if the angle between them is 90°.

When you are using a **protractor**, it is important that you:

- place the centre of the protractor *exactly* on the corner (vertex) of the angle
- lay the baseline of the protractor *exactly* along one side of the angle.

You must follow these two steps to obtain an accurate value for the angle you are measuring.

You should already have discovered how easy it is to measure acute angles and obtuse angles, using the common semicircular protractor.

EXAMPLE 1

Measure the angles ABC, DEF and GHI in the diagrams below.

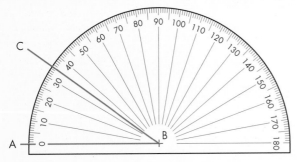
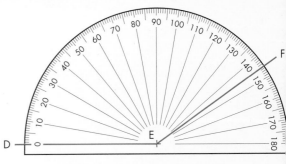

Acute angle ABC is 35° and obtuse angle DEF is 145°.

To measure reflex angles, such as angle GHI, it is easier to use a circular protractor if you have one.

Note the notation for angles.

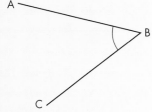

Angle ABC, or ∠ABC, means the angle at B between the lines AB and BC.

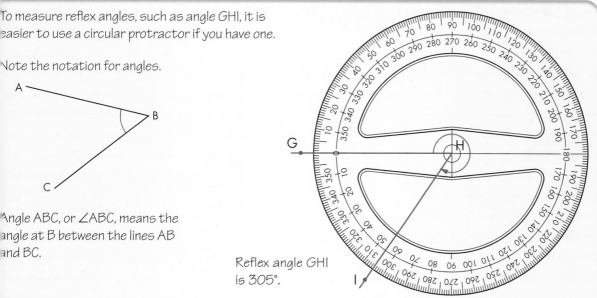

Reflex angle GHI is 305°.

EXERCISE 24A

1 Use a protractor to measure the size of each marked angle.

a

b

c

d

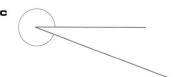

> **HINTS AND TIPS**
>
> Check that your answer is sensible. If an angle is reflex, the answer must be over 180°.

> **HINTS AND TIPS**
>
> A good way to measure a reflex angle is to measure the acute or obtuse angle and subtract it from 360°.

2 Use a protractor to draw angles of the following sizes.

    a 30°   **b** 125°   **c** 90°   **d** 212°   **e** 324°   **f** 19°   **g** 171°

3 Find three pairs of perpendicular lines from the following: AC, AD, AE, BE, CE, CF.

4 It is only safe to climb this ladder if the angle between the ground and the ladder is between 72° and 78°.

Is it safe for Oliver to climb the ladder?

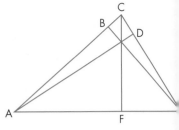

5 An obtuse angle is 10° more than an acute angle.
Write down a possible value for the size of the obtuse angle.

6 Use a ruler and a protractor to draw these triangles accurately. Then measure the unmarked angle in each one.

a

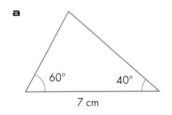

60°   40°
7 cm

b

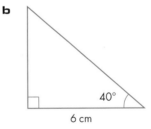

40°
6 cm

c

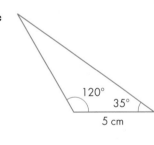

120°
35°
5 cm

24.2 Bearings

The **bearing** of a point B from a point A is the angle through which you turn *clockwise* as you change direction from *due north* to the direction of B.

For example, in this diagram the bearing of B from A is 060°.

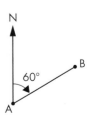

A bearing can have any value from 0° to 360°. It is usual to give all bearings as three figures. This is known as a **three-figure bearing**. So, in the example on the previous page, the bearing is written as 060°, using three figures. Here are three more examples.

D is on a bearing of 048° from C

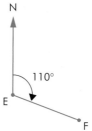

F is on a bearing of 110° from E

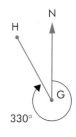

H is on a bearing of 330° from G

There are eight bearings which you should know. They are shown in the diagram.

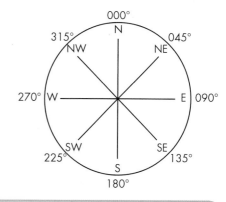

EXAMPLE 2

A, B and C are three towns.

Write down the bearing of B from A and the bearing of C from A.

The bearing of B from A is 070°.

The bearing of C from A is 360° − 115° = 245°.

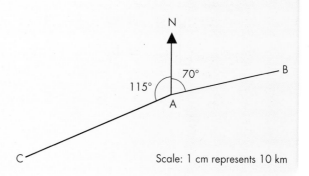

Scale: 1 cm represents 10 km

EXERCISE 24B

1. Look at this map. By measuring angles, find the following bearings.

   a T from D

   b D from E

   c M from D

   d G from A

   e M from G

   f T from M

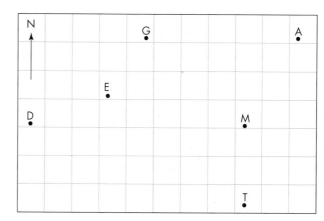

FOUNDATION

399

2 Draw sketches to illustrate the following situations.

   a C is on a bearing of 170° from H.   **b** B is on a bearing of 310° from W.

3 A is due north from C. B is due east from A. B is on a bearing of 045° from C. Sketch the layout of the three points, A, B and C.

4 The Captain decided to sail his ship around the four sides of a square kilometre.

   a Assuming he started sailing due north, write down the further three bearings he would use in order to complete the square in a clockwise direction.

   b Assuming he started sailing on a bearing of 090°, write down the further three bearings he would use in order to complete the square in an anticlockwise direction.

5 The map shows a boat journey around an island, starting and finishing at S. On the map 1 centimetre represents 10 kilometres. Measure the distance and bearing of each leg of the journey.

Copy and complete the table shown right.

Leg	Actual distance	Bearing
1		
2		
3		
4		
5		

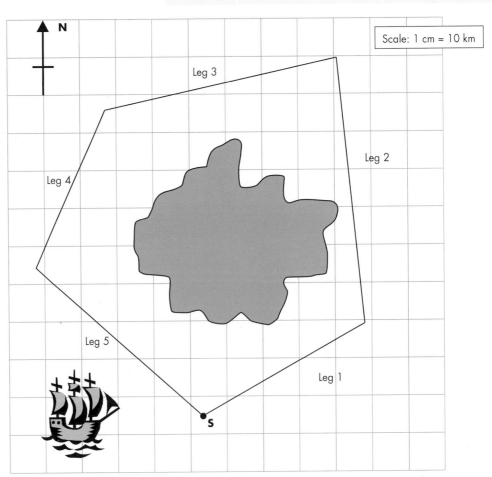

Scale: 1 cm = 10 km

6 The diagram shows a port P
and two harbours X and Y on
the coast.

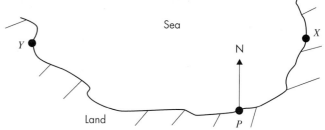

a A fishing boat sails to X from P.

What is the three-figure
bearing of X from P?

b A yacht sails to Y from P.

What is the three-figure bearing of Y from P?

7 Draw diagrams to solve the following problems.

a The three-figure bearing of A from B is 070°. Work out the three-figure bearing of B from A.

b The three-figure bearing of P from Q is 145°. Work out the three-figure bearing of Q from P.

c The three-figure bearing of X from Y is 324°. Work out the three-figure bearing of Y from X.

8 The diagram shows the position of Kim's house H and the college C.

Scale: 1 cm represents 200 m

a Use the diagram to work out the actual distance from Kim's house to the college.

b Measure and write down the three-figure bearing of the college from Kim's house.

c The supermarket S is 600 m from Kim's house on a bearing of 150°.

Mark the position of S on a copy of the diagram.

9 Chen is flying a plane on a bearing of 072°.

He is told to fly due south towards an airport.

Through what angle does he need to turn?

10 A, B and C are three villages in a bay.

They lie on the vertices of a square.

The bearing of B from A is 030°.

Work out the bearing of A from C.

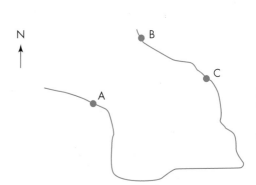

Two-dimensional shapes that are exactly the same size and shape are said to be **congruent**. For example, although they are in different positions, the triangles below are congruent, becau they are all exactly the same size and shape.

Congruent shapes fit exactly on top of each other. So, one way to see whether shapes are congruent is to trace one of them and check that it covers the other shapes exactly. For some c the shapes, you may have to turn your tracing paper over.

EXAMPLE 3

Which of these shapes is not congruent to the others?

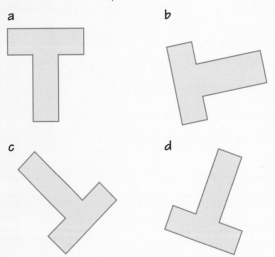

Trace shape **a** and check whether it fits exactly on top of the others.

You should find that shape **b** is not congruent to the others.

1 State whether the shapes in each pair, **a** to **f**, are congruent or not.

a

b

c

d

e

f

2 Which figure in each group, **a** to **c**, is not congruent to the other two?

a i   **ii**  **iii**

b i  **ii**   **iii**

c i  **ii**  **iii**

3 Draw a square PQRS. Draw in the diagonals PR and QS. Which triangles are congruent to each other?

4 Draw a rectangle EFGH. Draw in the diagonals EG and FH. Which triangles are congruent to each other?

5 Draw a parallelogram ABCD. Draw in the diagonals AC and BD. Which triangles are congruent to each other?

6 Draw an isosceles triangle ABC where AB = AC. Draw the line from A to the midpoint of BC. Which triangles are congruent to each other?

Two shapes are **similar** if one is an **enlargement** of the other.

Corresponding angles of similar shapes are equal.

These shapes are similar. Their corresponding angles are equal.

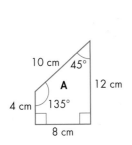
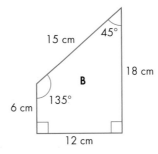

The ratio of their **corresponding sides** is 2 : 3.

The **linear scale factor** is $\frac{3}{2}$ because the lengths in shape B are $\frac{3}{2}$ of the lengths in shape A.

EXAMPLE 4

These two shapes are similar.

Find x and y.

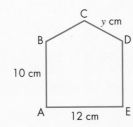
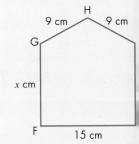

Look for two corresponding sides where we know the lengths.

AE corresponds to FJ and the lengths are 12 cm and 15 cm.

The scale factor is $\frac{15}{12}$ = 1.25.

To find x (GF) from the length given for BA:

$x = 1.25 \times AB$

$\quad = 1.25 \times 10$

$\quad = 12.5$

$x = 12.5$ cm

To find y (CD) from the length of HI:

$HI = 1.25 \times y$

$9 = 1.25 \times y$

$y = \frac{9}{1.25}$

$\quad = 7.2$

$y = 7.2$ cm

EXERCISE 24D

1 These diagrams are drawn to scale. What is the linear scale factor of the enlargement in each case?

a

b

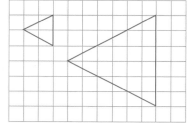

2 Are these pairs of shapes similar? If so, give the scale factor. If not, give a reason.

a

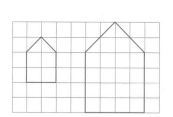

b

3 **a** Explain why these triangles are similar.

b Give the ratio of the sides.

c Which angle corresponds to angle C?

d Which side corresponds to side QP?

4 **a** Explain why these triangles are similar.

b Which angle corresponds to angle A?

c Which side corresponds to side AC?

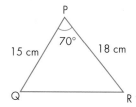

5 **a** Explain why triangle ABC is similar to triangle AQR.

b Which angle corresponds to the angle at B?

c Which side of triangle AQR corresponds to side AC of triangle ABC? Your answers to question **4** may help you.

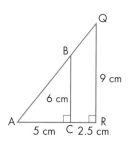

6 In the diagrams **a** to **d**, each pair of shapes is similar but not drawn to scale.

a Find x.

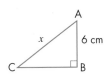 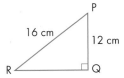

b Find x and y.

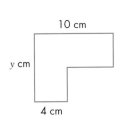 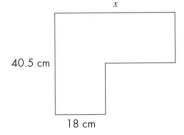

c Find x and y.

 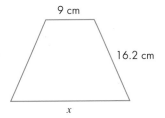

d Find the length of QR.

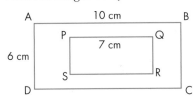

7 a Explain why all squares are similar.

b Are all rectangles similar? Explain your answer.

8 Sean is standing next to a tree.

His height is 1.6 m and he casts a shadow that has a length of 2.4 m.

The tree casts a shadow that has a length of 7.8 m.

Use what you know about similar triangles to work out the height of the tree, h.

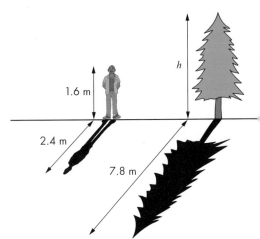

If two triangles have the same angles then they are similar.

Triangles ABC and DEF are similar.

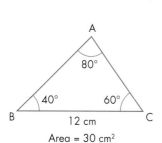

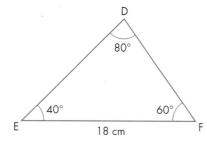

The linear scale factor is $\frac{18}{12} = 1.5$. The **area scale factor** is $1.5^2 = 2.25$.

If the area of triangle ABC is 30 cm^2 then the area of DEF is $30 \times 2.25 = 67.5$ cm^2.

If the linear scale factor is k, then the area scale factor is k^2.

EXERCISE 24E

1 These triangles are similar.

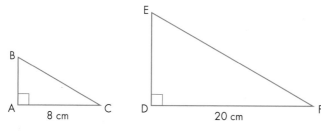

   a What is the linear scale factor?

   b The area of triangle ABC is 20 cm^2.

      Calculate the area of triangle DEF.

2 These are equilateral triangles.

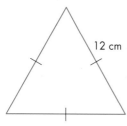

   a Explain why they are similar.

   b The area of the larger one is 62.4 cm^2 (to 3 significant figures).

      Calculate the area of the smaller one.

HIGHER

3 The area of triangle ABC is 7 cm².

Calculate the area of triangle ADE.

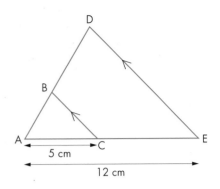

4 The area of triangle ABD is 25 cm².

Calculate the area of the trapezium CBDE.

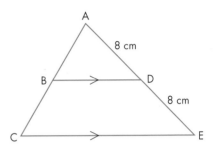

5 P is enlarged by a linear scale factor of 1.2 to make Q.

Q is enlarged by a linear scale factor of 1.2 to make R.

The area of Q is 100 cm².

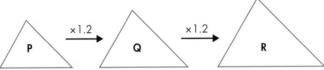

a What is the area of R?

b What is the area of P?

6 The area scale factor also applies to other similar shapes.

A and B are regular pentagons.

a Explain why they are similar.

b Calculate the area of B.

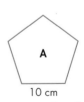
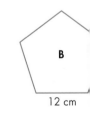

10 cm   12 cm

Area = 172 cm²

7 A, B and C are equilateral triangles.

B is four times the area of A.

a What is the linear scale factor?

b What is *x*?

c C is twice the area of A.
Calculate *y*.

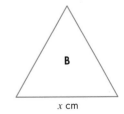
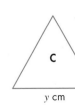

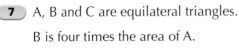

5 cm   *x* cm   *y* cm

8 These shapes are similar.

The smaller one has an area of 210 cm².

Find the area of the larger one.

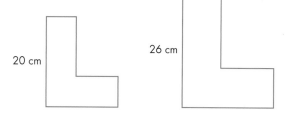

20 cm

26 cm

9 A photocopier has a setting for enlargements with a scale factor of 1.41.

What is special about this particular value?

10 All circles are similar.

If a circle with a diameter of 8 cm has an area of 50.3 cm², what is the area of a circle with diameter 6 cm?

24.6 Areas and volumes of similar shapes

You saw that if two shapes are similar and the linear scale factor is k then the area scale factor is k^2.

Two **solid shapes** are similar if corresponding lengths are in the same ratio and corresponding angles are equal. In that case the **volume scale factor** is k^3.

Generally, the relationship between similar shapes can be expressed as:

Length ratio $x : y$ Area ratio $x^2 : y^2$ Volume ratio $x^3 : y^3$

XAMPLE 5

A model yacht is made to a scale of $\frac{1}{20}$ of the size of the real yacht. The area of the sail of the model is 150 cm². What is the area of the sail of the real yacht?

At first sight, it may appear that you do not have enough information to solve this problem, but it can be done as follows.

Linear scale factor   = 1 : 20

Area scale factor   = 1 : 400 (square of the linear scale factor)

Area of real sail   = 400 × area of model sail

                      = 400 × 150 cm²

                      = 60 000 cm² = 6 m²

EXAMPLE 6

A bottle has a base radius of 4 cm, a height of 15 cm and a capacity of 650 cm^3.
A similar bottle has a base radius of 3 cm.

a What is the length ratio?

b What is the volume ratio?

c What is the volume of the smaller bottle?

a The length ratio is given by the ratio of the two radii, that is 4 : 3.

b The volume ratio is therefore $4^3 : 3^3 = 64 : 27$.

c Let v be the volume of the smaller bottle. Then the volume ratio is:

$$\frac{\text{volume of smaller bottle}}{\text{volume of larger bottle}} = \frac{v}{650} = \frac{27}{64}$$

$$\Rightarrow v = \frac{27 \times 650}{64} = 274 \text{ cm}^3 \text{ (3 significant figures)}$$

EXERCISE 24F

1 The length ratio between two similar solids is 2 : 5.

a What is the area ratio between the solids?

b What is the volume ratio between the solids?

2 The length ratio between two similar solids is 4 : 7.

a What is the area ratio between the solids?

b What is the volume ratio between the solids?

3 Copy and complete this table.

Linear scale factor	Linear ratio	Linear fraction	Area scale factor	Volume scale facto
2	1 : 2	$\frac{2}{1}$		
3				
$\frac{1}{4}$	4 : 1	$\frac{1}{4}$		$\frac{1}{64}$
			25	
				$\frac{1}{1000}$

4 A shape has an area of 15 cm^2. What is the area of a similar shape with lengths that are three times the corresponding lengths of the first shape?

5 A toy brick has a surface area of 14 cm^2. What would be the surface area of a similar to brick with lengths that are:

a twice the corresponding lengths of the first brick

b three times the corresponding lengths of the first brick?

6 A rug has an area of 12 m². What area would be covered by rugs with lengths that are:

    a twice the corresponding lengths of the first rug

    b half the corresponding lengths of the first rug?

7 A brick has a volume of 300 cm³. What would be the volume of a similar brick whose lengths are:

    a twice the corresponding lengths of the first brick

    b three times the corresponding lengths of the first brick?

8 A tin of paint, 6 cm high, holds a half litre of paint. How much paint would go into a similar tin which is 12 cm high?

9 A model statue is 10 cm high and has a volume of 100 cm³. The real statue is 2.4 m high. What is the volume of the real statue? Give your answer in m³.

10 A small tin of paint costs $0.75. What is the cost of a larger similar tin with height twice that of the smaller tin? Assume that the cost is based only on the volume of paint in the tin.

11 A small box of width 2 cm has a volume of 10 cm³. What is the width of a similar box with a volume of 80 cm³?

12 A cinema sells popcorn in two different-sized tubs that are similar in shape.

Show that it is true that the big tub is better value.

Popcorn

10 cm

20 cm

Small tub
$0.60

Large tub
$4

Better value if you buy the Big tub

13 The diameters of two ball bearings are given below. Work out:

    a the ratio of their radii

    b the ratio of their surface areas

    c the ratio of their volumes.

6 mm   8 mm

14 Cuboid A is similar to cuboid B.

The length of cuboid A is 10 cm and the length of cuboid B is 5 cm.

The volume of cuboid A is 720 cm³.

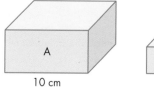

A

10 cm

B

5 cm

Zainab says that the volume of cuboid B must be 360 cm³.

Explain why she is wrong.

More complex problems using area and volume ratios

In some problems involving similar shapes, the length ratio is not given, so we have to start with the area ratio or the volume ratio. We usually then need to find the length ratio in order to proceed with the solution.

EXAMPLE 7

A manufacturer makes a range of clown hats that are all similar in shape. The smallest hat is 8 cm tall and uses 180 cm^2 of card. What will be the height of a hat made from 300 cm^2 of card?

The area ratio is 180 : 300.

Therefore, the length ratio is $\sqrt{180} : \sqrt{300}$ (do not calculate these yet):

Let the height of the larger hat be H, then:

$$\frac{H}{8} = \frac{\sqrt{300}}{\sqrt{180}} = \sqrt{\frac{300}{180}}$$

$$\Rightarrow H = 8 \times \sqrt{\frac{300}{180}} = 10.3 \text{ cm (1 decimal place)}$$

EXAMPLE 8

Two similar tins hold respectively 1.5 litres and 2.5 litres of paint. The area of the label on the smaller tin is 85 cm^2. What is the area of the label on the larger tin?

The volume ratio is 1.5 : 2.5.

Therefore, the length ratio is $\sqrt[3]{1.5} : \sqrt[3]{2.5}$ (do not calculate these yet).

So the area ratio is $(\sqrt[3]{1.5})^2 : (\sqrt[3]{2.5})^2$

Let the area of the label on the larger tin be A, then:

$$\frac{A}{85} = \frac{\sqrt[3]{2.5}^2}{\sqrt[3]{1.5}^2} = \left(\sqrt[3]{\frac{2.5}{1.5}}\right)^2$$

$$\Rightarrow A = 85 \times \left(\sqrt[3]{\frac{2.5}{1.5}}\right)^2 = 119 \text{ cm}^2 \qquad \text{(3 significant figures)}$$

EXERCISE 24G

1. A firm produces three sizes of similar-shaped labels for its products. Their areas are 150 cm^2, 250 cm^2 and 400 cm^2.

   The 250 cm^2 label fits around a can of height 8 cm. Find the heights of similar cans around which the other two labels would fit.

2 A firm makes similar boxes in three different sizes: small, medium and large. The areas of their lids are as follows.

Small: 30 cm^2 Medium: 50 cm^2 Large: 75 cm^2

The medium box is 5.5 cm high. Find the heights of the other two sizes.

3 A cone of height 8 cm can be made from a piece of card with an area of 140 cm^2. What is the height of a similar cone made from a similar piece of card with an area of 200 cm^2?

4 It takes 5.6 litres of paint to paint a chimney which is 3 m high. What is the tallest similar chimney that can be painted with 8 litres of paint?

5 A piece of card, 1200 cm^2 in area, will make a tube 13 cm long. What is the length of a similar tube made from a similar piece of card with an area of 500 cm^2?

6 If a television screen of area 220 cm^2 has a diagonal length of 21 cm, what will be the diagonal length of a similar screen of area 350 cm^2?

7 There are two similar bronze statues. One has a mass of 300 g, the other has a mass of 2 kg. The height of the smaller statue is 9 cm.

What is the height of the larger statue?

8 The sizes of the labels around three similar cans are as follows.

Small can: 24 cm^2 Medium can: 46 cm^2 Large can: 78 cm^2

The medium size can is 6 cm tall with a mass of 380 g. Calculate these quantities:

a The heights of the other two sizes.

b The masses of the other two sizes.

9 A statue has a mass of 840 kg. A similar statue was made out of the same material but two-fifths the height of the first one. What was the mass of the smaller statue?

10 A wooden model stands on a base of area 12 cm^2. A similar wooden model stands on a base of area 7.5 cm^2.

Calculate the mass of the smaller model if the larger one has a mass of 3.5 kg.

11 Steve fills two similar jugs with orange juice.

The first jug holds 1.5 litres of juice and has a base diameter of 8 cm.

The second jug holds 2 litres of juice. Work out the base diameter of the second jug.

12 The total surface areas of two similar cuboids are 500 cm^2 and 800 cm^2.

If the width of one of the cuboids is 10 cm, calculate the two possible widths for the other cuboid.

13 The volumes of two similar cylinders are 256 cm^3 and 864 cm^3.

Which of the following gives the ratio of their surface areas?

    a 2 : 3   **b** 4 : 9   **c** 8 : 27

Why this chapter matters

Engineers, town planners, architects, surveyors, builders and computer designers all need to work with great precision. Some projects involve working with very big lengths or distances; others involve very tiny ones. So how do they manage to work with these difficult measures?

The answer is that they all work with drawings drawn to scale. This allows them to represent lengths th cannot easily measure with standard equipment.

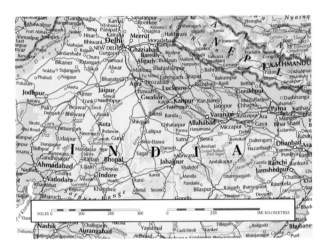

In a scale drawing, one length is used to represe another.

For example, a map cannot be drawn to the sam size as the area it represents. The measurements are scaled down, to make a map of a size that can be conveniently used by drivers, tourists an walkers.

Architects use scale drawings to show views of a planned house from different directions.

In computer design, people who design microchips need to scale up their drawings, as the dimensions they work with are so small. Use the Internet to research more about how scale drawings are used.

25

Geometrical constructions

ics	Level	Key words
Constructing shapes	**FOUNDATION**	construct, ruler, protractor, compasses, set square
Bisectors	**FOUNDATION**	bisect, angle bisector, perpendicular bisector
Scale drawings	**FOUNDATION**	scale drawing

What you need to be able to do in the examinations:

FOUNDATION

- Measure and draw lines to the nearest millimetre.
- Construct triangles and other two-dimensional shapes using a combination of a ruler, a protractor and compasses.
- Solve problems using scale drawings.
- Use straight edge and compasses to:
  - construct the perpendicular bisector of a line segment
  - construct the bisector of an angle.
- Use and interpret scale drawings.

Triangles and other polygons can be **constructed** using a **ruler**, a **protractor**, **compasses** and a **set square**.

When carrying out geometric constructions, always use a sharp pencil to give you thin, clear lines. The examiner will be marking your construction and will be looking for accuracy, which requires fine, clean lines and points as small as you can make them, while ensuring they are clearly visible.

EXAMPLE 1

Construct a triangle with sides that are 5 cm, 4 cm and 6 cm long.

- **Step 1:** Draw the longest side as the base. In this case, the base will be 6 cm, which you draw using a ruler. (The diagrams in this example are drawn at half-size.)

- **Step 2:** Deal with the second longest side, in this case the 5 cm side. Open the compasses to a radius of 5 cm (the length of the side), place the point on one end of the 6 cm line and draw a short faint arc, as shown here.

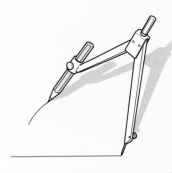

- **Step 3:** Deal with the shortest side, in this case the 4 cm side. Open the compasses to a radius of 4 cm, place the point on the other end of the 6 cm line and draw a second short faint arc to intersect the first arc, as shown here.

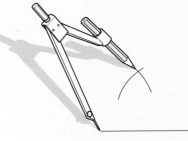

- **Step 4:** Complete the triangle by joining each end of the base line to the point where the two arcs intersect.

Note: The arcs are construction lines and so must be left in to show the examiner how you constructed the triangle.

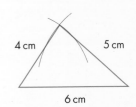

4 cm   5 cm

6 cm

EXAMPLE 2

Make an accurate drawing
of this trapezium.

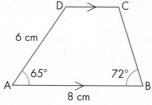

- **Step 1:** Draw the line AB 8 cm long.
  Place the protractor on AB with the
  centre at A and mark 65°.

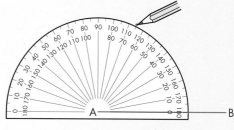

- **Step 2:** Draw a line from A through
  the 65° point. From A, using a pair of
  compasses, measure along this line.
  Label this point D.

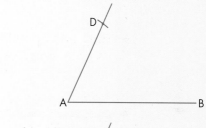

- **Step 3:** To draw a line through D parallel
  to AB, put one edge of a set square on
  AB and a ruler on the adjacent side of
  the set square.

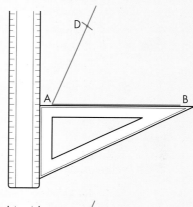

- **Step 4:** Hold the ruler still and slide the
  set square up to D. Draw a line from D
  along the top of the set square.

- **Step 5:** Use a protractor to draw an
  angle of 72° at B and mark C where the
  lines cross.

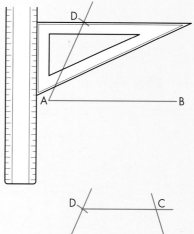

EXERCISE 25A

1 Draw the following triangles accurately and measure the sides and angles not given in the diagram.

a

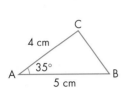

b

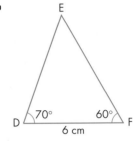

c

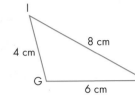

d

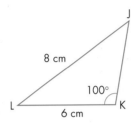

e

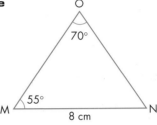

f

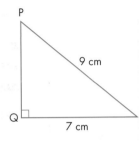

2 a Draw a triangle ABC, where AB = 7 cm, BC = 6 cm and AC = 5 cm.

b Measure the sizes of ∠ABC, ∠BCA and ∠CAB.

3 Draw an isosceles triangle that has two sides of length 7 cm and the included angle of 50°. Measure the length of the base of the triangle.

4 Make an accurate drawing of this quadrilateral.

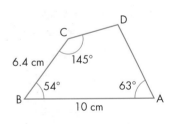

5 Make an accurate drawing of this diagram. Use a set square to draw the parallel lines.

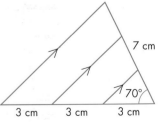

6 Make an accurate drawing of this trapezium.

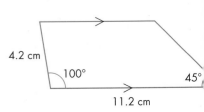

7 A triangle ABC has ∠ABC = 30°, AB = 6 cm and AC = 4 cm. There are two different triangles that can be drawn from this information.

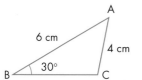

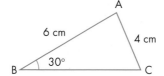

What are the two different lengths that BC can be?

8 Construct an equilateral triangle of side length 5 cm. Measure the height of the triangle.

9 Construct a parallelogram with sides of length 5 cm and 8 cm and with an angle of 120° between them. Measure the height of the parallelogram.

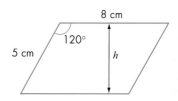

10 A rope has 12 equally-spaced knots. It can be laid out to give a triangle, like this.

It will always be a right-angled triangle.

Here are two more examples of such ropes.

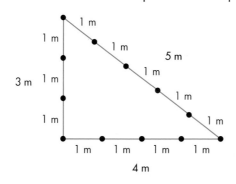

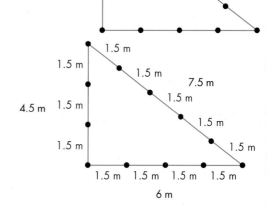

a Show, by constructing each of the above triangles (use a scale of 1 cm : 1 m), that each is a right-angled triangle.

b Choose a different triangle that you think might also be right-angled. Use the same knotted-rope idea to check.

11 Construct the triangle with the largest area which has a total perimeter of 12 cm.

12 Anil says that, as long as he knows all three angles of a triangle, he can draw it. Explain why Anil is wrong.

To **bisect** means to divide in half. So a bisector divides something into two equal parts.

- A **perpendicular bisector** divides a straight line into two equal lengths and is perpendicular to it.
- An **angle bisector** is the straight line that divides an angle into two equal angles.

To construct a perpendicular bisector of a line

It is usually more accurate to construct a perpendicular bisector than to measure its position (the midpoint of the line).

- **Step 1:** Here is a line to bisect.

- **Step 2:** Open your compasses to a radius of about three-quarters of the length of the line. Using each end of the line as a centre, and without changing the radius of your compasses, draw two intersecting arcs.

- **Step 3:** Join the two points at which the arcs intersect. This line is the perpendicular bisector of the original line.

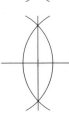

To construct an angle bisector

It is much more accurate to construct an angle bisector than to measure its position.

- **Step 1:** Here is an angle to bisect.

- **Step 2:** Open your compasses to any reasonable radius that is less than the length of the lines forming the angle. If in doubt, go for about 3 cm. With the vertex of the angle as centre, draw an arc through both lines.

- **Step 3:** With centres at the two points at which this arc intersects the lines, draw two more arcs so that they intersect.

- **Step 4:** Join the point at which these two arcs intersect to the vertex of the angle.

This line is the angle bisector.

EXERCISE 25B

1 Draw a line 7 cm long and bisect it. Check your accuracy by seeing if each half is 3.5 cm.

2 Draw a circle of about 4 cm radius.

Draw a triangle inside the circle so that the corners of the triangle touch the circle.

Bisect each side of the triangle.

The bisectors should all meet at the same point, which should be the centre of the circle.

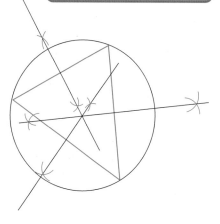

3 **a** Draw any triangle with sides that are between 5 cm and 10 cm.

b On each side construct the perpendicular bisector.

All your perpendicular bisectors should intersect at the same point.

c Using this point as the centre, draw a circle that goes through every vertex of the triangle.

4 Repeat question **3** with a different triangle and check that you get a similar result.

5 **a** Draw the quadrilateral on the right.

b Construct the perpendicular bisector of each side. These all should intersect at the same point.

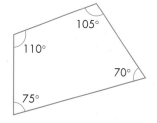

c Use this point as the centre of a circle that goes through the quadrilateral at each vertex. Draw this circle.

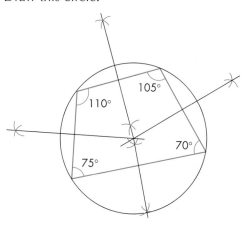

FOUNDATION

6 **a** Draw an angle of 50°.

b Construct the angle bisector.

c Check how accurate you have been by measuring each half. Both should be 25°.

7 Draw a circle with a radius of about 3 cm.

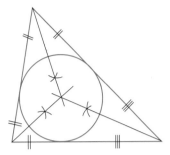

Draw a triangle so that the sides of the triangle are tangents to the circle.

Bisect each angle of the triangle.

The bisectors should all meet at the same point, which should be the centre of the circle.

8 **a** Draw any triangle with sides that are between 5 cm and 10 cm.

b At each angle construct the angle bisector.
All three bisectors should intersect at the same point.

c Use this point as the centre of a circle that just touches the sides of the triangle.

9 Repeat question **8** with a different triangle.

10 Draw a circle with radius about 4 cm.

Draw a quadrilateral, *not* a rectangle, inside the circle so that each vertex is on the circumference.

Construct the bisector of each side of the quadrilateral.

Where is the point where these bisectors all meet?

25.3 Scale drawings

A **scale drawing** is an accurate representation of a real object.

Scale drawings are usually smaller in size than the original objects. However, in certain cases, they have to be enlargements, typical examples of which are drawings of miniature electronic circuits and very small watch movements.

You will be told the scale being used, for example, '1 cm represents 20 m'.

EXAMPLE 3

The diagram shows the front of a kennel.
1 cm on the diagram represents a measurement
of 30 cm. Find:

a the actual width of the front

b the actual height of the doorway.

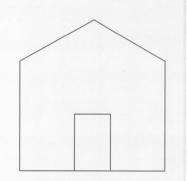

a The actual width of the front is
4 cm × 30 = 120 cm

b The actual height of the doorway is
1.5 cm × 30 = 45 cm

EXERCISE 25C

1 Look at this plan of a garden.

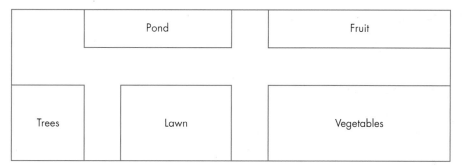

Scale: 1 cm represents 10 m

a State the actual dimensions of each plot of the garden.

b Calculate the actual area of each plot.

2 Below is a plan for a computer mouse mat.

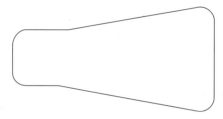

Scale: 1 cm represents 6 cm

HINTS AND TIPS

Remember to check the scale.

a How long is the actual mouse mat?

b How wide is the narrowest part of the mouse mat?

3 Below is a scale plan of the top of Ahmed's desk, where 1 cm represents 10 cm.

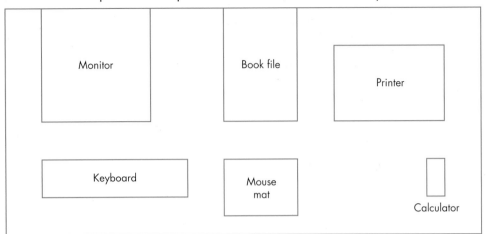

What are the actual dimensions of each of these objects?

a monitor   **b** keyboard   **c** mouse mat

d book file   **e** printer   **f** calculator

4 The diagram shows a sketch of a garden.

a Make an accurate scale drawing of the garden.

Use a scale of 1 cm to represent 2 m.

b Marie wants to plant flowers along the side marked x on the diagram. The flowers need to be planted 0.5 m apart. Use your scale drawing to work out how many plants she needs.

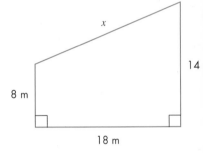

5 Look at the map below, drawn to a scale of 1 cm representing 2 km. Towns are shown with letters.

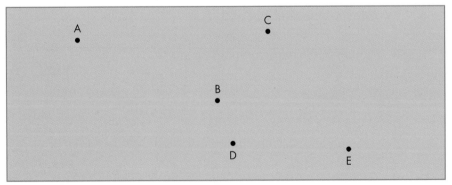

State the following actual distances to the nearest tenth of a kilometre.

a A to B   **b** B to C

c C to D   **d** D to E

e E to B   **f** B to D

6 This sketch shows the outline of a car park.

    a Make a scale drawing where 1 cm represents 5 m.

    b What is the length of AB in metres?

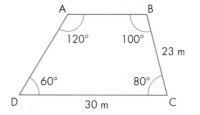

7 This map is drawn with a scale of 1 cm to 200 km.

Find the distances between these cities:

    a Paris and Berlin

    b Paris and Rome

    c Rome and Vienna

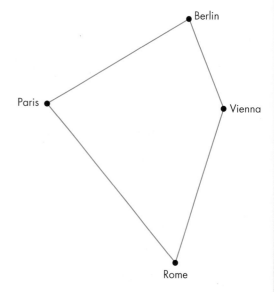

8 Here is a scale drawing of the Great Beijing Wheel in China.

The height of the wheel is 210 m.

Which of the following is the correct scale?

    a 1 cm represents 30 cm

    b 1 cm represents 7 m

    c 1 cm represents 30 m

    d 1 cm represents 300 m

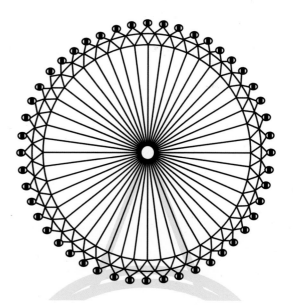

FOUNDATION

Why this chapter matters

How can you find the height of a mountain?

How do you draw an accurate map?

How can computers take an image and make it rotate so that you can view it from different directions?

How do Global Positioning Systems (GPS) work?

How can music be produced electronically?

The answer is by using the angles and sides of triangles and the connections between them. This important branch of mathematics is called trigonometry and is used in science, engineering, electronics and everyday life. This chapter gives a brief introduction to trigonometry.

The first major book of trigonometry was written by an astronomer called Ptolemy who lived in Alexandria, Egypt, over 1800 years ago.

It has tables of numbers, called 'trigonometric ratios', used in making calculations about the positions of stars and planets.

Trigonometry also helped Ptolemy to make a map of the world he knew. Today we no longer need to look up tables of values of trigonometric ratios because they are programmed into calculators and computers.

In the 19th century the French mathematician Jean Fourier showed how all musical sounds can be broken down into a combination of tones that can be described by trigonometry. His work makes it possible to imitate the sound of any instrument electronically.

26 Trigonometry

Topics	Level	Key words
1 Pythagoras' theorem	FOUNDATION	hypotenuse, Pythagoras' theorem
2 Trigonometric ratios	FOUNDATION	ratio, sine, cosine, tangent, opposite side, adjacent side
3 Calculating angles	FOUNDATION	inverse
4 Using sine, cosine and tangent functions	FOUNDATION	
5 Which ratio to use	FOUNDATION	
6 Solving problems using trigonometry	FOUNDATION	
7 Angles of elevation and depression	HIGHER	angle of elevation, angle of depression
8 Problems in three dimensions	HIGHER	
9 Sine, cosine and tangent of obtuse angles	HIGHER	obtuse angle
10 The sine rule and the cosine rule	HIGHER	sine rule, cosine rule, included angle
11 Using sine to find the area of a triangle	HIGHER	area sine rule

What you need to be able to do in the examinations:

FOUNDATION	HIGHER
<ul><li>Understand and use Pythagoras' Theorem in two dimensions.</li><li>Understand and use sine, cosine and tangent of acute angles to determine lengths and angles of a right-angled triangle.</li><li>Apply trigonometrical methods to solve problems in two dimensions.</li></ul>	<ul><li>Understand and use sine, cosine and tangent of obtuse angles.</li><li>Understand and use angles of elevation and depression.</li><li>Understand and use the sine and cosine rules for any triangle.</li><li>Use Pythagoras' Theorem in three dimensions.</li><li>Understand and use the formula $\frac{1}{2} ab \sin C$ for the area of a triangle.</li><li>Apply trigonometrical methods to solve problems in three dimensions, including finding the angle between a line and a plane.</li></ul>

Pythagoras, who was a philosopher as well as a mathematician, was born in 580BCE in Greece. He later moved to Italy, where he established the Pythagorean Brotherhood, which was a secret society devoted to politics, mathematics and astronomy.

This is his famous theorem.

Consider squares being drawn on each side of a right-angled triangle, with sides 3 cm, 4 cm and 5 cm.

The longest side is called the **hypotenuse** and is always opposite the right angle.

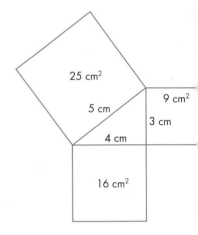

Pythagoras' theorem can then be stated as follows:

> *For any right-angled triangle, the area of the square drawn on the hypotenuse is equal to the sum of the areas of the squares drawn on the other two sides.*

The usual description is:

> *In any right-angled triangle, the square of the hypotenuse is equal to the sum of the squares of the other two sides.*

Pythagoras' theorem is more usually written as a formula:

$$c^2 = a^2 + b^2$$

Remember that Pythagoras' theorem can only be used in right-angled triangles.

Finding the hypotenuse

EXAMPLE 1

Find the length of the hypotenuse, marked x on the diagram.

Using Pythagoras' theorem gives:
$$x^2 = 8^2 + 5.2^2 \text{ cm}^2$$
$$= 64 + 27.04 \text{ cm}^2$$
$$= 91.04 \text{ cm}^2$$

So $x = \sqrt{91.04} = 9.5$ cm (1 decimal place)

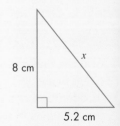

EXERCISE 26A

For each of the triangles in questions **1** to **9**, calculate the length of the hypotenuse, *x*, giving your answers to 1 decimal place.

1

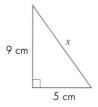

9 cm

x

5 cm

2

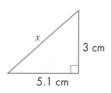

x

3 cm

5.1 cm

> **HINTS AND TIPS**
>
> In these examples you are finding the hypotenuse. The squares of the two short sides are added in every case.

3

4.8 cm

7 cm

x

4

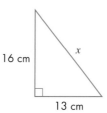

16 cm

x

13 cm

5

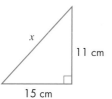

x

11 cm

15 cm

6

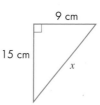

9 cm

15 cm

x

7

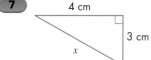

4 cm

3 cm

x

8

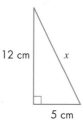

12 cm

x

5 cm

9

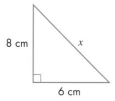

8 cm

x

6 cm

10 How does this diagram show that Pythagoras' theorem is true?

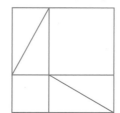

By rearranging the formula for Pythagoras' theorem, the length of one of the shorter sides can easily be calculated.

$$c^2 = a^2 + b^2$$

So, $a^2 = c^2 - b^2$ or $b^2 = c^2 - a^2$

EXAMPLE 2

Find the length x.

x is one of the shorter sides.

So using Pythagoras' theorem gives:
$$x^2 = 15^2 - 11^2 \, cm^2$$
$$= 225 - 121 \, cm^2$$
$$= 104 \, cm^2$$

So $x = \sqrt{104} = 10.2 \, cm$ (1 decimal place)

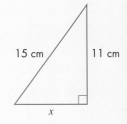

EXERCISE 26B

FOUNDATION

1 For each of the following triangles, calculate the length x, giving your answers to 1 decimal place.

a

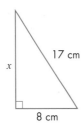

17 cm

x

8 cm

b

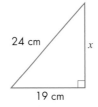

24 cm

x

19 cm

c

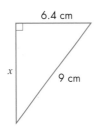

6.4 cm

x

9 cm

d

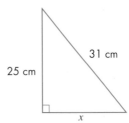

31 cm

25 cm

x

2 For each of the following triangles, calculate the length x, giving your answers to 1 decimal place.

a
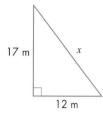
17 m
x
12 m

b
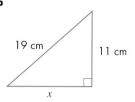
19 cm
11 cm
x

HINTS AND TIPS

These examples are a mixture. Make sure you combine the squares of the sides correctly. If you get it wrong in an examination, you get no marks.

c

17 m
x
23 m

d
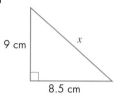
9 cm
x
8.5 cm

3 For each of the following triangles, find the length marked x.

a
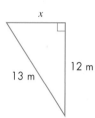
x
12 m
13 m

b
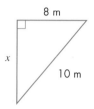
8 m
x
10 m

c
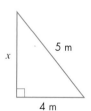
5 m
x
4 m

d
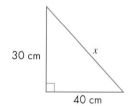
30 cm
x
40 cm

4 In question **3** you found sets of three numbers which satisfy $a^2 + b^2 = c^2$.

Can you find any more?

5 Calculate the value of x.

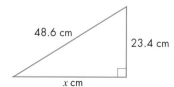
48.6 cm
23.4 cm
x cm

Trigonometry uses three important **ratios** to calculate sides and angles: **sine**, **cosine** and **tangent**. These ratios are defined in terms of the sides of a right-angled triangle and an angle. The angle is often written as θ.

In a right-angled triangle:

● the side opposite the right angle is called the hypotenuse and is the longest side

● the side opposite the angle θ is called the **opposite side**

● the other side next to both the right angle and the angle θ is called the **adjacent side**.

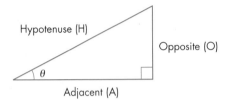

The sine, cosine and tangent ratios for θ are defined as:

$$\text{sine } \theta = \frac{\text{Opposite}}{\text{Hypotenuse}} \qquad \text{cosine } \theta = \frac{\text{Adjacent}}{\text{Hypotenuse}} \qquad \text{tangent } \theta = \frac{\text{Opposite}}{\text{Adjacent}}$$

These ratios are usually abbreviated as:

$$\sin \theta = \frac{O}{H} \qquad \cos \theta = \frac{A}{H} \qquad \tan \theta = \frac{O}{A}$$

These abbreviated forms are also used on calculator keys.

Using your calculator

You will need to use a calculator to find trigonometric ratios.

Different calculators work in different ways, so make sure you know how to use your model.

Angles are not always measured in degrees. Sometimes radians or grads are used instead. You do not need to learn about those in your IGCSE course. Calculators can be set to operate in any of these three units, so make sure your calculator is operating in degrees.

Use your calculator to find the sine of 60 degrees.

You will probably press the keys `sin` `6` `0` `=` in that order, but it might be different on your calculator.

The answer should be 0.8660…

3 cos 57° is a shorthand way of writing $3 \times \cos 57°$.

On most calculators you do not need to use the × button and you can just press the keys in the way it is written: `3` `cos` `5` `7` `=`

Check to see whether your calculator works this way.

The answer should be 1.63…

EXAMPLE 3

Find 5.6 sin 30°.

This means 5.6 × sine of 30 degrees.

Remember that you may not need to press the × button.

5.6 sin 30° = 2.8

EXERCISE 26C

1 Find these values, rounding off your answers to 3 significant figures.

    a sin 43°   **b** sin 56°   **c** sin 67.2°   **d** sin 90°

2 Find these values, rounding off your answers to 3 significant figures.

    a cos 43°   **b** cos 56°   **c** cos 67.2°   **d** cos 90°

3 **a** **i**   What is sin 35°?   **ii** What is cos 55°?

    b **i**   What is sin 12°?   **ii** What is cos 78°?

    c **i**   What is cos 67°?   **ii** What is sin 23°?

    d What connects the values in parts **a**, **b** and **c**?

    e Copy and complete these sentences.

        i   sin 15° is the same as cos …

        ii cos 82° is the same as sin …

        iii sin x is the same as cos …

4 Use your calculator to work out the values of the following.

    a tan 43°   **b** tan 56°   **c** tan 67.2°   **d** tan 90°

    e tan 45°   **f** tan 20°   **g** tan 22°   **h** tan 0°

5 What is so different about tan compared with both sin and cos?

6 Use your calculator to work out the values of the following.

    a 4 sin 63°   **b** 7 tan 52°   **c** 5 tan 80°   **d** 9 cos 8°

7 Use your calculator to work out the values of the following.

    a $\dfrac{5}{\sin 63°}$ **b** $\dfrac{6}{\cos 32°}$ **c** $\dfrac{3}{\tan 64°}$ **d** $\dfrac{7}{\tan 42°}$

8 Using the following triangles calculate sin x, cos x, and tan x. Leave your answers as fraction

a

b

c

26.3 Calculating angles

What angle has a cosine of 0.6? We can use a calculator to find out.

'The angle with a cosine of 0.6' is written as $\cos^{-1} 0.6$ and is called the **'inverse** cosine of 0.6'.

Find out where $\cos^{-1}$ is on your calculator.

You will probably find it on the same key as cos, but you will need to press

SHIFT or INV or 2ndF first.

Look to see if $\cos^{-1}$ is written above the cos key.

Check that $\cos^{-1} 0.6 = 53.1301… = 53.1°$ (1 decimal place)

Check that $\cos 53.1° = 0.600$ (3 decimal places)

Check that you can find the inverse sine and the inverse tangent in the same way.

EXAMPLE 4

What angle has a sine of $\frac{3}{8}$?

You need to find $\sin^{-1} \frac{3}{8}$.

You could use the fraction button on your calculator or you could calculate $\sin^{-1} (3 \div 8)$.

If you use the fraction key you may not need a bracket, or your calculator may put one in automatically.

Try to do it in both of these ways and then use whichever you prefer.

The answer should be 22.0°.

EXERCISE 26D

Use your calculator to find the answers to the following. Give your answers to 1 decimal place.

1 What angles have the following sines?

     a 0.5   **b** 0.785   **c** 0.64   **d** 0.877   **e** 0.999   **f** 0.707

2 What angles have the following cosines?

     a 0.5   **b** 0.64   **c** 0.999   **d** 0.707   **e** 0.2   **f** 0.7

3 What angles have the following tangents?

     a 0.6   **b** 0.38   **c** 0.895   **d** 1.05   **e** 2.67   **f** 4.38

4 What happens when you try to find the angle with a sine of 1.2? What is the largest value of sine you can put into your calculator without getting an error when you ask for the inverse sine? What is the smallest?

5 **a** **i** What angle has a sine of 0.3? (Keep the answer in your calculator memory.)

       ii What angle has a cosine of 0.3?

       iii Add the two accurate answers of parts **i** and **ii** together.

     b Will you always get the same answer to the above no matter what number you start with?

26.4   Using sine, cosine and tangent functions

Sine function

Remember sine $\theta = \dfrac{\text{Opposite}}{\text{Hypotenuse}}$

We can use the sine ratio to calculate the lengths of sides and angles in right-angled triangles.

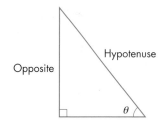

EXAMPLE 5

Find the angle θ, given that the opposite side is 7 cm and the hypotenuse is 10 cm.

Draw a diagram. (This is an essential step.)

From the information given, use sine.

$$\sin \theta = \frac{O}{H} = \frac{7}{10} = 0.7$$

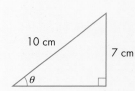

What angle has a sine of 0.7? To find out, use the inverse sine function on your calculator.

$$\sin^{-1} 0.7 = 44.4° \text{ (1 decimal place)}$$

EXAMPLE 6

Find the length of the side marked a in this triangle.

Side a is the opposite side, with 12 cm as the hypotenuse, so use sine.

$$\sin \theta = \frac{O}{H}$$

$$\sin 35° = \frac{a}{12}$$

So $a = 12 \sin 35° = 6.88$ cm (3 significant figures)

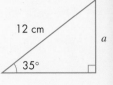

EXAMPLE 7

Find the length of the hypotenuse, h, in this triangle.

Note that although the angle is in the other corner, the opposite side is again given. So use sine.

$$\sin \theta = \frac{O}{H}$$

$$\sin 52° = \frac{8}{h}$$

So $h = \frac{8}{\sin 52°} = 10.2$ cm (3 significant figures)

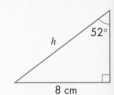

EXERCISE 26E

1 Find the angle marked x in each of these triangles.

a

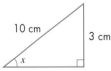

b

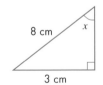

c

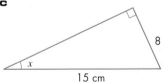

2 Find the side marked x in each of these triangles.

a

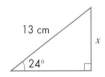

b

c

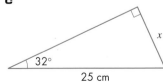

FOUNDATION

3 Find the side marked x in each of these triangles.

a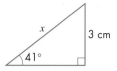

b

c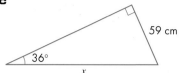

4 Find the side marked x in each of these triangles.

a

b

c

d

Cosine function

Remember cosine $\theta = \dfrac{\text{Adjacent}}{\text{Hypotenuse}}$

We can use the cosine ratio to calculate the lengths of sides and angles in right-angled triangles.

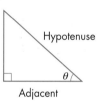

EXAMPLE 8

Find the angle θ, given that the adjacent side is 5 cm and the hypotenuse is 12 cm.

Draw a diagram. (This is an essential step.)

From the information given, use cosine.

$$\cos \theta = \frac{A}{H} = \frac{5}{12}$$

What angle has a cosine of $\frac{5}{12}$? To find out, use the inverse cosine function on your calculator.

$$\cos^{-1}\frac{5}{12} = 65.4° \text{ (1 decimal place)}$$

EXAMPLE 9

Find the length of the hypotenuse, h, in this triangle.

The adjacent side is given. So use cosine.

$$\cos \theta \quad = \frac{A}{H}$$

$$\cos 40° = \frac{20}{h}$$

So $h = \dfrac{20}{\cos 40°} = 26.1$ cm (3 significant figures)

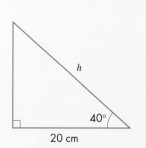

EXERCISE 26F

1 Find the angle marked x in each of these triangles.

a

8 cm
x
5 cm

b

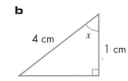

4 cm
x
1 cm

c

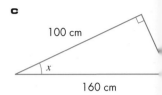

100 cm
x
160 cm

2 Find the side marked x in each of these triangles.

a

8 cm
48°
x

b

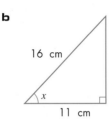

36°
x
12 cm

c

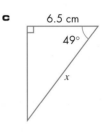

11 cm
24°
x

d

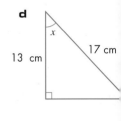

52° 14 cm
x

3 Find the value of x in each of these triangles.

a

10 cm
56°
x

b

16 cm
x
11 cm

c

6.5 cm
49°
x

d

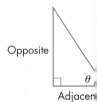

x
13 cm
17 cm

Tangent function

Remember tangent $\theta = \dfrac{\text{Opposite}}{\text{Adjacent}}$

We can use the tangent ratio to calculate the lengths of sides and angles in right-angled triangles.

Opposite
θ
Adjacen[t]

EXAMPLE 10

Find the length of the side marked x in this triangle.

Side x is the opposite side, with 9 cm as the adjacent side, so use tangent.

$\tan \theta = \dfrac{O}{A}$

$\tan 62° = \dfrac{x}{9}$

So $x = 9 \tan 62° = 16.9$ cm (3 significant figures)

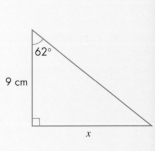

62°
9 cm
x

EXERCISE 26G

1 Find the angle marked x in each of these triangles.

a

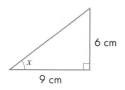

6 cm

x

9 cm

b

x

20 cm

15 cm

c

35 cm

45 cm

x

2 Find the side marked x in each of these triangles.

a

x

61°

5 cm

b

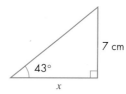

7 cm

43°

x

c

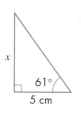

33°

11 cm

x

d

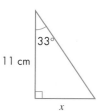

6 cm

34°

x

3 Find the value x in each of these triangles.

a

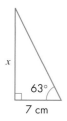

x

63°

7 cm

b

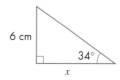

9 cm

x

8 cm

c

x

52°

9 cm

d

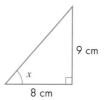

x

4 cm

3.5 cm

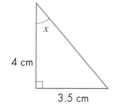

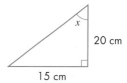

439

The difficulty with any trigonometric problem is knowing which ratio to use to solve it.

The following examples show you how to determine which ratio you need in any given situatio

EXAMPLE 11

Find the length of the side marked x in this triangle.

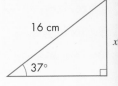

Step 1   Identify what information is given and what needs to be found. Namely, x is opposite the angle and 16 cm is the hypotenuse.

Step 2   Decide which ratio to use. Only one ratio uses opposite and hypotenuse: **sine**.

Step 3   Remember $\sin \theta = \dfrac{O}{H}$

Step 4   Put in the numbers and letters: $\sin 37° = \dfrac{x}{16}$

Step 5   Rearrange the equation and work out the answer:
$x = 16 \sin 37° = 9.629\,040\,371$ cm

Step 6   Give the answer to an appropriate degree of accuracy: $x = 9.63$ cm (3 significant figures)

In reality, you do not write down every step as in Example 11. Step 1 can be done by marking the triangle. Steps 2 and 3 can be done in your head. Steps 4 to 6 are what you write down.

Remember that examiners will want to see evidence of working. Any reasonable attempt at identifying the sides and using a ratio will probably get you some method marks, but only if th fraction is the right way round.

The next examples are set out in a way that requires the *minimum* amount of working but get *maximum* marks.

EXAMPLE 12

Find the length of the side marked x in this triangle.

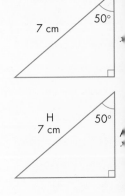

Mark on the triangle the side you know (H) and the side you want to find (A).

Recognise it is a **cosine** problem because you have A and H.

So $\cos 50° = \dfrac{x}{7}$

$x = 7 \cos 50° = 4.50$ cm (3 significant figures)

EXAMPLE 13

Find the angle marked x in this triangle.

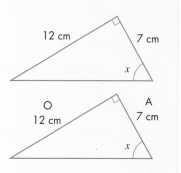

Mark on the triangle the sides you know.

Recognise it is a **tangent** problem because you have O and A.

So $\tan x = \dfrac{12}{7}$

$x = \tan^{-1}\dfrac{12}{7} = 59.7°$ (1 decimal place)

EXERCISE 26H

1 Find the length marked x in each of these triangles.

a

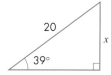

b

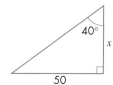

c

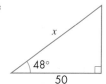

d

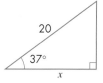

e

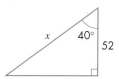

f

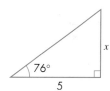

2 Find the angle marked x in each of these triangles.

a

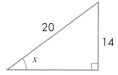

b

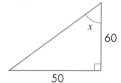

c

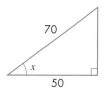

d

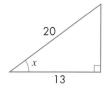

e

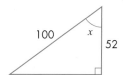

f

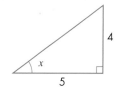

FOUNDATION

441

3 Find the angle or length marked x in each of these triangles.

a
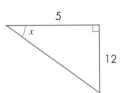

b
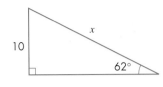

c
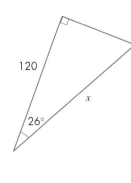

d
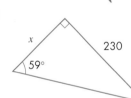

e
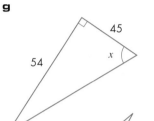

f
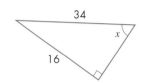

g
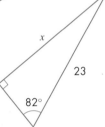

h
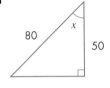

i
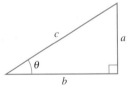

j
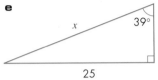

4 **a** How does this diagram show that $\tan \theta = \dfrac{\sin \theta}{\cos \theta}$?

b How does the diagram show that $(\sin \theta)^2 + (\cos \theta)^2 = 1$?

c Choose a value for θ and check the two results in parts **a** and **b** are true.

Solving problems using trigonometry

Many trigonometry problems are not straightforward triangles. Sometimes, solving a triangle is part of solving a practical problem. You should follow these steps when solving a practical problem using trigonometry.

● Draw the triangle required.

● Put on the information given (angles and sides).

● Put on x for the unknown angle or side.

● Mark on two of O, A or H as appropriate.

● Choose which ratio to use.

● Write out the equation with the numbers in.

● Rearrange the equation if necessary, then work out the answer.

● Give your answer to a sensible degree of accuracy. Answers given to 3 significant figures or to the nearest degree are acceptable in exams.

EXAMPLE 14

A window cleaner has a ladder which is 7 m long. The window cleaner leans it against a wall so that the foot of the ladder is 3 m from the wall. What angle does the ladder make with the wall?

Draw the situation as a right-angled triangle.

Then mark the sides and angle.

Recognise it is a sine problem because you have O and H.

So $\sin x = \dfrac{3}{7}$

$x = \sin^{-1}\dfrac{3}{7} = 25°$ (to the nearest degree)

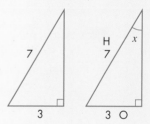

EXERCISE 26I

In these questions, give answers involving angles to the nearest degree.

1 A ladder, 6 m long, rests against a wall. The foot of the ladder is 2.5 m from the base of the wall. What angle does the ladder make with the ground?

2 The ladder in question **1** has a 'safe angle' with the ground of between 70° and 80°. What are the safe limits for the distance of the foot of this ladder from the wall? How high up the wall does the ladder reach?

3 A ladder, of length 10 m, is placed so that it reaches 7 m up the wall. What angle does it make with the ground?

4 A ladder is placed so that it makes an angle of 76° with the ground. The foot of the ladder is 1.7 m from the foot of the wall. How high up the wall does the ladder reach?

5 Calculate the angle that the diagonal makes with the long side of a rectangle which measures 10 cm by 6 cm.

6 This diagram shows a frame for a bookcase.

a What angle does the diagonal strut make with the long side?

b Use Pythagoras' theorem to calculate the length of the strut.

c Why might your answers be inaccurate in this case?

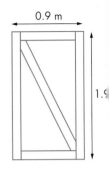

0.9 m

1.9

7 This diagram shows the side of a sloping roof.

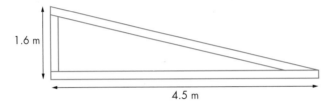

1.6 m

4.5 m

a What angle will the roof make with the horizontal?

b Calculate the length of the sloping support.

8 Alicia paces out 100 m from the base of a tall building. She then measures the angle to the top of the building as 23°. How would Alicia find the height of the building?

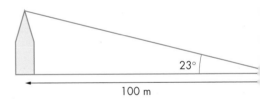

23°

100 m

9 A girl is flying a kite on a string 32 m long. The string, which is being held at 1 m above the ground, makes an angle of 39° with the horizontal. How high is the kite above the ground?

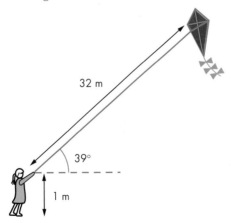

32 m

39°

1 m

10 Taipei is 800 km from Hong Kong on a bearing of 065°.

    a How far north of Hong Kong is Taipei?

    b How far west of Taipei is Hong Kong?

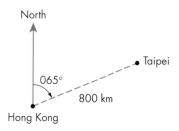

11 A ship is at S where it is 65 km from port P on a bearing of 132°.

It sails north to port T where it is east of the port P.

What is the distance from S to T?

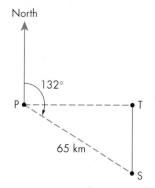

12 A plane at P is 160 km west of airport A.

It flies on a bearing of 070° to Q which is north of A.

How far does the plane fly?

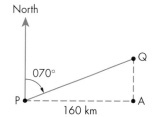

13 Seville is 200 km west of Madrid.

The bearing of Seville from Madrid is 208°.

How far apart are the cities?

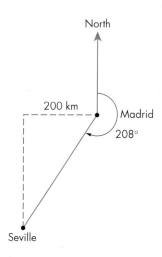

14 Helena is standing on one bank of a wide river.
She wants to find the width of the river.
She cannot get to the other side.
She asks if you can use trigonometry to find the width of the river.

What can you suggest?

When you look *up* at an aircraft in the sky, the angle through which your line of sight turns fr◦ looking straight ahead (the horizontal) is called the **angle of elevation**.

When you are standing on a high point and look *down* at a boat, the angle through which yo◦ line of sight turns from looking straight ahead (the horizontal) is called the **angle of depressio◦**

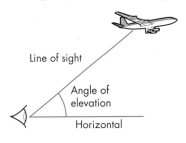

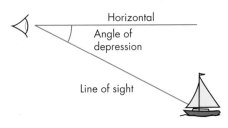

EXAMPLE 15

From the top of a vertical cliff, 100 m high, Ali sees a boat out at sea. The angle of depression from Ali to the boat is 42°. How far from the base of the cliff is the boat?

*The diagram of the situation is shown in figure **i**.*

*From this, you get the triangle shown in figure **ii**.*

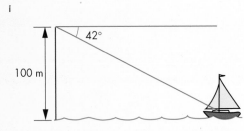

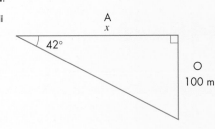

*From figure **ii**, you see that this is a tangent problem.*

$$\text{So } \tan 42° = \frac{100}{x}$$

$$x = \frac{100}{\tan 42°} = 111 \text{ m (3 significant figures)}$$

EXERCISE 26J

In these questions, give any answers involving angles to the nearest degree.

1 Eric sees an aircraft in the sky. The aircraft is at a horizontal distance of 25 km from Eric The angle of elevation is 22°. How high is the aircraft?

2 An aircraft is flying at an altitude of 4000 m and is 10 km from the airport. If a passeng◦ can see the airport, what is the angle of depression?

3 A man standing 200 m from the base of a television transmitter looks at the top of it and notices that the angle of elevation of the top is 65°. How high is the tower?

4 a From the top of a vertical cliff, 200 m high, a boat has an angle of depression of 52°. How far from the base of the cliff is the boat?

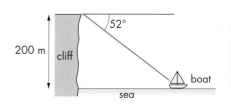

b The boat now sails away from the cliff so that the distance is doubled. Does that mean that the angle of depression is halved? Give a reason for your answer.

5 From a boat, the angle of elevation of the foot of a lighthouse on the edge of a cliff is 34°.

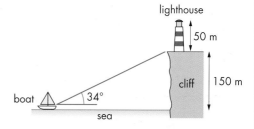

a If the cliff is 150 m high, how far from the base of the cliff is the boat?

b If the lighthouse is 50 m high, what would be the angle of elevation of the top of the lighthouse from the boat?

6 A bird flies from the top of a 12 m tall tree, at an angle of depression of 34°, to catch a worm on the ground.

a How far does the bird actually fly?

b How far was the worm from the base of the tree?

7 Sunil wants to work out the height of a building. He stands about 50 m away from a building. The angle of elevation from Sunil to the top of the building is about 15°. How tall is the building?

8 The top of a ski run is 100 m above the finishing line. The run is 300 m long. What is the angle of depression of the ski run?

9 Nessie and Cara are standing on opposite sides of a tree.

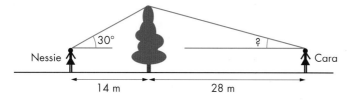

Nessie is 14 m away and the angle of elevation of the top of the tree is 30°.

Cara is 28 m away. She says the angle of elevation for her must be 15° because she is twice as far away.

Is she correct?

What do you think the angle of elevation is?

To find the value of an angle or side in a three dimensional figure you need to find a right-ang
triangle in the figure which contains it. This triangle also has to contain two known values tha
you can use in the calculation.

You must redraw this triangle separately as a plain, right-angled triangle. Add the known value
and the unknown value you want to find. Then use the trigonometric ratios and Pythagoras'
theorem to solve the problem.

EXAMPLE 16

A, B and C are three points at ground level. They are in the same horizontal plane. C is
50 km east of B. B is north of A. C is on a bearing of 050° from A.

An aircraft, flying east, passes over B and over C at the same height. When it passes ove
B, the angle of elevation from A is 12°. Find the angle of elevation of the aircraft from A
when it is over C.

First, draw a diagram containing
all the known information.

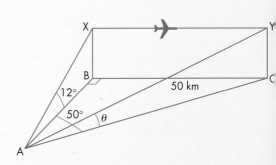

Next, use the right-angled triangle ABC to calculate AB and AC.

$$AB = \frac{50}{\tan 50°} = 41.95 \text{ km} \qquad (4 \text{ significant figures})$$

$$AC = \frac{50}{\sin 50°} = 65.27 \text{ km} \qquad (4 \text{ significant figures})$$

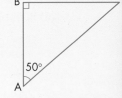

Then use the right-angled triangle ABX to calculate BX,
and hence CY.

$$BX = 41.95 \tan 12° = 8.917 \text{ km} \quad (4 \text{ significant figures})$$

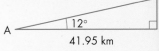

Finally, use the right-angled triangle ACY to
calculate the required angle of elevation, θ.

$$\tan \theta = \frac{8.917}{65.27} = 0.1366$$

$$\Rightarrow \theta = \tan^{-1} 0.1366 = 7.8° \quad (1 \text{ decimal place})$$

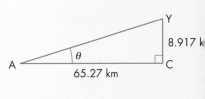

Always write down intermediate working values to at least 4 significant figures, or use the
answer on your calculator display to avoid inaccuracy in the final answer.

EXERCISE 26K

1 A vertical flagpole AP stands at the corner of a rectangular courtyard ABCD.

Calculate the angle of elevation of P from C.

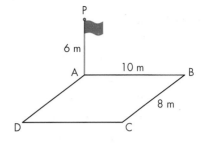

2 The diagram shows a pyramid. The base is a horizontal rectangle ABCD, 20 cm by 15 cm. The length of each sloping edge is 24 cm. The apex, V, is over the centre of the rectangular base. Calculate:

a the length of the diagonal AC

b the size of the angle VAC

c the height of the pyramid.

3 The diagram shows the roof of a building. The base ABCD is a horizontal rectangle 7 m by 4 m. The triangular ends are equilateral triangles. Each side of the roof is an isosceles trapezium. The length of the top of the roof, EF, is 5 m. Calculate:

a the length EM, where M is the midpoint of AB

b the size of angle EBC

c the size of the angle between EM and the base ABCD.

4 ABCD is a vertical rectangular plane. EDC is a horizontal triangular plane. Angle CDE = 90°, AB = 10 cm, BC = 4 cm and ED = 9 cm. Calculate:

a angle AED   **b** angle DEC

c EC   **d** angle BEC.

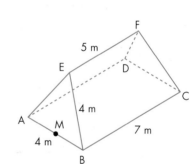

5 In the diagram, XABCD is a pyramid with a rectangular base.

Revina says that the angle between the edge XD and the base ABCD is 56.3°.

Work out the correct answer to show that Revina is wrong.

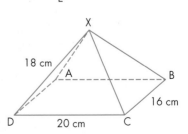

So far you have only used sines and cosines of right-angled triangles.
A calculator also gives the sine and cosine of **obtuse angles**.
Check that sin 115° = 0.906 and cos 115° = –0.423.

We cannot have a right-angled triangle with an obtuse angle
so how can we calculate sin and cos of an obtuse angle?

Imagine a rod OP lying on the x-axis as shown.

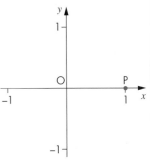

It rotates anticlockwise about the origin O.

ONP is a right-angled triangle.

The hypotenuse OP = 1

If the angle which OP makes with the x-axis is θ:

the adjacent side ON = OP × cos θ = 1 × cos θ = cos θ

So the x-coordinate of P is cos θ.

Similarly:

the opposite side NP = OP × sin θ = 1 × sin θ = sin θ

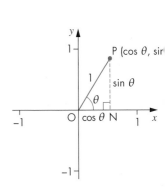

So the y-coordinate of P is sin θ.
Therefore the coordinates of P are (cos θ, sin θ).

Imagine that OP continues to rotate so that angle θ
becomes obtuse.
We still define cos θ and sin θ as the coordinates of P.
We can see from the diagram that the x-coordinate for P
must now be negative.
The y-coordinate is still positive.

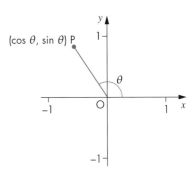

For example, if θ = 115°, the diagram looks like this:

You can see that the angle adjoining θ on the x-axis must
have the same sine as θ. When adjacent angles add up to
180° their sines are the same. So:

If angle θ is obtuse then sin θ = sin (180 – θ)

Their cosines have the same numeric value but the cosine of
the obtuse angle is negative:

If angle θ is obtuse cos θ = –cos (180 – θ)

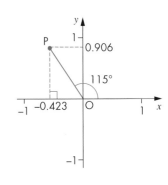

For example, if θ = 115 then 180 – θ = 65
sin 115 = sin 65 = 0.906

$\cos 115 = -\cos 65 = -0.423$

A calculator will also find the tangent of an obtuse angle.

Check that $\tan 115° = -2.144$

In an earlier exercise you saw that $\tan \theta = \dfrac{\sin \theta}{\cos \theta}$ for any angle.

Check that $\tan 115° = \dfrac{\sin 115°}{\cos 115°}$

EXERCISE 26L

1 **a** Copy and complete this table.

Angle	10°	30°	50°	85°	90°	95°	130°	150°	170°
Sine		0.5			1				

b Draw a graph of $y = \sin x$ for $0 \leqslant x \leqslant 180$

c Describe the symmetry of the graph.

d Find two examples from the table to show that if two angles add up to 180 degrees they have the same sine.

2 Find two angles that have a sine of 0.5.

3 Find two angles that have a sine of 0.72.

4 x is an obtuse angle and $\sin x = 0.84$. Find x.

5 **a** Copy and complete this table.

Angle	15°	35°	60°	80°	90°	100°	120°	145°	165°
Cosine									

b Draw a graph of $y = \cos x$ for $0 \leqslant x \leqslant 180$

c Describe the symmetry of the graph.

6 Find the following angles.

a $\cos^{-1} 0.85$ **b** $\cos^{-1}(-0.85)$ **c** $\cos^{-1}(-0.5)$

d $\cos^{-1} 0$ **e** $\cos^{-1} 0.125$ **f** $\cos^{-1}(-0.125)$

7 Solve these equations where $0 \leqslant x \leqslant 180$. Give your answers to the nearest degree. There may be more than one solution.

a $\cos x = 0.6$ **b** $\cos x = -0.25$ **c** $\sin x = \dfrac{3}{4}$ **d** $\sin x = 1$

e $\cos x = 0$ **f** $\sin x = 0.95$ **g** $\sin x = 2$ **h** $\sin x = \cos x$

8 Copy and complete this table.

Angle	30°	55°	80°	100°	125°	150°
Tangent	0.577					

9 Given tan 10° = 0.176, write down tan 170°.

10 **a** Check that $\tan 134° = \dfrac{\sin 134°}{\cos 134°}$

b Why can you not find tan 90°?

11 What obtuse angle has a tangent of −1?

26.10 The sine rule and the cosine rule

Any triangle has six measurements: three sides and three angles. To find any unknown angles or sides), we need to know at least three of the measurements. Any combination of three measurements – except that of all three angles – is enough to work out the rest.

When we need to find a side or an angle in a triangle which contains no right angle, we can use one of two rules, depending on what is known about the triangle. These are the **sine rule** and the **cosine rule**.

The sine rule

Take a triangle ABC and draw the perpendicular from A to the opposite side BC.

From right-angled triangle ADB, $h = c \sin B$

From right-angled triangle ADC, $h = b \sin C$

Therefore:

$$c \sin B = b \sin C$$

which can be rearranged to give:

$$\frac{c}{\sin C} = \frac{b}{\sin B}$$

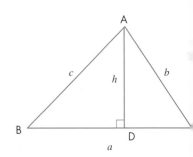

By drawing a perpendicular from each of the other two vertices to the opposite side (or by algebraic symmetry), we see that:

$$\frac{a}{\sin A} = \frac{c}{\sin C} \quad \text{and that} \quad \frac{a}{\sin A} = \frac{b}{\sin B}$$

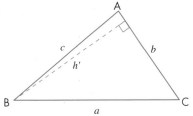

These are usually combined in the form:

$$\frac{a}{\sin A} = \frac{b}{\sin B} = \frac{c}{\sin C}$$

which can be inverted to give:

$$\frac{\sin A}{a} = \frac{\sin B}{b} = \frac{\sin C}{c}$$

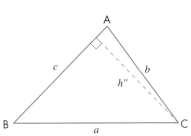

Remember when using the sine rule: take each side in turn, divide it by the sine of the angle opposite and then equate the results.

Note:

- When you are calculating a *side*, use the rule with the *sides on top*.
- When you are calculating an *angle*, use the rule with the *sines on top*.

EXAMPLE 17

In triangle ABC, find the value of x.

Use the sine rule with sides on top, which gives:

$$\frac{x}{\sin 84°} = \frac{25}{\sin 47°}$$

$$\Rightarrow x = \frac{25 \sin 84°}{\sin 47°} = 34.0 \text{ cm} \quad \text{(3 significant figures)}$$

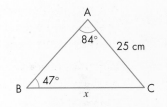

EXAMPLE 18

In the triangle ABC, find the value of the acute angle x.

Use the sine rule with sines on top, which gives:

$$\frac{\sin x}{7} = \frac{\sin 40°}{6}$$

$$\Rightarrow \quad \sin x = \frac{7 \sin 40°}{6} = 0.7499$$

$$\Rightarrow \quad x = \sin^{-1} 0.7499 = 48.6° \quad \text{(3 significant figures)}$$

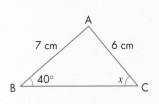

The sine rule works even if the triangle has an obtuse angle, because we can find the sine of an obtuse angle.

EXERCISE 26M

1 Find the length x in each of these triangles.

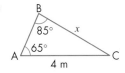

a

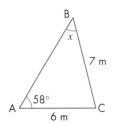

b

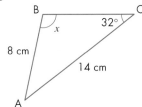

c

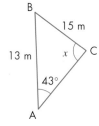

2 Find the angle x in each of these triangles.

a

b

c

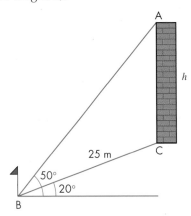

3 To find the height of a tower standing on a small hill, Maria made some measurements (see diagram).

From a point B, the angle of elevation of C is 20°, the angle of elevation of A is 50°, and the distance BC is 25 m.

a Calculate these angles.

i ABC

ii BAC

b Using the sine rule and triangle ABC, calculate the height h of the tower.

4 Use the information on this sketch to calculate the width, *w*, of the river.

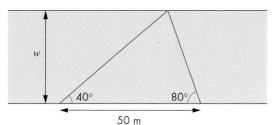

5 An old building is unsafe and is protected by a fence. A company is going to demolish the building and has to work out its height BD, marked *h* on the diagram.

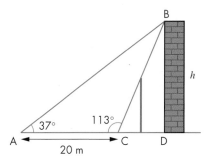

Calculate the value of *h*, using the given information.

6 A mass is hung from a horizontal beam using two strings. The shorter string is 2.5 m long and makes an angle of 71° with the horizontal. The longer string makes an angle of 43° with the horizontal. What is the length of the longer string?

7 A rescue helicopter is based at an airfield at A.

It is sent out to rescue a man from a mountain at M, due north of A.

The helicopter then flies on a bearing of 145° to a hospital at H as shown on the diagram.

Calculate the direct distance from the mountain to the hospital.

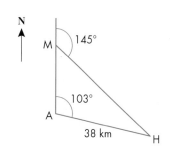

8 Triangle ABC has an obtuse angle at B.

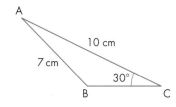

Calculate the size of angle ABC.

The cosine rule

Take the triangle, shown on the right, where D is the foot of the perpendicular to BC from A.

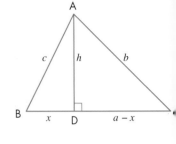

Using Pythagoras' theorem on triangle BDA:

$$h^2 = c^2 - x^2$$

Using Pythagoras' theorem on triangle ADC:

$$h^2 = b^2 - (a - x)^2$$

Therefore,

$$c^2 - x^2 = b^2 - (a - x)^2$$
$$c^2 - x^2 = b^2 - a^2 + 2ax - x^2$$
$$\Rightarrow c^2 = b^2 - a^2 + 2ax$$

From triangle BDA, $x = c \cos B$.

So,

$$c^2 = b^2 - a^2 + 2ac \cos B$$

Rearranging gives:

$$b^2 = a^2 + c^2 - 2ac \cos B$$

By algebraic symmetry:

$$a^2 = b^2 + c^2 - 2bc \cos A \quad \text{and} \quad c^2 = a^2 + b^2 - 2ab \cos C$$

This is the cosine rule, which can be best remembered by the diagram on the right, where:

$$a^2 = b^2 + c^2 - 2bc \cos A$$

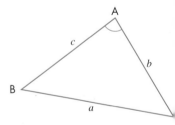

Note the symmetry of the rule and how the rule works using two adjacent sides and the angle between them (the **included angle**).

The formula can be rearranged to find any of the three angles.

$$\cos A = \frac{b^2 + c^2 - a^2}{2bc}$$

$$\cos B = \frac{a^2 + c^2 - b^2}{2ac}$$

$$\cos C = \frac{a^2 + b^2 - c^2}{2ab}$$

EXAMPLE 19

Find x in this triangle.

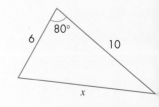

By the cosine rule:
$$x^2 = 6^2 + 10^2 - 2 \times 6 \times 10 \times \cos 80°$$
$$x^2 = 115.16$$
$$\Rightarrow x = 10.7 \quad \text{(3 significant figures)}$$

EXAMPLE 20

Find x in this triangle.

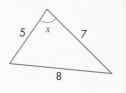

By the cosine rule:

$$\cos x = \frac{5^2 + 7^2 - 8^2}{2 \times 5 \times 7} = 0.1428$$

$$\Rightarrow x = 81.8° \qquad \text{(3 significant figures)}$$

It is possible to find the cosine of an angle that is greater than 90°. For example, cos 120° = –0.5.

EXERCISE 26N

1 Find the length x in each of these triangles.

a
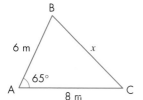

b
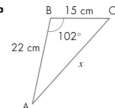

c
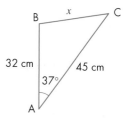

2 Find the angle x in each of these triangles.

a
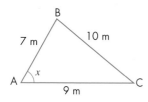

b
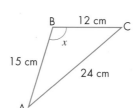

c
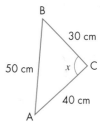

d Explain the significance of the answer to part **c**.

3 In triangle ABC, AB = 5 cm, BC = 6 cm and angle ABC = 55°. Find AC.

4 A triangle has two sides of length 40 cm and an angle of 110°. Work out the length of the third side of the triangle.

5 The diagram shows a trapezium ABCD.
AB = 6.7 cm, AD = 7.2 cm, CB = 9.3 cm and angle DAB = 100°.

Calculate:

a the length DB   **b** angle DBA

c angle DBC   **d** the length DC

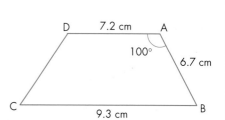

HIGHER

6 A ship sails from a port on a bearing of 050° for 50 km then turns on a bearing of 150° for 40 km. A crewman is taken ill, so the ship drops anchor. What course and distance should a rescue helicopter from the port fly to reach the ship in the shortest possible time?

7 The three sides of a triangle are given as 3*a*, 5*a* and 7*a*. Calculate the smallest angle in the triangle.

8 Two ships, X and Y, leave a port at 9 am.

Ship X travels at an average speed of 20 km/h on a bearing of 075° from the port.

Ship Y travels at an average speed of 25 km/h on a bearing of 130° from the port.

Calculate the distance between the two ships at 11 am.

9 Calculate the size of the largest angle in the triangle ABC.

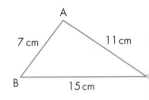

Choosing the correct rule

When finding the unknown sides and angles in a triangle, there are several situations that can occur.

Two sides and the included angle

1 Use the cosine rule to find the third side.

2 Use the sine rule to find either of the other angles.

3 Use the sum of the angles in a triangle to find the third angle.

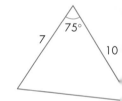

Two angles and a side

1 Use the sum of the angles in a triangle to find the third angle.

2, 3 Use the sine rule to find the other two sides.

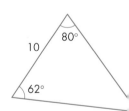

Three sides

1 Use the cosine rule to find one angle.

2 Use the sine rule to find another angle.

3 Use the sum of the angles in a triangle to find the third angle.

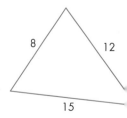

HIGHER

1 Find the length or angle x in each of these triangles.

a
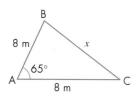

b
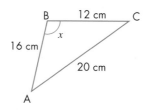

c
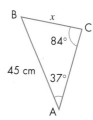

d
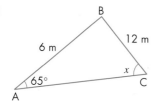

e
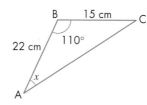

f
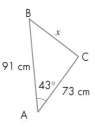

g
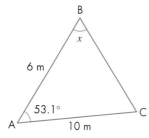

h
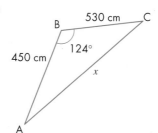

i
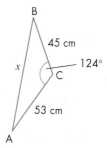

2 The hands of a clock have lengths 3 cm and 5 cm. Find the distance between the tips of the hands at 4 o'clock.

3 A spacecraft is seen hovering at a point which is in the same vertical plane as two towns, X and F, which are on the same level. Its distances from X and F are 8.5 km and 12 km respectively. The angle of elevation of the spacecraft when observed from F is 43°. Calculate the distance between the two towns.

4 Triangle ABC has sides with lengths a, b and c, as shown in the diagram.

   a What can you say about the angle BAC, if $b^2 + c^2 - a^2 = 0$?

   b What can you say about the angle BAC, if $b^2 + c^2 - a^2 > 0$?

   c What can you say about the angle BAC, if $b^2 + c^2 - a^2 < 0$?

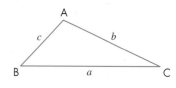

5 The diagram shows a sketch of a field ABCD.
A farmer wants to put a new fence round the perimeter of the field.

Calculate the perimeter of the field.

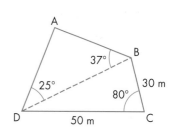

In triangle ABC, the vertical height is BD and the base is AC.

Let $BD = h$ and $AC = b$, then the area of the triangle is given by:

$$\tfrac{1}{2} \times AC \times BD = \tfrac{1}{2}bh$$

However, in triangle BCD:

$$h = BC \sin C = a \sin C$$

where $BC = a$.

Substituting into $\tfrac{1}{2}bh$ gives:

$$\tfrac{1}{2}b \times (a \sin C) = \tfrac{1}{2}ab \sin C$$

as the area of the triangle.

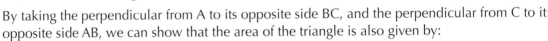

By taking the perpendicular from A to its opposite side BC, and the perpendicular from C to it opposite side AB, we can show that the area of the triangle is also given by:

$$\tfrac{1}{2}ac \sin B \quad \text{and} \quad \tfrac{1}{2}bc \sin A$$

Note the pattern: the area is given by the product of two sides multiplied by the sine of the included angle. This is the **area sine rule**. Starting from any of the three forms, it is also possib to use the sine rule to establish the other two.

EXAMPLE 21

Find the area of triangle ABC.

Area $= \tfrac{1}{2}ab \sin C$

Area $= \tfrac{1}{2} \times 5 \times 7 \times \sin 38° = 10.8 \text{ cm}^2$

(3 significant figures)

EXAMPLE 22

Find the area of triangle ABC.

You have all three sides but no angle. So first you must find an angle in order to apply the area sine rule.

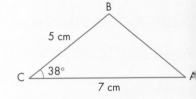

Find angle C, using the cosine rule.

$$\cos C = \frac{a^2 + b^2 - c^2}{2ab}$$

$$= \frac{13^2 + 19^2 - 8^2}{2 \times 13 \times 19} = 0.9433$$

$$\Rightarrow C = \cos^{-1} 0.9433 = 19.4°$$

(Keep the exact value in your calculator memory.)

Now apply the area sine rule.

$$\tfrac{1}{2}ab \sin C = \tfrac{1}{2} \times 13 \times 19 \times \sin 19.4°$$

$$= 41.0 \text{ cm}^2 \quad \text{(3 significant figures)}$$

EXERCISE 26P

1 Find the area of each of the following triangles.

   a Triangle ABC where BC = 7 cm, AC = 8 cm and angle ACB = 59°

   b Triangle ABC where angle BAC = 86°, AC = 6.7 cm and AB = 8 cm

   c Triangle PQR where QR = 27 cm, PR = 19 cm and angle QRP = 109°

   d Triangle XYZ where XY = 231 cm, XZ = 191 cm and angle YXZ = 73°

   e Triangle LMN where LN = 63 cm, LM = 39 cm and angle NLM = 85°

2 The area of triangle ABC is 27 cm². If BC = 14 cm and angle BCA = 115°, find AC.

3 The area of triangle LMN is 113 cm², LM = 16 cm and MN = 21 cm. Angle LMN is acute. Calculate these angles.

   a LMN   **b** MNL

4 A board is in the shape of a triangle with sides 60 cm, 70 cm and 80 cm. Find the area of the board.

5 Two circles, centres P and Q, have radii of 6 cm and 7 cm respectively. The circles intersect at X and Y. Given that PQ = 9 cm, find the area of triangle PXQ.

6 Sanjay is making a kite.
The diagram shows a sketch of his kite.

Calculate the area of the material required to make the kite.

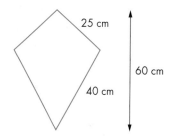

7 The triangular area ABC is to be made a national park.

The bearing of B from A is 324°.
The bearing of C from A is 42°.

Calculate the area of the park.

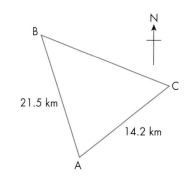

8 The diagram shows the dimensions of a four-sided field.

   a Show that the length of the diagonal BD is 66 metres to the nearest metre.

   b Calculate the size of angle C.

   c Calculate the area of the field.

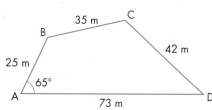

Why this chapter matters

People have always needed to measure areas and volumes.

In everyday life, you will, for instance, need to find the area to work out how many tiles to buy to cover a floor; or you will need to find the volume to see how much water is needed to fill a swimming pool. You can do this quickly using formulae.

Measuring the world

From earliest times, farmers have wanted to know the area of their fields to see how many crops they could grow or animals they could support. When land is bought and sold, the cost depends on the area.

Volumes are important too. Volumes tell us how much space there is inside any structure. Whether it is a house, barn, aeroplane, car or office, the volume is important. In some countries there are regulations about the number of people who can use an office, based on the volume of the room.

Volumes of containers for liquids also need to be measured. Think, for example, of a car fuel tank, the water tank in a building, or a reservoir. It is important to be able to calculate the capacity of all these things.

So how do we measure areas and volumes? In this chapter, you will learn formulae that can be used to calculate areas and volumes of different shapes, based on a few measureme

Many of these formulae were first worked ou thousands of years ago. They are still in use today because they are important in everyda life.

The process of calculating areas and volume using formulae is called **mensuration**.

27

Mensuration

Topics	Level	Key words
1 Perimeter and area of a rectangle	**FOUNDATION**	length, width, perimeter, area
2 Area of a triangle	**FOUNDATION**	base, perpendicular height
3 Area of a parallelogram	**FOUNDATION**	parallelogram
4 Area of a trapezium	**FOUNDATION**	trapezium
5 Circumference and area of a circle	**FOUNDATION**	circumference, diameter, radius, π, semicircle
6 Surface area and volume of a cuboid	**FOUNDATION**	face, volume, cuboid, vertices, edge, vertex, cube, surface area, litre
7 Volume of a prism	**FOUNDATION**	cross-section, prism, pyramid, cone, sphere
8 Volume and surface area of a cylinder	**FOUNDATION**	curved surface, cylinder
9 Arcs and sectors	**HIGHER**	arc, sector, subtended
10 Volume and surface area of a cone	**HIGHER**	cone, vertical height, slant height
11 Volume and surface area of a sphere	**HIGHER**	sphere

What you need to be able to do in the examinations:

FOUNDATION	HIGHER
• Find the perimeter of shapes made from triangles and rectangles. • Find the area of parallelograms, trapezia and simple shapes. • Find circumferences and areas of circles. • Find perimeters and areas of semicircles. • Recognise 3-D solids and understand the terms face, edge and vertex. • Find the surface area of a cylinder and of simple shapes. • Find the volume of right prisms using an appropriate formula.	• Find the perimeters and areas of sectors of circles. • Find the surface area and volume of a sphere and a right circular cone. • Convert between volume measures.

Perimeter and area of a rectangle

The **perimeter** of a rectangle is the total distance around the outside.

$$\text{perimeter} = l + w + l + w$$
$$= 2(l + w)$$

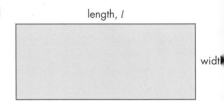

length, *l*

width

The **area** of a rectangle is **length** × **width**.

$$\text{area} = lw$$

EXAMPLE 1

Calculate the area and perimeter of this rectangle.

11 cm

4 cm

Area of rectangle = length × width
$$= 11\,\text{cm} \times 4\,\text{cm}$$
$$= 44\,\text{cm}^2$$

Perimeter = 2 × 11 + 2 × 4
$$= 30\,\text{cm}$$

Some two-dimensional shapes are made up of two or more rectangles.

These shapes can be split into simpler shapes, which makes it easy to calculate their areas.

EXAMPLE 2

Find the area and perimeter of the shape shown on the right.

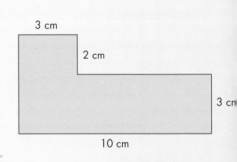

3 cm

2 cm

3 cm

10 cm

First, split the shape into two rectangles, A and B, and find the missing lengths.

The perimeter is

$$3 + 2 + 7 + 3 + 10 + 5 = 30\,\text{cm}$$

area of A = 2 × 3 = 6 cm^2

area of B = 10 × 3 = 30 cm^2

The area of the shape is:

area of A + area of B = 6 + 30 = 36 cm^2

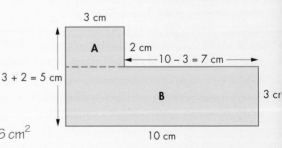

3 cm

A

2 cm

10 − 3 = 7 cm

3 + 2 = 5 cm

B

3 cm

10 cm

EXERCISE 27A

1 Calculate the area and the perimeter for each of the rectangles.

a

7 cm

5 cm

b

11 cm

3 cm

c
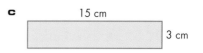
15 cm

3 cm

d

10 cm

7 cm

e

8 cm

7 cm

f

5 cm

2 cm

2 Calculate the area and the perimeter for each of the rectangles.

a

8.2 cm

6.5 cm

b
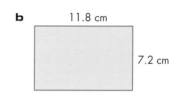
11.8 cm

7.2 cm

3 A rectangular field is 150 m long and 45 m wide.

Fencing is needed to go all the way around the field.

The fencing is sold in 10-metre long pieces.

How many pieces are needed?

4 A soccer pitch is 160 m long and 70 m wide.

a Before a game, the players have to run about 1500 m to help them loosen up. How many times will they need to run round the perimeter of the pitch to do this?

b The groundsman waters the pitch at the rate of 100 m^2 per minute. How long will it take him to water the whole pitch?

5 What is the perimeter of a square with an area of 100 cm^2?

6 Which rectangle has the largest area? Which has the largest perimeter?

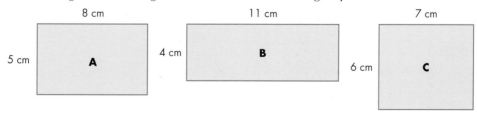

8 cm

5 cm

A

11 cm

4 cm

B

7 cm

6 cm

C

Explain your answers.

7 Doubling the length and width of a rectangle doubles the area of the rectangle.

Is this statement:
- always true
- sometimes true
- never true?

Explain your answer.

8 Calculate the perimeter and area of each of the compound shapes below.

a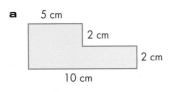
5 cm, 2 cm, 2 cm, 10 cm

b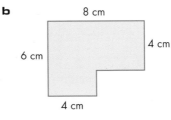
8 cm, 4 cm, 6 cm, 4 cm

c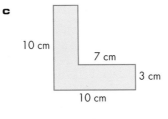
10 cm, 7 cm, 3 cm, 10 cm

d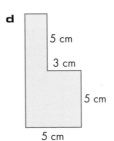
5 cm, 3 cm, 5 cm, 5 cm

e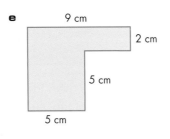
9 cm, 2 cm, 5 cm, 5 cm

f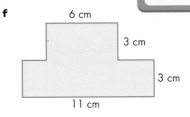
6 cm, 3 cm, 3 cm, 11 cm

g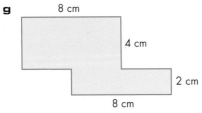
8 cm, 4 cm, 2 cm, 8 cm

h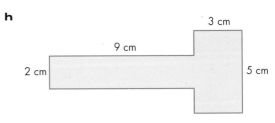
3 cm, 9 cm, 2 cm, 5 cm

9 This compound shape is made from four rectangles that are all the same size.

Work out the area of the compound shape.

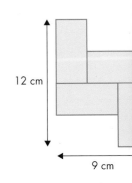
12 cm, 3 cm, 9 cm

10 This shape is made from five squares that are all the same size.

It has an area of 80 cm^2.

Work out the perimeter of the shape.

Area of a triangle

The area of any triangle is given by the formula:

area = $\frac{1}{2}$ × **base** × **perpendicular height**

As an algebraic formula, this is written as:

$A = \frac{1}{2}bh$

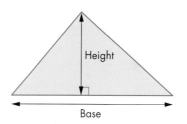

This diagram shows why the area for a triangle is half the area of a rectangle with the same base and height.

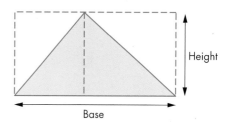

EXAMPLE 3

Calculate the area of this triangle.

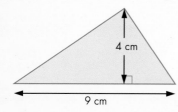

Area = $\frac{1}{2}$ × 9 cm × 4 cm

= $\frac{1}{2}$ × 36 cm^2

= 18 cm^2

EXAMPLE 4

Calculate the area of the shape shown below.

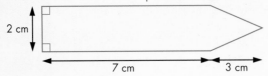

This shape can be split into a rectangle (R) and a triangle (T).

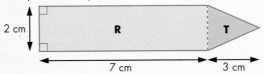

Area of the shape = area of R + area of T

$$= 7 \times 2 + \tfrac{1}{2} \times 2 \times 3$$
$$= 14 + 3$$
$$= 17 \, cm^2$$

EXERCISE 27B

1 Calculate the area of each of these triangles.

a

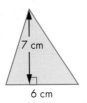

7 cm

6 cm

b

3 cm

8 cm

c

7 cm

4 cm

d

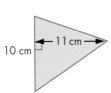

10 cm   11 cm

e

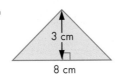

12 cm

15 cm

f

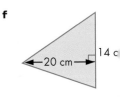

20 cm   14 c

2 Copy and complete the following table for triangles **a** to **f**.

	Base	Perpendicular height	Area
a	8 cm	7 cm	
b		9 cm	36 cm
c		5 cm	10 cm
d	4 cm		6 cm^2
e	6 cm		21 cm
f	8 cm	11 cm	

3 This regular hexagon has an area of 48 cm².

What is the area of the square that surrounds the hexagon?

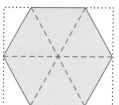

4 Find the area of each of these shapes.

a

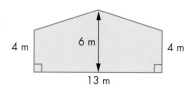

6 cm

5 cm

10 cm

> **HINTS AND TIPS**
>
> Refer to Example 4 on how to find the area of a compound shape.

b

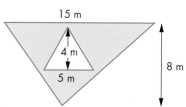

4 m   6 m   4 m

13 m

c

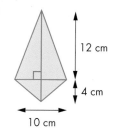

12 cm

4 cm

10 cm

5 Find the area of each shaded shape.

a

4 cm   ←6 cm→   7 cm

11 cm

> **HINTS AND TIPS**
>
> Find the area of the outer shape and subtract the area of the inner shape.

b

15 m

4 m

5 m

8 m

6 Write down the dimensions of two different-sized triangles that have the same area of 50 cm².

7 Which triangle is the odd one out? Give a reason for your answer.

a

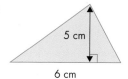

5 cm

6 cm

b

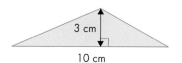

3 cm

10 cm

c

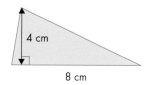

4 cm

8 cm

A **parallelogram** can be changed into a rectangle by moving a triangle.

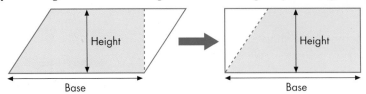

This shows that the area of the parallelogram is the area of a rectangle with the same base and height. The formula is:

area of a parallelogram = base × height

As an algebraic formula, this is written as:

$A = bh$

EXAMPLE 5

Find the area of this parallelogram.

Area = 8 cm × 6 cm
= 48 cm²

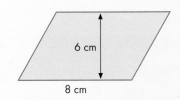

EXERCISE 27C

1 Calculate the area of each parallelogram below.

a
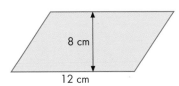
8 cm
12 cm

b

10 cm
7 cm

c
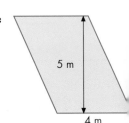
5 m
4 m

d
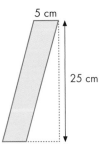
5 cm
25 cm

e
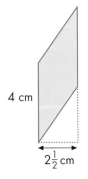
4 cm
$2\frac{1}{2}$ cm

f
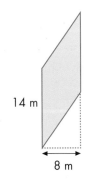
14 m
8 m

2 Sandeep says that the area of this parallelogram is 30 cm².

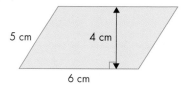

Is she correct? Give a reason for your answer.

3 This shape is made from four parallelograms that are all the same size. The area of the shape is 120 cm².

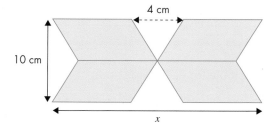

Work out the length marked x on the diagram.

4 This logo, made from two identical parallelograms, is cut from a sheet of card.

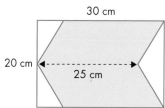

a Calculate the area of the logo.

b How many logos can be cut from a sheet of card that measures 1 m by 1 m?

27.4 Area of a trapezium

The area of a **trapezium** is calculated by finding the average of the lengths of its parallel sides and multiplying this by the perpendicular height between them.

The area of a trapezium is given by this formula:

$A = \frac{1}{2}(a + b)h$

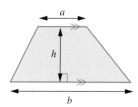

EXAMPLE 6

Find the area of the trapezium ABCD.

Area $= \frac{1}{2}(4 + 7) \times 3$

$= \frac{1}{2} \times 11 \times 3$

$= 16.5 \text{ cm}^2$

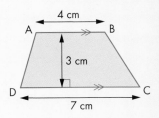

EXERCISE 27D

1 Copy and complete the following table for each trapezium.

	Parallel side 1	Parallel side 2	Perpendicular height	Area
a	8 cm	4 cm	5 cm	
b	10 cm	12 cm	7 cm	
c	7 cm	5 cm	4 cm	
d	5 cm	9 cm	6 cm	
e	3 cm	13 cm	5 cm	
f	4 cm	10 cm		42 cm²
g	7 cm	8 cm		22.5 cm²

2 Calculate the perimeter and the area of each trapezium.

a

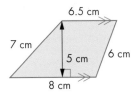

b

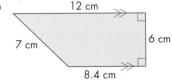

c

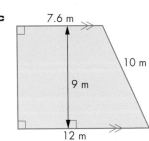

HINTS AND TIPS

Make sure you use the right measurement for the height. Some questions may include the slant side length, which is not used for the area.

3 A trapezium has an area of 25 cm². Its vertical height is 5 cm.
Work out a possible pair of lengths for the two parallel sides.

4 Which of the following shapes has the largest area?

a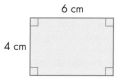
6 cm
4 cm

b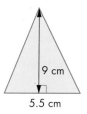
9 cm
5.5 cm

c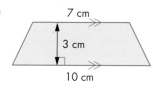
7 cm
3 cm
10 cm

5 Which of the following shapes has the smallest area?

a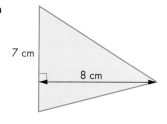
7 cm
8 cm

b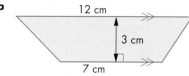
12 cm
3 cm
7 cm

c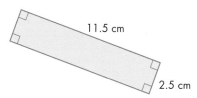
11.5 cm
2.5 cm

6 Which of the following is the area of this trapezium?

a 45 cm^2 **b** 65 cm^2 **c** 70 cm^2

You must show your workings.

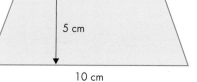
8 cm
5 cm
10 cm

7 Work out the value of a so that the square and the trapezium have the same area.

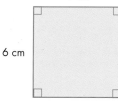
6 cm
6 cm

7 cm
8 cm
a

8 The side of a ramp is a trapezium, as shown in the diagram. Calculate its area, giving your answer in square metres.

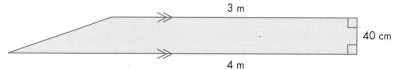
3 m
40 cm
4 m

HINTS AND TIPS

Change the height into metres first.

The perimeter of a circle is called the **circumference**.

You calculate the circumference, C, of a circle by multiplying its **diameter**, d, by π.

The value of π is found on all scientific calculators, with $\pi = 3.141\,592\,654$, but if it is not on your calculator, then take $\pi = 3.142$.

The circumference of a circle is given by the formula:

circumference = $\pi \times$ diameter *or* $C = \pi d$

As the diameter is twice the **radius**, r, this formula can also be written as $C = 2\pi r$.

EXAMPLE 7

Calculate the circumference of the circle with a diameter of 4 cm.

Use the formula:

$C = \pi d$

$\quad = \pi \times 4$

$\quad = 12.6$ cm (rounded to 1 decimal place)

Remember The length of the radius of a circle is half the length of its diameter. So, when you are given a radius, in order to find a circumference you must first double the radius to get the diameter.

EXAMPLE 8

Calculate the diameter of a circle that has a circumference of 40 cm.

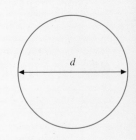

$C = \pi \times d$

$40 = \pi \times d$

$d = \dfrac{40}{\pi} = 12.7$ cm (rounded to 1 decimal place)

The area, A, of a circle is given by the formula:

area = $\pi \times$ radius^2 *or* $A = \pi \times r \times r$ *or* $A = \pi r^2$

Remember This formula uses the radius of a circle. So, when you are given the diameter of a circle, you must *halve* it to get the radius.

Half of a circle is called a **semicircle**.

EXAMPLE 9

a Calculate the area of a circle of diameter 12 cm.

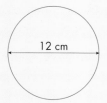
12 cm

b Calculate
i the perimeter
ii the area of a semicircle of diameter 12 cm.

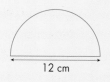
12 cm

a First, halve the diameter to get the radius:

radius = 12 ÷ 2 = 6 cm

Then, find the area:

$$area = \pi r^2$$
$$= \pi \times 6^2$$
$$= \pi \times 36$$
$$= 113.1 \text{ cm}^2 \text{ (rounded to 1 decimal place)}$$

b i The perimeter of the semicircle is half the circumference of the whole circle + the diameter

Circumference of whole circle = 12 × π = 37.7 cm
Perimeter of semicircle = 37.7 ÷ 2 + 12 = 30.8 cm

ii The semicircle has half the area of the circle.
Area = π × 36 ÷ 2 = 56.5 cm² (rounded to 1 decimal place)

EXERCISE 27E

1 Calculate **i** the circumference **ii** the area of each circle illustrated below. Give your answers to 1 decimal place.

a
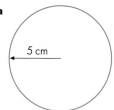
5 cm

b

3 cm

c

1.5 cm

d

4 cm

2 Calculate **i** the circumference **ii** the area of each circle illustrated below. Give your answers to 1 decimal place.

a
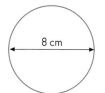
8 cm

b

5 cm

c

9.2 cm

d

4.7 cm

3 Calculate **i** the circumference **ii** the area of these semicircles.

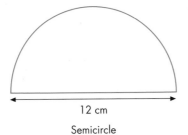

a 30 cm

b 8 cm

c 2.4 m

d 0.5 m

4 A circle has a circumference of 60 cm.

a Calculate the diameter of the circle to 1 decimal place.

b What is the radius of the circle to 1 decimal place?

c Calculate the area of the circle to 1 decimal place.

5 Calculate the area of a circle with a circumference of 110 cm.

6 Calculate **i** the circumference **ii** the area of the following shapes.

a

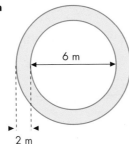

12 cm

Semicircle

b

5 cm

Quadrant

7 Calculate the area of the shaded part of each of these diagrams.

a

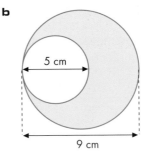

6 m

2 m

b

5 cm

9 cm

c

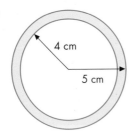

4 cm

5 cm

8 The diagram shows a circular photograph frame.

Work out the area of the frame. Give your answer to 1 decimal place.

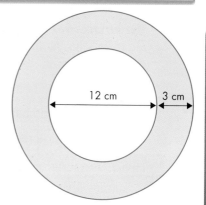

9 A square has sides of length a and a circle has radius r.

The area of the square is equal to the area of the circle.

Show that $r = \dfrac{a}{\sqrt{\pi}}$

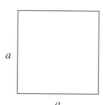

10 A circle fits exactly inside a square of side 10 cm.

Calculate the area of the shaded region. Give your answer to 1 decimal place.

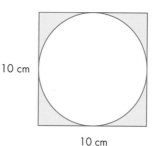

27.6 Surface area and volume of a cuboid

A cuboid is a box shape, all six **faces** of which are rectangles.

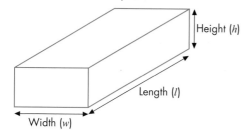

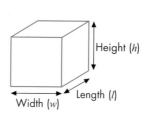

A **cube** is a cuboid where the length, width and height are all the same size. All the faces are squares.

A cuboid has 8 corners or **vertices**. Each face has 4 **edges**. A cuboid has a total of 12 edges.

Every day you will come across many examples of **cuboids**, such as food packets, DVD players – and even this book.

The **volume** of a cuboid is given by the formula:

volume = length × width × height *or* $V = l \times w \times h$ *or* $V = lwh$

HINTS AND TIPS

The correct name of a corner is a **vertex**. The plural of vertex is vertices.

The **surface area** of a cuboid is calculated by finding the total area of the six faces, which are rectangles. Notice that each pair of opposite rectangles have the same area. So, from the diagram above:

area of top and bottom rectangles = 2 × length × width = $2lw$

area of front and back rectangles = 2 × height × width = $2hw$

area of two side rectangles = 2 × height × length = $2hl$

Hence, the surface area of a cuboid is given by the formula:

surface area = $A = 2lw + 2hw + 2hl$

EXAMPLE 10

Calculate the volume and surface area of this cuboid.

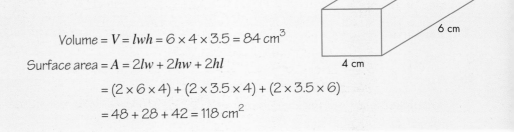

Volume = $V = lwh = 6 \times 4 \times 3.5 = 84\ cm^3$

Surface area = $A = 2lw + 2hw + 2hl$

$= (2 \times 6 \times 4) + (2 \times 3.5 \times 4) + (2 \times 3.5 \times 6)$

$= 48 + 28 + 42 = 118\ cm^2$

Note:

$1\ cm^3 = 1000\ mm^3$ and $1\ m^3 = 1\,000\,000\ cm^3$

$1000\ cm^3 = 1$ **litre**

$1\ m^3 = 1000$ litres

EXERCISE 27F

1 Find **i** the volume and **ii** the surface area of each of these cuboids.

a
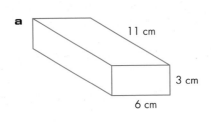

b
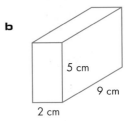

c
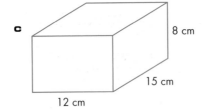

d
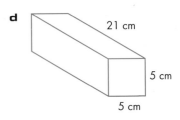

2 Find the capacity of a fish-tank with dimensions: length 40 cm, width 30 cm and height 20 cm. Give your answer in litres.

3 Find the volume of the cuboid in each of the following cases.

a The area of the base is 40 cm^2 and the height is 4 cm.

b The base has one side 10 cm and the other side 2 cm longer, and the height is 4 cm.

c The area of the top is 25 cm^2 and the depth is 6 cm.

4 Calculate **i** the volume and **ii** the surface area of each of the cubes with these edge lengths.

a 4 cm **b** 7 cm **c** 10 mm **d** 5 m **e** 12 m

5 Safety regulations say that in a room where people sleep there should be at least 12 m^3 for each person. A dormitory is 20 m long, 13 m wide and 4 m high. What is the greatest number of people who can safely sleep in the dormitory?

6 A tank contains 32 000 litres of water. The base of the tank measures 6.5 m by 3.1 m. Find the depth of water in the tank. Give your answer to one decimal place.

7 A room contains 168 m^3 of air. The height of the room is 3.5 m. What is the area of the floor?

8 What are the dimensions of cubes with these volumes?

a 27 cm^3 **b** 125 m^3 **c** 8 mm^3 **d** 1.728 m^3

9 Calculate the volume of each of these shapes.

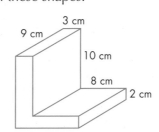

a

2 cm
3 cm
6 cm
2 cm
10 cm
7 cm

b

3 cm
9 cm
10 cm
8 cm
2 cm

> **HINTS AND TIPS**
>
> Split the solid into two separate cuboids and work out the dimensions of each of them from the information given.

10 A cuboid has a volume of 125 cm^3 and a total surface area of 160 cm^2.

Is it possible that this cuboid is a cube? Give a reason for your answer.

11 The volume of a cuboid is 1000 cm^3. What is the smallest surface area it could have?

A **prism** is a three dimensional shape which has the same **cross-section** running all the way through it.

Name:	Cuboid	Triangular prism	Cylinder	Cuboid	Hexagonal prism	Cube
Cross-section:	Rectangle	Triangle	Circle	Square	Hexagon	Square

These three dimensional shapes are not prisms. They do not have the same cross section all the way through.

Name:	**Pyramid**	**Cone**	**Sphere**

The volume of a prism is found by multiplying the area of its cross-section by the length of the prism (or height if the prism is stood on end).

That is, volume of prism = area of cross-section × length
or $V = Al$

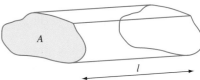

EXAMPLE 11

Find the volume of the triangular prism.

The area of the triangular cross-section = $A = \dfrac{5 \times 7}{2} = 17.5 \text{ cm}^2$

The volume is the area of its cross-section × length = Al

$= 17.5 \times 9 = 157.5 \text{ cm}^3$

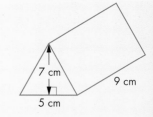

EXERCISE 27G

1 For each prism shown:

   i calculate the area of the cross-section   **ii** calculate the volume.

a

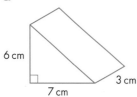

6 cm

7 cm

3 cm

b

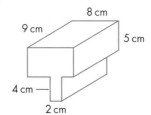

8 cm

9 cm

5 cm

4 cm

2 cm

c

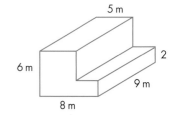

5 m

6 m

2

8 m

9 m

2 Calculate the volume of each of these prisms.

a
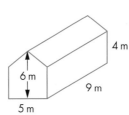

b
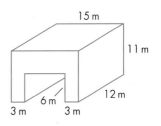

c
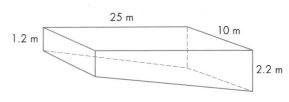

3 A swimming pool is 10 m wide and 25 m long.

It is 1.2 m deep at one end and 2.2 m deep at the other end. The floor slopes uniformly from one end to the other.

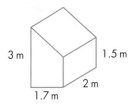

a Explain why the shape of the pool is a prism.

b The pool is filled with water at a rate of 2 m³ per minute. How long will it take to fill the pool?

4 A building is in the shape of a prism. Calculate the volume of air (in litres) inside the building with the dimensions shown in the diagram.

5 Each of these prisms has a uniform cross-section in the shape of a right-angled triangle.

a Find the volume of each prism. **b** Find the total surface area of each prism.

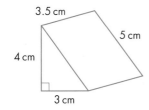

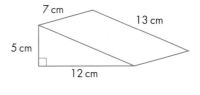

6 The top and bottom of the container shown here are the same size, both consisting of a rectangle, 4 cm by 9 cm, with a semi-circle at each end. The depth is 3 cm. Find the volume of the container.

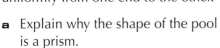

7 A horse trough is in the shape of a semi-circular prism as shown.

What volume of water will the trough hold when it is filled to the top? Give your answer in litres.

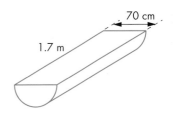

8 **a** Calculate the volume of this prism.

b For this prism, find the number of **i** faces **ii** vertices **iii** edges.

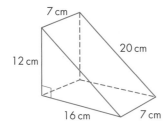

7 cm

20 cm

12 cm

16 cm   7 cm

27.8 # Volume and surface area of a cylinder

Volume

Since a **cylinder** is an example of a prism, its volume is found by multiplying the area of one of its circular ends by the height.

That is, volume $= \pi r^2 h$

where r is the radius of the cylinder and h is its height or length.

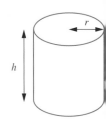

EXAMPLE 12

What is the volume of a cylinder having a radius of 5 cm and a height of 12 cm?

Volume = area of circular base × height

$= \pi r^2 h$

$= \pi \times 5^2 \times 12 \text{ cm}^3$

$= 942 \text{ cm}^3$ (3 significant figures)

Surface area

The total surface area of a cylinder is made up of the area of its **curved surface** plus the area of its two circular ends.

The curved surface area, when opened out, is a rectangle with length equal to the circumference of the circular end.

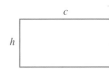

curved surface area = circumference of end × height of cylinder

$= 2\pi rh$ or πdh

area of one end $= \pi r^2$

Therefore, total surface area $= 2\pi rh + 2\pi r^2$ or $\pi dh + 2\pi r^2$

EXAMPLE 13

What is the total surface area of a cylinder with a radius of 15 cm and a height of 2.5 m?

First, you must change the dimensions to a *common unit*. Use centimetres in this case.

$$\text{Total surface area} = \pi dh + 2\pi r^2$$
$$= \pi \times 30 \times 250 + 2 \times \pi \times 15^2 \text{ cm}^2$$
$$= 23\,562 + 1414 \text{ cm}^2$$
$$= 24\,976 \text{ cm}^2$$
$$= 25\,000 \text{ cm}^2 \text{ (3 significant figures)}$$

EXERCISE 27H

1 For the cylinders below find:

   i   the volume   **ii**   the total surface area.

Give your answers to 3 significant figures.

a   **b**   **c**   **d**

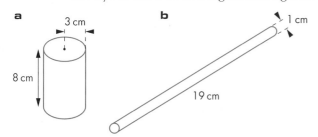
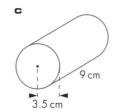
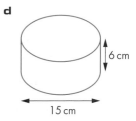

2 For each of these cylinder dimensions find:

   i   the volume   **ii**   the curved surface area.

Give your answers in terms of π.

  a   Base radius 3 cm and height 8 cm   **b**   Base diameter 8 cm and height 7 cm

  c   Base diameter 12 cm and height 5 cm   **d**   Base radius of 10 m and length 6 m

3 A solid cylinder has a diameter of 8.4 cm and a height of 12.0 cm. Calculate the volume of the cylinder.

4 A cylindrical food can has a height of 10.5 cm and a diameter of 7.4 cm.

What can you say about the size of the paper label around the can?

5 A cylindrical container is 65 cm in diameter. Water is poured into the container until it is 1 m deep.

How much water is in the container? Give your answer in litres.

6 A drinks manufacturer plans a new drink in a can. The quantity in each can must be 330 m

Suggest a suitable height and diameter for the can. (You might like to look at the dimensions of a real drinks can.)

7 Wire is commonly made by putting hot metal through a hole in a plate.

What length of wire of diameter 1 mm can be made from a 1 cm cube of metal?

8 The engine size of a car is measured in litres. This tells you the total volume of its cylinders. Cylinders with the same volume can be long and thin or short and fat.

In a racing car, the diameter of a cylinder is twice its length. Suggest possible dimensions for a 0.4 litre racing car cylinder.

27.9 Arcs and sectors H

A **sector** is part of a circle, bounded by two radii of the circle and one of the **arcs** formed by the intersections of these radii with the circumference.

The angle at O is the angle of the sector. We say it is **subtended** at the centre of the circle by the arc.

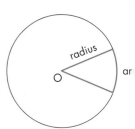

Length of an arc and area of a sector

A sector is a fraction of the whole circle, the size of the fraction being determined by the size of angle of the sector. The angle is often written as θ, a Greek letter pronounced *theta*. For example,

the sector shown in the diagram represents the fraction $\dfrac{\theta}{360}$.

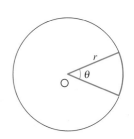

This applies to both its arc length and its area. Therefore:

$$\text{arc length} \ = \frac{\theta}{360} \times 2\pi r \quad \text{or} \quad \frac{\theta}{360} \times \pi d$$

$$\text{sector area} \ = \frac{\theta}{360} \times \pi r^2$$

EXAMPLE 14

Find the arc length and the area of the sector in the diagram.

The sector angle is 28° and the radius is 5 cm. Therefore:

$$\text{arc length} \ = \frac{28}{360} \times \pi \times 2 \times 5 = 2.4 \text{ cm (1 decimal place)}$$

$$\text{sector area} \ = \frac{28}{360} \times \pi \times 5^2 = 6.1 \text{ cm}^2 \text{ (1 decimal place)}$$

EXERCISE 27I

1 For each of these sectors, calculate: **i** the arc length **ii** the sector area.

a
40°
8 cm

b
95°
5 cm

c
78°
12 cm

d
130°
7 cm

2 Calculate the arc length and the area of a sector whose arc subtends an angle of 60° at the centre of a circle with a diameter of 12 cm. Give your answers in terms of π.

3 Calculate the total perimeter of each of these sectors. **a**
11 cm

b
22°
8.5 cm

4 Calculate the area of each of these sectors.

a
110°
7 cm

b
50°
8 cm

5 O is the centre of a circle of radius 12.5 cm.

Calculate the length of the arc ACB.

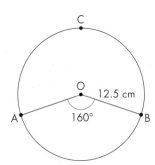

6 The diagram shows quarter of a circle. Calculate the area of the shaded shape, giving your answer in terms of π.

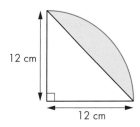

7 ABCD is a square of side length 8 cm. APC and AQC are arcs of the circles with centres D and B. Calculate the area of the shaded part.

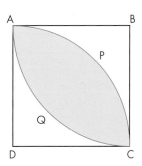

8 Find:

   a the perimeter

   b the area

of this shape.

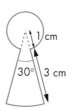

1 cm

30° 3 cm

27.10 **Volume and surface area of a cone** **H**

A **cone** can be treated as a pyramid with a circular base. Therefore, the formula for the volume of a cone is the same as that for a pyramid.

   volume $= \frac{1}{3} \times$ base area $\times$ vertical height

     $V = \frac{1}{3}\pi r^2 h$

where r is the radius of the base and h is the **vertical height** of the cone.

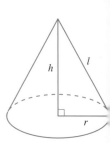

The curved surface area of a cone is given by:

   curved surface area $= \pi \times$ radius $\times$ slant height

       $S = \pi r l$

where l is the **slant height** of the cone.

So the total surface area of a cone is given by the curved surface area plus the area of its circular base.

   $A = \pi r l + \pi r^2$

EXAMPLE 15

For the cone in the diagram, calculate:

   i its volume

   ii its total surface area.

Give your answers in terms of π.

   i The volume is given by $V = \frac{1}{3}\pi r^2 h$

                   $= \frac{1}{3} \times \pi \times 36 \times 8 = 96\pi \text{ cm}^3$

   ii The total surface area is given by $A = \pi r l + \pi r^2$

                                $= \pi \times 6 \times 10 + \pi \times 36 = 96\pi \text{ cm}^2$

10 cr

8 cm

6 cm

EXERCISE 27J

1 For each cone, calculate:

   i its volume   **ii** its total surface area.

Give your answers to 3 significant figures.

a 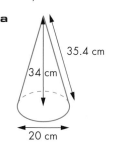   **b** 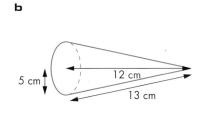   **c**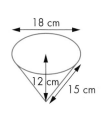

2 Find the total surface area of a cone whose base radius is 3 cm and slant height is 5 cm. Give your answer in terms of π.

3 Calculate the volume of each of these shapes. Give your answers in terms of π.

a 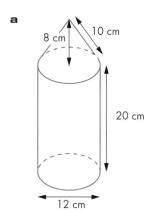   **b**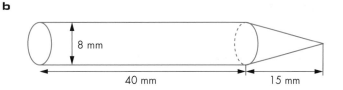

4 You could work with a partner on this question.

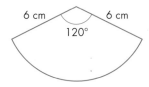

A sector of a circle, as in the diagram, can be made into a cone (without a base) by sticking the two straight edges together.

  a What would be the diameter of the base of the cone in this case?

  b What is the diameter if the angle is changed to 180°?

  c Investigate other angles.

5 A cone has the dimensions shown in the diagram.

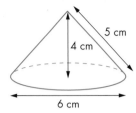

Calculate the total surface area, leaving your answer in terms of π.

6 If the slant height of a cone is equal to the base diameter, show that the area of the curved surface is twice the area of the base.

7 The model shown below is made from aluminium.

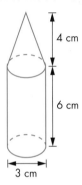

What is the mass of the model, given that the density of aluminium is 2.7 g/cm^3? (This means that 1 cm^3 of aluminium has a mass of 2.7 g.)

8 A container in the shape of a cone, base radius 10 cm and vertical height 19 cm, is full of water. The water is poured into an empty cylinder of radius 15 cm. How high is the water in the cylinder?

Volume and surface area of a sphere

The volume of a **sphere**, radius r, is given by:

$$V = \frac{4}{3}\pi r^3$$

Its surface area is given by:

$$A = 4\pi r^2$$

EXAMPLE 16

For a sphere of radius of 8 cm, calculate **i** its volume and **ii** its surface area.

i The volume is given by:

$$V = \frac{4}{3}\pi r^3$$

$$= \frac{4}{3} \times \pi \times 8^3 = \frac{2048}{3} \times \pi = 2140 \text{ cm}^3 \quad \text{(3 significant figures)}$$

ii The surface area is given by:

$$A = 4\pi r^2$$

$$= 4 \times \pi \times 8^2 = 256 \times \pi = 804 \text{ cm}^2 \quad \text{(3 significant figures)}$$

EXERCISE 27K

1 Calculate the volume and surface area of each of these spheres. Give your answers in terms of π.

    a Radius 3 cm   **b** Radius 6 cm   **c** Diameter 20 cm

2 Calculate the volume and the surface area of a sphere with a diameter of 50 cm.

3 A sphere fits exactly into an open cubical box of side 25 cm. Calculate the following.

    a The surface area of the sphere   **b** The volume of the sphere

4 A metal sphere of radius 15 cm is melted down and recast into a solid cylinder of radius 6 cm. Calculate the height of the cylinder.

5 Lead has a density of 11.35 g/cm^3. Calculate the maximum number of lead spheres of radius 1.5 mm which can be made from 1 kg of lead.

6 A sphere has a radius of 5.0 cm. A cone has a base radius of 8.0 cm. The sphere and the cone have the same volume.

    Calculate the height of the cone.

7 A sphere of diameter 10 cm is carved out of a wooden block in the shape of a cube of side 10 cm. What percentage of the wood is wasted?

Why this chapter matters

If you look carefully, you will be able to spot symmetry all around you. It is present in the natural world and in objects made by man. But, does it have a purpose and why do we need it?

Symmetry in nature, art and literature

Symmetry is everywhere you look in nature. Plants and animals have symmetrical body shapes and patterns. For example, if you divide a leaf in half, you will see that one half is the same shape as the other half.

Where is the symmetry in this butterfly, star fish and peacock? What effect does this symmetry have?

Pegasus.

This painting by a Dutch artist called M.C. Escher (1898–1972) uses line symmetry and rotational symmetry. Why do you think Escher used symmetry in his paintings?

Every tiger has its own unique pattern of stripes. These appear on the tiger's skin as well as its fur. What purpose do these symmetrical stripes serve?

Can you identify symmetry in the face of the tiger?

Symmetry in structures

St Peter's Basilica, Rome.

St Peter's Basilica, in the Vatican City in Rome, was started in 1506 and completed in 1626. It is a very symmetrical structure – see if you can identify all the symmetry that is present.

Why do you think that the designers of this building used symmetry?

These examples show some of the uses of symmetry in the world. Now think about where symmetry occurs in your own life – how important is it to you?

28

Symmetry

pics		Level	Key words
1	Lines of symmetry	**FOUNDATION**	line of symmetry, mirror line
2	Rotational symmetry	**FOUNDATION**	rotational symmetry, order of rotational symmetry
3	Symmetry of special two-dimensional shapes	**FOUNDATION**	rectangle, square, parallelogram, kite, rhombus, trapezium, isosceles triangle, equilateral triangle

What you need to be able to do in the examinations:

FOUNDATION

- Recognise line and rotational symmetry.
- Identify any lines of symmetry and the order of rotational symmetry of a given two-dimensional figure.

Many two-dimensional shapes have one or more **lines of symmetry**.

A line of symmetry is a line that can be drawn through a shape so that what can be seen on one side of the line is the mirror image of what is on the other side. This is why a line of symmetry is sometimes called a **mirror line**.

It is also the line along which a shape can be folded exactly onto itself.

EXAMPLE 1

Find the number of lines of symmetry for this cross.

There are four altogether.

EXERCISE 28A

1 Copy these shapes and draw on the lines of symmetry for each one. If it will help you, u[se] tracing paper or a mirror to check your results.

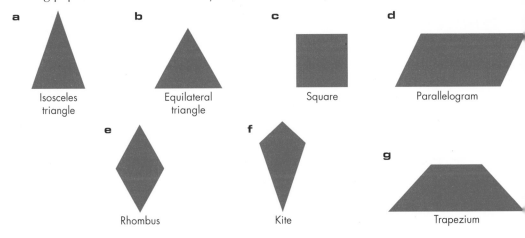

a Isosceles triangle

b Equilateral triangle

c Square

d Parallelogram

e Rhombus

f Kite

g Trapezium

2 **a** Find the number of lines of symmetry for each of these regular polygons.

i

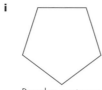

Regular pentagon

ii

Regular hexagon

iii

Regular octagon

b How many lines of symmetry do you think a regular decagon has? (A decagon is a ten-sided polygon.)

3 Write down the number of lines of symmetry for each of these flags.

Austria   Canada   Iceland   Switzerland   Greece

4 These road signs all have lines of symmetry. Copy them and draw on the lines of symmetry for each one.

5 The animal and plant kingdoms are full of symmetry. Four examples are given below. State the number of lines of symmetry for each one.

a

b

c

d

Can you find other examples? Find suitable pictures, copy them and state the number of lines of symmetry each one has.

6 Copy this diagram.

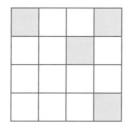

On your copy, shade in four more squares so that the diagram has four lines of symmetry.

Rotational symmetry

A two-dimensional shape has **rotational symmetry** if it can be rotated about a point to look exactly the same in a new position.

The **order of rotational symmetry** is the number of different positions in which the shape looks the same when it is rotated 360° about the point (that is, one complete turn).

The easiest way to find the order of rotational symmetry for any shape is to trace it and count the number of times that the shape stays the same as you turn the tracing paper through one complete turn.

EXAMPLE 2

Find the order of rotational symmetry for this shape.

First, hold the tracing paper on top of the shape and trace the shape. Then rotate the tracing paper and count the number of times the tracing matches the original shape in one complete turn.

You will find three different positions.

So, the order of rotational symmetry for the shape is 3.

EXERCISE 28B

1 Copy these shapes and write below each one the order of rotational symmetry. If it will help you, use tracing paper.

a

Square

b

Rectangle

c

Parallelogram

d

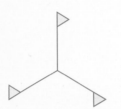

Equilateral triangle

e

Regular hexagon

2 Find the order of rotational symmetry for each of these shapes.

a   b   c  d   e

3 The following are Greek capital letters. Write down the order of rotational symmetry for each one.

ᵃ Φ   **ᵇ H**   **ᶜ Z**   **ᵈ Θ**   **ᵉ Ξ**

4 Here is a star pattern.

Inside the star there are two patterns that have rotational symmetry.

a What is the order of rotational symmetry of the whole star?

b What is the order of rotational symmetry of the two patterns inside the star?

5 Copy the grid on the right. On your copy, shade in four squares so that the shape has rotational symmetry of order 2.

28.3 Symmetry of special two-dimensional shapes

Some three and four-sided shapes have special names such as **isosceles triangle** or **parallelogram**. You need to know the symmetry of these shapes.

For example:

- An isosceles triangle has one line of symmetry and no rotational symmetry.

- A parallelogram has no lines of symmetry and rotational symmetry of order 2.

EXERCISE 28C

1 Draw diagrams to show all the lines of symmetry on these shapes:

    a a **rectangle**

    b a **kite**

    c a **square**

    d an **equilateral triangle**

    e a **rhombus**.

2 **a** Which shape in question 1 has no rotational symmetry?

    b Find the order of rotational symmetry for each of the others.

3 **a** What do we call a triangle with one line of symmetry?

    b Can you draw a triangle with exactly two lines of symmetry?

4 What shape are the following?

    a A quadrilateral with no lines of symmetry and rotational symmetry of order 2.

    b A quadrilateral with rotational symmetry of order 4.

5 **a** Name two different quadrilaterals which have two lines of symmetry and rotational symmetry of order 2.

    b Can you draw a quadrilateral which has two lines of symmetry but no rotational symmetry?

6 The dotted line is a line of symmetry.

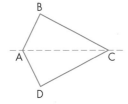

    a Which angles *must* be equal?

    b Which sides *must* be equal?

    c What is the name of this shape?

7 **a** What is the special name for a line of symmetry of a circle?

b How many lines of symmetry does a circle have?

c What is the order of rotational symmetry of a circle?

8 This shape has rotational symmetry of order 2.

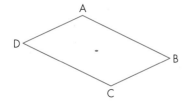

a Which angles *must* be equal?

b Which sides *must* be equal?

c What is the name of this shape?

9 If a **trapezium** has a line of symmetry, what can you say about its angles?

Why this chapter matters

Vectors are used to represent any quantity that has both magnitude and direction. The velocity of a speeding car is its direction *and* its speed. Velocity is a vector.

To understand how a force acts on an object, you need to know the magnitude of the force and the direction in which it moves – that is, its vector.

In science, vectors are used to describe displacement, acceleration and momentum.

But are vectors used in real life? Yes! Here are some examples.

In the 1950s, a group of talented Brazilian footballers invented the swerving free kick. By kicking the ball in just the right place, they managed to make it curl around the wall of defending players and go into the goal. When a ball is in flight, it is acted upon by various forces which can be described by vectors.

Formula One teams always employ physicists and mathematicians to help them build the perfect racing car. Since vectors describe movements and forces, they are used as the basis of a car's design.

Pilots have to consider wind speed and direction when they plan to land an aircraft at an airport. Vectors are an important part of the computerised landing system.

Vectors play a key role in the design of aircraft wings, where an upward force or lift is needed to enable the aircraft to fly.

Vectors are used extensively in computer graphics. Software f animations uses the mathematics of vectors.

29 Vectors

Topics	Level	Key words
Introduction to vectors	HIGHER	magnitude, direction, vector, column, scalar, coordinate grid
Using vectors	HIGHER	position vector
The magnitude of a vector	HIGHER	magnitude, Pythagoras' theorem

What you need to be able to do in the examinations:

HIGHER

- Understand that a vector has both magnitude and direction.
- Understand and use vector notation.
- Multiply vectors by scalar quantities.
- Add and subtract vectors.
- Calculate the modulus (magnitude) of a vector.
- Find the resultant of two or more vectors.
- Apply vector methods for simple geometrical proofs.

A **vector** is something that has both **magnitude** and **direction** and can be represented by an arrow.

Examples are velocity, acceleration, force and momentum.

Vectors can be written down in several ways:

$\overrightarrow{AB}$ – giving the start and end points with an arrow over the top.

a – as a lower case letter printed in bold. When you are writing vectors **a**, **b** ... by hand yo cannot show them in bold type. Instead write them with a line underneath as a, b ... to sho that they stand for a vector and not a number.

When vectors are drawn on a **coordinate grid** they can be represented by two numbers in brackets in a **column**.

The top number shows how far the line moves from one side to the other between its start and end points, and the bottom number shows how far it moves up or down.

For example, on this grid:

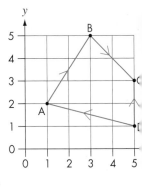

$\overrightarrow{AB} = \begin{pmatrix} 2 \\ 3 \end{pmatrix}$ means move 2 right and 3 upwards to get to point B from point A.

If a line moves either downwards or left the coordinate is negative.
So:

$\overrightarrow{BC} = \begin{pmatrix} 2 \\ -2 \end{pmatrix}$ means move 2 right and 2 downwards to get to point C from point B.

$\overrightarrow{DA} = \begin{pmatrix} -4 \\ 1 \end{pmatrix}$ means move 4 left and 1 upwards to get to point A from point D.

$\overrightarrow{DC} = \begin{pmatrix} 0 \\ 2 \end{pmatrix}$ means move neither left or right, just upwards to get to point C from point D.

Notice that the line joining A and B can be written as AB and BA and these both have the san magnitude (in this case length).

But the vectors $\overrightarrow{AB}$ and $\overrightarrow{BA}$ are not the same because their directions are different:

$\overrightarrow{AB} = \begin{pmatrix} 2 \\ 3 \end{pmatrix}$ and $\overrightarrow{BA} = \begin{pmatrix} -2 \\ -3 \end{pmatrix}$

Therefore $\overrightarrow{BA} = -\overrightarrow{AB}$

The vectors on this grid show that if $\mathbf{a} = \begin{pmatrix} 1 \\ 2 \end{pmatrix}$ then $-\mathbf{a} = \begin{pmatrix} -1 \\ -2 \end{pmatrix}$.

It is a vector with the same length (magnitude) but in the opposite direction.

HINTS AND TIPS

Do not forget the arrow t indicate a vector.

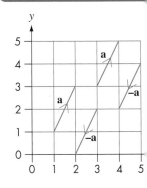

Adding, subtracting and multiplying vectors

Vectors are added together by placing them end to end.

On this grid $\mathbf{a} + \mathbf{b} = \mathbf{c}$

or $\begin{pmatrix} 3 \\ 2 \end{pmatrix} + \begin{pmatrix} 1 \\ -3 \end{pmatrix} = \begin{pmatrix} 4 \\ -1 \end{pmatrix}$

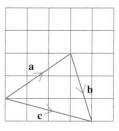

Notice that you add the top figures together $(3 + 1 = 4)$ and the bottom figures together $(2 + -3 = -1)$.

Vectors can be subtracted too.

$\mathbf{a} - \mathbf{b} = \mathbf{d}$

$\mathbf{a} - \mathbf{b}$ is the same as $\mathbf{a} + (-\mathbf{b})$

$\begin{pmatrix} 3 \\ 2 \end{pmatrix} - \begin{pmatrix} 1 \\ -3 \end{pmatrix} = \begin{pmatrix} 3 \\ 2 \end{pmatrix} + \begin{pmatrix} -1 \\ 3 \end{pmatrix} = \begin{pmatrix} 2 \\ 5 \end{pmatrix}$

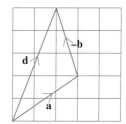

Notice that $3 - 1 = 2$ and $2 - -3 = 5$.

Vectors can be multiplied by a number.

If $\mathbf{a} = \begin{pmatrix} 3 \\ 2 \end{pmatrix}$

$3\mathbf{a} = 3 \times \begin{pmatrix} 3 \\ 2 \end{pmatrix} = \begin{pmatrix} 9 \\ 6 \end{pmatrix}$

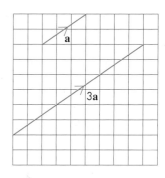

Notice that $3 \times 3 = 9$ and $3 \times 2 = 6$.

Here 3 is called a **scalar**, to distinguish it from a vector.

If k is a scalar, then $k\begin{pmatrix} x \\ y \end{pmatrix} = \begin{pmatrix} kx \\ ky \end{pmatrix}$.

EXERCISE 29A

1

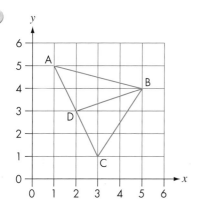

a Write these as column vectors:

   i $\overrightarrow{AB}$ **ii** $\overrightarrow{DB}$ **iii** $\overrightarrow{CB}$ **iv** $\overrightarrow{CA}$ **v** $\overrightarrow{AC}$ **vi** $\overrightarrow{DA}$

b Show that $\overrightarrow{AD} = \overrightarrow{DC}$. What does this tell you about the position of D on line AC?

HIGHER

2   F has the coordinates (4, 2), G has the coordinates (2, 6), M is the midpoint of FG and C the origin.

     a   Mark F, G and M on a coordinate grid.

        Write these as column vectors:

        i $\overrightarrow{FG}$ **ii** $\overrightarrow{GF}$ **iii** $\overrightarrow{OM}$ **iv** $\overrightarrow{MG}$ **v** $\overrightarrow{GO}$

3   A has the coordinates (3, 4).

$$\overrightarrow{AP} = \begin{pmatrix} 2 \\ 1 \end{pmatrix}, \ \overrightarrow{AQ} = \begin{pmatrix} 1 \\ -3 \end{pmatrix}, \ \overrightarrow{AR} = \begin{pmatrix} -2 \\ 0 \end{pmatrix}$$

     Mark A, P, Q and R on a coordinate grid.

4 $\mathbf{a} = \begin{pmatrix} 2 \\ 3 \end{pmatrix}$ and $\mathbf{b} = \begin{pmatrix} 1 \\ -4 \end{pmatrix}$.

     Draw diagrams to show:

     a $\mathbf{a} + \mathbf{b}$ **b** $-\mathbf{b}$ **c** $\mathbf{a} - \mathbf{b}$ **d** $\mathbf{b} + \mathbf{a}$ **e** $\mathbf{b} - \mathbf{a}$ **f** $2\mathbf{b}$

5 $\mathbf{e} = \begin{pmatrix} -3 \\ -2 \end{pmatrix}$ and $\mathbf{f} = \begin{pmatrix} 2 \\ 4 \end{pmatrix}$.

     Find:

     a $\mathbf{e} + \mathbf{f}$ **b** $3\mathbf{f}$ **c** $\mathbf{e} - \mathbf{f}$ **d** $\mathbf{f} - \mathbf{e}$ **e** $4\mathbf{e}$ **f** $2\mathbf{f} + \mathbf{e}$

6

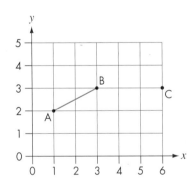

     a   Copy this diagram.

     b $\overrightarrow{AE} = 2\overrightarrow{AB}$. Mark E on the grid.

     c $\overrightarrow{CD} = -2\overrightarrow{AB}$. Mark D on the grid.

     d $\overrightarrow{AB} = 2\overrightarrow{AM}$. Mark M on the grid.

     e $\overrightarrow{CN} = 2\overrightarrow{CB}$. Mark N on the grid.

     f   If $\overrightarrow{DC} = k\overrightarrow{AM}$, what is the value of k?

Using vectors

This grid is made of identical parallelograms.

O is the origin.

The position vector of A is

$$\overrightarrow{OA} = \mathbf{a}$$

The **position vector** of B is

$$\overrightarrow{OB} = \mathbf{b}$$

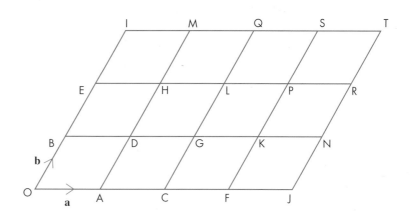

We can give the position vectors of the other points in terms of **a** and **b**. For example:

The position vector of G = $\overrightarrow{OG}$ = 2**a** + **b**.

The position vector of S = $\overrightarrow{OS}$ = 3**a** + 3**b**.

We can write other vectors in terms of **a** and **b**. For example:

$$\overrightarrow{CL} = 2\mathbf{b}$$
$$\overrightarrow{CP} = \overrightarrow{CL} + \overrightarrow{LP} = 2\mathbf{b} + \mathbf{a} \text{ or } \mathbf{a} + 2\mathbf{b}$$
$$\overrightarrow{CH} = \overrightarrow{CL} + \overrightarrow{LH} = 2\mathbf{b} + -\mathbf{a} = 2\mathbf{b} - \mathbf{a}$$

EXAMPLE 1

a Using the grid above, write down the following vectors in terms of **a** and **b**.

i $\overrightarrow{BH}$ ii $\overrightarrow{HP}$ iii $\overrightarrow{GT}$

iv $\overrightarrow{TI}$ v $\overrightarrow{FH}$ vi $\overrightarrow{BQ}$

b What is the relationship between the following vectors?

i $\overrightarrow{BH}$ and $\overrightarrow{GT}$ ii $\overrightarrow{BQ}$ and $\overrightarrow{GT}$ iii $\overrightarrow{HP}$ and $\overrightarrow{TI}$

c Show that B, H and Q lie on the same straight line.

a i **a** + **b**   ii 2**a**   iii 2**a** + 2**b**   iv −4**a**   v −2**a** + 2**b**   vi 2**a** + 2**b**

b i $\overrightarrow{BH}$ and $\overrightarrow{GT}$ are parallel and $\overrightarrow{GT}$ is twice the length of $\overrightarrow{BH}$.

ii $\overrightarrow{BQ}$ and $\overrightarrow{GT}$ are equal.

iii $\overrightarrow{HP}$ and $\overrightarrow{TI}$ are in opposite directions and $\overrightarrow{TI}$ is twice the length of $\overrightarrow{HP}$.

c $\overrightarrow{BH}$ and $\overrightarrow{BQ}$ are parallel and start at the same point B. Therefore, B, H and Q must lie on the same straight line.

EXERCISE 29B

1 On this grid O is the origin, $\overrightarrow{OA}$ is **a** and $\overrightarrow{OB}$ is **b**.

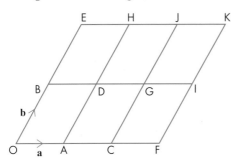

  a Name three other vectors equivalent to **a**.

  b Name three other vectors equivalent to **b**.

  c Name three vectors equivalent to –**a**.

  d Name three vectors equivalent to –**b**.

2 Using the same grid as in question **1**, give the following vectors in terms of **a** and **b**.

  a The position of vector C.   **b** The position of vector E.

  c The position of vector K.   **d** $\overrightarrow{OH}$

  e $\overrightarrow{AG}$ **f** $\overrightarrow{AK}$ **g** $\overrightarrow{BK}$

3 On the grid in question **1**, there are three vectors equivalent to $\overrightarrow{OG}$. Name all three.

4 On the grid in question **1**, there are three vectors that are three times the magnitude of and in the same direction. Name all three.

5 On a copy of this grid, mark on the points C to G to show the following.

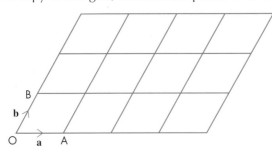

  a $\overrightarrow{OC} = 2\mathbf{a} + 3\mathbf{b}$ **b** $\overrightarrow{OD} = 2\mathbf{a} + \mathbf{b}$

  c $\overrightarrow{OE} = 4\mathbf{a}$ **d** $\overrightarrow{OF} = 4\mathbf{a} + 2\mathbf{b}$

  e $\overrightarrow{OG} = \frac{1}{2}\mathbf{a} + 2\mathbf{b}$

6 On this grid, $\overrightarrow{OA}$ is **a** and $\overrightarrow{OB}$ is **b**.

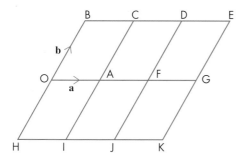

Give the following vectors in terms of **a** and **b**.

a $\overrightarrow{OH}$

b $\overrightarrow{OK}$

c $\overrightarrow{OJ}$

d $\overrightarrow{OI}$

e $\overrightarrow{OC}$

f $\overrightarrow{CO}$

g $\overrightarrow{AK}$

h $\overrightarrow{DI}$

i $\overrightarrow{JE}$

j $\overrightarrow{AB}$

k $\overrightarrow{CK}$

l $\overrightarrow{DK}$

7 **a** On the grid in question **6**, there are two vectors that are twice the size of $\overrightarrow{AB}$ and in the opposite direction. Name both of them.

b On the grid in question **6**, there are three vectors that are three times the size of $\overrightarrow{OA}$ and in the opposite direction. Name all three.

8 On a copy of this grid, mark on the points C to P to show the following.

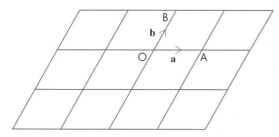

a $\overrightarrow{OC} = 2\mathbf{a} - \mathbf{b}$

b $\overrightarrow{OD} = 2\mathbf{a} + \mathbf{b}$

c $\overrightarrow{OE} = \mathbf{a} - 2\mathbf{b}$

d $\overrightarrow{OF} = \mathbf{b} - 2\mathbf{a}$

e $\overrightarrow{OG} = -\mathbf{a}$

f $\overrightarrow{OH} = -\mathbf{a} - 2\mathbf{b}$

g $\overrightarrow{OI} = 2\mathbf{a} - 2\mathbf{b}$

h $\overrightarrow{OJ} = -\mathbf{a} + \mathbf{b}$

i $\overrightarrow{OK} = -\mathbf{a} - \mathbf{b}$

j $\overrightarrow{OM} = -\mathbf{a} - \frac{3}{2}\mathbf{b}$

k $\overrightarrow{ON} = -\frac{1}{2}\mathbf{a} - 2\mathbf{b}$

l $\overrightarrow{OP} = \frac{3}{2}\mathbf{a} - \frac{3}{2}\mathbf{b}$

9 The diagram shows two sets of parallel lines. O is the origin.

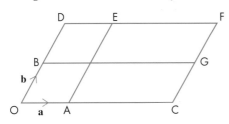

$\overrightarrow{OA} = \mathbf{a}$ and $\overrightarrow{OB} = \mathbf{b}$

$\overrightarrow{OC} = 3\overrightarrow{OA}$ and $\overrightarrow{OD} = 2\overrightarrow{OB}$

a Write down the following vectors in terms of **a** and **b**.

  i $\overrightarrow{OF}$ **ii** $\overrightarrow{OG}$ **iii** $\overrightarrow{EG}$ **iv** $\overrightarrow{CE}$

b Write down two vectors that can be written as $3\mathbf{a} - \mathbf{b}$.

10 This grid shows the vectors $\overrightarrow{OA} = \mathbf{a}$ and $\overrightarrow{OB} = \mathbf{b}$. O is the origin.

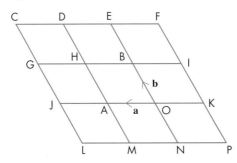

Give the position vectors of the following points:

a G   **b** F

c The midpoint of DH   **d** The centre of OAHB

e The centre of DCGH   **f** The centre of AMLJ

11 The diagram shows the vectors $\overrightarrow{OA} = \mathbf{a}$ and $\overrightarrow{OB} = \mathbf{b}$. M is the midpoint of AB.

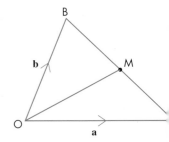

a **i** Work out the vector $\overrightarrow{AB}$ in terms of **a** and **b**.

    ii Work out the vector $\overrightarrow{AM}$.

    iii Explain why $\overrightarrow{OM} = \overrightarrow{OA} + \overrightarrow{AM}$.

    iv Using your answers to parts **ii** and **iii**, work out $\overrightarrow{OM}$ in terms of **a** and **b**.

b Copy the diagram and show on it the vector $\overrightarrow{OC}$ which is equal to $\mathbf{a} + \mathbf{b}$.

c Describe in geometrical terms the position of M in relation to O, A, B and C.

12 The diagram shows the vectors $\overrightarrow{OA}$ = **a** and $\overrightarrow{OB}$ = **b**.
The point C divides the line AB in the ratio 1:2.

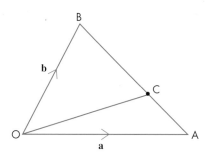

a **i** Work out the vector $\overrightarrow{AB}$.

   ii Work out the vector $\overrightarrow{AC}$.

   iii Work out the vector $\overrightarrow{OC}$ in terms of **a** and **b**.

b If C now divides the line AB in the ratio 1 : 3, write down the vector that represents $\overrightarrow{OC}$.

13 The diagram shows the vectors $\overrightarrow{OA}$ = **a** and $\overrightarrow{OB}$ = **b**.

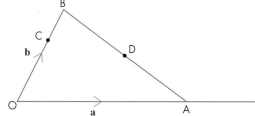

The point C divides OB in the ratio 2 : 1. The point E is such that $\overrightarrow{OE} = 2\overrightarrow{OA}$.
D is the midpoint of AB.

a Write down (or work out) these vectors in terms of **a** and **b**.

   i $\overrightarrow{OC}$ **ii** $\overrightarrow{OD}$ **iii** $\overrightarrow{CO}$

b The vector $\overrightarrow{CD}$ can be written as $\overrightarrow{CD} = \overrightarrow{CO} + \overrightarrow{OD}$. Use this fact to work out $\overrightarrow{CD}$ in terms of **a** and **b**.

c Write down a similar rule to that in part **b** for the vector $\overrightarrow{DE}$. Use this rule to work out $\overrightarrow{DE}$ in terms of **a** and **b**.

d Explain why C, D and E lie on the same straight line.

The **magnitude** of a vector is represented by two vertical lines which stand for 'magnitude of' 'length of', eg $|\overrightarrow{AB}|$ or $|\mathbf{a}|$.

If a vector is drawn on a rectangular coordinate grid we can calculate the magnitude using **Pythagoras' theorem**.

For example,

if $\overrightarrow{AB} = \begin{pmatrix} 4 \\ -3 \end{pmatrix}$

we know that it can form the hypotenuse of a triangle with sides of lengths 3 and 4 as shown on the grid below.

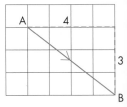

The square of the hypotenuse is equal to the sum of the squares of the other two sides so:

$$|\overrightarrow{AB}| = \sqrt{4^2 + 3^2}$$

$$= \sqrt{25}$$

$$= 5$$

In general, if $\mathbf{a} = \begin{pmatrix} x \\ y \end{pmatrix}$ then $|\mathbf{a}| = \sqrt{x^2 + y^2}$.

EXAMPLE 2

If A has coordinates $(3, -2)$ and B has coordinates $(-3, 5)$, find $|\overrightarrow{AB}|$.

$$\overrightarrow{AB} = \begin{pmatrix} -6 \\ 7 \end{pmatrix}$$

$$|\overrightarrow{AB}| = \sqrt{(-6)^2 + 7^2}$$

$$= \sqrt{36 + 49}$$

$$= \sqrt{85}$$

$$= 9.22 \text{ to 2 decimal places}$$

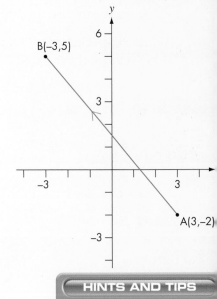

HINTS AND TIPS

Remember that $(-6)^2 = 6^2$

EXERCISE 29C

1 O is the origin.

P, Q and R have position vectors $\begin{pmatrix} 3 \\ 5 \end{pmatrix}$, $\begin{pmatrix} 6 \\ -2 \end{pmatrix}$, $\begin{pmatrix} 0 \\ -4 \end{pmatrix}$.

a Show O, P, Q and R on a diagram.

b Find $|\overrightarrow{OP}|$, $|\overrightarrow{OQ}|$, and $|\overrightarrow{OR}|$. You can leave square root signs in your answers.

c Find $|\overrightarrow{PQ}|$.

d Find $|\overrightarrow{QR}|$.

2 $\mathbf{a} = \begin{pmatrix} 6 \\ 8 \end{pmatrix}$ and $\mathbf{b} = \begin{pmatrix} 5 \\ -12 \end{pmatrix}$

a Find $|\mathbf{a}|$ and $|\mathbf{b}|$.

b Find $\mathbf{a} + \mathbf{b}$.

c Find $|\mathbf{a} + \mathbf{b}|$.

d Is it true that $|\mathbf{a} + \mathbf{b}| = |\mathbf{a}| + |\mathbf{b}|$? Give a reason for your answer.

e Find $|\mathbf{a} - \mathbf{b}|$.

f Find $|\mathbf{b} - \mathbf{a}|$.

g Is it always true that $|\mathbf{a} - \mathbf{b}| = |\mathbf{b} - \mathbf{a}|$? Give a reason for your answer.

3 A, B, C and D have position vectors $\begin{pmatrix} 4 \\ 4 \end{pmatrix}$, $\begin{pmatrix} 10 \\ 12 \end{pmatrix}$, $\begin{pmatrix} -4 \\ 10 \end{pmatrix}$ and $\begin{pmatrix} 4 \\ -6 \end{pmatrix}$.

a Find $|\overrightarrow{AB}|$, $|\overrightarrow{AC}|$ and $|\overrightarrow{AD}|$.

b Explain why B, C and D must lie on a circle with centre A and state the radius of the circle.

4 $\mathbf{c} = \begin{pmatrix} 1 \\ 4 \end{pmatrix}$ and $\mathbf{d} = \begin{pmatrix} -2 \\ 5 \end{pmatrix}$. Find:

a $|\mathbf{c}|$

b $|3\mathbf{d}|$

c $|2\mathbf{c} + \mathbf{d}|$

d $|4\mathbf{c} - 2\mathbf{d}|$

How many sides does a strip of paper have? Two or one?

Take a strip of paper about 20 cm by 2 cm.

How many sides does it have? Easy! You can see that this has two sides, a topside and an underside. If you were to draw a line along one side of the strip, you would have one side with a line 20 cm long on it and one side blank.

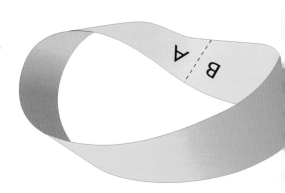

Now mark the ends A and B, put a single twist in the strip of paper and tape (or glue) the two ends together, as shown.

How many sides does this strip of paper have now?

Take a pen and draw a line on the paper, starting at any point you like. Continue the line along the length of the paper – you will eventually come back to your starting point. Your strip has only one side now! There is no blank side.

You have transformed a two-sided piece of paper into a one-sided piece of paper.

This shape is called a Möbius strip. It is named after Aug Ferdinand Möbius, a 19th-century German mathematici. Möbius caused a revolution in geometry.

Möbius strips have a number of applications that use its property of one-sidedness, including conveyor belts in industry and in vacuum cleaners.

The Möbius strip has become the universal symbol of recycling. The symbol was created in 1970 by Gary Anderson at the University of Southern California, as part of a contest sponsored by a paper company.

The Möbius strip is a form of transformation. In this chapter, you will look at some other transformations of shapes.

30 Transformations

ics	Level	Key words
1 Translations	FOUNDATION	transformation, translation, vector
2 Reflections	FOUNDATION	reflection, object, image, mirror line
3 Further reflections	FOUNDATION	
4 Rotations	FOUNDATION	rotation, centre of rotation, angle of rotation, clockwise, anticlockwise, positive, negative
5 Further rotations	FOUNDATION	
6 Enlargements	FOUNDATION	scale factor, enlargement, centre of enlargement, ray method, coordinate method, fractional enlargement

What you need to be able to do in the examinations:

FOUNDATION

- Understand that rotations are specified by a centre and an angle.
- Rotate a shape about a point through a given angle.
- Recognise that an anti-clockwise rotation is a *positive* angle of rotation and a clockwise rotation is a *negative* angle of rotation.
- Understand that reflections are specified by a mirror line.
- Construct a mirror line given an object and reflect a shape given a mirror line.
- Understand that translations are specified by a distance and direction and translate a shape.
- Understand that rotations, reflections and translations preserve length and angle so that a transformed shape under any of these transformations remains congruent to the original shape.
- Understand that enlargements are specified by a centre and a scale factor.
- Understand that enlargements preserve angles and not lengths.
- Enlarge a shape given the scale factor.
- Identify and give complete descriptions of transformations.

A **transformation** changes the position or the size of a shape.

There are four basic ways of changing the position and size of two-dimensional shapes: a **translation**, a reflection, a rotation or an enlargement. All of these transformations, except enlargement, keep shapes congruent.

A translation is the 'movement' of a shape from one place to another without reflecting it or rotating it. It is sometimes called a glide, since the shape appears to glide from one place to another. Every point in the shape moves in the same direction and through the same distance.

We describe translations by using **vectors**. A vector is represented by the combination of a horizontal shift and a vertical shift.

EXAMPLE 1

Use vectors to describe the translations of the following triangles.

a A to B

b B to C

c C to D

d D to A

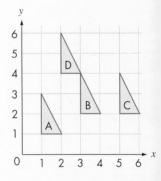

a The vector describing the translation from A to B is $\begin{pmatrix} 2 \\ 1 \end{pmatrix}$.

b The vector describing the translation from B to C is $\begin{pmatrix} 2 \\ 0 \end{pmatrix}$.

c The vector describing the translation from C to D is $\begin{pmatrix} -3 \\ 2 \end{pmatrix}$.

d The vector describing the translation from D to A is $\begin{pmatrix} -1 \\ -3 \end{pmatrix}$.

Note:

● The top number in the vector describes the horizontal movement. To the right +, to the left −.

● The bottom number in the vector describes the vertical movement. Upwards +, downwards −.

● These vectors are also called *direction vectors*.

EXERCISE 30A

1 Use vectors to describe the following translations of the shapes on the grid below.

a **i** A to B
  ii A to C
  iii A to D

b **i** B to E
  ii B to F
  iii B to G

c **i** C to A
  ii C to E
  iii C to G

d **i** G to D
  ii F to G
  iii G to E

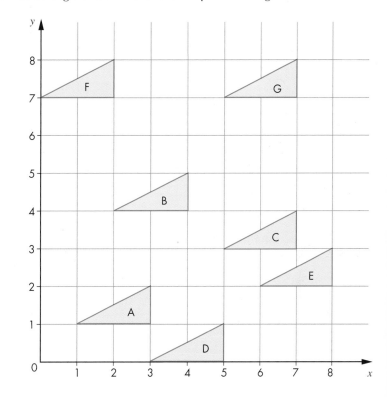

2 **a** Draw a set of coordinate axes and on it the triangle with coordinates A(1, 1), B(2, 1) and C(1, 3).

b Draw the image of ABC after a translation with vector $\begin{pmatrix} 2 \\ 3 \end{pmatrix}$. Label this triangle P.

c Draw the image of ABC after a translation with vector $\begin{pmatrix} -1 \\ 2 \end{pmatrix}$. Label this triangle Q.

d Draw the image of ABC after a translation with vector $\begin{pmatrix} 3 \\ -2 \end{pmatrix}$. Label this triangle R.

e Draw the image of ABC after a translation with vector $\begin{pmatrix} -2 \\ -4 \end{pmatrix}$. Label this triangle S.

3 Using your diagram from question **2**, use vectors to describe the translation that will move:

a P to Q   **b** Q to R   **c** R to S   **d** S to P

e R to P   **f** S to Q   **g** R to Q   **h** P to S.

4 If a translation is given by:

$$\begin{pmatrix} x \\ y \end{pmatrix}$$

describe the translation that would take the image back to the original position.

5 A boat travels between three jetties X, Y and Z on a lake. It uses direction vectors, with distance in kilometres.

The direction vector from X to Y is $\begin{pmatrix} 3 \\ -1 \end{pmatrix}$ and the direction vector from Y to Z is $\begin{pmatrix} -2 \\ -3 \end{pmatrix}$.

Using centimetre-squared paper, draw a diagram to show journeys between X, Y and Z. Use a scale of 1 cm represents 1 km. Work out the direction vector for the journey from Z to X.

30.2 Reflections

A **reflection** transforms a shape so that it becomes a mirror image of itself.

EXAMPLE 2

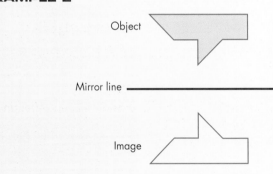

Object

Mirror line

Image

Notice the reflection of each point in the original shape, called the **object**, is perpendicular to the mirror line. So if you 'fold' the whole diagram along the **mirror line**, the object will coincide with its reflection, called its **image**.

EXERCISE 30B

1 Copy the diagram below and draw the reflection of the given triangle in the following lines.

a $x = 2$

b $x = -1$

c $x = 3$

d $y = 2$

e $y = -1$

f y-axis

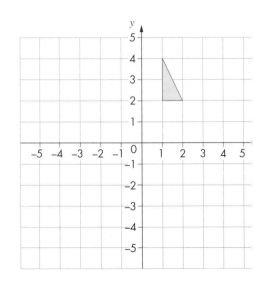

2 **a** Draw a pair of axes. Label the *x*-axis from –5 to 5 and the *y*-axis from –5 to 5.

b Draw the triangle with coordinates A(1, 1), B(3, 1), C(4, 5).

c Reflect the triangle ABC in the *x*-axis. Label the image P.

d Reflect triangle P in the *y*-axis. Label the image Q.

e Reflect triangle Q in the *x*-axis. Label the image R.

f Describe the reflection that will move triangle ABC to triangle R.

3 **a** Draw a pair of axes. Label the *x*-axis from –5 to +5 and the *y*-axis from –5 to +5.

b Reflect the points A(2, 1), B(5, 0), C(–3, 3), D(3, –2) in the *x*-axis.

c What do you notice about the values of the coordinates of the reflected points?

d What would the coordinates of the reflected point be if the point (*a*, *b*) were reflected in the *x*-axis?

4 **a** Draw a pair of axes. Label the *x*-axis from –5 to +5 and the *y*-axis from –5 to +5.

b Reflect the points A(2, 1), B(0, 5), C(3, –2), D(–4, –3) in the *y*-axis.

c What do you notice about the values of the coordinates of the reflected points?

d What would the coordinates of the reflected point be if the point (*a*, *b*) were reflected in the *y*-axis?

5 By using the middle square as a starting square ABCD, describe how to keep reflecting the square to obtain the final shape in the diagram.

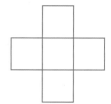

6 Triangle A is drawn on a grid.

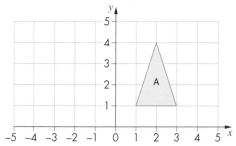

Triangle A is reflected to form a new triangle B.
The coordinates of B are (–4, 4), (–3, 1) and (–5, 1).

Work out the equation of the mirror line.

Up to now we have just looked at reflections in horizontal or vertical lines.

We can reflect a shape in any line.

EXAMPLE 3

Draw the reflection of triangle t in the line with equation $y = x$.

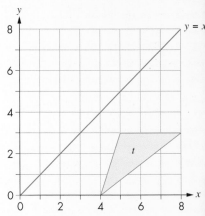

To find the image of each vertex of the triangle, draw lines perpendicular to the mirror.

Each vertex and its image are the same distance from the mirror but on opposite sides. Use the grid to help you find the new vertices.

Join the new vertices to draw the reflection M(t) of the triangle t.

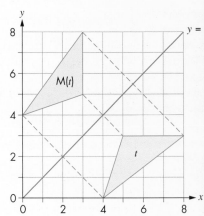

1. A designer used the following instructions to create a design.

   • Start with any rectangle ABCD.

   • Reflect the rectangle ABCD in the line AC.

   • Reflect the rectangle ABCD in the line BD.

   Draw a rectangle and use the above to create a design.

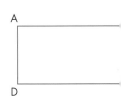

2 Draw each of these triangles on squared paper, leaving plenty of space on the opposite side of the given mirror line. Then draw the reflection of each triangle.

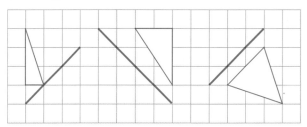

> **HINTS AND TIPS**
>
> Turn the page around so that the mirror lines are vertical or horizontal.

3 **a** Draw a pair of axes and the lines $y = x$ and $y = -x$, as shown.

b Draw the triangle with coordinates A(2, 1), B(5, 1), C(5, 3).

c Draw the reflection of triangle ABC in the x-axis and label the image P.

d Draw the reflection of triangle P in the line $y = -x$ and label the image Q.

e Draw the reflection of triangle Q in the y-axis and label the image R.

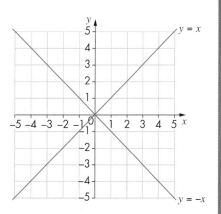

f Draw the reflection of triangle R in the line $y = x$ and label the image S.

g Draw the reflection of triangle S in the x-axis and label the image T.

h Draw the reflection of triangle T in the line $y = -x$ and label the image U.

i Draw the reflection of triangle U in the y-axis and label the image W.

j What single reflection will move triangle W to triangle ABC?

4 Copy the diagram and reflect the triangle in the following lines.

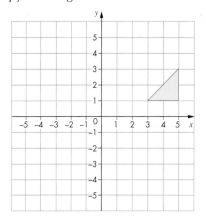

a $y = x$ **b** $x = 1$

c $y = -x$ **d** $y = -1$

5 **a** Draw a pair of axes. Label the *x*-axis from −5 to +5 and the *y*-axis from −5 to +5.

b Draw the line *y* = *x*.

c Reflect the points A(2, 1), B(5, 0), C(−3, 2), D(−2, −4) in the line *y* = *x*.

d What do you notice about the values of the coordinates of the reflected points?

e What would the coordinates of the reflected point be if the point (*a*, *b*) were reflected in the line *y* = *x*?

6 **a** Draw a pair of axes. Label the *x*-axis from −5 to +5 and the *y*-axis from −5 to +5.

b Draw the line *y* = −*x*.

c Reflect the points A(2, 1), B(0, 5), C(3, −2), D(−4, −3) in the line *y* = −*x*.

d What do you notice about the values of the coordinates of the reflected points?

e What would the coordinates of the reflected point be if the point (*a*, *b*) were reflected in the line *y* = −*x*?

30.4 Rotations

A **rotation** transforms a shape to a new position by turning it about a fixed point called the **centre of rotation**.

EXAMPLE 4

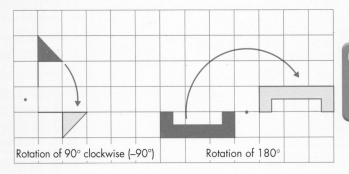

Rotation of 90° clockwise (−90°) Rotation of 180°

> **HINTS AND TIPS**
>
> Use tracing paper to check rotations.

Note:

● The direction of turn or the **angle of rotation** is expressed as **clockwise** or **anticlockwise** or as **negative** and **positive**.

● Positive rotations are anticlockwise. Negative rotations are clockwise.

● The position of the centre of rotation is always specified.

● The rotations 180° clockwise and 180° anticlockwise are the same.

The rotations that most often appear in examination questions are 90° and 180°.

FOUNDATION

EXERCISE 30D

1 On squared paper, draw each of these shapes and its centre of rotation, leaving plenty of space all round the shape.

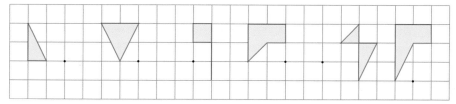

a Rotate each shape about its centre of rotation:

    i first by 90° clockwise (call the image A)

    ii then by 90° anticlockwise (call the image B).

b Describe, in each case, the rotation that would take:

    i A back to its original position   **ii** A to B.

2 A graphics designer came up with the following routine for creating a design.

- Start with a triangle ABC.

- Reflect the triangle in the line AB.

- Rotate the whole shape about point C 90°, then a further 90°, then a further 90°.

From any triangle of your choice, create a design using the above routine.

3 By using the middle square as a starting square ABCD, describe how to keep rotating the square to obtain the final shape in the diagram.

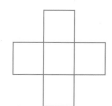

4 Copy the diagram and rotate the given triangle by the following.

a −90° about (0, 0)

b 180° about (0, 0)

c 90° about (1, 4)

d 180° about (1, 3)

e −90° about (2, 2)

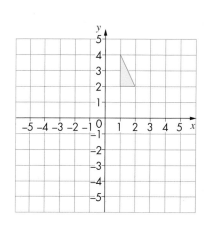

FOUNDATION

5

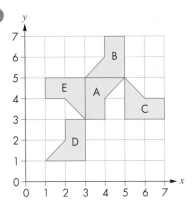

Give the centre and the angle for the rotations that will take:

a A onto B

b A onto C

c A onto D

d A onto E

6

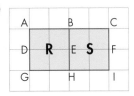

a A 180° rotation will take R onto S. Where is the centre of rotation?

b A –90° rotation will take R onto S. Where is the centre?

7 **a** Draw a pair of axes where both the x-values and y-values are from –5 to 5.

b Draw the triangle ABC, where A = (1, 2), B = (2, 4) and C = (4, 1).

c **i** Rotate triangle ABC 90° clockwise about the origin (0, 0) and label the image A′, C′, where A′ is the image of A, etc.

ii Write down the coordinates of A′, B′, C′.

iii What connection is there between A, B, C and A′, B′, C′?

iv Will this connection always be so for a 90° clockwise rotation about the origin?

8 Repeat question **7**, but rotate triangle ABC through 180°.

9 Show that a reflection in the x-axis followed by a reflection in the y-axis is equivalent to rotation of 180° about the origin.

FOUNDATION

1 Draw x and y axes from 0 to 12.

Draw the triangle with vertices at (5, 5), (7, 5) and (7, 8). Label it T.

a Rotate T 180° about (4, 5). Label the new triangle A.

b Rotate T –90° about (7, 4). Label the new triangle B.

c Rotate T 90° about (7, 9). Label the new triangle C.

d What rotation will take triangle B onto triangle C?

2 **a** A 180° rotation will take square A onto square B.
Where is the centre of the rotation?

b A 90° *clockwise* rotation will take A onto B.
Where is the centre of the rotation?

c A 90° *anticlockwise* rotation will take A onto B.
Where is the centre of the rotation?

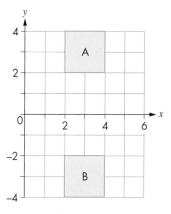

3 Give the centre and angle for the following rotations:

a A onto B　　　　**b** B onto C

c C onto D　　　　**d** D onto A

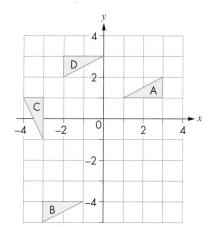

4 Show that a reflection in the line $y = x$ followed by a reflection in the line $y = -x$ is equivalent to a rotation of 180° about the origin.

5 **a** Draw a regular hexagon ABCDEF with centre O. The letters should go round the hexagon clockwise.

b Using O as the centre of rotation, describe a transformation that will result in the following movements.

   i Triangle AOB to triangle BOC   **ii** Triangle AOB to triangle COD

   iii Triangle AOB to triangle DOE   **iv** Triangle AOB to triangle EOF

c Describe the transformations that will move the rhombus ABCO to these positions.

   i Rhombus BCDO   **ii** Rhombus DEFO

6 Triangle A, as shown on the grid, is rotated to form a new triangle B.

The coordinates of the vertices of B are (0, –2), (–3, –2) and (–3, –4).

Describe fully the rotation that maps triangle A onto triangle B.

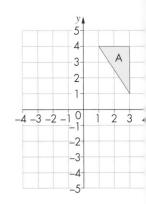

30.6 Enlargements

An **enlargement** changes the size of an object but all the angles stay the same. The lengths of sides all multiply by the same **scale factor**.

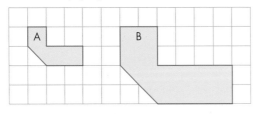

B is an enlargement of A with a scale factor of 2. Every side of B is twice as long as the corresponding side of A.

Very often an enlargement will have a **centre of enlargement** as well as a scale factor.

Every length of the enlarged shape will be:

original length × scale factor

The distance of each image point on the enlargement from the centre of enlargement will be:

distance of original point from centre of enlargement × scale factor

EXAMPLE 5

The diagram shows the enlargement of triangle ABC by scale factor 3 about the centre of enlargement X.

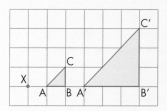

Note:

● Each length on the enlargement A′B′C′ is three times the corresponding length on the original shape.

This means that the corresponding sides are in the same ratio:

AB : A′B′ = AC : A′C′ = BC : B′C′ = 1 : 3

● The distance of any point on the enlargement from the centre of enlargement is three times the distance from the corresponding point on the original shape to the centre of enlargement.

There are two distinct ways to enlarge a shape: the **ray method** and the **coordinate method** (counting squares).

Ray method

This is the *only* way to construct an enlargement when the diagram is not on a grid.

EXAMPLE 6

Enlarge triangle ABC by scale factor 3 about the centre of enlargement X.

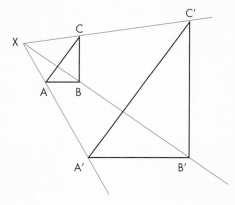

Notice that the rays have been drawn from the centre of enlargement to each vertex and beyond.

The distance from X to each vertex on triangle ABC is measured and multiplied by 3 to give the distance from X to each vertex A′, B′ and C′ for the enlarged triangle A′B′C′.

Once each image vertex has been found, the whole enlarged shape can then be drawn.

Check the measurements and see for yourself how the calculations have been done.

Notice again that the length of each side on the enlarged triangle is three times the length of the corresponding side on the original triangle.

Coordinate method

In this method, you use the coordinates of the vertices to 'count squares'.

EXAMPLE 7

Enlarge the triangle ABC by scale factor 3 from the centre of enlargement (1, 2).

To find the coordinates of each image vertex, first work out the horizontal and vertical distances from each original vertex to the centre of enlargement.

Then multiply each of these distances by 3 to find the position of each image vertex.

For example, to find the coordinates of C′ work out the distance from the centre of enlargement (1, 2) to the point C(3, 5).

    horizontal distance = 2

    vertical distance = 3

Make these 3 times longer to give:

    new horizontal distance = 6

    new vertical distance = 9

So the coordinates of C′ are: (1 + 6, 2 + 9) = (7, 11)

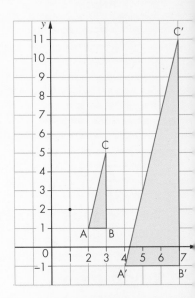

Notice again that the length of each side is three times as long in the enlargement.

Fractional enlargement

Strange but true … you can have an enlargement in mathematics that is actually smaller than the original shape! This happens when the scale factor is a fraction and is called **fractional enlargement**.

EXAMPLE 8

Triangle ABC has been enlarged by a scale factor of $\frac{1}{2}$ about the centre of enlargement 0 to give triangle A′B′C′.

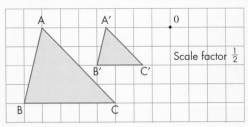

Scale factor $\frac{1}{2}$

EXERCISE 30F

1 Use squared paper to draw an enlargement of each of these shapes. Use the given scale factor.

a   **b**   **c**   **d**

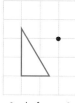 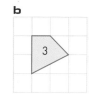 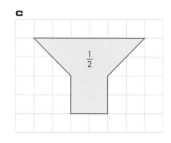 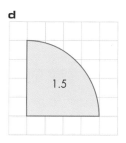

2 Copy each of these figures with its centre of enlargement. Then enlarge it by the given scale factor, using the ray method.

a   **b**   **c**   **d**

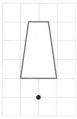 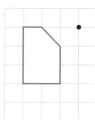 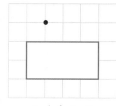

     Scale factor 2   Scale factor 3   Scale factor 2   Scale factor 3

3 Copy each of these diagrams onto squared paper and enlarge it by scale factor 2, using the origin as the centre of enlargement.

a   **b**

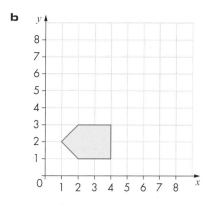

c

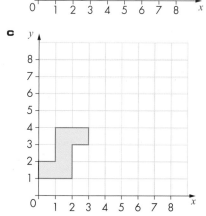

> ### HINTS AND TIPS
>
> Even if you are using a counting square method, you can always check by using the ray method.

4 Enlarge each of these shapes by a scale factor of $\frac{1}{2}$ about its closest centre of enlargement

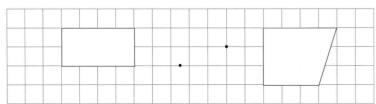

5 Copy this diagram onto squared paper.

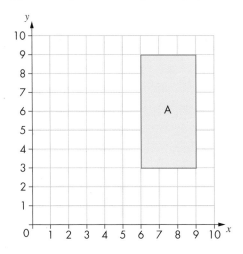

a Enlarge the rectangle A by scale factor $\frac{1}{3}$ about the origin. Label the image B.

b Write down the ratio of the lengths of the sides of rectangle A to the lengths of the si of rectangle B.

c Work out the ratio of the perimeter of rectangle A to the perimeter of rectangle B.

d Work out the ratio of the area of rectangle A to the area of rectangle B.

6 Copy each of these diagrams onto squared paper and enlarge it by scale factor 2, using the given centre of enlargement.

a

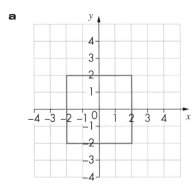

Centre of enlargement (−1, 1)

b

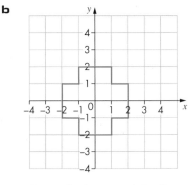

Centre of enlargement (−2, −3)

7 Copy the diagram onto squared paper.

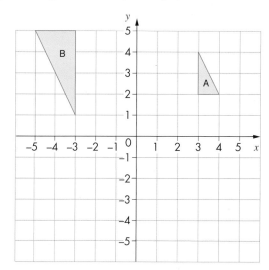

a Enlarge A by a scale factor of 3 about a centre (4, 5).

b Enlarge B by a scale factor $\frac{1}{2}$ about a centre (–1, –3).

FOUNDATION

1 The diagram shows a shape on a centimetre grid.

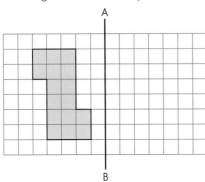

a Write down the order of rotational symmetry of the shape. [1]

b Work out the perimeter of the shape. [1]

c Work out the area of the shape. [1]

d Reflect the shape in the line *AB*. [2]

Edexcel Limited Paper 1F Q9 Jan 15

2 The diagram shows a shape with one line of symmetry.

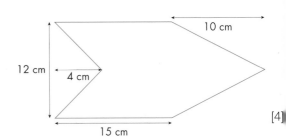

Work out the area of the shape. [4]

Edexcel Limited Paper 1F Q16 Jan 15

3 A steam engine for pulling trains has wheels of diameter 1.5 metres.

a Calculate the circumference of a wheel.

Give your answer correct to 3 significant figures. [2]

The steam engine travels 1000 metres along a test track.

b Work out the number of complete turns of a wheel. [2]

Edexcel Limited Paper 1F Q18 May 15

4

Diagram NOT
accurately drawn

A solid cylinder has a radius of 5.1 cm and a height of 3.7 cm.

Work out the **total** surface area of the cylinder.

Give your answer to 3 significant figures. [3]

Edexcel Limited Paper 1F Q19 May 13

5

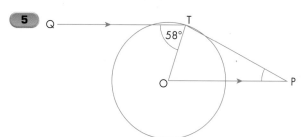

Edexcel Limited Paper 1F Q19 Jan 14

T is a point on a circle, centre *O*.

Q is a point such that angle *QTO* = 58°

P is the point such that *OP* is parallel to *QT* and *PT* is a tangent to the circle.

Work out the size of angle *OPT*.　　[3]

6

A cylinder has radius 5.4 cm and height 16 cm.

a Work out the volume of the cylinder.

Give your answer correct to the nearest whole number.　　[2]

The radius 5.4 cm is correct to 2 significant figures.

b i Write down the upper bound of the radius.　　[1]

　　ii Write down the lower bound of the radius.　　[1]

Edexcel Limited Paper 1F Q19 Jan 16

7 Here is a prism.

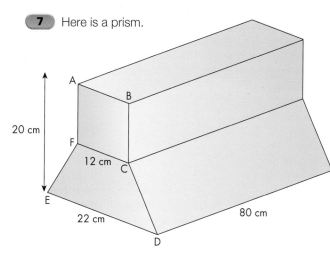

ABCDEF is a cross section of the prism.

ABCF is a square of side 12 cm.

FCDE is a trapezium.

ED = 22 cm.

The height of the prism is 20 cm.

The length of the prism is 80 cm.

Work out the total volume of the prism.　[5]

Edexcel Limited Paper 1F Q23 May 15

8

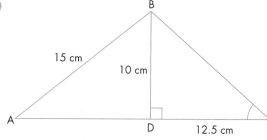

ABC is a triangle.

The point *D* lies on *AC*.

Angle *BDC* = 90°

BD = 10 cm, *AB* = 15 cm and *DC* = 12.5 cm.

a Calculate the length of *AD*.

Give your answer correct to 3 significant figures.　　[3]

b Calculate the size of angle *BCD*.

Give your answer correct to 1 decimal place.　　[3]

Edexcel Limited Paper 1F Q21 Jan 15

FOUNDATION

PAPER 2F

1

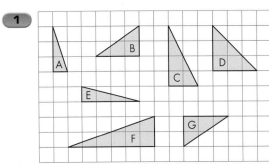

a Write down the letters of the two triangles that are congruent. [1]

b One of the triangles is similar to triangle A. Write down the letter of this triangle. [1]

c One of the triangles is isosceles. Write down the letter of this triangle. [1]

Edexcel Limited Paper 2F Q7 May 14

2

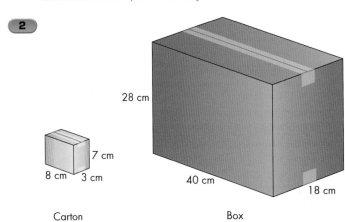

Carton   Box

A carton measures 8 cm by 3 cm by 7 cm.

Cartons are packed into boxes.

A box measures 40 cm by 18 cm by 28 cm.

Work out the number of cartons that can completely fill one box. [3]

Edexcel Limited Paper 2F Q12 Jan 15

3

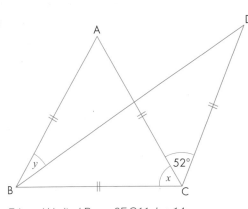

The diagram shows an equilateral triangle *ABC* and an isosceles triangle *BCD*.

$AB = AC = BC = CD$.

Angle $ACD = 52°$

Angle $ACB = x°$

a Find the value of x. [1]

Angle $ABD = y°$

b Work out the value of y. [3]

Edexcel Limited Paper 2F Q11 Jan 14

4

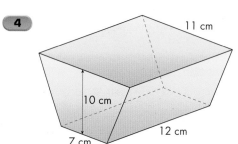

The diagram shows a solid prism.

The cross section of the prism is a trapezium.

The lengths of the parallel sides of the trapezium are 11 cm and 7 cm.

The perpendicular distance between the parallel sides of the trapezium is 10 cm.

The length of the prism is 12 cm.

a How many faces has the prism? [1]

b How many vertices has this prism? [1]

c Work out the area of the trapezium. [2]

d Work out the volume of the prism. [2]

Edexcel Limited Paper 2F Q14 Jan 14

5

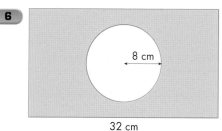

Work out the area of this shape. [4]

Edexcel Limited Paper 2F Q10 Jan 16

6

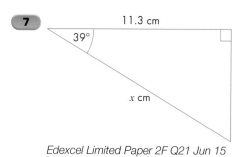

The diagram shows a circle inside a rectangle.

Work out the area of the shaded region.

Give your answer to 3 significant figures.

Edexcel Limited Paper 2F Q15 Jan 15

7

Work out the value of x.

Give your answer to 2 decimal places. [3]

Edexcel Limited Paper 2F Q21 Jun 15

PAPER 3H

1 **a** Find the sum of the interior angles of a polygon with 7 sides. [2]

The diagram shows a regular polygon with 7 sides. [2]

b Work out the value of x.

Give your answer to 1 decimal place. [2]

Edexcel Limited Paper 3H Q9 Jan 15

2

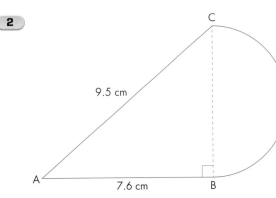

The diagram shows a shape made from triangle ABC and a semicircle with diameter BC.

Triangle ABC is right-angled at B.

$AB = 7.6$ cm and $AC = 9.5$ cm.

Calculate the area of the shape.

Give your answer correct to 3 significant figures. [5]

Edexcel Limited Paper 3H Q12 Jan 14

3 The diagram shows a metal plate.

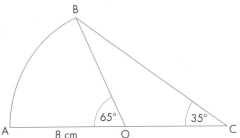

The metal plate is made from a sector OAB of a circle, center O, and a triangle OCB.

Angle $AOB = 65°$ Angle $OCB = 35°$

$OA = OB = 8$ cm.

AOC is a straight line.

a Calculate the length of BC.

Give your answer correct to 3 significant figures. [3]

b Calculate the area of the metal plate.

Give your answer correct to 3 significant figures. [3]

Edexcel Limited Paper 3H Q22 May 14

4

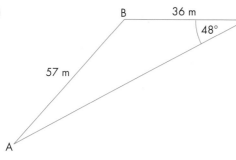

Work out the area of triangle ABC.

Give your answer correct to 3 significant figures. [4]

Edexcel Limited Paper 3H Q20 Jan 16

5

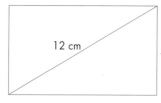

The diagram shows a triangle ABC.

$AB = (2x + 1)$ cm, $AC = (2x - 1)$ cm and $BC = 2\sqrt{7}$ cm.

Angle $BAC = 60°$

Work out the value of x.

Show clear algebraic working. [3]

Edexcel Limited Paper 3H Q22 Jan 15

6 The diagram shows a rectangle.

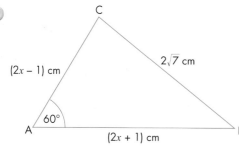

The width of the rectangle is x cm.

The length of a diagonal of the rectangle is 12 cm.

The perimeter of the rectangle is 28 cm.

Find the possible values of x.

Give your values correct to 3 significant figures.

Show your working clearly. [7]

Edexcel Limited Paper 3H Q22 Jan 16

7

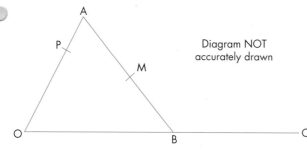

Diagram NOT accurately drawn

OAB is a triangle.

P is the point on OA such that $OP : PA = 2 : 1$

C is the point such that B is the midpoint of OC.

M is the midpoint of AB.

$\overrightarrow{OA} = 6\mathbf{a}$ $\overrightarrow{OB} = 4\mathbf{b}$

Show that PMC is a straight line. [5]

Edexcel Limited Paper 3H Q23 Jan 16

HIGHER

PAPER 4H

1

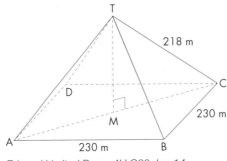

The diagram shows part of a regular polygon.

The interior angle and the exterior angle at a vertex are marked.

The size of the interior angle is 7 times the size of the exterior angle.

Work out the number of sides of the polygon. [3]

Edexcel Limited Paper 4H Q19 Jan 14

2

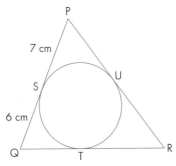

The sides of triangle *PQR* are tangents to a circle.

The tangents touch the circle at the points *S*, *T* and *U*.

QS = 6 cm. *PS* = 7 cm.

a **i** Write down the length of *QT*. [1]

   ii Give a reason for your answer. [1]

The perimeter of triangle *PQR* is 42 cm.

b Calculate the size of angle *PQR*.

Give your answer to 1 decimal place. [4]

Edexcel Limited Paper 4H Q19 May 13

3

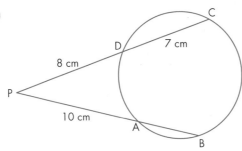

A, *B*, *C* and *D* are points on a circle.

PAB and *PDC* are straight lines.

PA = 10 cm, *PD* = 8 cm and *DC* = 7 cm.

Calculate the length of *AB*.

Edexcel Limited Paper 4H Q20 Jan 16

4

A pyramid has a horizontal square base *ABCD* with sides of length 230 metres.

M is the midpoint of *AC*.

The vertex, *T*, is vertically above *M*.

The slant edges of the pyramid are of length 218 metres.

Calculate the height, *MT*, of the pyramid.

Give your answer correct to 3 significant figures. [5]

Edexcel Limited Paper 4H Q23 Jan 14

5 A sphere has a surface area of 81π cm^2.

Work out the volume of the sphere.

Give your answer correct to 3 significant figures. [4]

Edexcel Limited Paper 4H Q21 May 14

6 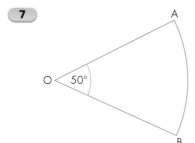 The diagram shows a solid cylinder.

The cylinder has radius $4\sqrt{3}$ cm and height h cm.

The total surface area of the cylinder is $56\pi\sqrt{6}$ cm^2

Find the exact value of h.

Give your answer in the form $a\sqrt{2} + b\sqrt{3}$, where a and b are integers.

Show your working clearly. [5]

h cm

Edexcel Limited Paper 4H Q23 Jan 16

7  The diagram shows sector OAB of a circle, centre O.

Angle $AOB = 50°$

Sector OAB has area 20π cm^2

Calculate the perimeter of sector OAB.

Give your answer correct to 3 significant figures. [5]

Edexcel Limited Paper 4H Q24 Jan 16

8 The diagram shows a cylinder and a sphere.

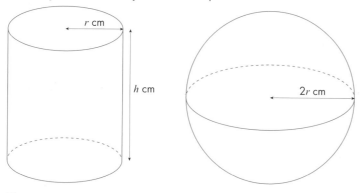

The cylinder has radius r cm and height h cm.

The sphere has radius $2r$ cm.

The volume of the cylinder is equal to the volume of the sphere.

Find an expression for h in terms of r.

Give your answer in its simplest form. [3]

Edexcel Limited Paper 4H Q21 Jan 15

Why this chapter matters

Statistical graphs such as bar charts and line graphs are used in many areas of life from science to politics. They help us to analyse and interpret information.

One of the best ways to analyse information is to present it in a visual form. Some of the earliest types of statistical diagram were line graphs, bar charts and pie charts. They all show information in different ways.

Take the owner of a bookshop.

He might use a graph like the one below to show how his sales go up and down over the year. Graphs like these are particularly good at showing trends in figures over time (see chapter 14).

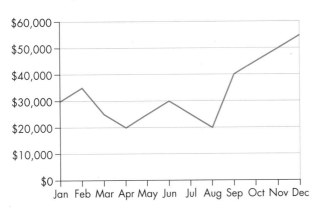

He might use a bar chart like the one on the right to show how many books they sell in different categories. Bar charts are very good at showing actual numbers.

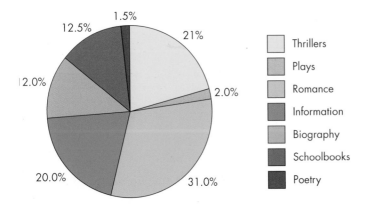

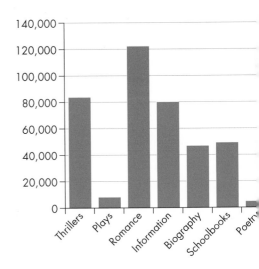

And he can get an idea of the percenta of different types of books he sells out total sales by using a pie chart like the one on the left. Pie charts are good for analysing a whole (100%) by its parts.

This chapter introduces you to some o the most common forms of statistical representation. They fall into two grou graphical diagrams such as bar charts pie charts and quantitative diagrams s as frequency tables.

31

Statistical representation

ics	Level	Key words
Frequency tables	FOUNDATION	tally chart, frequency, frequency table, classes, class interval, grouped frequency table
Pictograms	FOUNDATION	pictogram, symbol, key
Bar charts	FOUNDATION	bar chart, axis, dual bar chart
Pie charts	FOUNDATION	pie chart, angle, sector
Histograms	HIGHER	histogram, frequency density

hat you need to be able to do in the examinations:

FOUNDATION	HIGHER
Use pictograms, bar charts and pie charts to represent data.	• Construct and interpret histograms.
Use appropriate methods of tabulation to enable the construction of statistical diagrams.	
Interpret statistical diagrams.	

Frequency tables

Statistics is concerned with the collection and organisation of data, the representation of data diagrams and the interpretation of data.

When you are collecting data for simple surveys, it is usual to use a **tally chart**. For example, data collection sheets are used to gather information on how people travel to work, how students spend their free time and the amount of time people spend watching TV.

It is easy to record the data by using tally marks, as shown in Example 1. Counting up the tally marks in each row of the chart gives the **frequency** of each category. By listing the frequencies a column on the right-hand side of the chart, you can make a **frequency table** (see Example 1 Frequency tables are an important part of making statistical calculations.

EXAMPLE 1

Sandra wanted to find out about the ways in which students travelled to school. She carried out a survey. Her frequency table looked like this:

Method of travel	Tally	Frequency
Walk	HHT HHT HHT HHT HHT III	28
Car	HHT HHT II	12
Bus	HHT HHT HHT HHT III	23
Bicycle	HHT	5
Taxi	II	2

By adding together all the frequencies, you can see that 70 students took part in the survey. The frequencies also show you that more students travelled to school on foot than by any other method of transport.

Grouped data

Many surveys produce a lot of data that covers a wide range of values. In these cases, it is sensible to put the data into groups before attempting to compile a frequency table. These gro of data are called **classes** or **class intervals**.

Once the data has been grouped into classes, a **grouped frequency table** can be completed. The method is shown in Example 2.

EXAMPLE 2

These marks are for 36 students in a Year 10 mathematics examination.

31	49	52	79	40	29	66	71	73	19	51	47
81	67	40	52	20	84	65	73	60	54	60	59
25	89	21	91	84	77	18	37	55	41	72	38

a Construct a frequency table, using classes of 1–20, 21–40 and so on.

b What was the most frequent interval of marks?

a Draw the grid of the table shown below and put in the headings.

Next, list the classes, in order, in the column headed 'Marks'.

Using tally marks, indicate each student's score against the class to which it belongs. For example, 81, 84, 89 and 91 belong to the class 81–100, giving five tally marks, as shown below.

Finally, count the tally marks for each class and enter the result in the column headed 'Frequency'. The table is now complete.

Marks	Tally	Frequency
1–20	|||	3
21–40	HHT |||	8
41–60	HHT HHT		11
61–80	HHT ||||	9
81–100	HHT	5

b From the grouped frequency table, you can see that the highest number of students obtained a mark in the 41–60 interval.

EXERCISE 31A

1 Kurt kept a record of the number of goals scored by his local team in the last 20 matches. These are his results:

0   1   1   0   2   0   1   3   2   1

0   1   0   3   2   1   0   2   1   1

a Draw a frequency table for his data.   **b** Which was the most frequent score?

c How many goals were scored in total for the 20 matches?

2 Monique was doing a geography project on the weather. As part of her work, she kept a record of the daily midday temperatures in June.

a Copy and complete the grouped frequency table for her data.

b In which interval do the most temperatures lie?

c Describe what the weather was probably like throughout the month.

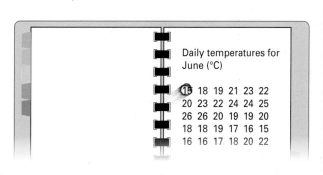

Daily temperatures for June (°C)

15 18 19 21 23 22
20 23 22 24 24 25
26 26 20 19 19 20
18 18 19 17 16 15
16 16 17 18 20 22

Temperature (°C)	Tally	Frequency
14–16		
17–19		
20–22		
23–25		
26–28		

FOUNDATION

3 In a game, Mitesh used a six-sided dice. He decided to keep a record of his scores to see whether the dice was fair. His scores were:

2 4 2 6 1 5 4 3 3 2 3 6 2 1 3

5 4 3 4 2 1 6 5 1 6 4 1 2 3 4

a Draw a frequency table for his data.

b How many throws did Mitesh have during the game?

c Do you think the dice was a fair one? Explain why.

4 The data shows the heights, in centimetres, of a sample of 32 students.

172 158 160 175 180 167 159 180

167 166 178 184 179 156 165 166

184 175 170 165 164 172 154 186

167 172 170 181 157 165 152 164

a Draw a grouped frequency table for the data, using class intervals 151–155, 156–160, …

b In which interval do the most heights lie?

c Does this agree with a survey of the students in your class?

5 A student used a stopwatch to time how long it took her rabbit to find food left in its hutch.

The following is her record in seconds.

7	30	14	27	8	31	8	28	10	41	51	37	15	21	37	16	38
23	20	9	11	55	9	33	8	35	45	35	25	25	49	23	43	55
45	8	13	9	39	12	57	16	37	26	32	19	48	29	37		

Find the best way to put this data into a frequency chart to illustrate the length of time it took the rabbit to find the food.

6 A student was doing a survey to find the ages of people at a football competition.

He said that he would make a frequency table with the regions 15–20, 20–25, 25–30.

Explain what difficulty he could have with these class divisions.

Data collected from a survey can be presented in pictorial or diagrammatic form to help people to understand it more quickly. You see plenty of examples of this in newspapers and magazines and on TV, where every type of visual aid is used to communicate statistical information.

Pictograms

A **pictogram** is a frequency table in which frequency is represented by a repeated **symbol**. The symbol itself usually represents a number of items, as Example 4 on the next page shows. However, sometimes it is more sensible to let a symbol represent just a single unit, as in Example 3 below. The **key** tells you how many items are represented by a symbol.

EXAMPLE 3

The pictogram shows the number of phone calls made by Nurul from her mobile phone during a week.

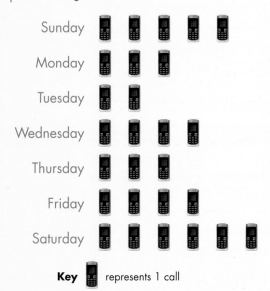

How many calls did Nurul make in the week?

From the pictogram, you can see that Nurul made a total of 27 calls.

Although pictograms can have great visual impact (particularly as used in advertising) and are easy to understand, they have a serious drawback. Apart from a half, fractions of a symbol cannot usually be drawn accurately and so frequencies are often represented only approximately by symbols.

Example 4 on the next page highlights this difficulty.

EXAMPLE 4

The pictogram shows the number of students who were late for school during a week.

Monday

Tuesday

Wednesday

Thursday

Friday

Key represents 5 students

How many students were late on:

a Monday

b Thursday?

Precisely how many students were late on Monday and Thursday respectively?

If you assume that each 'limb' of the symbol represents one student and its 'body' also represents one student, then the answers are:

a 19 students were late on Monday

b 13 on Thursday.

EXERCISE 31B

1 The frequency table shows the numbers of cars parked in a shop's car park at various times of the day. Draw a pictogram to illustrate the data. Use a key of 1 symbol = 5 cars.

Time	9 am	11 am	1 pm	3 pm	5 pm
Frequency	40	50	70	65	45

2 A milkman kept a record of how many pints of milk he delivered to 10 apartments on a particular morning. Draw a pictogram for the data. Use a key of 1 symbol = 1 pint.

Flat 1	Flat 2	Flat 3	Flat 4	Flat 5	Flat 6	Flat 7	Flat 8	Flat 9	Flat 10
2	3	1	2	4	3	2	1	5	1

3 The pictogram, taken from a Suntours brochure, shows the average daily hours of sunshine for five months in Tenerife.

a Write down the average daily hours of sunshine for each month.

b Give a reason why pictograms are useful in holiday brochures.

May

June

July

August

September

Key represents 2 hours

4　The pictogram shows the amounts of money collected by six students after they had completed a sponsored walk for charity.

Anthony	$ $ $ $ $
Ben	$ $ $ $ $ $
Emma	$ $ $ $ $
Leanne	$ $ $ $
Reena	$ $ $ $ $ $
Simon	$ $ $ $ $ $ $

Key $ represents $5

a Who raised the most money?

b How much money was raised altogether by the six students?

c Robert also took part in the walk and raised $32. Why would it be difficult to include him on the pictogram?

5　A newspaper showed the following pictogram about a family and the number of emails each family member received during one Sunday.

Key ⊠ represents 4 emails

		Frequency
Dad	⊠ ⊠ ⊠	
Mum	⊠ ▷	
Teenage son	⊠ ⊠ ⊠ ▷	
Teenage daughter		23
Young son		9

a How many emails did:

　i Dad receive　　**ii** Mum receive　　**iii** the teenage son receive?

b Copy and complete the pictogram.

c How many emails were received altogether?

31.3　Bar charts

A **bar chart** consists of a series of bars or blocks of the *same* width, drawn either vertically or horizontally from an **axis**.

Sometimes, the bars are separated by narrow gaps of equal width, which makes the chart easier to read.

EXAMPLE 5

The grouped frequency table shows the marks of 24 students in a test. Draw a bar chart for the data.

Marks	1–10	11–20	21–30	31–40	41–50
Frequency	2	3	5	8	6

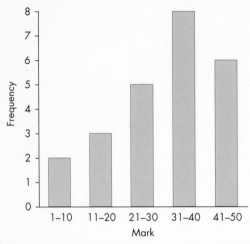

Note:

- Both axes are labelled.
- The class intervals are written under the middle of each bar.
- The bars are separated by equal spaces.

By using a **dual bar chart**, it is easy to compare two sets of related data, as Example 6 shows.

EXAMPLE 6

This dual bar chart shows the average daily maximum temperatures for England and Turkey over a five-month period.

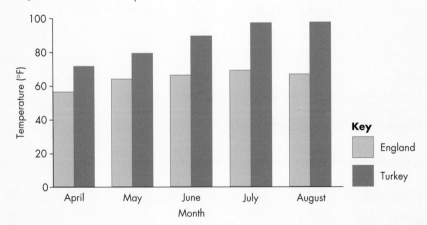

In which month was the *difference* between temperatures in England and Turkey the greatest?

The largest difference can be seen in August.

Note: You must always include a key to identify the two different sets of data.

EXERCISE 31C

1 For her survey on fitness, Samina asked a sample of people, as they left a sports centre, which activity they had taken part in. She then drew a bar chart to show her data.

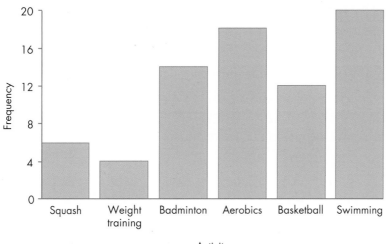

a Which was the most popular activity?

b How many people took part in Samina's survey?

2 The frequency table below shows the levels achieved by 100 students in their practice IGCSE examinations.

Grade	F	E	D	C	B	A
Frequency	12	22	24	25	15	2

a Draw a suitable bar chart to illustrate the data.

b What fraction of the students achieved a grade C or grade B?

c Give one advantage of drawing a bar chart rather than a pictogram for this data.

3 This table shows the number of points Amir and Hasrul were each awarded in eight rounds of a general knowledge quiz.

Round	1	2	3	4	5	6	7	8
Amir	7	8	7	6	8	6	9	4
Hasrul	6	7	6	9	6	8	5	6

a Draw a dual bar chart to illustrate the data.

b Comment on how well each of them did in the quiz.

4 Mira did a survey on the time it took students in her class to get to school on a particular morning. She wrote down their times to the nearest minute.

15 23 36 45 8 20 34 15 27 49

10 60 5 48 30 18 21 2 12 56

49 33 17 44 50 35 46 24 11 34

a Draw a grouped frequency table for Mira's data, using class intervals 1–10, 11–20, …

b Draw a bar chart to illustrate the data.

c What conclusions can Mira draw from the bar chart?

5 This table shows the number of accidents at a dangerous road junction over a six-year period.

Year	2005	2006	2007	2008	2009	2010
No. of accidents	6	8	7	9	6	4

a Draw a pictogram for the data.

b Draw a bar chart for the data.

c Which diagram would you use if you were going to argue that traffic lights should be installed at the junction? Explain why.

6 The diagram below shows the minimum and maximum temperatures for one day in August in five cities.

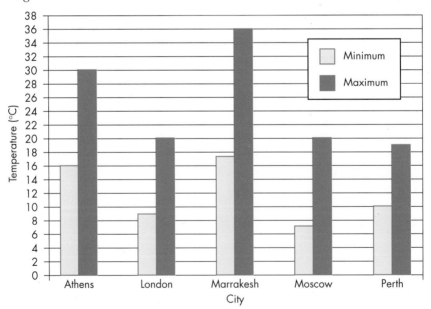

Lee says that the minimum temperature is always about half the maximum temperature for most cities.

Is Lee correct?

Give reasons to justify your answer.

Pictograms, bar charts and line graphs are easy to draw but they can be difficult to interpret when there is a big difference between the frequencies or there are only a few categories. In these cases, it is often more convenient to illustrate the data on a **pie chart**.

In a pie chart, the whole of the data is represented by a circle (the 'pie') and each category of it is represented by a **sector** of the circle (a 'slice of the pie'). The **angle** of each sector is proportional to the frequency of the category it represents.

So, a pie chart cannot show individual frequencies, like a bar chart can, for example. It can only show proportions.

Sometimes the pie chart will be marked off in equal sections rather than angles. In these cases, the numbers are always easy to work with.

EXAMPLE 7

20 people were surveyed about their preferred drink. Their replies are shown in the table.

Drink	Tea	Coffee	Milk	Cola
Frequency	6	7	4	3

Show the results on the pie chart given.

You can see that the pie chart has 10 equally-spaced divisions.

As there are 20 people, each division is worth two people. So the sector for tea will have 3 of these divisions. In the same way, coffee will have $3\frac{1}{2}$ divisions, milk will have 2 divisions and cola will have $1\frac{1}{2}$ divisions.

The finished pie chart will look like the one in the diagram below.

Preferred drinks

Note:

- You should always label the sectors of the pie chart (use shading and a separate key if there is not enough space to write on the pie chart).

- Give your chart a title.

EXAMPLE 8

In a survey on holidays, 120 people were asked to state which type of transport they used on their last holiday. This table shows the results of the survey. Draw a pie chart to illustrate the data.

Type of transport	Train	Bus	Car	Ship	Plane
Frequency	24	12	59	11	14

You need to find the angle for the fraction of 360° that represents each type of transport. This is usually done in a table, as shown below.

Type of transport	Frequency	Calculation	Angle
Train	24	$\frac{24}{120} \times 360° = 72°$	72°
Bus	12	$\frac{12}{120} \times 360° = 36°$	36°
Car	59	$\frac{59}{120} \times 360° = 177°$	177°
Ship	11	$\frac{11}{120} \times 360° = 33°$	33°
Plane	14	$\frac{14}{120} \times 360° = 42°$	42°
Totals	120		360°

Draw the pie chart, using the calculated angle for each sector.

Note:

- Use the frequency total (120 in this case) to calculate each fraction.

- Check that the sum of all the angles is 360°.

- Label each sector.

- The angles or frequencies do not have to be shown on the pie chart.

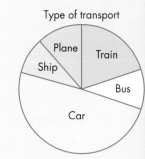

Type of transport

EXERCISE 31D

1 Copy the diagram on the right and draw a pie chart to show each of the following sets of data.

a The favourite pets of 10 children.

Pet	Bird	Cat	Rabbit
Frequency	4	5	1

b The makes of cars of 20 teachers.

Make of car	Ford	Toyota	BMW	Nissan	Peugeot
Frequency	4	5	2	3	6

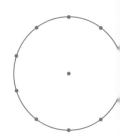

c The newspaper read by 40 office workers.

Newspaper	*The Post*	*Today*	*The Mail*	*The Times*
Frequency	14	8	6	12

2 Draw a pie chart to represent each of the following sets of data.

> **HINTS AND TIPS**
>
> Remember to complete a table as shown in the examples. Check that all angles add up to 360°.

a The number of children in 40 families.

No. of children	0	1	2	3	4
Frequency	4	10	14	9	3

b How 90 students get to school.

Journey to school	Walk	Car	Bus	Cycle
Frequency	42	13	25	10

3 Mariam asked 24 of her friends which sport they preferred to play. Her data is shown in this frequency table.

Sport	Rugby	Football	Tennis	Baseball	Basketball
Frequency	4	11	3	1	5

Illustrate her data in a pie chart.

4 Ameer wrote down the number of lessons he had per week in each subject on his school timetable.

Mathematics 5	English 5	Science 8	History 6
Geography 6	Arts 4	Sport 2	

a How many lessons did Ameer have on his timetable?

b Draw a pie chart to show the data.

c Draw a bar chart to show the data.

d Which diagram better illustrates the data? Give a reason for your answer.

5 A market researcher asked 720 people which new brand of tinned beans they preferred. The results are given in the table.

A	248
B	264
C	152
D	56

a Draw a pie chart to illustrate the data.

b Why do you think pie charts are used to show this sort of information?

6 This pie chart shows the proportions of the different shoe sizes worn by 144 pupils in one year group in a school.

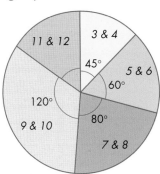

a What is the angle of the sector representing shoe sizes 11 and 12?

b How many pupils had a shoe size of 11 or 12?

7 The table below shows the numbers of candidates, at each grade, taking music examinations in Strings and Brass.

	Grades					Total number of candidates
	3	**4**	**5**	**6**	**7**	
Strings	300	980	1050	600	70	3000
Brass	250	360	300	120	70	1100

a Draw a pie chart to represent each of the two examinations.

b Compare the pie charts to decide which group of candidates, Strings or Brass, did better overall. Give reasons to justify your answer.

8 In a survey, a rail company asked passengers whether their service had improved.

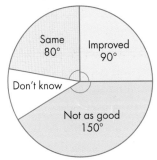

What is the probability that a person picked at random from this survey answered "Don't know"?

You should already be familiar with bar charts like the one on the right in which the vertical axis represents frequency, and each bar has a label to show what it represents. (Sometimes it is more convenient to have the axes the other way round.)

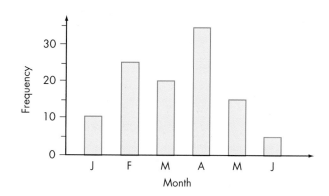

A **histogram** looks similar to a bar chart, but there are *three* fundamental differences.

- There are no gaps between the bars.
- The horizontal axis has a continuous scale.
- The area of each bar represents the class or group frequency of the bar.

Here is a histogram. It shows how long some people waited to see a doctor.

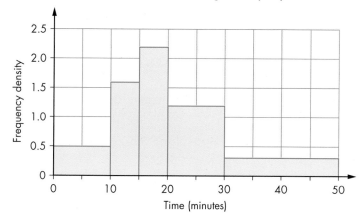

The *area* of each column gives the frequency.

The number who waited less than 10 minutes is the area of the first column = 0.5 × 10 = 5 people.
Between 10 and 15 minutes is 1.6 × 5 = 8 people (the width is 15 − 10 = 5).
Between 15 and 20 minutes is 2.2 × 5 = 11 people.
Between 20 and 30 minutes is 1.2 × 10 = 12 people (the width is 30 − 20 = 10).
Between 30 and 50 minutes is 0.3 × 20 = 6 people.

The number on the vertical axis of a histogram is called the **frequency density**.

Frequency = frequency density × class width

$$\text{Frequency density} = \frac{\text{frequency}}{\text{class width}}$$

EXAMPLE 9

The heights of a group of girls were measured. The results were classified as shown in the table.

Height, h (cm)	$151 \leqslant h < 153$	$153 \leqslant h < 154$	$154 \leqslant h < 155$	$155 \leqslant h < 159$	$159 \leqslant h < 160$
Frequency	64	43	47	96	12

It is convenient to write the table vertically and add two columns, class width and frequency density.

The class width is found by subtracting the lower class boundary from the upper class boundary. The frequency density is found by dividing the frequency by the class width.

Height, h (cm)	Frequency	Class width	Frequency density
$151 \leqslant h < 153$	64	2	32
$153 \leqslant h < 154$	43	1	43
$154 \leqslant h < 155$	47	1	47
$155 \leqslant h < 159$	96	4	24
$159 \leqslant h < 160$	12	1	12

The histogram can now be drawn.

The horizontal scale should be marked off as normal, from a value below the lowest value in the table to a value above the largest value in the table. In this case, mark the scale from 150 cm to 160 cm.

The vertical scale is always frequency density and is marked up to at least the largest frequency density in the table. In this case, 50 is a sensible value.

Each bar is drawn between the lower class interval and the upper class interval horizontally, and up to the frequency density vertically.

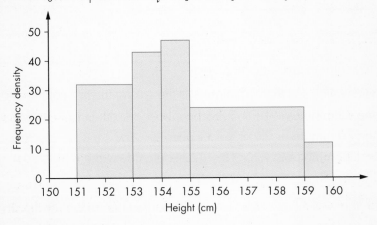

We can check that the area of each column is equal to the frequency.

151–153 is 2 × 32 = 64

153–154 is 1 × 43 = 43

and so on.

If the bars are of equal width, the frequency density and the frequency will be proportional. In that case we can use frequency on the vertical axis, as we did in section 31.3.

EXERCISE 31E

1 Draw histograms for these grouped frequency distributions.

a

Temperature, t (°C)	$8 \leqslant t < 10$	$10 \leqslant t < 12$	$12 \leqslant t < 15$	$15 \leqslant t < 17$	$17 \leqslant t < 20$	$20 \leqslant t < 24$
Frequency	5	13	18	4	3	6

b

Wage, w ($1000)	$6 \leqslant w < 10$	$10 \leqslant w < 12$	$12 \leqslant w < 16$	$16 \leqslant w < 24$
Frequency	16	54	60	24

c

Age, a (years)	$11 \leqslant a < 14$	$14 \leqslant a < 16$	$16 \leqslant a < 17$	$17 \leqslant a < 20$
Frequency	51	36	12	20

d

Pressure, p (mm)	$745 \leqslant p < 755$	$755 \leqslant p < 760$	$760 \leqslant p < 765$	$765 \leqslant p < 775$
Frequency	4	6	14	10

e

Time, t (min)	$0 \leqslant t < 8$	$8 \leqslant t < 12$	$12 \leqslant t < 16$	$16 \leqslant t < 20$
Frequency	72	84	54	36

2 The following information was gathered about the weekly pocket money given to 14-year-olds.

Pocket money, p ($)	$0 \leqslant p < 2$	$2 \leqslant p < 4$	$4 \leqslant p < 5$	$5 \leqslant p < 8$	$8 \leqslant p < 10$
Girls	8	15	22	12	4
Boys	6	11	25	15	6

Represent the information about the boys and girls on separate histograms.

3 The sales of the *Star* newspaper over 70 years are recorded in this table.

Years	1940–60	1961–80	1981–90	1991–2000	2001–05	2006–2010
Copies	62 000	68 000	71 000	75 000	63 000	52 000

Illustrate this information on a histogram. Take the class boundaries as 1940, 1960, 1980, 1990, 2000, 2005, 2010.

4 The Madrid trains were always late, so one month a survey was undertaken to find how many trains were late, and by how many minutes. The results are illustrated by this histogram.

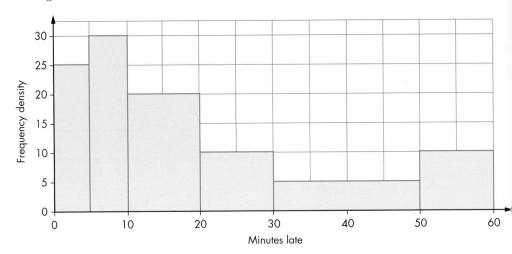

a How many trains were in the survey?

b How many trains were delayed for longer than 15 minutes?

5 For each of the frequency distributions illustrated in the histograms write down the grouped frequency table.

a

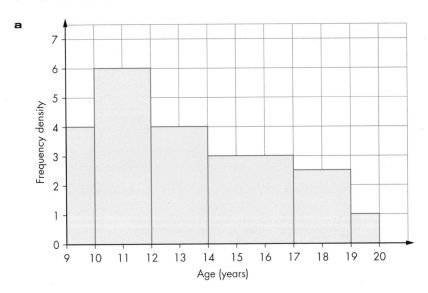

b

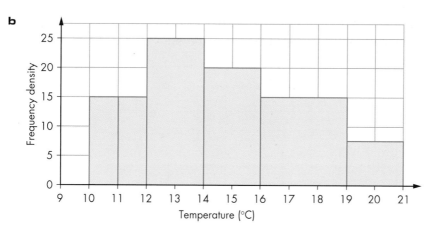

c

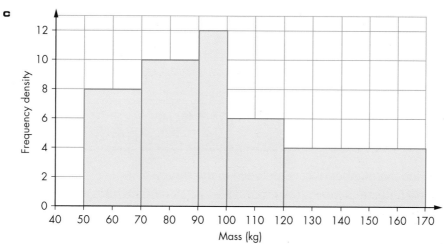

6 All the patients in a hospital were asked how long it was since they last saw a doctor. The results are shown in the table.

Hours, h	$0 \leqslant h < 2$	$2 \leqslant h < 4$	$4 \leqslant h < 6$	$6 \leqslant h < 10$	$10 \leqslant h < 16$	$16 \leqslant h < 24$
Frequency	8	12	20	30	20	10

a Draw a histogram to illustrate the data.

b Estimate how many people waited more than 8 hours.

> **HINTS AND TIPS**
>
> Find the area to the right of $h = 8$.

7 One summer, Albert monitored the mass of the tomatoes grown on each of his plants. His results are summarised in this table.

Mass, m (kg)	$6 \leqslant m < 10$	$10 \leqslant m < 12$	$12 \leqslant m < 16$	$16 \leqslant m < 20$	$20 \leqslant m < 25$
Frequency	8	15	28	16	10

a Draw a histogram for this distribution.

b Estimate how many plants produced more than 15 kg.

8 A survey was carried out to find the speeds of cars passing a particular point on a road. The histogram illustrates the results of the survey.

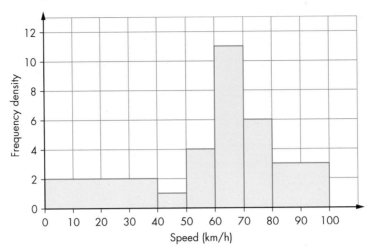

a Copy and complete this table.

Speed, v (km/h)	$0 < v \leqslant 40$	$40 < v \leqslant 50$	$50 < v \leqslant 60$	$60 < v \leqslant 70$	$70 < v \leqslant 80$	$80 < v \leqslant 1$
Frequency		10	40	110		

b Find the number of cars included in the survey.

9 The histogram shows the test scores for 320 students in a school.

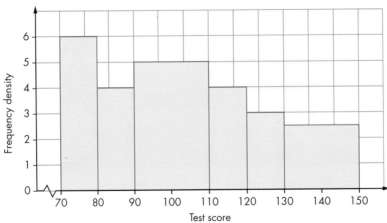

a How many students scored more than 120?

b The pass mark was 90. What percentage of students failed the test?

10 Adrienne and Bernice collected the same data about journey times but grouped it differently.

Here are Adrienne's figures:

Journey time (*t* minutes)	$0 \leqslant t < 5$	$5 \leqslant t < 10$	$10 \leqslant t < 15$	$15 \leqslant t < 20$	$20 \leqslant t < 25$	$25 \leqslant t < 30$
Frequency	10	15	30	12	6	3

Here are Bernice's figures:

Journey time (*t* minutes)	$0 \leqslant t < 10$	$10 \leqslant t < 15$	$15 \leqslant t < 30$
Frequency	25	30	21

a Draw a histogram for each set of figures. Use frequency density on the vertical axis each time.

b Describe any similarities or differences between the histograms.

Why this chapter matters

The idea of 'average' is important in statistics. But there are several ways of working out an average which have been developed over a long period of time.

Mean in Ancient India

There is a story about Rituparna who was born in India around 5000 BCE. He wanted to estimate the amount of fruit on a tree:

- He counted how much fruit was on one branch, then estimated the number of branches on the tree.

- He multiplied the estimated number of branches by the counted fruit on one branch.

He was amazed that the total was very close to the actual counted number of fruit when it was picked.

Rituparna was one of the first to use the arithmetic **mean**. The branch he chose was an average one representing all the branches. So the number of fruit on that branch would have been in the middle of the smallest and largest number of fruit on other branches on the tree.

Mode in Ancient Greece

This story comes from a war in Ancient Greece (431–404 BCE). It is about a battle between the Spartans and the Athenians.

The Athenians had to get over the Spartan Wall so they needed to work out its height. They started by counting the layers of bricks. This was done by hundreds of soldiers at the same time because many of them would get it wrong – but the majority would get it about right.

This is seen as an early use of the **mode**: the number of layers that occurred the most in the counting was taken as the one most likely to be correct.

They then had to guess the height of one brick and so calculate the total height of the wall. They could then make ladders long enough to reach the top of the wall.

The other average that we use is the **median**, and there is no record of any use of this (which finds the middle value) being used until the early 17th century.

These ancient examples demonstrate that we do not always work out the average in the same way – we must choose a method that is appropriate to the situation.

32

Statistical measures

ics		Level	Key words
①	The mode	FOUNDATION	average, mode, frequency, modal value
②	The median	FOUNDATION	median, middle value
③	The mean	FOUNDATION	average, mean
④	The range	FOUNDATION	range, spread, consistency
⑤	Which average to use	FOUNDATION	representative, appropriate, extreme values
⑥	Using frequency tables	FOUNDATION	frequency table
⑦	Grouped data	FOUNDATION	grouped data, estimated, modal class, continuous data, discrete data
⑧	Measuring spread	HIGHER	interquartile range, spread, lower quartile, upper quartile
⑨	Cumulative frequency diagrams	HIGHER	cumulative frequency, quartile, cumulative frequency diagram

What you need to be able to do in the examinations:

FOUNDATION	HIGHER
• Understand the concept of average. • Calculate the mean, median, mode and range for a discrete data set. • Calculate an estimate for the mean for grouped data. • Identify the modal class for grouped data.	• Estimate the median and interquartile range from a cumulative frequency diagram. • Understand the concept of a measure of spread. • Find the interquartile range from a discrete data set. • Use cumulative frequency diagrams and construct them from tabulated data.

Average is a term often used when describing or comparing sets of data, for example, average rainfall over a year or the average mark in an examination for a group of students.

In each of the above examples, you are representing the whole set of many values (rainfall on every day of the year or the marks of all the students) by just a single, 'typical' value, which is called the average.

The idea of an average is extremely useful, because it enables you to compare one set of data with another set by comparing just two values – their averages.

There are several ways of expressing an average, but the most commonly used averages are the **mode**, the median and the mean.

The mode is the value that occurs the most in a set of data. That is, it is the value with the highest **frequency**.

The mode is a useful average because it is very easy to find and it can be applied to non-numerical data (qualitative data). For example, you could find the modal style of skirts sold in a particular month.

EXAMPLE 1

Suhail scored the following number of goals in 12 football matches:

1 2 1 0 1 0 0 1 2 1 0 2

What is the mode of his scores?

The number which occurs most often in this list is 1. So, the mode is 1.

You can also say that the modal score or **modal value** is 1.

EXERCISE 32A

1 Find the mode for each set of data.

a 3, 4, 7, 3, 2, 4, 5, 3, 4, 6, 8, 4, 2, 7

b 47, 49, 45, 50, 47, 48, 51, 48, 51, 48, 52, 48

c −1, 1, 0, −1, 2, −2, −2, −1, 0, 1, −1, 1, 0, −1, 2, −1, 2

d $\frac{1}{2}, \frac{1}{4}, 1, \frac{1}{2}, \frac{3}{4}, \frac{1}{4}, 0, 1, \frac{3}{4}, \frac{1}{4}, 1, \frac{1}{4}, \frac{3}{4}, \frac{1}{4}, \frac{1}{2}$

e 100, 10, 1000, 10, 100, 1000, 10, 1000, 100, 1000, 100, 10

f 1.23, 3.21, 2.31, 3.21, 1.23, 3.12, 2.31, 1.32, 3.21, 2.31, 3.21

2 Find the mode for each set of data.

a red, green, red, amber, green, red, amber, green, red, amber

b rain, sun, cloud, sun, rain, fog, snow, rain, fog, sun, snow, sun

c α, γ, α, β, γ, α, α, γ, β, α, β, γ, β, β, α, β, γ, β

d ✳, ☆, ★, ★, ☆, ✳, ★, ☆, ★, ☆, ★, ✳, ✪, ☆, ★, ★, ☆

> **HINTS AND TIPS**
>
> It helps to put the data in order or group all the same things together.

3 Halima did a survey to find the shoe sizes of students in her class. The bar chart illustrates her data.

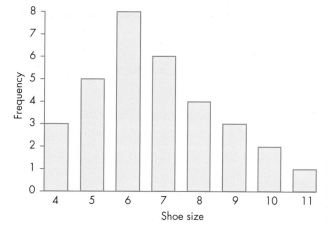

a How many students are in Halima's class?

b What is the modal shoe size?

c Can you tell which are the boys' shoes sizes and which are the girls' shoe sizes?

d Halima then decided to draw a bar chart to show the shoe sizes of the boys and the girls separately. Do you think that the mode for the boys and the mode for the girls will be the same as the mode for the whole class? Explain your answer.

4 The frequency table shows the marks that a class obtained in a spelling test.

Mark	3	4	5	6	7	8	9	10
Frequency	1	2	6	5	5	4	3	4

a Write down the mode for their marks.

b Do you think this is a typical mark for the class? Explain your answer.

5 Explain why the mode is often referred to as the 'shopkeeper's average'.

6 This table shows the colours of eyes of the students in a class.

a How many students are in the class?

	Blue	Brown	Green
Boys	4	8	1
Girls	8	5	2

b What is the modal eye colour for:

  i boys   **ii** girls   **iii** the whole class?

c After two students join the class the modal eye colour for the whole class is blue. Which of the following statements is true?

- Both students had green eyes.
- Both students had brown eyes.
- Both students had blue eyes.
- You cannot tell what their eye colours were.

7 Here is a large set of raw data.

5 6 8 2 4 8 9 8 1 3 4 2 7 2 4 6 7 5 3 8

9 1 3 1 5 6 2 5 7 9 4 1 4 3 3 5 6 8 6 9

8 4 8 9 3 4 6 7 7 4 5 4 2 3 4 6 7 6 5 5

a What problems may occur if you attempted to find the mode by counting individual numbers?

b Explain a method that would make finding the mode more efficient and accurate.

c Use your method to find the mode of the data.

The **median** is the **middle value** of a list of values when they are put in *order of size*, from lowest to highest.

The advantage of using the median as an average is that half the data values are below the median value and half are above it. Therefore, the average is only slightly affected by the presence of any particularly high or low values that are not typical of the data as a whole.

EXAMPLE 2

Find the median for the following list of numbers:

2, 3, 5, 6, 1, 2, 3, 4, 5, 4, 6

Putting the list in numerical order gives:

1, 2, 2, 3, 3, **4**, 4, 5, 5, 6, 6

There are 11 numbers in the list, so the middle of the list is the 6th number. Therefore, the median is 4.

EXAMPLE 3

Find the median of the data shown in the frequency table.

Value	2	3	4	5	6	7
Frequency	2	4	6	7	8	3

First, add up the frequencies to find out how many pieces of data there are.

The total is 30 so the median value will be between the 15th and 16th values.

Now, add up the frequencies to give a running total, to find out where the 15th and 16th values are.

Value	2	3	4	5	6	7
Frequency	2	4	6	7	8	3
Running total	2	6	12	19	27	30

There are 12 data-values up to the value 4 and 19 up to the value 5.

Both the 15th and 16th values are 5, so the median is 5.

To find the median in a list of n values, written in order, use the rule:

$$\text{median} = \frac{n + 1}{2}\text{th value}$$

EXERCISE 32B

1 Find the median for each set of data.

a 7, 6, 2, 3, 1, 9, 5, 4, 8

b 26, 34, 45, 28, 27, 38, 40, 24, 27, 33, 32, 41, 38

c 4, 12, 7, 6, 10, 5, 11, 8, 14, 3, 2, 9

d 12, 16, 12, 32, 28, 24, 20, 28, 24, 32, 36, 16

e 10, 6, 0, 5, 7, 13, 11, 14, 6, 13, 15, 1, 4, 15

f −1, −8, 5, −3, 0, 1, −2, 4, 0, 2, −4, −3, 2

g 5.5, 5.05, 5.15, 5.2, 5.3, 5.35, 5.08, 5.9, 5.25

> **HINTS AND TIPS**
>
> Remember to put the data in order before finding the median.

> **HINTS AND TIPS**
>
> If there is an even number of pieces of data, the median will be halfway between the two middle values.

2 A group of 15 students had lunch in the school's cafeteria. Given below are the amounts that they spent.

$2.30, $2.20, $2, $2.50, $2.20, $3.50, $2.20, $2.25, $2.20, $2.30, $2.40, $2.20, $2.30, $2, $2.35

a Find the mode for the data.

b Find the median for the data.

c Which is the better average to use? Explain your answer.

3 **a** Find the median of 7, 4, 3, 8, 2, 6, 5, 2, 9, 8, 3.

b Without putting them in numerical order, write down the median for each of these sets.

i 17, 14, 13, 18, 12, 16, 15, 12, 19, 18, 13

ii 217, 214, 213, 218, 212, 216, 215, 212, 219, 218, 213

iii 12, 9, 8, 13, 7, 11, 10, 7, 14, 13, 8

iv 14, 8, 6, 16, 4, 12, 10, 4, 18, 16, 6

> **HINTS AND TIPS**
>
> Look for a connection between the original data and the new data. For example, in **i**, the numbers are each 10 more than those in part **a**.

4 Given below are the age, height and mass of each of the seven players in a netball team

	Ella	Linda	Pat	Marion	Amina	Martha	Elisa
Age (yr)	13	15	12	11	11	15	14
Height (cm)	161	165	162	158	154	168	169
Mass (kg)	61	62	57	52	55	62	60

a Find the median age of the team. Which player has the median age?

b Find the median height of the team. Which player has the median height?

c Find the median mass of the team. Which player has the median mass?

d Who would you choose as the average player in the team? Give a reason for your answer

5 The table shows the number of sandwiches sold in a shop over 25 days.

Sandwiches sold	10	11	12	13	14	15	16
Frequency	2	3	6	4	3	4	3

a What is the modal number of sandwiches sold?

b What is the median number of sandwiches sold?

6 **a** Write down a list of nine numbers that has a median of 12.

b Write down a list of 10 numbers that has a median of 12.

c Write down a list of nine numbers that has a median of 12 and a mode of 8.

d Write down a list of 10 numbers that has a median of 12 and a mode of 8.

7 A list contains seven even numbers. The largest number is 24. The smallest number is half the largest. The mode is 14 and the median is 16. Two of the numbers add up to 42 What are the seven numbers?

8 Look at this list of numbers.

4, 4, 5, 8, 10, 11, 12, 15, 15, 16, 20

a Add four numbers to make the median 12.

b Add six numbers to make the median 12.

c What is the least number of numbers to add that will make the median 4?

9 Here are five payments.

$3, $5, $8, $100, $3000

Explain why the median is not a good average to use in this set of payments.

The mean

The **mean** of a set of data is the sum of all the values in the set divided by the total number of values in the set. That is:

$$\text{mean} = \frac{\text{sum of all values}}{\text{total number of values}}$$

This is what most people mean when they use the term '**average**'.

Another name for this average is the arithmetic mean.

The advantage of using the mean as an average is that it takes into account all the values in the set of data.

XAMPLE 4

The ages of 11 players in a football squad are:

21, 23, 20, 27, 25, 24, 25, 30, 21, 22, 28

What is the mean age of the squad?

Sum of all the ages = 266

Total number in squad = 11

Therefore, mean age = $\frac{266}{11}$ = 24.1818... = 24.2 (1 decimal place)

EXERCISE 32C

1. Find the mean for each set of data.

   a. 7, 8, 3, 6, 7, 3, 8, 5, 4, 9

   b. 47, 3, 23, 19, 30, 22

   c. 42, 53, 47, 41, 37, 55, 40, 39, 44, 52

   d. 1.53, 1.51, 1.64, 1.55, 1.48, 1.62, 1.58, 1.65

   e. 1, 2, 0, 2, 5, 3, 1, 0, 1, 2, 3, 4

2. Calculate the mean for each set of data, giving your answers correct to 1 decimal place.

   a. 34, 56, 89, 34, 37, 56, 72, 60, 35, 66, 67

   b. 235, 256, 345, 267, 398, 456, 376, 307, 282

   c. 50, 70, 60, 50, 40, 80, 70, 60, 80, 40, 50, 40, 70

   d. 43.2, 56.5, 40.5, 37.9, 44.8, 49.7, 38.1, 41.6, 51.4

   e. 2, 3, 1, 0, 2, 5, 4, 3, 2, 0, 1, 3, 4, 5, 0, 3, 1, 2

3 The table shows the marks that 10 students obtained in Mathematics, English and Science in their examinations.

Student	Ahmed	Badru	Camille	Dayar	Evrim	Fatima	George	Helga	Imran	Josie
Maths	45	56	47	77	82	39	78	32	92	62
English	54	55	59	69	66	49	60	56	88	44
Science	62	58	48	41	80	56	72	40	81	52

    a Work out the mean mark for Mathematics.

    b Work out the mean mark for English.

    c Work out the mean mark for Science.

    d Which student obtained marks closest to the mean in all three subjects?

    e How many students were above the average mark in all three subjects?

4 Suni kept a record of the amount of time she spent on her homework over 10 days:

$\frac{1}{2}$h, 20 min, 35 min, $\frac{1}{4}$h, 1 h, $\frac{1}{2}$h, 1$\frac{1}{2}$h, 40 min, $\frac{3}{4}$h, 55 min

Calculate the mean time, in minutes, that Suni spent on her homework.

> **HINTS AND TIPS**
>
> Convert all times to minutes, for example, $\frac{1}{4}$h = 15 minutes.

5 The weekly wages of 10 people working in an office are:

$350 $200 $180 $200 $350 $200 $240 $480 $300 $280

    a Find the modal wage.

    b Find the median wage.

    c Calculate the mean wage.

    d Which of the three averages best represents the office staff's wages? Give a reason for your answer.

> **HINTS AND TIPS**
>
> Remember that the mean can be distorted by extreme values.

6 The ages of five people in a group of walkers are 38, 28, 30, 42 and 37.

    a Calculate the mean age of the group.

    b Steve, who is 41, joins the group. Calculate the new mean age of the group.

7 **a** Calculate the mean of 3, 7, 5, 8, 4, 6, 7, 8, 9 and 3.

    b Calculate the mean of 13, 17, 15, 18, 14, 16, 17, 18, 19 and 13. What do you notice?

    c Write down, without calculating, the mean for each of the following sets of data.

       i 53, 57, 55, 58, 54, 56, 57, 58, 59, 53

      ii 103, 107, 105, 108, 104, 106, 107, 108, 109, 103

     iii 4, 8, 6, 9, 5, 7, 8, 9, 10, 4

> **HINTS AND TIPS**
>
> Look for a connection between the original data and the new data. For example in **i** the numbers are 50 more.

8 Two families were in a competition.

Speed family		Roberts family	
Brian	aged 59	Frank	aged 64
Kath	aged 54	Marylin	aged 62
James	aged 34	David	aged 34
Helen	aged 34	James	aged 32
John	aged 30	Tom	aged 30
Joseph	aged 24	Helen	aged 30
Joy	aged 19	Evie	aged 16

Each family had to choose four members with a mean age of between 35 and 36.

Choose two teams, one from each family, that have this mean age between 35 and 36.

9 Asif had an average batting score of 35 runs. He had scored 315 runs in nine games of cricket.

What is the least number of runs he needs to score in the next match if he is to get a higher average score?

10 The mean age of a group of eight walkers is 42. Joanne joins the group and the mean age changes to 40. How old is Joanne?

32.4 The range

The **range** for a set of data is the highest value of the set minus the lowest value.

The range is *not* an average. It shows the **spread** of the data. It is, therefore, used when comparing two or more sets of similar data. You can also use it to comment on the **consistency** of two or more sets of data.

EXAMPLE 5

Rachel's marks in 10 mental arithmetic tests were 4, 4, 7, 6, 6, 5, 7, 6, 9 and 6.

Therefore, her mean mark is $60 \div 10 = 6$ and the range is $9 - 4 = 5$.

Adil's marks in the same tests were 6, 7, 6, 8, 5, 6, 5, 6, 5 and 6.

Therefore, his mean mark is $60 \div 10 = 6$ and the range is $8 - 5 = 3$.

Although the means are the same, Adil has a smaller range. This shows that Adil's results are more consistent.

EXERCISE 32D

1 Find the range for each set of data.

a 3, 8, 7, 4, 5, 9, 10, 6, 7, 4

b 62, 59, 81, 56, 70, 66, 82, 78, 62, 75

c 1, 0, 4, 5, 3, 2, 5, 4, 2, 1, 0, 1, 4, 4

d 3.5, 4.2, 5.5, 3.7, 3.2, 4.8, 5.6, 3.9, 5.5, 3.8

e 2, –1, 0, 3, –1, –2, 1, –4, 2, 3, 0, 2, –2, 0, –3

2 The table shows the maximum and minimum temperatures at midday for five cities in England during a week in August.

	Birmingham	Leeds	London	Newcastle	Sheffield
Maximum temperature (°C)	28	25	26	27	24
Minimum temperature (°C)	23	22	24	20	21

a Write down the range of the temperatures for each city.

b What do the ranges tell you about the weather for England during the week?

3 Over a three-week period, a school sweet shop took the following amounts.

	Monday	Tuesday	Wednesday	Thursday	Friday
Week 1	$32	$29	$36	$30	$28
Week 2	$34	$33	$25	$28	$20
Week 3	$35	$34	$31	$33	$32

a Calculate the mean amount taken each week.

b Find the range for each week.

c What can you say about the total amounts taken for each of the three weeks?

4 In a golf tournament, the club chairperson had to choose either Maria or Fay to play in first round. In the previous eight rounds, their scores were as follows.

Maria's scores: 75, 92, 80, 73, 72, 88, 86, 90

Fay's scores: 80, 87, 85, 76, 85, 79, 84, 88

a Calculate the mean score for each golfer.

b Find the range for each golfer.

c Which golfer would you choose to play in the tournament? Explain why.

> **HINTS AND TIPS**
>
> The best person to choose may not be the one with the biggest mean but could be the most consistent player.

5 Dan has a choice of two buses to get to school: Number 50 or Number 63. Over a month, he kept a record of the number of minutes each bus was late when it set off from his home bus stop.

No. 50: 4, 2, 0, 6, 4, 8, 8, 6, 3, 9

No. 63: 3, 4, 0, 10, 3, 5, 13, 1, 0, 1

a For each bus, calculate the mean number of minutes late.

b Find the range for each bus.

c Which bus would you advise Dan to catch? Give a reason for your answer.

6 The table gives the ages and heights of 10 children.

Name	Age (years)	Height (cm)
Evrim	9	121
Isaac	4	73
Lilla	8	93
Lewis	10	118
Evie	3	66
Badru	6	82
Oliver	4	78
Halima	2	69
Isambard	9	87
Chloe	7	82

a Chloe is having a party. She wants to invite as many children as possible but does not want the range of ages to be more than 5. Who will she invite?

b This is a sign at a theme park:

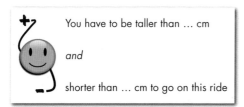

You have to be taller than … cm

and

shorter than … cm to go on this ride

Isaac is the shortest person who can go on the ride and Isambard is the tallest.

What are the smallest and largest missing values on the sign?

7 a The age range of a school quiz team is 20 years and the mean age is 34. Who would you expect to be in this team? Explain your answer.

b Another team has an average age of $15\frac{1}{2}$ and a range of 1. Who would you expect to be in this team? Explain your answer.

An average must be truly **representative** of a set of data. So, when you have to find an average is crucial to choose the **appropriate** type of average for this particular set of data.

If you use the wrong average, your results will be distorted and give misleading information.

This table, which compares the advantages and disadvantages of each type of average, will hel you to make the correct decision.

	Mode	Median	Mean
Advantages	Very easy to find Not affected by **extreme values** Can be used for non-numerical data	Easy to find for ungrouped data Not affected by extreme values	Easy to find Uses all the values The total for a given number of values can be calculated from it
Disadvantages	Does not use all the values May not exist	Does not use all the values Often not understood	Extreme values can distort it Has to be calculated
Use for	Non-numerical data Finding the most likely value	Data with extreme values	Data with values that are spread in a balanced way

EXERCISE 32E

1. The ages of the members of a hockey team were:

    29 26 21 24 26 28 35 23 29 28 29

    a Give:

    i the modal age   **ii** the median age   **iii** the mean age.

    b What is the range of the ages?

2. **a** For each set of data, find the mode, the median and the mean.

    i 6, 10, 3, 4, 3, 6, 2, 9, 3, 4   **ii** 6, 8, 6, 10, 6, 9, 6, 10, 6, 8

    iii 7, 4, 5, 3, 28, 8, 2, 4, 10, 9

    b For each set of data, decide which average is the best one to use and give a reason.

3. A shop sold the following numbers of copies of *The Evening Star* on 12 consecutive evenings during a promotion exercise organised by the newspaper's publisher.

    65 73 75 86 90 112 92 87 77 73 68 62

    a Find the mode, the median and the mean for the sales.

    b The shopkeeper had to report the average sales to the publisher after the promotion. Which of the three averages would you advise the shopkeeper to use? Explain why.

4 The mean age of a group of 10 young people is 15.

   a What do all their ages add up to?

   b What will be their mean age in five years' time?

5 Decide which average you would use for each of the following. Give a reason for your answer.

   a The average mark in an examination.

   b The average pocket money for a group of 16-year-old students.

   c The average shoe size for all the girls in one year at school.

   d The average height for all the artistes on tour with a circus.

   e The average hair colour for students in your school.

   f The average mass of all newborn babies in a hospital's maternity ward.

6 A pack of matches consisted of 12 boxes. The contents of each box were counted as:

   34   31   29   35   33   30   31   28   29   35   32   31

On the box it stated 'Average contents 32 matches'. Is this correct?

7 Mr Brennan told each student their test mark and only gave the test statistics to the whole class. He gave the class the modal mark, the median mark and the mean mark.

   a Which average would tell a student whether they were in the top half or the bottom half of the class?

   b Which average tells the students nothing really?

   c Which average allows a student to gauge how well they have done compared with everyone else?

8 Three players were hoping to be chosen for the basketball team.

The following table shows their scores in the last few games they played.

Tom	16, 10, 12, 10, 13, 8, 10
David	16, 8, 15, 25, 8
Mohaned	15, 2, 15, 3, 5

The teacher said they would be chosen by their best average score.

Which average would each boy want to be chosen by?

9 **a** Find five numbers that have *both* the properties below:
    ● a range of 5
    ● a mean of 5.

   b Find five numbers that have *all* the properties below:
    ● a range of 5
    ● a median of 5
    ● a mean of 5.

10 What is the average pay at a factory with 10 employees?

The boss said: "$43 295"

A worker said: "$18 210"

They were both correct.

Explain how this can be.

11 A list of nine numbers has a mean of 7.6. What number must be added to the list to give new mean of 8?

12 A dance group of 17 people had a mean weight of 54.5 kg. To enter a competition there needed to be 18 people with an average weight of 54.4 kg or less. What is the maximum weight that the eighteenth person must be?

32.6 Using frequency tables

When a lot of information has been gathered, it is often convenient to put it together in a **frequency table**. From this table you can then find the values of the three averages and the rang

EXAMPLE 6

A survey was done on the number of people in each car leaving a shopping centre. The results are summarised in the table below.

Number of people in each car	1	2	3	4	5	6
Frequency	45	198	121	76	52	13

For the number of people in a car, calculate:

a the mode   **b** the median   **c** the mean.

a The modal number of people in a car is easy to spot. It is the number with the largest frequency, which is 198. Hence, the modal number of people in a car is 2.

b The median number of people in a car is found by working out where the middle of the set of numbers is located. First, add up frequencies to get the total number of cars surveyed, which comes to 505. Next, calculate the middle position.

$(505 + 1) \div 2 = 253$

Now add the frequencies across the table to find which group contains the 253rd item. The 243rd item is the end of the group with 2 in a car. Therefore, the 253rd item must be in the group with 3 in a car. Hence, the median number of people in a car is 3.

c To calculate the mean number of people in a car, multiply the number of people in the car by the frequency. This is best done in an extra column. Add these to find the total number of people and divide by the total frequency (the number of cars surveyed).

Number in car	Frequency	Number in these cars
1	45	$1 \times 45 = 45$
2	198	$2 \times 198 = 396$
3	121	$3 \times 121 = 363$
4	76	$4 \times 76 = 304$
5	52	$5 \times 52 = 260$
6	13	$6 \times 13 = 78$
Totals	505	1446

Hence, the mean number of people in a car is $1446 \div 505 = 2.9$ (to 1 decimal place).

EXERCISE 32F

1 Find **i** the mode, **ii** the median and **iii** the mean from each frequency table below.

a A survey of the shoe sizes of all the boys in one year of a school gave these results.

Shoe size	4	5	6	7	8	9	10
Number of students	12	30	34	35	23	8	3

b A survey of the number of eggs laid by hens over a period of one week gave these results.

Number of eggs	0	1	2	3	4	5	6
Frequency	6	8	15	35	48	37	12

c This is a record of the number of babies born each week over one year in a small maternity unit.

Number of babies	0	1	2	3	4	5	6	7	8	9	10	11	12	13	14
Frequency	1	1	1	2	2	2	3	5	9	8	6	4	5	2	1

d A school did a survey on how many times in a week students arrived late at school. These are the findings.

Number of times late	0	1	2	3	4	5
Frequency	481	34	23	15	3	4

FOUNDATION

2 A survey of the number of children in each family of a school's intake gave these results.

Number of children	1	2	3	4	5
Frequency	214	328	97	26	3

a Assuming each child at the school is shown in the data, how many children are at the school?

b Calculate the mean number of children in a family.

c How many families have this mean number of children?

d How many families would consider themselves average from this survey?

3 A dentist kept records of how many teeth he extracted from his patients.

In 1989 he extracted 598 teeth from 271 patients.

In 1999 he extracted 332 teeth from 196 patients.

In 2009 he extracted 374 teeth from 288 patients.

a Calculate the average number of teeth taken from each patient in each year.

b Explain why you think the average number of teeth extracted falls each year.

4 One hundred cases of apples delivered to a supermarket were inspected and the number of bad apples were recorded.

Bad apples	0	1	2	3	4	5	6	7	8	9
Frequency	52	29	9	3	2	1	3	0	0	1

Give:

a the modal number of bad apples per case

b the mean number of bad apples per case.

5 Two dice are thrown together 60 times. The sums of the scores are shown below.

Score	2	3	4	5	6	7	8	9	10	11	12
Frequency	1	2	6	9	12	15	6	5	2	1	1

Find:

a the modal score

b the median score

c the mean score.

6 During a one-month period, the number of days off taken by 100 workers in a factory were noted as follows.

Number of days off	0	1	2	3	4
Number of workers	35	42	16	4	3

Calculate:

a the modal number of days off

b the median number of days off

c the mean number of days off.

7 Two friends often played golf together. They recorded the number of shots they made to get their balls into each hole over the last five games to compare who was more consistent and who was the better player. Their results are summarised in the following table.

Number of shots	1	2	3	4	5	6	7	8	9
Roger	0	0	0	14	37	27	12	0	0
Brian	5	12	15	18	14	8	8	8	8

a What is the modal score for each player?

b What is the range of scores for each player?

c What is the median score for each player?

d What is the mean score for each player?

e Which player is the more consistent and why?

f Who would you say is the better player and why?

8 A tea stain on a newspaper removed four numbers from the following frequency table of goals scored in 40 football matches one weekend.

Goals	0	1	2			5
Frequency	4	6	9			3

The mean number of goals scored is 2.4.

What could the missing four numbers be?

9 Manju made day trips to Mumbai frequently during a year.

The table shows how many days in a week she travelled.

Days	0	1	2	3	4	5
Frequency	17	2	4	13	15	1

Explain how you would find the median number of days Manju travelled in a week to Mumbai.

Sometimes the information you are given is grouped in some way (called **grouped data**), as in Example 7, which shows the range of weekly pocket money given to Year 12 students in a particular class.

Normally, grouped tables use **continuous data**, which is data that can have any value within a range of values, for example, height, mass, time, area and capacity. In these situations, the mean can only be **estimated** as you do not have all the information.

Discrete data is data that consists of separate numbers, for example, goals scored, marks in a test, number of children and shoe sizes.

In both cases, when using a grouped table to estimate the mean, first find the midpoint of the interval by adding the two end-values and then dividing by two.

EXAMPLE 7

Pocket money, p ($)	$0 < p \leqslant 1$	$1 < p \leqslant 2$	$2 < p \leqslant 3$	$3 < p \leqslant 4$	$4 < p \leqslant 5$
No. of students	2	5	5	9	15

a Write down the **modal class**.

b Calculate an estimate of the mean weekly pocket money.

a The modal class is easy to pick out, since it is simply the one with the largest frequency. Here the modal class is $4 to $5.

b To estimate the mean, assume that each person in each class has the 'midpoint' amount, then build up the following table.

To find the midpoint value, the two end-values are added together and then divided by two.

Pocket money, p ($$)	Frequency (f)	Midpoint (m)	$f \times m$
$0 < p \leqslant 1$	2	0.50	1.00
$1 < p \leqslant 2$	5	1.50	7.50
$2 < p \leqslant 3$	5	2.50	12.50
$3 < p \leqslant 4$	9	3.50	31.50
$4 < p \leqslant 5$	15	4.50	67.50
Totals	36		120

The estimated mean will be $120 ÷ 36 = $3.33 (rounded to the nearest cent).

If you had written 0.01–1.00, 1.01–2.00 and so on for the groups, then the midpoints would have been 0.505, 1.505 and so on. This would not have had a significant effect on the final answer as it is only an estimate.

Note that you *cannot* find the median or the range from a grouped table as you do not know the actual values.

EXERCISE 32G

1 For each table of values given below, find:

    i the modal group

    ii an estimate for the mean.

HINTS AND TIPS

When you copy the tables, draw them vertically as in Example 7.

a

x	$0 < x \leqslant 10$	$10 < x \leqslant 20$	$20 < x \leqslant 30$	$30 < x \leqslant 40$	$40 < x \leqslant 50$
Frequency	4	6	11	17	9

b

y	$0 < y \leqslant 100$	$100 < y \leqslant 200$	$200 < y \leqslant 300$	$300 < y \leqslant 400$	$400 < y \leqslant 500$	$500 < y \leqslant 600$
Frequency	95	56	32	21	9	3

c

z	$0 < z \leqslant 5$	$5 < z \leqslant 10$	$10 < z \leqslant 15$	$15 < z \leqslant 20$
Frequency	16	27	19	13

d

Weeks	1–3	4–6	7–9	10–12	13–15
Frequency	5	8	14	10	7

2 Jason brought 100 pebbles back from the beach and found their masses, recording each mass to the nearest gram. His results are summarised in the table below.

Mass, m (g)	$40 < m \leqslant 60$	$60 < m \leqslant 80$	$80 < m \leqslant 100$
Frequency	5	9	22

Mass, m (g)	$100 < m \leqslant 120$	$120 < m \leqslant 140$	$140 < m \leqslant 160$
Frequency	27	26	11

Find:

a the modal class of the pebbles

b an estimate of the total mass of all the pebbles

c an estimate of the mean mass of the pebbles.

3 A gardener measured the heights of all his daffodils to the nearest centimetre and summarised his results as follows.

Height (cm)	10–14	15–18	19–22	23–26	27–40
Frequency	21	57	65	52	12

a How many daffodils did the gardener have?

b What is the modal class of the daffodils?

c What is the estimated mean height of the daffodils?

4 A survey was created to see how quickly an emergency service got to cars which had broken down. The following table summarises the results.

Time (min)	1–15	16–30	31–45	46–60	61–75	76–90	91–105
Frequency	2	23	48	31	27	18	11

  a How many calls were used in the survey?

  b Estimate the mean time taken per call.

  c Which average would the emergency service use for the average call-out time?

  d What percentage of calls do the emergency service get to within the hour?

5 One hundred light bulbs were tested by their manufacturer to see whether the average life-span of the manufacturer's bulbs was over 200 hours. The following table summarises the results.

Life span, h (hours)	$150 < h \leqslant 175$	$175 < h \leqslant 200$	$200 < h \leqslant 225$	$225 < h \leqslant 250$	$250 < h \leqslant 27$
Frequency	24	45	18	10	3

  a What is the modal length of time a bulb lasts?

  b What percentage of bulbs last longer than 200 hours?

  c Estimate the mean life-span of the light bulbs.

  d Do you think the test shows that the average life-span is over 200 hours? Fully explain your answer.

6 Three shops each claimed to have the lowest average price increase over the year. The following table summarises their price increases.

Price increase (p)	1–5	6–10	11–15	16–20	21–25	26–30	31–35
Soundbuy	4	10	14	23	19	8	2
Springfields	5	11	12	19	25	9	6
Setco	3	8	15	31	21	7	3

Using their average price increases, make a comparison of the supermarkets and say which one has the lowest price increases over the year. Do not forget to justify your answers.

7 The table shows the distances run, over a month, by an athlete who is training for a half-marathon.

Distance, d (km)	$0 < d \leqslant 5$	$5 < d \leqslant 10$	$10 < d \leqslant 15$	$15 < d \leqslant 20$	$20 < d \leqslant 2$
Frequency	3	8	13	5	2

A half-marathon is 21 kilometres. It is recommended that an athlete's daily average distance should be at least a third of the distance of the race for which they are training. this athlete doing enough training?

8 The table shows the points scored in a general-knowledge competition by all the players.

Points	0–9	10–19	20–29	30–39	40–49
Frequency	8	5	10	5	2

Balvir noticed that two frequencies were the wrong way round and that this made a difference of 1.7 to the arithmetic mean.

Which two frequencies were the wrong way round?

9 The profit made each week by a charity shop is shown in the table below.

Profit	$0–$500	$501–$1000	$1001–$1500	$1501–$2000
Frequency	15	26	8	3

Estimate the mean profit made each week.

10 The table shows the number of members of 100 football clubs.

Members	20–29	30–39	40–49	50–59	60–69
Frequency	16	34	27	18	5

Roger claims that the median number of members is 39.5.

Is he correct? Explain your answer.

32.8 Measuring spread

Here are the ages of eleven players in a football team:

17, 17, 18, 18, 20, 23, 23, 24, 26, 27, 35

The range is 35 – 17 = 18 years.

The median is the 6th age in the list, that is, 23 years ($\frac{11 + 1}{2}$th = 6th).

$$17 \quad 17 \quad 18 \quad 18 \quad 20 \quad (23) \quad 23 \quad 24 \quad 26 \quad 27 \quad 35$$
$$\uparrow \qquad\qquad\qquad \text{median} \qquad\qquad \uparrow$$

There are five ages on each side of the median.

The median of each of these sets of five is marked with an arrow:

18 is the **lower quartile**.

26 is the **upper quartile**.

The median and the quartiles together divide the ages into four equal groups.

The difference between the quartiles is the **interquartile range**. The interquartile range here is 8 years:

$$26 - 18 = 8$$

Half the ages are within the interquartile range.

The interquartile range is often used to measure the **spread** of a set of data, that is, how spread out they are. It is better than using the range for measuring spread because it is less affected by extreme values at each end of a set of numbers.

EXAMPLE 8

Here are 20 test marks.

Find the median and the interquartile range.

0	12	12	13	15	15	18	20	20	20
21	24	24	25	25	27	27	27	27	30

Because there are 20 marks there is no middle number ($\frac{20 + 1}{2} = 10\frac{1}{2}$).

The median is between the 10th mark (20) and the 11th mark (21):

$$\text{median} = \frac{20 + 21}{2} = 20.5$$

There are ten numbers less than the median:

0   12   12   13   15   ↑   15   18   20   20   20   ↑   21

             lower quartile   median

The lower quartile is the mean of the 5th and 6th marks. These are both 15 so the lower quartile is 15.

The upper quartile is the mean of the 15th and 16th marks:

$$\frac{25 + 27}{2} = 26$$

The interquartile range is:

$$26 - 15 = 11 \text{ marks}$$

EXERCISE 32H

1 Find the median, the quartiles and the interquartile range for each of these sets of numbers:

a 3, 3, 3, 5, 8, 10, 13, 13, 14, 15, 16, 20, 21, 29, 40

b 17, 17, 18, 21, 28, 29, 32, 33, 36

c 100, 99, 97, 96, 92, 88, 83, 76, 72, 70

d 16, 12, 8, 5, 4, 10, 7

HIGHER

2 Here are the results of 20 throws of a dice.

Score	Frequency
1	4
2	8
3	4
4	2
5	1
6	1

a Find the median.

b Find the interquartile range.

c Do you think the dice was thrown fairly?

3 Here are the expenses (in dollars) claimed by 15 employees.

0	2	8	15	16
21	21	28	41	47
51	58	75	101	132

a Find the range.

b Find the median.

c Find the interquartile range.

4 Here are the times, to the nearest minute, of 11 runners in a race:

30, 41, 41, 42, 45, 46, 49, 49, 52, 60, 78

a Find the median, the range and the interquartile range.

b If you remove the times of the fastest and slowest runners, what are the median, range and interquartile range of the remaining nine?

5 Here are the masses, in kilograms, of the baggage of 18 passengers on an aircraft.

11.5	11.8	12.3	12.4	13.2	14.0
14.9	15.1	15.2	15.5	15.5	15.5
17.3	18.2	19.4	20.3	20.7	20.8

a Find the median mass.

b Find the upper quartile.

c Find the interquartile range.

d The scales were faulty. Each piece of baggage has a mass which was 0.1 kg heavier than they showed. What are the correct median and interquartile range?

This section will show how to find the interquartile range and the median of a set of data by drawing a **cumulative frequency diagram**.

Look at the marks of 50 students in a mathematics test, which have been put into a grouped table as shown below. Note that it includes a column for the **cumulative frequency**, which is the sum of all the frequencies up to that point.

Mark	No. of students	Cumulative frequency
21 to 30	1	1
31 to 40	3	4
41 to 50	6	10
51 to 60	10	20
61 to 70	13	33
71 to 80	6	39
81 to 90	4	43
91 to 100	3	46
101 to 110	2	48
111 to 120	2	50

This data can then be used to plot a graph of the top value of each group against its cumulative frequency. The points to be plotted are (30, 1), (40, 4), (50, 10), (60, 20), etc., which will give the graph shown below. Note that the cumulative frequency is *always* the vertical (*y*) axis.

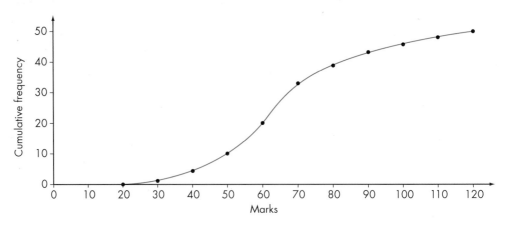

Also note that the scales on both axes are labelled at each graduation mark, in the usual way. *Do not* label the scales as shown below. It is *wrong* because a cumulative frequency graph must have a continuous horizontal scale.

21–30	31–40	41–50

The plotted points can be joined by a freehand curve, to give a cumulative frequency diagram.

The median

The median is the middle item of data once all the items have been put in order of size, from lowest to highest. So, if you have n items of data plotted as a cumulative frequency diagram, you can find the median from the middle value of the cumulative frequency, that is the $\frac{1}{2}n$th value.

But remember, if you want to find the median from a simple list of discrete data, you *must* use the $\frac{1}{2}(n + 1)$th value. The reason for the difference is that the cumulative frequency diagram treats the data as continuous, even when using data such as examination marks, which are discrete. You can use the $\frac{1}{2}n$th value when working with cumulative frequency diagrams because you are only looking for an *estimate* of the median.

There are 50 values in the table on the previous page. To find the median:

● The middle value will be the 25th value.

● Draw a horizontal line from the 25th value to meet the graph.

● Now go down to the horizontal axis.

This will give an estimate of the median. In this example, the median is about 64 marks.

The interquartile range

By dividing the cumulative frequency into four parts, you can obtain **quartiles** and the interquartile range. You have already met these for discrete sets of data.

The lower quartile is the item one-quarter of the way up the cumulative frequency axis and is given by the $\frac{1}{4}n$th value.

The upper quartile is the item three-quarters of the way up the cumulative frequency axis and is given by the $\frac{3}{4}n$th value.

The interquartile range is the difference between the lower and upper quartiles.

These are illustrated on the graph below.

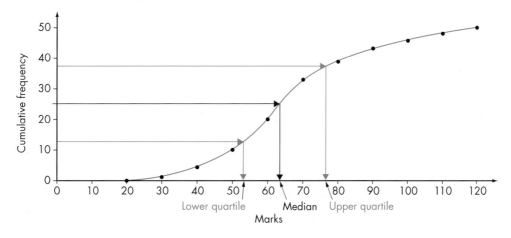

The quarter and three-quarter values out of 50 values are the 12.5th value and the 37.5th value. Draw lines across to the cumulative frequency curve from these values and down to the horizontal axis. These give the lower and upper quartiles. In this example, the lower quartile is 54 marks, the upper quartile is 77 marks and the interquartile range is 77 − 54 = 23 marks.

EXAMPLE 9

This table shows the marks of 100 students in a mathematics test.

a   Draw a cumulative frequency curve.

b   Use your graph to find the median and the interquartile range.

c   Students who score less than 44 have to have extra teaching. How many students will have to have extra teaching?

The groups are given in a different way to those in the table on page 583. You will meet several ways of giving groups (for example, 21–30, $20 < x \leqslant 30$, $21 \leqslant x \leqslant 30$) but the important thing to remember is to plot the top point of each group against the corresponding cumulative frequency.

Mark	No. of students	Cumulative frequency
$21 \leqslant x \leqslant 30$	3	3
$31 \leqslant x \leqslant 40$	9	12
$41 \leqslant x \leqslant 50$	12	24
$51 \leqslant x \leqslant 60$	15	39
$61 \leqslant x \leqslant 70$	22	61
$71 \leqslant x \leqslant 80$	16	77
$81 \leqslant x \leqslant 90$	10	87
$91 \leqslant x \leqslant 100$	8	95
$101 \leqslant x \leqslant 110$	3	98
$111 \leqslant x \leqslant 120$	2	100

a and b Draw the graph and put on the lines for the median (50th value), lower and upper quartiles (25th and 75th values).

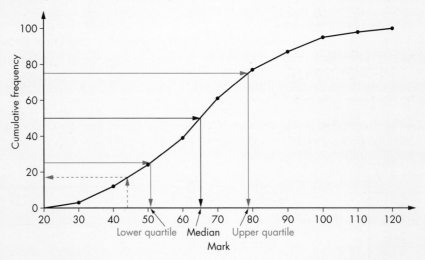

The required answers are read from the graph.

> Median = 65 marks
> Lower quartile = 51 marks
> Upper quartile = 79 marks
> Interquartile range = 79 – 51 = 28 marks

c   At 44 on the mark axis, draw a perpendicular line to intersect the graph, and at the point of intersection draw a horizontal line across to the cumulative frequency axis, a shown. The number of students needing extra teaching is 18.

Note: An alternative way in which the table in Example 9 could have been set out is shown below. This arrangement has the advantage that the points to be plotted are taken straight from the last two columns. You have to decide which method you prefer.

Mark	No. of students	Less than	Cumulative frequency
$21 \leqslant x \leqslant 30$	3	30	3
$31 \leqslant x \leqslant 40$	9	40	12
$41 \leqslant x \leqslant 50$	12	50	24
$51 \leqslant x \leqslant 60$	15	60	39
$61 \leqslant x \leqslant 70$	22	70	61
$71 \leqslant x \leqslant 80$	16	80	77
$81 \leqslant x \leqslant 90$	10	90	87
$91 \leqslant x \leqslant 100$	8	100	95
$101 \leqslant x \leqslant 110$	3	110	98
$111 \leqslant x \leqslant 120$	2	120	100

EXERCISE 32I

1 A class of 30 students was asked to guess when one minute had passed. The table on the right shows the results.

Time (seconds)	No. of students
$20 < x \leqslant 30$	1
$30 < x \leqslant 40$	3
$40 < x \leqslant 50$	6
$50 < x \leqslant 60$	12
$60 < x \leqslant 70$	3
$70 < x \leqslant 80$	3
$80 < x \leqslant 90$	2

   a Copy the table and complete a cumulative frequency column.

   b Draw a cumulative frequency diagram.

   c Use your diagram to estimate the median time and the interquartile range.

2 A group of 50 pensioners was given the task in question **1**. The results are shown in the table on the right.

Time (seconds)	No. of pensioners
$10 < x \leqslant 20$	1
$20 < x \leqslant 30$	2
$30 < x \leqslant 40$	2
$40 < x \leqslant 50$	9
$50 < x \leqslant 60$	17
$60 < x \leqslant 70$	13
$70 < x \leqslant 80$	3
$80 < x \leqslant 90$	2
$90 < x \leqslant 100$	1

   a Copy the table and complete a cumulative frequency column.

   b Draw a cumulative frequency diagram.

   c Use your diagram to estimate the median time and the interquartile range.

   d Which group, the students or the pensioners, was better at estimating time? Give a reason for your answer.

HIGHER

3 The sizes of 360 senior schools are recorded in the table on the right.

a Copy the table and complete a cumulative frequency column.

b Draw a cumulative frequency diagram.

c Use your diagram to estimate the median size of the schools and the interquartile range.

d Schools with fewer than 350 students are threatened with closure. About how many schools are threatened with closure?

No. of students	No. of schools
100–199	12
200–299	18
300–399	33
400–499	50
500–599	63
600–699	74
700–799	64
800–899	35
900–999	11

4 The temperature at a seaside town was recorded for 50 days. It was recorded to the nearest degree. The table on the right shows the results.

a Copy the table and complete a cumulative frequency column.

b Draw a cumulative frequency diagram. Note that as the temperature is to the nearest degree the top values of the groups are 7.5 °C, 10.5 °C, 13.5 °C, 16.5 °C, etc.

Temperature (°C)	No. of days
5–7	2
8–10	3
11–13	5
14–16	6
17–19	6
20–22	9
23–25	8
26–28	6
29–31	5

c Use your diagram to estimate the median temperature and the interquartile range.

d Use your diagram to estimate the 10th percentile.

5 A game consists of throwing three darts and recording the total score. The results of the first 80 people to throw are recorded in the table on the right.

a Draw a cumulative frequency diagram to show the data.

b Use your diagram to estimate the median score and the interquartile range.

c People who score over 90 get a prize. About what percentage of the people get a prize?

Total score	No. of players
$1 \leqslant x \leqslant 20$	9
$21 \leqslant x \leqslant 40$	13
$41 \leqslant x \leqslant 60$	23
$61 \leqslant x \leqslant 80$	15
$81 \leqslant x \leqslant 100$	11
$101 \leqslant x \leqslant 120$	7
$121 \leqslant x \leqslant 140$	2

6 One hundred children were asked to say how much pocket money they got in a week. The results are in the table on the right.

Amount of pocket money (cents)	No. of children
51–100	6
101–150	10
151–200	20
201–250	28
251–300	18
301–350	11
351–400	5
401–450	2

a Copy the table and complete a cumulative frequency column.

b Draw a cumulative frequency diagram.

c Use your diagram to estimate the median amount of pocket money and the interquartile range.

7 Johan set his class an end-of-course test with two papers, A and B. He produced the cumulative frequency graphs on the right.

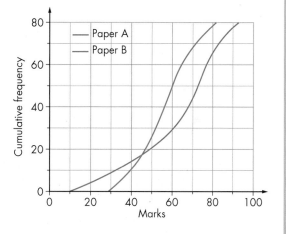

a What is the median score for each paper?

b What is the interquartile range for each paper?

c Which is the harder paper? Explain how you know.

Johan wanted 80% of the students to pass each paper and 20% of the students to get a top grade in each paper.

d What marks for each paper give:

i a pass   **ii** the top grade?

8 The lengths of time taken by 60 helpline telephone calls were recorded.

A cumulative frequency diagram of this data is shown on the right.

Estimate the percentage of calls lasting more than 10 minutes.

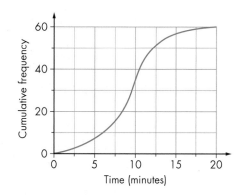

9 Byron was given a cumulative frequency diagram showing the marks obtained by students in a mental maths test.

He was told the top 10% were given the top grade.

How would he find the marks needed to gain this top award?

Why this chapter matters

Chance is a part of everyday life. Judgements are frequently made based on the probability of something happening.

For example:

- there is an 80% chance that my team will win the game tomorrow
- there is a 40% chance of rain tomorrow
- she has a 50–50 chance of having a baby girl
- there is a 10% chance of the bus being on time tonight.

In everyday life we talk about the probability of something happening. Two people might give different probabilities to the same events because of their different views. For example, some people might not agree that there is an 80% chance of your team winning the game. They might say that there is only a 70% chance of them winning tomorrow. A lot depends on what people believe or have experienced.

When people first started to predict the weather scientifically over 150 years ago, they used probabilities to do it. For example, meteorologists looked for three important indicators of rain:

- the number of nimbus clouds in the sky
- falling pressure on a barometer
- the direction of the wind and whether it was blowing from a part of the country with high rainfall.

If all three of these things occurred together rain would almost certainly follow soon.

Now, in the 21st century, probability theory is used to control the flow of traffic through road systems (below left) or the running of telephone exchanges (below right), and to look at patterns of the spread of infections.

33 Probability

Topics	Level	Key words
The probability scale	FOUNDATION	chance, outcome, event, probability, impossible, certain, probability scale
Calculating probabilities	FOUNDATION	equally likely, probability fraction, random
Probability that an event will not happen	FOUNDATION	
Addition rule for probabilities	FOUNDATION	mutually exclusive
Probability from data	FOUNDATION	Venn diagram
Expected frequency	FOUNDATION	expected frequency
Combined events	FOUNDATION	sample space
Tree diagrams	HIGHER	tree diagram, conditional probability

What you need to be able to do in the examinations:

FOUNDATION	HIGHER
• Understand the language of probability. • Understand and use the probability scale. • Understand and use estimates or measures of probability from theoretical models. • Understand the concepts of a sample space and an event, and how the probability of an event happening can be determined. • From the sample space list all the outcomes for single events and for two successive events in a systematic way. • Estimate probabilities from previously collected data. • Find probabilities from a Venn diagram. • Calculate the probability of the complement of an event happening. • Use the addition rule of probability for mutually exclusive events. • Understand and use the term *expected frequency*.	• Draw and use tree diagrams. • Determine the probability that two or more independent events will both occur. • Use simple conditional probability when combining events. • Apply probability to simple problems.

Almost daily, you hear somebody talking about the probability of whether this or that will happen. They usually use words such as '**chance**', 'likelihood' or 'risk' rather than 'probability'. For example:

"What is the likelihood of rain tomorrow?"

"What chance does she have of winning the 100 metre sprint?"

"Is there a risk that his company will go bankrupt?"

You can give a value to the chance of any of these **outcomes** or **events** happening – and millio of others, as well. This value is called the **probability**.

It is true that some things are certain to happen and that some things cannot happen; that is, the chance of something happening can be anywhere between **impossible** and **certain**. This situation is represented on a sliding scale called the **probability scale**, as shown below.

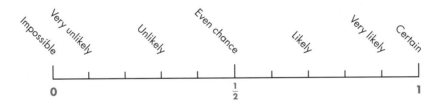

Note: All probabilities lie somewhere in the range of **0** to **1**.

An outcome or an event that cannot happen (is impossible) has a probability of 0. For example the probability that donkeys will fly is 0.

An outcome or an event that is certain to happen has a probability of 1. For example, the probability that the sun will rise tomorrow is 1.

EXAMPLE 1

Put arrows on the probability scale to show the probability of each of the outcomes of these events.

a   You will get a head when throwing a coin.

b   You will get a six when throwing a dice.

c   You will have maths homework this week.

a   This outcome is an even chance. (Commonly described as a fifty-fifty chance.)

b   This outcome is fairly unlikely.

c   This outcome is likely.

The arrows show the approximate probabilities on the probability scale.

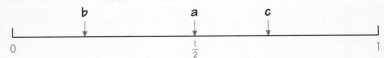

EXERCISE 33A

1 State whether each of the following events is impossible, very unlikely, unlikely, even chance, likely, very likely or certain.

a Someone in your class is left-handed.

b You will live to be 100.

c You get a score of seven when you throw a dice.

d You will watch some TV this evening.

e A new-born baby will be a girl.

2 Draw a probability scale and put an arrow to show the approximate probability of each of the following events happening.

a The next car you see will have been made in Japan.

b A person in your class will have been born in the 20th century.

c It will rain tomorrow.

d In the next Olympic Games, someone will run the 1500 m race in 3 minutes.

e During this week, you will have noodles with a meal.

3 **a** Draw a probability scale and mark an arrow to show the approximate probability of each of the following events.

A The next person to come into the room will be male.

B The person sitting next to you in mathematics is over 16 years old.

C Someone in the class will have a mobile phone.

b What number on your scale corresponds to each arrow?

4 **a** Give two events of your own for which you think the probability of an outcome is as follows:

A impossible   B very unlikely

C evens   D likely

E certain

b Draw a probability scale numbered from 0 to 1 and put an arrow for each of your events.

c What number on your scale corresponds to each arrow?

5 "The train was late yesterday so it is very likely that it will be late today." Is this true?

In Exercise 33A, you may have had difficulty in knowing exactly where to put some of the arrows on the probability scale. It would have been easier for you if each result of the event could have been given a value, from 0 to 1, to represent the probability for that result.

For some events, this can be done by first finding all the possible results, or outcomes, for a particular event. For example, when you throw a coin there are two **equally likely** outcomes: i lands heads up or tails up. (The 'head' of a coin is the side which often shows a head, the 'tail' the side which shows the value of the coin.)

If you want to calculate the probability of getting a head, there is only one outcome that is possible. So, you can say that there is a 1 in 2, or 1 out of 2, chance of getting a head. This is usually given as a **probability fraction**, namely $\frac{1}{2}$. So, you would write the event as:

$P(\text{head}) = \frac{1}{2}$

Probabilities can also be written as decimals or percentages, so that:

$P(\text{head}) = \frac{1}{2}$ or 0.5 or 50%

The probability of an outcome is defined as:

$$P(\text{event}) = \frac{\text{number of ways the outcome can happen}}{\text{total number of possible outcomes}}$$

This definition always leads to a fraction, which should be cancelled to its simplest form.

Another probability term you will meet is at **random**. This means that the outcome cannot be predicted or affected by anyone.

EXAMPLE 2

The spinner shown here is spun and the score on the side on which it lands is recorded.

What is the probability that the score is:

a   2

b   odd

c   less than 5?

a   There are two 2s out of six sides, so $P(2) = \frac{2}{6} = \frac{1}{3}$

b   There are four odd numbers, so $P(\text{odd}) = \frac{4}{6} = \frac{2}{3}$

c   All of the numbers are less than 5, so this is a certain event.

$P(\text{less than 5}) = 1$

EXAMPLE 3

Bernice is always early, just on time or late for work.

The probability that she is early is 0.1, the probability she is just on time is 0.5.

What is the probability that she is late?

As all the possibilities are covered – that is 'early', 'on time' and 'late' – the total of the three probabilities is 1. So,

P(early) + P(on time) = 0.1 + 0.5 = 0.6

So, the probability of Bernice being late is 1 – 0.6 = 0.4.

EXERCISE 33B

1 There are ten balls in a bag. One is red, two are blue, three are yellow and four are green. A ball is taken out without looking.

What is the probability that it is:

a red **b** green

c green or yellow **d** red or green

e white?

> **HINTS AND TIPS**
>
> If an event is impossible, just write the probability as 0, not as a fraction such as $\frac{0}{6}$. If it is certain, write the probability as 1, not as a fraction such as $\frac{6}{6}$.

2 An 8-sided spinner has the numbers 1, 2, 3, 4, 5, 6, 7 and 8 on it. It is spun once.

What is the probability that the score is:

a 3 **b** more than 3

c an even number?

> **HINTS AND TIPS**
>
> Remember to cancel the fractions if possible.

3 A bag contains only blue balls. If I take one out at random, what is the probability of each of these outcomes?

a I get a black ball. **b** I get a blue ball.

4 Number cards with the numbers 1 to 10 inclusive are placed in a hat. Amir takes a number card out of the bag without looking. What is the probability that he draws:

a the number 7 **b** an even number

c a number greater than 6 **d** a number less than 3

e a number between 3 and 8?

5 A pencil case contains six red pens and five blue pens. Paulo takes out a pen without looking at what it is. What is the probability that he takes out:

a a red pen **b** a blue pen **c** a pen that is not blue?

FOUNDATION

6 A bag contains 50 balls. 10 are green, 15 are red and the rest are white. Galenia takes a ball from the bag at random. What is the probability that she takes:

  a a green ball

  b a white ball

  c a ball that is not white

  d a ball that is green or white?

7 There are 500 students in a school and 20 students in Ali's class. One person is chosen at random to welcome a special visitor.

What is the probability the person is in Ali's class?

8 Anton, Bianca, Charlie, Debbie and Elisabeth are in the same class. Their teacher wants two students to do a special job.

  a Write down all the possible combinations of two people, for example, Anton and Bianca, Anton and Charlie. (There are 10 combinations altogether.)

  b How many pairs give two boys?

  c What is the probability of choosing two boys?

  d How many pairs give a boy and a girl?

  e What is the probability of choosing a boy and a girl?

  f What is the probability of choosing two girls?

> **HINTS AND TIPS**
>
> Try to be systematic when writing out all the pairs.

9 A bag contains 25 coloured balls. 12 are red, 7 are blue and the rest are green. Ravi takes a ball at random from the bag.

  a Find:

  i P(he takes a red)   **ii** P(he takes a blue)   **iii** P(he takes a green).

  b Add together the three probabilities. What do you notice?

  c Explain your answer to part **b**.

10 The weather tomorrow will be sunny, cloudy or raining.

If P(sunny) = 40%, P(cloudy) = 25%, what is P(raining)?

11 At morning break, Priya has a choice of coffee, tea or hot chocolate.

If P(she chooses coffee) = 0.3 and P(she chooses hot chocolate) = 0.2, what is P(she chooses tea)?

12 The following information is known about the classes at Bradway School.

Year	Y1		Y2		Y3		Y4		Y5		Y6	
Class	P	Q	R	S	T	U	W	X	Y	Z	K	
Girls	7	8	8	10	10	10	9	11	8	12	14	
Boys	9	10	9	10	12	13	11	12	10	8	16	

A class representative is chosen at random from each class.

Which class has the best chance of choosing a boy as the representative?

13 The teacher chooses, at random, a student to ring the school bell.

Tom says: "It's even chances that the teacher chooses a boy or a girl."

Explain why Tom might not be correct.

33.3 Probability that an event will not happen

In some questions in Exercise 33B, you were asked for the probability of something not happening. For example, in question 5 you were asked for the probability of picking a pen that is not blue. You could answer this because you knew how many pens were in the case. However, sometimes you do not have this type of information.

The probability of throwing a six on a fair, six-sided dice is $P(6) = \frac{1}{6}$.

There are five outcomes that are not sixes: 1, 2, 3, 4, 5.

So, the probability of *not* throwing a six on a dice is:

$P(\text{not a } 6) = \frac{5}{6}$

Notice that:

$P(6) = \frac{1}{6}$ and $P(\text{not a } 6) = \frac{5}{6}$

So:

$P(6) + P(\text{not a } 6) = 1$

If you know that $P(6) = \frac{1}{6}$, then $P(\text{not a } 6)$ is:

$1 - \frac{1}{6} = \frac{5}{6}$

So, if you know P(outcome happening), then:

$P(\text{outcome not happening}) = 1 - P(\text{outcome happening})$

XAMPLE 4

A box of coloured pencils has 20 different pencils. There are four red pencils, five blue, one green, three yellow, two brown, one black, and four other colours.

A pencil is chosen at random. What is the probability that it is not red?

There are 4 red pencils out of 20.

The probability that a red is chosen is $= \frac{4}{20} = \frac{1}{5}$

The probability that a red is **not** chosen is $1 - \frac{1}{5} = \frac{4}{5}$

EXERCISE 33C

FOUNDATION

1 **a** The probability that a football team will win their next match is $\frac{1}{4}$. What is the probability that the team will not win?

b The probability that snow will fall during the winter holidays is 0.45. What is the probability that it will not snow?

c The probability that Paddy wins a game of chess is 0.7 and the probability that he draws the game is 0.1. What is the probability that he loses the game?

2 Look at Example 4.

What is the probability that the pencil is:

a not blue   **b** not yellow   **c** not black?

3 The following letter cards are put into a bag.

| M | A | T | H | E | M |

| A | T | I | C | A | L |

a Lee takes a letter card at random.
   i What is the probability he takes a letter A?
   ii What is the probability he does not take a letter A?

b Ziad picks an M and keeps it. Tasnim now takes a letter from those remaining.
   i What is the probability she takes a letter A?
   ii What is the probability she does not take a letter A?

4 Hamzah is told: "The chance of you winning this game is 0.3."

Hamzah says: "So I have a chance of 0.7 of losing."

Explain why Hamzah might be wrong.

33.4 Addition rule for probabilities

Mutually exclusive events are ones that cannot happen at the same time, such as throwing an odd number and an even number on a roll of a dice.

When two events are mutually exclusive, you can work out the probability of either of them occurring by adding up the separate probabilities.

For example, a plane arriving at an airport could be early, on time or late. These are mutually exclusive events. It cannot be both early and late!

Suppose the probability that a plane arrives early is 0.4 and the probability that it arrives on time is 0.25:

The probability that it is either early or on time is $0.4 + 0.25 = 0.65$.

The probability that the plane is not early or on time (in other words, it is late) is $1 - 0.65 = 0.35$.

Notice that because just one of these three events must happen the probabilities add up to 1:

$0.4 + 0.25 + 0.35 = 1$.

EXAMPLE 5

A bag contains 12 red balls, 8 green balls, 5 blue balls and 15 black balls. A ball is drawn at random. What is the probability that it is the following:

a red   **b** black   **c** red or black   **d** not green?

a $P(\text{red}) = \frac{12}{40} = \frac{3}{10}$

b $P(\text{black}) = \frac{15}{40} = \frac{3}{8}$

c $P(\text{red or black}) = P(\text{red}) + P(\text{black}) = \frac{3}{10} + \frac{3}{8} = \frac{27}{40}$

d $P(\text{not green}) = \frac{32}{40} = \frac{4}{5}$

EXERCISE 33D

1 Iqbal throws an ordinary dice. What is the probability that he throws:

    a a 2   **b** a 5   **c** a 2 or a 5?

2 A bag contains a large number of coloured counters. One is taken out at random.

    The probability that it is red is 0.1.

    The probability that it is blue is 0.15.

    The probability that it is green is 0.2.

    What is the probability that it is:

    a red or blue   **b** red, blue or green   **c** not blue or green?

3 A letter is chosen at random from the letters in the word PROBABILITY. What is the probability that the letter will be:

    a B   **b** a vowel   **c** B or a vowel?

4 A bag contains 10 white balls, 12 black balls and 8 red balls. A ball is drawn at random from the bag. What is the probability that it will be:

    a white   **b** black

    c black or white   **d** not red

    e not red or black?

HINTS AND TIPS

You can only add fractions with the same denominator.

FOUNDATION

5 As part of a computer game, the player is taken to a different country.

The probability that it is in Europe is 0.12.

The probability that it is in Africa is 0.35.

The probability that it is in Asia is 0.24.

What is the probability that the country is:

a In Africa or Asia   **b** In Europe or Asia   **c** Not in Europe, Africa or As

6 John needs his calculator for his mathematics lesson. It is always in his pocket, bag or locker. The probability it is in his pocket is 0.35 and the probability it is in his bag is 0.4 What is the probability that:

a he will have the calculator for the lesson   **b** his calculator is in his locke

7 Aneesa has 20 unlabelled CDs, 12 of which are rock, 5 are pop and 3 are classical. She picks a CD at random. What is the probability that it will be:

a rock or pop   **b** pop or classical   **c** not pop?

8 The probability that it rains on Monday is 0.5. The probability that it rains on Tuesday is 0.5 and the probability that it rains on Wednesday is 0.5. Kelly argues that it is certain to rain on Monday, Tuesday or Wednesday because 0.5 + 0.5 + 0.5 = 1.5, which is bigger than 1 so it is a certain event. Explain why she is wrong.

9 In a TV game show, contestants throw darts at the dartboard shown.

The angle at the centre of each black sector is 15°.

If a dart lands in a black sector the contestant loses.

Any dart missing the board is rethrown.

What is the probability that a contestant throwing a dart at random does not lose?

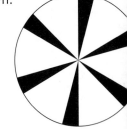

10 There are 45 patients sitting in the hospital waiting room.

8 patients are waiting for Dr Speed.
12 patients are waiting for Dr Mayne.
9 patients are waiting for Dr Kildare.
10 patients are waiting for Dr Pattell.
6 patients are waiting for Dr Stone.

A patient suddenly has to go home.

What is the probability that the patient who left was due to see Dr Speed?

11 The probability of it snowing on any one day in February is $\frac{1}{4}$.

One year, there was no snow for the first 14 days.

Ciara said: "The chance of it snowing on any day in the rest of February must now be $\frac{1}{2}$."

Explain why Ciara is wrong.

FOUNDATION

Probability from data

If you drop a drawing pin, what is the probability that it will land point up?

There is no theoretical way to answer this. However we could do an experiment and use the results to find the probability.

Suppose we drop a drawing pin 50 times and it lands point up 17 times.
The probability of point up is:

$$\frac{17}{50} = 0.34$$

EXAMPLE 6

The frequency table shows the speeds of 160 vehicles that pass a radar speed check on a fast road.

Speed (km/h)	20–29	30–39	40–49	50–59	60–69	70+
Frequency	14	23	28	35	52	8

a   What is the probability that a car is travelling faster than 70 km/h?

b   What is the probability that a car is travelling slower than 50 km/h?

a   The probability is $\frac{8}{160} = \frac{1}{20}$ or 0.05.

b   28 + 23 + 14 = 65 cars were slower than 50 km/h.
   The probability is $\frac{65}{160} = 0.41$

Information in **Venn diagrams** can be used to calculate probabilities.

EXAMPLE 7

A group of students are asked if they have a calculator (C) or a dictionary (D).

The results are shown in this Venn diagram.

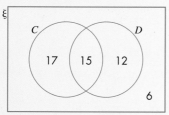

Find the probability that a student:

a   has both a calculator and a dictionary   b   does not have a calculator.

a   The total number of students is 17 + 15 + 12 + 6 = 50
   The number who have both is 15
   The probability is $\frac{15}{50} = 0.3$

b   The number who do not have a calculator is 12 + 6 = 18
   The probability is $\frac{18}{50} = 0.36$

FOUNDATION

EXERCISE 33E

1 This table shows the weather on 1st April over the last 40 years in my town.

Weather	Wet	Dry but cloudy	Sunny
Number of years	10	18	12

What is the probability that the next 1st April will be:

a sunny   **b** not wet?

2 Marta and Maria have played each other at badminton 25 times. Maria has won 10 times.
What is the probability that Marta will win the next game?

3 A survey asks people if they can speak Spanish (S) or Chinese (S). The results are in this Venn diagram.

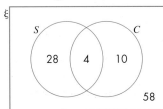

Find the probability that a person can speak:

a Spanish   **b** Chinese   **c** neither language.

4 Zia catches a train to work each day.

He compiles these statistics over a two month period.

Train	On time	Less than 5 minutes late	5–10 minutes late	More than 10 minutes late
Frequency	23	10	3	4

What is the probability that tomorrow's train will be:

a over 10 minutes late   **b** late?

5 A dice is biased. The numbers are not all equally likely to fall face up.

Here are the results of 500 throws.

Score	1	2	3	4	5	6
Frequency	84	123	62	91	47	93

What is the probability of throwing:

a an odd number   **b** a '4' or more?

6 A group of 80 people are asked if they have been to the theatre (T) or a concert (C) in the past three months. Here are the results.

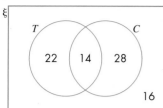

Find the probability that someone went to the theatre or a concert but not both.

7 A number of cars were given a safety check. Here are the results.

		Pass	Fail	**Total**
Age of car	Less than 5 years	27	12	**39**
	Over 5 years	26	35	**61**
	Total	53	47	**100**

a My car is less than 5 years old. What is the probability that it passed?

b My friend's car failed. What is the probability that it is over 5 years old?

8 Here are the results of a survey of whether people use an internet website.

Age group	Under 18	18–60	Over 60
Use it	24	31	7
Do not use it	10	21	18

a What is the probability that someone under 18 uses the site?

b What is the probability that someone over 60 uses the site?

c Sami is one of the people in the survey who use the site. What is the probability that Sami is under 18?

9 There are 150 students.

The Venn diagram shows how many are studying chemistry (C), physics (P) or biology (B).

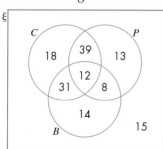

Find the probability that a student is studying:

a all three subjects   **b** two of the three subjects

c one of the three subjects   **d** none of them.

Expected frequency

When you know the probability of an event you can predict how many times it will happen in a certain number of trials. This is the **expected frequency**.

EXAMPLE 8

A bag contains 20 balls, 9 of which are black, 6 white and 5 yellow. A ball is drawn at random from the bag, its colour is noted and then it is put back in the bag. This is repeated 500 times.

a How many times would you expect a black ball to be drawn?

b How many times would you expect a yellow ball to be drawn?

c How many times would you expect a black or a yellow ball to be drawn?

a P(black ball) = $\frac{9}{20}$

Expected number of black balls = $\frac{9}{20} \times 500 = 225$

b P(yellow ball) = $\frac{5}{20} = \frac{1}{4}$

Expected number of yellow balls = $\frac{1}{4} \times 500 = 125$

c Expected number of black or yellow balls = 225 + 125 = 350

EXERCISE 33F

1 **a** What is the probability of throwing a 6 with an ordinary dice?

b I throw an ordinary dice 150 times. How many times can I expect to get a score of 6?

2 **a** What is the probability of tossing a head with a coin?

b I toss a coin 2000 times. How many times can I expect to get a head?

3 When Yusef plays chess with his father, Yusef's probability of winning is 0.3 and of losing is 0.6.

a If they play 20 games, how many can Yusef expect to:

i win

ii lose

iii draw?

b If they play 30 games, how many can Yusef's father expect to:

i win

ii lose

iii draw?

4 Anita plays golf. She estimates the probabilities of the number of strokes it will take to put the ball into one particular hole:

   3 strokes or less: 0.25

   4 strokes: 0.3

   5 strokes: 0.15

If she plays the hole 20 times, how often does she expect to take:

a 4 strokes

b 5 strokes

c Less than 5 strokes

d More than 5 strokes?

5 In a bag there are 30 balls, 15 of which are red, 5 yellow, 5 green, and 5 blue. A ball is taken out at random and then replaced. This is done 300 times. How many times would I expect to get:

a a red ball

b a yellow or blue ball

c a ball that is not blue

d a pink ball?

6 The experiment described in question 5 is carried out 1000 times. Approximately how many times would you expect to get:

a a green ball

b a ball that is not blue?

7 A sampling bottle contains red and white balls. Balls are tipped out one at a time and then replaced. It is known that the probability of getting a red ball is 0.3. If 1500 samples are taken, how many of them would you expect to give a white ball?

8 Josie said, "When I throw a dice, I expect to get a score of 3.5."

   "Impossible," said Paul, "You can't score 3.5 with a dice."

   "Do this and I'll prove it," said Josie.

a An ordinary dice is thrown 60 times. Fill in the table for the expected number of times each score will occur.

Score	1	2	3	4	5	6
Expected occurrences						

b Now work out the average score that is expected over 60 throws.

c There is an easy way to get an answer of 3.5 for the expected average score. Can you see what it is?

9 The probabilities of some cloud types being seen on any day are given below.

Cumulus	0.3
Stratocumulus	0.25
Stratus	0.15
Altocumulus	0.11
Cirrus	0.05
Cirrcocumulus	0.02
Nimbostratus	0.005
Cumulonimbus	0.004

a What is the probability of *not* seeing one of the above clouds in the sky?

b On how many days of the year would you expect to see altocumulus clouds in the sky?

10 Every evening Anne and Chris draw cards out of a pack to see who washes up. There are 52 cards in a pack, including 4 Jacks, 4 Queens and 4 Kings.

If they draw a King or a Jack, Chris washes up.

If they draw a Queen, Anne washes up.

Otherwise, they wash up together.

In a year of 365 days, how many days would you expect them to wash up together?

11 A market gardener is supplied with tomato plant seedlings and knows that the probability that any plant will develop a disease is 0.003.

How will she find out how many of the tomato plants she should expect to develop a disease?

33.7 Combined events

There are many situations where two events occur together. Four examples are given below.

Throwing two dice

Imagine that two dice, one red and one blue, are thrown.

The red dice can land with any one of six scores: 1, 2, 3, 4, 5 or 6.

The blue dice can also land with any one of six scores. This gives a total of 36 possible combinations.

These are shown in the left-hand diagram, where combinations are given as (2, 3) and so on. The first number is the score on the blue dice and the second number is the score on the red dice.

The combination (2, 3) gives a total of 5. The total scores for all the combinations are shown in the diagram on the right-hand side. Diagrams that show all the outcomes of combined events are called **sample spaces**.

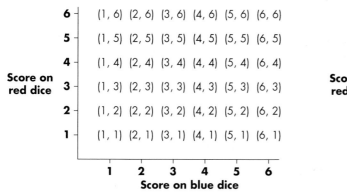

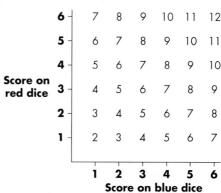

From the diagram on the right above, you can see that there are two ways to get a score of 3. This gives a probability of scoring 3 as:

$$P(3) = \frac{2}{36} = \frac{1}{18}$$

From the diagram on the left, you can see that there are six ways to get a 'double'. This gives a probability of scoring a double as:

$$P(\text{double}) = \frac{6}{36} = \frac{1}{6}$$

Dice and coins

Throwing a dice and a coin

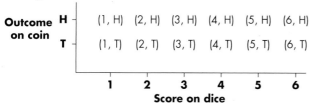

Hence: P (head and an even number) = 3 ways out of 12 = $\frac{3}{12} = \frac{1}{4}$

1 To answer these questions, use the diagram at the top of the page for all the possible scores when two fair dice are thrown together.

   a What is the most likely score?

   b Which two scores are least likely?

   c Write down the probabilities of throwing all the scores from 2 to 12.

   d What is the probability of a score that is:

     i bigger than 10   **ii** between 3 and 7   **iii** even

     iv a square number   **v** a prime number   **vi** a triangular number?

FOUNDATION

2 Use the diagram on the previous page that shows the outcomes when two fair, six-sided dice are thrown together as coordinates. What is the probability that:

a the score is an even 'double'

b at least one of the dice shows 2

c the score on one dice is twice the score on the other dice

d at least one of the dice shows a multiple of 3?

3 Use the diagram on the previous page that shows the outcomes when two fair, six-sided dice are thrown together as coordinates. What is the probability that:

a both dice show a 6

b at least one of the dice will show a 6

c exactly one dice shows a 6?

4 The diagram shows the scores for the event 'the difference between the scores when two fair, six-sided dice are thrown'.

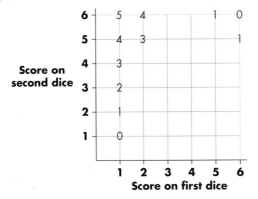

Copy and complete the diagram.

For the event described above, what is the probability of a difference of:

a 1   **b** 0

c 4   **d** 6

e an odd number?

5 When two fair coins are thrown together, what is the probability of:

a two heads

b a head and a tail

c at least one tail

d no tails?

Use a diagram of the outcomes when two coins are thrown together.

6 Two five-sided spinners are spun together and the total score of the sides that they land on is worked out. Copy and complete the diagram shown.

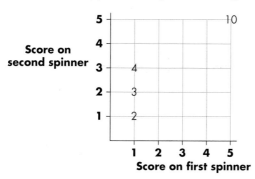

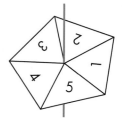

a What is the most likely score?

b When two five-sided spinners are spun together, what is the probability that:

   i the total score is 5

   ii the total score is an even number

   iii the score is a 'double'

   iv the total score is less than 7?

7 Two eight-sided spinners showing the numbers 1 to 8 were thrown at the same time.

a Draw a diagram to show the product of the two scores.

b What is the probability that the product of the two scores is an even square number?

8 Isaac rolls two dice and multiplies both numbers to give their product. He wants to know the probability of rolling two dice that will give him a product between 19 and 35.

Draw a suitable diagram and use it to answer the question.

9 Nic went to a garden centre to buy some roses.

She found they came in six different colours – white, red, orange, yellow, pink and copper.

She also found they came in five different sizes – dwarf, small, medium, large climbing and rambling.

a Draw a diagram to show all the options.

b She buys a random rose for Auntie Janet. Auntie Janet only likes red and pink roses and she does not like climbing or rambling roses.

What is the probability that Nic has bought for Auntie Janet a rose:

   i that she likes

   ii that she does not like?

Imagine you have to draw two cards from this pack of six cards, but you must replace the first card before you select the second card.

You can tackle problems like this by using a **tree diagram**.

When you pick the first card, there are three possible outcomes: a square, a triangle or a circle. For a single event:

$$P(\text{square}) = \frac{3}{6} \qquad\qquad P(\text{triangle}) = \frac{2}{6} \qquad\qquad P(\text{circle}) = \frac{1}{6}$$

You can show this by depicting each event as a branch and writing its probability on the branch.

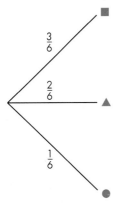

The diagram can then be extended to take into account a second choice. Because the first card has been replaced, you can still pick a square, a triangle or a circle. This is true no matter what is chosen the first time. You can demonstrate this by adding three more branches to the 'square' branch in the diagram.

Here is the complete tree diagram.

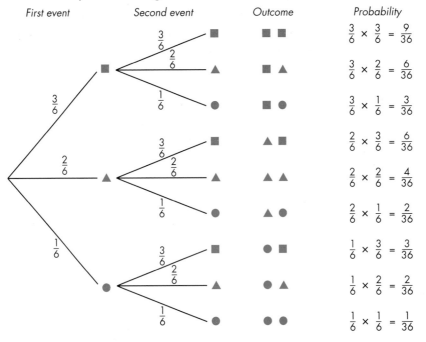

The probability of any outcome is calculated by multiplying all the probabilities on its branches. For instance:

$$P(\text{two squares}) = \frac{3}{6} \times \frac{3}{6} = \frac{9}{36}$$

$$P(\text{triangle followed by circle}) = \frac{2}{6} \times \frac{1}{6} = \frac{2}{36}$$

EXAMPLE 9

Using the tree diagram on the previous page, what is the probability of obtaining:

a two triangles

b a circle followed by a triangle

c a square and a triangle, in any order

d two circles

e two shapes the same?

a $P(\text{two triangles}) = \frac{4}{36}$

b $P(\text{circle followed by triangle}) = \frac{2}{36}$

c There are two places in the outcome column that have a square and a triangle. These are the second and fourth rows. The probability of each is $\frac{1}{6}$. Their combined probability is given by the addition rule.

$$P(\text{square and triangle, in any order}) = \frac{6}{36} + \frac{6}{36} = \frac{12}{36}$$

d $P(\text{two circles}) = \frac{1}{36}$

e There are three places in the outcome column that have two shapes the same. These are the first, fifth and last rows. The probabilities are respectively $\frac{1}{4}, \frac{1}{9}$ and $\frac{1}{36}$. Their combined probability is given by the addition rule.

$$P(\text{two shapes the same}) = \frac{9}{36} + \frac{4}{36} + \frac{1}{36} = \frac{14}{36}$$

The term **conditional probability** is used to describe the situation when the probability of an event is dependent on the outcome of another event. For instance, if a card is taken from a pac and not returned, then the probabilities for the next card drawn will be altered. The following example illustrates this situation.

EXAMPLE 10

A bag contains nine balls, of which five are white and four are black.

A ball is taken out and not replaced. Another is then taken out. If the first ball removed is black, what is the probability that:

a the second ball will be black

b both balls will be black?

When a black ball is removed, there are five white balls and three black balls left, reducing the total to eight.

Hence, when the second ball is taken out:

a P(second ball black) = $\frac{3}{8}$

b P(both balls black) = $\frac{4}{9} \times \frac{3}{8} = \frac{1}{6}$

We could show this on a tree diagram and put in other probabilities too.

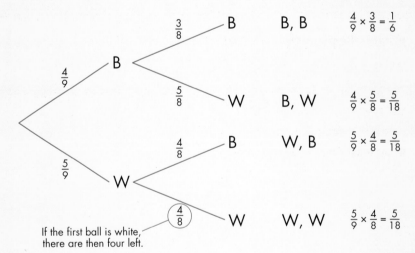

EXERCISE 33H

1 A coin is tossed twice.

Copy and complete this tree diagram to show all the outcomes.

First event	Second event	Outcome	Probability

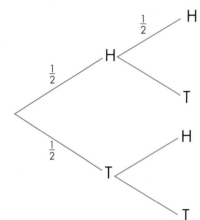

H (H, H) $\frac{1}{2} \times \frac{1}{2} = \frac{1}{4}$

Use your tree diagram to work out the probability of each of these outcomes.

a Getting two heads

b Getting a head and a tail

c Getting at least one tail

2 A bag contains lots of counters. There are equal numbers of green, gold, black and red counters.

A counter is taken out at random. It is replaced. Then a second counter is taken.

a What is the probability that the first counter is gold?

b What is the probability that the second counter is gold?

c Draw a tree diagram to show the outcome of each counter being gold or not. Use it to work out the probability that:

i both counters are gold

ii at least one counter will be gold.

3 On my way to work, I drive through two sets of road works with traffic lights which only show green or red. I know that the probability of the first set being green is $\frac{1}{3}$ and the probability of the second set being green is $\frac{1}{2}$.

a What is the probability that the first set of lights will be red?

b What is the probability that the second set of lights will be red?

c Copy and complete this tree diagram, showing the possible outcomes when passing through both sets of lights.

First event	Second event	Outcome	Probability

$$\frac{1}{2} \quad G \qquad (G, G) \qquad \frac{1}{3} \times \frac{1}{2} = \frac{1}{6}$$

G

$\frac{1}{3}$

R

G

R

R

d Using the tree diagram above, what is the probability of each of the following outcomes?

i I do not get held up at either set of lights.

ii I get held up at exactly one set of lights.

iii I get held up at least once.

e Over a school term I make 90 journeys to work. On how many days can I expect to g two green lights?

4 Six out of every 10 cars in Britain are made in other countries.

a What is the probability that any car will be British made?

b Two cars can be seen approaching in the distance. Draw a tree diagram to work out the probability of each of these outcomes.

i Both cars will be British made.

ii One car will be British and the other car will be made elsewhere.

5 Three coins are tossed. Copy and complete the tree diagram below and use it to answer the questions.

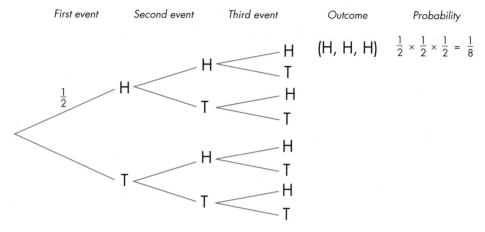

If a coin is tossed three times, what is the probability of each of these outcomes?

a Three heads

b Two heads and a tail

c At least one tail

6 Thomas has to take a three-part language examination paper.

- The first part is speaking. He has a 0.4 chance of passing this part.
- The second is listening. He has a 0.5 chance of passing this part.
- The third part is writing. He has a 0.7 chance of passing this part.

a Draw a tree diagram covering three events, where the first event is passing or failing the speaking part of the examination, the second event is passing or failing the listening part and the third event is passing or failing the writing part.

b If he passes all three parts, his father will give him $20. What is the probability that he gets the money?

c If he passes two parts only, he can resit the other part. What is the chance he will have to resit?

d If he fails all three parts, he will be thrown off the course. What is the chance he is thrown off the course?

7 In a group of 10 girls, six like the pop group Smudge and four like the pop group Mirage. Two girls are to be chosen for a pop quiz.

   a What is the probability that the first girl chosen will be a Smudge fan?

   b If the first girl chosen is a Smudge fan, explain why the probability that the second girl chosen is a Smudge fan is $\frac{5}{9}$.

   c Copy and complete this tree diagram showing what the group likes when two girls are chosen.

   d Use your tree diagram to work out the probability that:

      i both girls chosen will like Smudge

      ii both girls chosen will like the same group

      iii both girls chosen will like different groups.

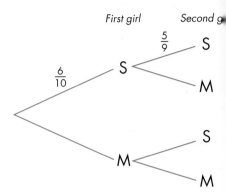

8 There are five white eggs and one brown egg in an egg box. Kate decides to make a two-egg omelette. She takes each egg from the box without looking at its colour.

   a What is the probability that the first egg taken is brown?

   b If the first egg taken is brown, what is the probability that the second egg taken will be brown?

   c Copy this tree diagram and put probabilities on the branches.

   d What is the probability that Kate gets an omelette made from:

      i two white eggs

      ii one white and one brown egg

      iii two brown eggs?

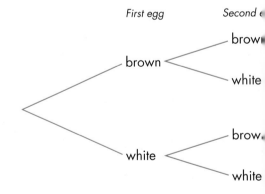

9 A box contains 10 red and 15 yellow balls. One is taken out and not replaced. Another is taken out.

   a If the first ball taken out is red, what is the probability that the second ball is:

      i red

      ii yellow?

   b If the first ball taken out is yellow, what is the probability that the second ball is:

      i red

      ii yellow?

10 A fruit bowl contains six oranges and eight lemons. Kevin takes two pieces of fruit at random.

    a If the first piece is an orange, what is the probability that the second is:

        i an orange

        ii a lemon?

    b What is the probability that:

        i both are oranges

        ii both are lemons?

11 A bag contains three black balls and seven red balls. A ball is taken out and not replaced. This is repeated twice. What is the probability that:

    a all three are black

    b exactly two are black

    c exactly one is black

    d none are black?

12 On my way to work, I pass two sets of traffic lights. The probability that the first is green is $\frac{1}{3}$. If the first is green, the probability that the second is green is $\frac{1}{3}$. If the first is red, the probability that the second is green is $\frac{2}{3}$. What is the probability that:

    a both are green

    b none are green

    c exactly one is green

    d at least one is green?

13 An engineering test is in two parts, a written test and a practical test. 90% of those who take the written test pass. When a person passes the written test, the probability that he or she will also pass the practical test is 60%. When a person fails the written test, the probability that he or she will pass the practical test is 20%.

    a What is the probability that someone passes both tests?

    b What is the probability that someone passes one test?

    c What is the probability that someone fails both tests?

    d What is the combined probability of the answers to parts **a**, **b** and **c**?

FOUNDATION

1 The bar chart shows information about the number of candidates for an examination from each of five countries.

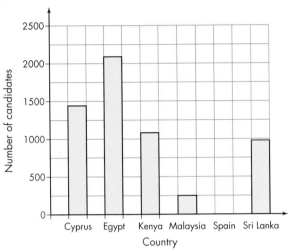

a Which of the five countries had the greatest number of candidates? [1]

b Write down the number of candidates from Malaysia. [1]

c The number of candidates from one country was 1086.
Which country was this? [1]

d The number of candidates from Spain was 727.
Draw a bar on the bar chart to show this information. [1]

Edexcel Limited Paper 1F Q2 May 13

2 In a game, a fair coin is spun and a fair 6-sided dice is rolled.

A score is given according to the rules below.

coin lands on **heads**	coin lands on **tails**
score = 2 × number on the dice	score = 1 + number on the dice

a Complete the table to show all the possible scores. [2]

		Dice					
		1	2	3	4	5	6
Coin	Heads					10	
	Tails			4			

Peter plays the game once.

b Find the probability that Peter's score is 4. [2]

George plays the game 60 times.

c Work out an estimate for the number of times George's score is 10. [2]

Edexcel Limited Paper 1F Q12 May 14

3 Rohan plays for his village cricket team.

Here are the number of runs he scored in each of six games.

12   4   35   67   32   54

a Find the range of the number of runs Rohan scored. [2]

b Find the mean of the number of runs Rohan scored. [2]

One of the six games is picked at random.

c Find the probability that Rohan scored more than 50 runs in this game. [2]

Edexcel Limited Paper 1F Q11 Jan 15

4 w, x, y and z are 4 integers written in order of size, starting with the smallest.

The mean of w, x, y and z is 13.

The sum of w, x and y is 33.

a Find the value of z. [2]

Given also that the range of w, x, y and z is 10,

b Work out the median of w, x, y and z. [2]

Edexcel Limited Paper 1F Q16 May 15

5 The pie chart gives information about the amounts spent by a gas company in one year.

The amount spent on materials was 225.5 million euros.

The amount spent on services was the same as the amount spent on wages.

Work out the amount spent on services. [3]

Edexcel Limited Paper 1F Q20 May 15

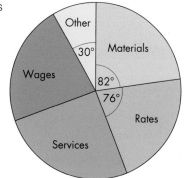

6 Zara must take 5 tests.

Each test is out of 100.

After 4 tests her mean score is 64%.

What score must Zara get in her 5th test to increase her mean score in all 5 tests to 70%? [3]

Edexcel Limited Paper 1F Q17 May 14

7 There are 32 students in Mr Newton's class.

20 are boys and 12 are girls.

The mean height of the boys is 151 cm.

The mean height of the girls is 148 cm.

Calculate the mean height of all the students in Mr Newton's class. [3]

Edexcel Limited Paper 1F Q24 May 15

PAPER 2F

1 The pictogram shows information about the number of books sold in a shop on each of six days.

Monday	
Tuesday	
Wednesday	
Thursday	
Friday	
Saturday	

a On which day was the least number of books sold in the shop? [1]

The number of books sold in the shop on one of these days was twice the number of books sold on Wednesday.

b On which day was this? [1]

The number of books sold in the shop on Tuesday was 18.

c Work out the number of books sold in the shop on Friday. [2]

Edexcel Limited Paper 2F Q3 Jan 16

2 The mean of four numbers is 2.6

One of the four numbers is 5

Find the mean of the other three numbers. [3]

Edexcel Limited Paper 2F Q22 Jan 14

3 Here are 6 cards.

Each card has a number on it.

(3) (1) (5) (5) (2) (3)

a Find the median of the numbers on the cards. [2]

Uzma places two extra cards next to the six cards.

(3) (1) (5) (5) (2) (3) (   ) (   )

She wants the mean of the numbers on the 8 cards to be 4

She wants the range of the numbers on the 8 cards to be 9

b Find the numbers that she should write on the two extra cards. [3]

Edexcel Limited Paper 2F Q11 Jan 16

4

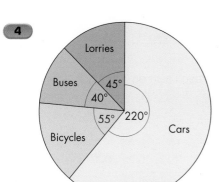

The pie chart shows information about the types of vehicles that went past a school, in one hour, on Monday morning.

a 16 buses went past the school.

Work out the number of bicycles that went past the school. [2]

b The table shows the numbers of vehicles that went past the school, in one hour, on Tuesday morning.

Vehicles	Frequency
Cars	41
Bicycles	15
Buses	7
Lorries	9
Total	72

A pie chart is to be drawn to show this information.

Work out the size of the angle in the pie chart for the 9 lorries. [2]

Edexcel Limited Paper 2F Q14 May 14

5 A bag contains only red counters, blue counters and yellow counters.

The number of red counters in the bag is the same as the number of blue counters.

Mikhail takes at random a counter from the bag.

The probability that the counter is yellow is 0.3.

Work out the probability that the counter Mikhail takes is red. [3]

Edexcel Limited Paper 2F Q14 Jan 15

6 Three positive whole numbers are all different.

The numbers have a median of 8 and a mean of 6.

Find the three numbers. [2]

Edexcel Limited Paper 2F Q21 Jan 15

PAPER 3H

1 Kim asked 40 people how many text messages they each sent on Monday.

The table shows her results.

Number of text messages sent	Frequency
0 to 4	6
5 to 9	3
10 to 14	5
15 to 19	12
20 to 24	14

a Write down the modal class. [1]

b Calculate an estimate for the mean number of text messages sent. [4]

c What percentage of these 40 people sent 20 or more text messages. [2]

Edexcel Limited Paper 3H Q5 Jan 16

2 Here are the marks that James scored in eleven maths tests.

16   12   19   18   17   13   13   20   11   19   17

a Find the interquartile range of these marks. [3]

Sunil did the same eleven maths tests.

The median mark Sunil scored in his tests is 17

The interquartile range is 8

b Which one of Sunil or James has the more consistent marks?
Give a reason for your answer. [1]

Sunil did four more maths tests.

His scores in these four tests were 16, 20, 18 and 10.

c How does his new median mark for the fifteen tests compare with his median mark of 17 for the eleven tests?
Tick (✓) one box.
new median is lower ☐
new median is 17 ☐
new median is higher ☐

Explain your answer. [1]

Edexcel Limited Paper 3H Q13 Jan 16

3 The cumulative frequency graph gives information about the intelligence quotients (IQ) of a random sample of 100 adults.

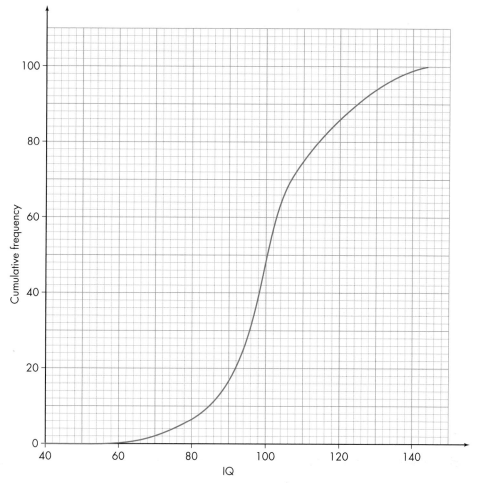

a Use the cumulative frequency graph to find an estimate for the number of adults in the sample who have an IQ between 85 and 115. [2]

b Find an estimate for the upper quartile of the IQ of adults in the sample. [2]

Edexcel Limited Paper 3H Q13 Jan 15

4 Leonidas has a fair dice.

He throws the dice twice.

a Work out the probability that he gets the number 5 both times. [2]

Alicia has a fair dice.

She throws the dice 3 times.

b Work out the probability that she gets the number 5 exactly once. [3]

Edexcel Limited Paper 3H Q20 May 15

PAPER 4H

1 A box contains toy cars.
Each car is red or blue or black or silver.
Emily takes at random a car from the box.
The table shows the probabilities that Emily takes a red car or a blue car or a black car.

Colour of car	Probability
Red	0.20
Blue	0.05
Black	0.15
Silver	

a Work out the probability that Emily takes a silver car. [2]

Emily puts the car back in the box.
There are 6 blue cars in the box.

b Work out the total number of cars in the box. [2]

Edexcel Limited Paper 4H Q8 Jan 16

2 The table gives information about the numbers of goals scored by a football team in 30 matches.

Number of goals scored	Frequency
0	2
1	10
2	7
3	6
4	3
5	2

Find the mean number of goals scored. [3]

Edexcel Limited Paper 4H Q4 Jan 14

3 Chris and Sunil each take a driving test.

The probability that Chris passes the driving test is 0.9.

The probability that Sunil passes the driving test is 0.65.

a Complete the probability tree diagram.

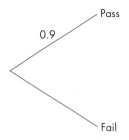

Chris   *Sunil*

0.9   Pass

Fail

b Work out the probability that exactly one of Chris or Sunil passes the driving test. [3]

Edexcel Limited Paper 4H Q16 Jan 15

4 The histogram shows information about the heights of some tomato plants.

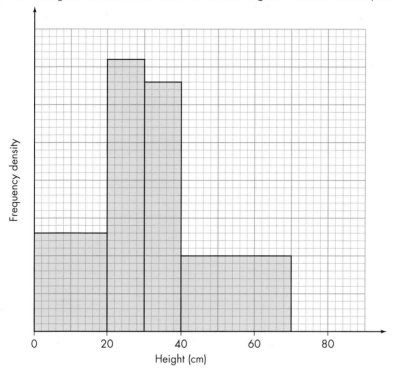

26 plants have a height of less than 20 cm.

Work out the total number of tomato plants. [3]

Edexcel Limited Paper 4H Q20 Jan 15

5 There are 6 milk chocolates and 4 plain chocolates in a box.

Rob takes at random a chocolate from the box and eats it.

Then Alison takes at random a chocolate from the box and eats it.

a Complete the probability tree diagram. [3]

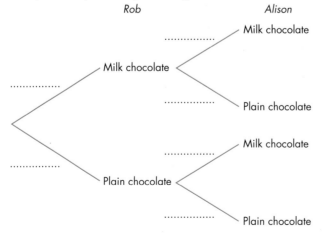

b Work out the probability that there are now exactly 3 plain chocolates in the box. [3]

Edexcel Limited Paper 4H Q15 Jun 15

Answers to Chapter 1

1.1 Multiples of whole numbers

Exercise 1A

1 a 3, 6, 9, 12, 15   **b** 7, 14, 21, 28, 35
  c 9, 18, 27, 36, 45   **d** 11, 22, 33, 44, 55
  e 16, 32, 48, 64, 80

2 a 72, 132, 216   **b** 161, 91   **c** 72, 102, 132, 78, 216

3 a 98   **b** 99   **c** 96   **d** 95   **e** 98   **f** 96

4 4 or 5 (as 2, 10 and 20 are not realistic answers)

5 a 18   **b** 28   **c** 15

6 5 numbers: 18, 36, 54, 72, 90

1.2 Factors of whole numbers

Exercise 1B

1 a 1, 2, 5, 10   **b** 1, 2, 4, 7, 14, 28
  c 1, 2, 3, 6, 9, 18   **d** 1, 17
  e 1, 5, 25   **f** 1, 2, 4, 5, 8, 10, 20, 40
  g 1, 2, 3, 5, 6, 10, 15, 30   **h** 1, 3, 5, 9, 15, 45
  i 1, 2, 3, 4, 6, 8, 12, 24   **j** 1, 2, 4, 8, 16

2 a 55   **b** 67   **c** 29
  d 39   **e** 65   **f** 80

3 a 1, 2   **b** 1, 2   **c** 1, 3   **d** 1, 5
  e 1, 3   **f** 1, 3   **g** 1, 7   **h** 1, 5
  i 1, 2, 5, 10   **j** 1, 11

4 5

1.3 Prime numbers

Exercise 1C

1 23 and 29

2 97

3 All these numbers are not prime.

4 3, 5, 7

5 Only if all 31 bars are in a single row, as 31 is a prime number and its only factors are 1 and 31.

1.4 Square numbers and cube numbers

Exercise 1D

1 36, 49, 64, 81, 100, 121, 144, 169, 196, 225, 256, 289, 324, 361, 400

2 4, 9, 16, 25, 36, 49

3 a 50, 65, 82   **b** 98, 128, 162
  c 51, 66, 83   **d** 48, 63, 80

4 a 25, 169, 625, 1681
  b Answers in each row are the same

5 a 125   **b** 216   **c** 1000

6 $1331 = 11^3$

7 10^3 is larger. Difference is 100.

8 a student's proof   **b** yes

9 Three

10 36 and 49

Exercise 1E

1 a 12, 24, 36   **b** 20, 40, 60   **c** 15, 30, 45
  d 18, 36, 54   **e** 35, 70, 105

2

	Square number	Factor of 70
Even number	16	14
Multiple of 7	49	35

3 4761 (69^2) or 1764 (42^2)

4 24 seconds

5 30 seconds

6 a 12   **b** 9   **c** 6   **d** 13   **e** 1?
  f 16   **g** 10   **h** 17   **i** 8   **j** 1?

7 2197 (13^3)

1.5 Products of prime numbers

Exercise 1F

1 a 48   **b** 1323   **c** 100 000
  d 1215   **e** 10 000   **f** 82 944

2 a $2^3 \times 3^2$ **b** $2^2 \times 5^2$ **c** $2^2 \times 3^2 \times 7$
  d $2^4 \times 5 \times 7$ **e** $3 \times 5 \times 19$ **f** 3^6
  g $2^2 \times 3 \times 37$ **h** $2^7 \times 7$ **i** $3^3 \times 5^2$
  j $3^3 \times 7^2$

3 a The digits repeat.
  b Yes, because $7 \times 11 \times 13 = 1001$

1.6 HCF and LCM

Exercise 1G

1 a $10 = 2 \times 5, 20 = 2 \times 10$ **b** No. The HCF is 10.

2 a 4   **b** 3   **c** 4
  d 3   **e** 10   **f** 50

3 a 6   **b** 12   **c** 24   **d** 12
  e 9   **f** 6   **g** 28   **h** 55

4 a $60 = 2 \times 30, 60 = 3 \times 20$ **b** No. The LCM is 6.

5 a 10   **b** 14   **c** 15   **d** 21

6 a 60   **b** 48   **c** 84   **d** 200
  e 126   **f** 240   **g** 96   **h** 770

7 12

8 420

9 60

Answers to Chapter 2

Equivalent fractions

a $\frac{8}{20}$ **b** $\frac{3}{12}$ **c** $\frac{15}{40}$

d $\times 6$, $\frac{12}{18}$ **e** $\times 3$, $\frac{9}{12}$ **f** $\times 5$, $\frac{25}{40}$

a $\frac{2}{3}$ **b** $\frac{4}{5}$ **c** $\frac{5}{7}$ **d** $\div 6$, $\frac{2}{3}$

e $25 \div 5$, $\frac{3}{5}$ **f** $\div 3$, $\frac{7}{10}$

a $\frac{2}{3}$ **b** $\frac{1}{3}$ **c** $\frac{2}{3}$ **d** $\frac{3}{4}$ **e** $\frac{1}{3}$

f $\frac{1}{2}$ **g** $\frac{7}{8}$ **h** $\frac{4}{5}$ **i** $\frac{1}{2}$ **j** $\frac{1}{4}$

a $\frac{1}{2}, \frac{2}{3}, \frac{5}{6}$ **b** $\frac{1}{2}, \frac{5}{8}, \frac{3}{4}$ **c** $\frac{2}{5}, \frac{1}{2}, \frac{7}{10}$

d $\frac{7}{12}, \frac{2}{3}, \frac{3}{4}$ **e** $\frac{1}{6}, \frac{1}{4}, \frac{1}{3}$ **f** $\frac{3}{4}, \frac{4}{5}, \frac{9}{10}$

a $\frac{1}{2}$ **b** $\frac{1}{4}$ **c** $\frac{1}{5}$ **d** $\frac{3}{4}$ **e** $\frac{3}{10}$

a $\frac{3}{4}$ **b** $\frac{2}{3}$ **c** $\frac{3}{5}$ **d** $\frac{1}{3}$ **e** $\frac{3}{8}$

a $2\frac{1}{3}$ **b** $2\frac{2}{3}$ **c** $2\frac{1}{4}$ **d** $1\frac{3}{7}$ **e** $2\frac{2}{5}$ **f** $1\frac{2}{5}$

a $\frac{10}{3}$ **b** $\frac{35}{6}$ **c** $\frac{9}{5}$ **d** $\frac{37}{7}$ **e** $\frac{41}{10}$ **f** $\frac{17}{3}$

g $\frac{5}{2}$ **h** $\frac{13}{4}$ **i** $\frac{43}{6}$ **j** $\frac{29}{8}$ **k** $\frac{19}{3}$ **l** $\frac{89}{9}$

Students check their own answers.

$\frac{27}{4} = 6\frac{3}{4}$, $\frac{31}{5} = 6\frac{1}{5}$, $\frac{13}{2} = 6\frac{1}{2}$, so $\frac{27}{4}$ is the biggest since $\frac{1}{5}$ is less than $\frac{1}{2}$ and $\frac{3}{4}$ is greater than $\frac{1}{2}$

Any mixed number which is between 7.7272 and 7.9. For example $7\frac{4}{5}$

Fractions and decimals

a $\frac{7}{10}$ **b** $\frac{2}{5}$ **c** $\frac{1}{2}$ **d** $\frac{3}{100}$ **e** $\frac{3}{50}$

f $\frac{13}{100}$ **g** $\frac{1}{4}$ **h** $\frac{19}{50}$ **i** $\frac{11}{20}$ **j** $\frac{16}{25}$

a 0.5   **b** 0.75   **c** 0.6   **d** 0.9

e 0.125   **f** 0.625   **g** 0.875   **h** 0.35

a 0.3, $\frac{1}{2}$, 0.6   **b** 0.3, $\frac{2}{5}$, 0.8   **c** 0.15, $\frac{1}{4}$, 0.35

d $\frac{7}{10}$, 0.71, 0.72   **e** 0.7, $\frac{3}{4}$, 0.8   **f** $\frac{1}{20}$, 0.08, 0.1

g 0.4, $\frac{1}{2}$, 0.55   **h** 1.2, 1.23, $1\frac{1}{4}$

a 0.333...   **b** 0.666...   **c** 0.111...

d 0.444...   **e** 0.0909...   **f** 0.7272...

a terminating   **b** recurring   **c** terminating

d recurring   **e** recurring

$\frac{7}{8}$ (= 0.875)

$\frac{2}{3}$ (= 0.67)

2.3 Recurring decimals

1 0.6666... or $0.\dot{6}$

2 a $\frac{2}{9}$ **b** $\frac{7}{9}$ **c** $\frac{4}{9}$ **d** 1

3 a $\frac{3}{11}$ **b** $\frac{1}{11}$ **c** $\frac{7}{11}$

   d Other 11ths follow a similar pattern

4 a $\frac{5}{9}$ **b** $\frac{8}{33}$ **c** $\frac{16}{33}$

5 a $\frac{1}{30}$ **b** $\frac{1}{15}$

2.4 Percentages, fractions and decimals

1 a $\frac{2}{25}$ **b** $\frac{1}{2}$ **c** $\frac{1}{4}$ **d** $\frac{7}{20}$ **e** $\frac{9}{10}$ **f** $\frac{3}{4}$

2 a 0.27   **b** 0.85   **c** 0.13   **d** 0.06   **e** 0.8   **f** 0.32

3 a $\frac{3}{25}$ **b** $\frac{2}{5}$ **c** $\frac{9}{20}$ **d** $\frac{17}{25}$ **e** $\frac{1}{4}$ **f** $\frac{5}{8}$

4 a 29%   **b** 55%   **c** 3%   **d** 16%   **e** 60%   **f** 125%

5 a 28%   **b** 30%   **c** 95%   **d** 34%   **e** 27.5% **f** 87.5%

6 a 0.6   **b** 0.075   **c** 0.76   **d** 0.3125 **e** 0.05   **f** 0.125

7 a 63%, 83%, 39%, 62%, 77%   **b** English

8 34%, 0.34, $\frac{17}{50}$; 85%, 0.85, $\frac{17}{20}$; 7.5%, 0.075, $\frac{3}{40}$; 45%, 0.45, $\frac{9}{20}$; 30%, 0.3, $\frac{3}{10}$; 67%, 0.67, $\frac{2}{3}$; 84%, 0.84, $\frac{21}{25}$; 45%, 0.45, $\frac{9}{20}$; 37.5%, 0.375, $\frac{3}{8}$

2.5 Calculating a percentage

1 a 0.88   **b** 0.3   **c** 0.25   **d** 0.08   **e** 1.15

2 a 78%   **b** 40%   **c** 75%   **d** 5%   **e** 110%

3 a $45   **b** $6.30   **c** 128.8 kg   **d** 1.125 kg

   e 1.08 h   **f** 37.8 cm   **g** $0.12   **h** 2.94 m

   i $7.60   **j** 33.88 min **k** 136 kg   **l** $162

4 $2410

5 a 86%   **b** 215

6 8520

7 287

8 990

9 Mon: 816, Tue: 833, Wed: 850, Thu: 799, Fri: 748

10 a $3.25   **b** 2.21 kg   **c** $562.80

    d $6.51   **e** 42.93 m   **f** $24

11 480 cm^3 nitrogen, 120 cm^3 oxygen

12 13

13 $270

14 More this year as it was 3% of a higher amount than last year.

2.6 Increasing or decreasing quantities by a percentage

Exercise 2F

1 **a** 1.1 **b** 1.03 **c** 1.2 **d** 1.07 **e** 1.12
2 **a** $62.40 **b** 12.96 kg **c** 472.5 g **d** 599.5 m
e $38.08 **f** $90 **g** 391 kg **h** 824.1 cm
i 253.5 g **j** $143.50 **k** 736 m **l** $30.24
3 $29 425
4 1 690 200
5 **a** Caretaker: $17 325, Driver: $18 165, Supervisor:
$20 475, Manager: $26 565
b 5% of different amounts is not a fixed amount. The more
pay to start with, the more the increase (5%) will be.
6 $411.95
7 193 800
8 575 g
9 918
10 60
11 TV: $287.88, microwave: $84.60, CD: $135.13, stereo: $34.66
12 $10

Exercise 2G

1 **a** 0.92 **b** 0.85 **c** 0.75 **d** 0.91 **e** 0.88
2 **a** $9.40 **b** 23 kg **c** 212.4 g **d** 339.5 m
e $4.90 **f** 39.6 m **g** 731 m **h** 83.52 g
i 360 cm **j** 117 min **k** 81.7 kg **l** $37.70
3 $5525
4 **a** 52.8 kg **b** 66 kg **c** 45.76 kg
5 Mr Patel $176, Mrs Patel $297.50,
Sandeep $341, Priyanka $562.50
6 448
7 705
8 **a** 66.5 km/h **b** 73.5 km/h
9 No, as the total is $101. She will save $20.20, which is less
than the $25 it would cost to join the club.
10 Offer A gives 360 grams for $1.40, i.e. 0.388 cents per gram.
Offer B gives 300 grams for $1.12, i.e 0.373 cents per
gram, so Offer B is the better offer.
Or Offer A is 360 for 1.40 = 2.6 grams per cent, offer B is
300 for 1.12 = 2.7 grams per cent, so offer B is better.

2.7 Expressing one quantity as a percentage of another

Exercise 2H

1 **a** 25% **b** 60.6% **c** 46.3% **d** 12.5%
e 41.7% **f** 60% **g** 20.8% **h** 10%
i 1.9% **j** 8.3% **k** 45.5% **l** 10.5%
2 32%
3 6.5%
4 33.7%
5 **a** 49.2% **b** 64.5% **c** 10.6%
6 17.9%
7 4.9%

8 90.5%
9 **a** Brit Com: 20.9%, USA: 26.5%, France: 10.3%, Other 42.
b Total 100%, all imports
10 Nadia had the greater percentage increase.
Nadia: (20 – 14) × 100 ÷ 14 = 42.9%.
Imran: (17 – 12) × 100 ÷ 12 = 41.7%
11 Yes, as 38 out of 46 is over 80% (82.6%)
12 Vase 20% loss, radio 25% profit, doll 175% profit, toy trai
64% loss

2.8 Reverse percentage

Exercise 2I

1 **a** 800 g **b** 250 m **c** 60 cm
d $3075 **e** $200 **f** $400
2 80
3 T shirt: $8.40, Tights: $1.20, Shorts: $5.20, Sweater:
$10.74, Trainers: $24.80, Boots: $32.40
4 $833.33
5 $300
6 240
7 537.63 dollars
8 4750 blue bottles
9 2200 dollars.
10 $1440
11 $2450
12 95 dollars
13 $140
14 $945
15 $1325
16 $1300
17 Lee has assumed that 291.2 is 100% instead of 112%. He
rounded his wrong answer to the correct answer of $260.

2.9 Interest and depreciation

Exercise 2J

1 **a** $2060 **b** $2121.80
2 **a** $2120 **b** $2247.20
3 **a** $819 **b** $69
4 **a** £6897.85 **b** $397.85
5 **a** $11250 **b** $9562.50
6 **a** $30800 **b** $ 27104 **c** $23851.52
7 **a i** $11000 **ii** $12100 **iii** $13310
b i $1000 **ii** $1100 **iii** $1210
8 **a i** $20000 **ii** $16000 **iii** $12800
b i $5000 **ii** $4000 **iii** $3200
9 With Axel she has $5712 and with Barco she has $5724
so Barco is better.
10 No. After 2 years it will be worth $3375.

10 Compound interest problems

ercise 2K

$884.32

a $3649.96 **b** $649.96

a $1229.87 **b** $1225.04 **c** $1276.28

a $100 **b** $552.56 **c** $1257.79

a $551.91 **b** $607.75 **c** $667.73 **d** $732.05

After 9 years she has $6205.31

11 Repeated percentage change

ercise 2L

a 3.402 kg **b** 13.4%

2 56%

3 19.1% to 1 d.p.

4 a $367.50 **b** 51%

5 52.1% to 1 d.p.

6 a 40% **b** 61.6%

7 a 32.25% **b** 101.1% to one d.p.

8 $2 \times 1.1^{15} = 8.35$ which is more than 8.

9 An increase of 8.73%

10 a $1.2 \times 1.2 = 1.44$ the multiplier for a 44% increase
   b $0.8 \times 0.8 = 0.64$ the multiplier for a 36% decrease
   c a 4% decrease

11 a a decrease of 1% **b** a decrease of 6.25%
   c a decrease of 56.25%

12 They are the same. In both cases the multiplier is 0.87×1.42 $= 1.2354$ an increase of 23.54%

Answers to Chapter 3

Order of operations

ercise 3A

a 11 **b** 6 **c** 10 **d** 12 **e** 11 **f** 13

g 11 **h** 12 **i** 12 **j** 4 **k** 13 **l** 3

a 16 **b** 2 **c** 10 **d** 10 **e** 6 **f** 18

g 6 **h** 15 **i** 9 **j** 12 **k** 3 **l** 8

a (4 + 1) **b** No brackets needed

c (2 + 1) **d** No brackets needed

e (4 + 4) **f** (16 − 4)

g No brackets needed **h** No brackets needed

i (20 − 10) **j** No brackets needed

k (5 + 5) **l** (4 + 2)

m (15 − 5) **n** (7 − 2)

o (3 + 3) **p** No brackets needed

q No brackets needed **r** (8 − 2)

No, correct answer is 5 + 42 = 47

a $2 \times 3 + 5 = 11$ **b** $2 \times (3 + 5) = 16$

c $2 + 3 \times 5 = 17$ **d** $5 − (3 − 2) = 4$

e $5 \times 3 − 2 = 13$ **f** $5 \times 3 \times 2 = 30$

$4 + 5 \times 3 = 19$

$(4 + 5) \times 3 = 27$. So $4 + 5 \times 3$ is smaller

$(5 − 2) \times 6 = 18$

$8 \div (5 − 3) = 4$

Choosing the correct operation

ercise 3B

a 6000

b 5 cans cost $1.95, so 6 cans cost $1.95. 32 − 5 × 6 = 2.
Cost is $10.53.

a 288 **b** 16

a 38

b Coach price for adults = $8, coach price for juniors = $4,
money for coaches raised by tickets = $12 400, cost of
coaches = $12 160, profit = $240

4 $34.80

5 (18.81…) Kirsty can buy 18 models.

6 (7.58 …) Michelle must work for 8 weeks.

7 $8.40 per year, 70 cents per copy

8 $450

9 15

10 Gavin pays 2296.25 − 1840 = $456.25

3.3 Finding a fraction of a quantity

Exercise 3C

1 a 18 **b** 10 **c** 18 **d** 28

2 a $1800 **b** 128 g **c** 160 kg
   d $116 **e** 65 litres **f** 90 min

3 a $\frac{5}{8}$ of 40 = 25 **b** $\frac{3}{4}$ of 280 = 210
   c $\frac{4}{5}$ of 70 = 56 **d** $\frac{5}{6}$ of 72 = 60

4 $6080

5 $31 500

6 52 kg

7 a 856 **b** 187 675

8 a $50 **b** $550

9 a $120 **b** $240

10 Lion Autos

11 Offer B

Exercise 3D

1 Both equal 45.

2 a $5\frac{1}{3}$ **b** $5\frac{1}{4}$ **c** $9\frac{3}{4}$ **d** $1\frac{3}{5}$ **e** $3\frac{3}{5}$ **f** $3\frac{1}{8}$

3 a $\frac{9}{10}$ **b** $\frac{9}{10}$ **c** $\frac{8}{15}$

4 $6\frac{2}{3}, 7\frac{1}{2}, 8\frac{1}{3}, 13\frac{1}{3}, 15, 16\frac{2}{3}$

5 a $18\frac{3}{4}$ **b** $26\frac{2}{3}$ **c** $9\frac{3}{5}$ **d** $5\frac{1}{10}$ **e** $18\frac{1}{8}$ **f** $26\frac{1}{4}$

3.4 Adding and subtracting fractions

Exercise 3E

1 a $\frac{5}{7}$ **b** $\frac{7}{9}$ **c** $\frac{4}{5}$ **d** $\frac{6}{7}$

2 a $\frac{3}{7}$ **b** $\frac{1}{9}$ **c** $\frac{4}{11}$ **d** $\frac{7}{13}$

3 a $\frac{6}{8} = \frac{3}{4}$ **b** $\frac{4}{10} = \frac{2}{5}$ **c** $\frac{6}{9} = \frac{2}{3}$ **d** $\frac{2}{4} = \frac{1}{2}$

4 a $\frac{4}{8} = \frac{1}{2}$ **b** $\frac{4}{10} = \frac{2}{5}$ **c** $\frac{4}{6} = \frac{2}{3}$ **d** $\frac{8}{10} = \frac{4}{5}$

5 a $\frac{12}{10} = \frac{6}{5} = 1\frac{1}{5}$ **b** $\frac{9}{8} = 1\frac{1}{8}$ **c** $\frac{9}{8} = 1\frac{1}{8}$

  d $\frac{13}{8} = 1\frac{5}{8}$ **e** $\frac{11}{8} = 1\frac{3}{8}$ **f** $\frac{7}{6} = 1\frac{1}{6}$

  g $\frac{9}{6} = \frac{3}{2} = 1\frac{1}{2}$ **h** $\frac{5}{4} = 1\frac{1}{4}$

6 a $\frac{10}{8} = \frac{5}{4} = 1\frac{1}{4}$ **b** $\frac{6}{4} = \frac{3}{2} = 1\frac{1}{2}$

  c $\frac{5}{5} = 1$ **d** $\frac{16}{10} = \frac{8}{5} = 1\frac{3}{5}$

7 a $\frac{5}{8}$ **b** $\frac{5}{10} = \frac{1}{2}$ **c** $\frac{1}{4}$ **d** $\frac{3}{8}$

  e $\frac{1}{4}$ **f** $\frac{3}{8}$ **g** $\frac{4}{10} = \frac{2}{5}$ **h** $\frac{5}{16}$

Exercise 3F

1 a $\frac{8}{15}$ **b** $\frac{7}{12}$ **c** $\frac{3}{10}$ **d** $\frac{11}{12}$ **e** $\frac{7}{8}$ **f** $\frac{1}{2}$

  g $\frac{1}{6}$ **h** $\frac{1}{20}$ **i** $\frac{1}{10}$ **j** $\frac{1}{8}$ **k** $\frac{1}{12}$ **l** $\frac{1}{3}$

  m $\frac{1}{6}$ **n** $\frac{7}{9}$ **o** $\frac{5}{8}$ **p** $\frac{3}{8}$ **q** $\frac{1}{15}$ **r** $1\frac{13}{24}$

  s $\frac{59}{80}$ **t** $\frac{22}{63}$ **u** $\frac{37}{54}$

2 a $3\frac{5}{14}$ **b** $10\frac{3}{5}$ **c** $2\frac{1}{6}$ **d** $3\frac{31}{45}$

  e $4\frac{47}{60}$ **f** $\frac{41}{72}$ **g** $\frac{29}{48}$ **h** $1\frac{43}{48}$

  i $1\frac{109}{120}$ **j** $1\frac{23}{30}$ **k** $1\frac{31}{84}$

3 $\frac{1}{20}$

4 a $\frac{1}{6}$ **b** 30, must be divisible by 2 and 3

3.5 Multiplying and dividing fractions

Exercise 3G

1 a $\frac{1}{6}$ **b** $\frac{1}{10}$ **c** $\frac{3}{8}$ **d** $\frac{3}{14}$ **e** $\frac{8}{15}$

  f $\frac{1}{5}$ **g** $\frac{2}{7}$ **h** $\frac{3}{10}$ **i** $\frac{1}{2}$ **j** $\frac{2}{5}$

2 a $\frac{3}{32}$ **b** $\frac{3}{8}$ **c** $\frac{7}{20}$

  d $\frac{16}{45}$ **e** $\frac{3}{5}$ **f** $\frac{5}{8}$

3 $\frac{1}{12}$

4 $\frac{3}{8}$

5 a $\frac{5}{12}$ **b** $2\frac{1}{12}$ **c** $6\frac{1}{4}$ **d** $2\frac{11}{12}$

  e $3\frac{9}{10}$ **f** $3\frac{1}{3}$ **g** $12\frac{1}{2}$ **h** 30

6 $\frac{2}{5}$ of $6\frac{1}{2} = 2\frac{3}{5}$

Exercise 3H

1 a $\frac{3}{4}$ **b** $1\frac{2}{5}$ **c** $1\frac{1}{15}$ **d** $1\frac{1}{14}$ **e** 4

  f 4 **g** 5 **h** $1\frac{5}{7}$ **i** $\frac{4}{9}$ **j** $1\frac{1}{8}$

2 18

3 40

4 15

5 16

6 a $2\frac{2}{15}$ **b** 38 **c** $1\frac{7}{8}$ **d** $\frac{9}{32}$ **e** $\frac{1}{16}$ **f** $\frac{25}{62}$

Answers to Chapter 4

4.1 Introduction to directed numbers

Exercise 4A

1 a 0 °C **b** 5 °C **c** –2 °C **d** –5 °C **e** –1 °C

2 a 11 degrees **b** 9 degrees

3 8 degrees

4.2 Everyday use of directed numbers

Exercise 4B

1 –$5

2 –200 m

3 above

4 –5 h

5 –2 °C

6 – 70 km

7 +5 minutes

8 –5 km/h

9 –2

10 a –11 °C **b** 6 degrees

11 1.54 am

3 The number line

Exercise 4C

1. **a** < **b** > **c** < **d** < **e** > **f** <
   g < **h** > **i** > **j** < **k** < **l** >
2. **a** < **b** < **c** < **d** > **e** < **f** <
3. **a**
   $$-5 \quad -4 \quad -3 \quad -2 \quad -1 \quad 0 \quad 1 \quad 2 \quad 3 \quad 4 \quad 5$$
   b
   $$-25 \quad -20 \quad -15 \quad -10 \quad -5 \quad 0 \quad 5 \quad 10 \quad 15 \quad 20 \quad 25$$
   c
   $$-10 \quad -8 \quad -6 \quad -4 \quad -2 \quad 0 \quad 2 \quad 4 \quad 6 \quad 8 \quad 10$$
   d
   $$-50 \quad -40 \quad -30 \quad -20 \quad -10 \quad 0 \quad 10 \quad 20 \quad 30 \quad 40 \quad 50$$
4. 6 °C   −2 °C   −4 °C   2 °C

4 Adding and subtracting directed numbers

Exercise 4D

1. **a** −2° **b** −3° **c** −2° **d** −3° **e** −2° **f** −3°
   g 3 **h** 3 **i** −1 **j** −1 **k** 2 **l** −3
   m −4 **n** −6 **o** −6 **p** −1 **q** −5 **r** −4
   s 4 **t** −1 **u** −5 **v** −4 **w** −5 **x** −5
2. **a** 7 degrees **b** −6 °C
3. **a** 2 − 8
   b 2 + 5 − 8 or 2 + 4 − 7 or 8 − 4 − 5 or 8 − 2 − 7 or 5 − 4 − 2
   c 2 − 5 − 7 − 8
   d 2 + 5 − 4 − 7 − 8
4. 250 metres

Exercise 4E

1. **a** −8 **b** −10 **c** −11 **d** −3 **e** 2 **f** −5
   g 1 **h** 4 **i** 7 **j** −8 **k** −5 **l** −11
   m 11 **n** 6 **o** 8 **p** 8 **q** −2 **r** −1
   s −9 **t** −5

2. **a** 10 degrees Celsius **b** 7 degrees Celsius
   c 9 degrees Celsius
3. **a** 2 **b** −3 **c** −5 **d** −7 **e** −10 **f** −20
4. **a** 2 **b** 4 **c** −1 **d** −5 **e** −11 **f** 8
5. **a** 13 **b** 2 **c** 5 **d** 4 **e** 11 **f** −2
6. **a** −10 **b** −5 **c** −2 **d** 4 **e** 7 **f** −4
7. **a** +6 + +5 = 11 **b** +6 + −9 = −3
   c +6 − −9 = 15 **d** +6 + +5 = 1
8. It may not come on as the thermometer inaccuracy might be between 0° and 2° or 2° and 4°
9. −1 and 6

4.5 Multiplying and dividing directed numbers

Exercise 4F

1. **a** −15 **b** −14 **c** −24 **d** 6 **e** 14 **f** 2
   g −2 **h** −8 **i** −4 **j** 3 **k** −24 **l** −10
   m −18 **n** 16 **o** 36 **p** −4 **q** −12 **r** −4
   s 7 **t** 25 **u** 18
2. **a** −9 **b** 16 **c** −3 **d** −32 **e** 18 **f** 18
   g 6 **h** −4 **i** 20 **j** 16 **k** 8 **l** −48
   m 13 **n** −13 **o** −8 **p** 0 **q** 16 **r** −42
3. **a** −2 **b** 30 **c** 15 **d** −27 **e** −7
4. **a** 4 **b** −9 **c** −3 **d** 6 **e** −4
5. **a** −9 **b** 3 **c** 1
6. **a** 16 **b** −2 **c** −12
7. **a** 24 **b** 6 **c** −4 **d** −2
8. For example: 1 × (−12), −1 × 12, 2 × (−6), 6 × (−2), 3 × (−4), 4 × (−3)
9. For example: 4 ÷ (−1), 8 ÷ (−2), 12 ÷ (−3), 16 ÷ (−4), 20 ÷ (−5), 24 ÷ (−6)
10. −5 × 4, 3 × −6, −20 ÷ 2, −16 ÷ −4
11. **a** 4 **b** 25 **c** 12 **d** 1

Answers to Chapter 5

Squares and square roots

Exercise 5A

1. **a** 49 **b** 100 **c** 1.44 **d** 6.25 **e** 256 **f** 400
   g 9.61 **h** 20.25 **i** 9 **j** 64 **k** 0.25 **l** 0.25
2. **a** 3 and −3 **b** 10 and −10 **c** 11 and −11
   d 1.2 and −1.2 **e** 20 and −20 **f** 3.5 and −3.5
   g 1 and −1 **h** 100 and −100
3. **a** 5 **b** 6 **c** 10 **d** 7 **e** 8
   f 1.5 **g** 5.5 **h** 1.2 **i** 20 **j** 0.5
4. **a** 81 **b** 40 **c** 100 **d** 14 **e** 36
   f 15 **g** 49 **h** 12 **i** 25 **j** 21
5. **a** 24 **b** 31 **c** 45 **d** 40 **e** 67
   f 101 **g** 3.6 **h** 6.5 **i** 13.9 **j** 22.2
6. √50, 3², √90, 4²
7. **a** 6² is 36 and 7² is 49; 40 is between 36 and 49
   b 6.3245553......

8. 4 and 5
9. **a** 8 and 9 **b** 9 and 10
   c 12 and 13 **d** 15 and 16
10. √324 = 18
11. 15

5.2 Cubes and cube roots

Exercise 5B

1. **a** 8 **b** 27 **c** 512 **d** 1000
   e 1.331 **f** 15.625 **g** −27 **h** −125
   i 8000 **j** 68.921 **k** −68.921
2. **a** 2 **b** 5 **c** 9 **d** 1 **e** 3
   f −3 **g** 10 **h** 1.5 **i** 4.5 **j** 0.5
3. **a** 5 and 6 **b** 6 and 7
   c 7 and 8 **d** −8 and −7

4 2^3 because it equals 8, the rest equal 9

5 One possible answer is $8^2 = 4^3$

6 $\sqrt[3]{2000}$, $\sqrt{225}$, 2.5^3, 4^2

7

Number	Square	Cube
10	100	1000
5	25	125
4	16	64
11	121	1331
9	81	729

8 0.8^3, 0.8^2, $\sqrt{0.8}$, $\sqrt[3]{0.8}$.

5.3 Surds

Exercise 5C

1 a $\sqrt{6}$ **b** $\sqrt{15}$ **c** 2 **d** 4 **e** $\sqrt{14}$ **f** 6
g 6 **h** $\sqrt{30}$

2 a 2 **b** $\sqrt{5}$ **c** $\sqrt{6}$ **d** $\sqrt{3}$ **e** 2 **f** $\sqrt{6}$
g 1 **h** 3

3 a $2\sqrt{3}$ **b** 15 **c** $4\sqrt{2}$ **d** $4\sqrt{3}$ **e** $2\sqrt{7}$ **f** $6\sqrt{5}$
g $6\sqrt{3}$ **h** 30

4 a $\sqrt{3}$ **b** 1 **c** $2\sqrt{2}$ **d** $\sqrt{2}$ **e** $\sqrt{5}$ **f** $\sqrt{3}$
g $\sqrt{2}$ **h** $\sqrt{7}$ **i** $\sqrt{7}$ **j** $2\sqrt{3}$ **k** $2\sqrt{3}$ **l** 1

5 a a **b** 1 **c** $\sqrt{a}$

6 a $3\sqrt{2}$ **b** $2\sqrt{6}$ **c** $2\sqrt{3}$ **d** $5\sqrt{2}$ **e** $2\sqrt{2}$ **f** $3\sqrt{3}$
g $4\sqrt{3}$ **h** $5\sqrt{3}$ **i** $3\sqrt{5}$ **j** $3\sqrt{7}$ **k** $4\sqrt{2}$ **l** $10\sqrt{2}$

7 a 36 **b** $16\sqrt{30}$ **c** 54 **d** 32 **e** $48\sqrt{6}$ **f** $48\sqrt{6}$
g $18\sqrt{15}$ **h** 84 **i** 64 **j** 100 **k** 50 **l** 56

8 a $20\sqrt{6}$ **b** $6\sqrt{15}$ **c** 24 **d** 16 **e** $12\sqrt{10}$ **f** 18
g $20\sqrt{3}$ **h** $10\sqrt{21}$ **i** $6\sqrt{14}$ **j** 36 **k** 24 **l** $12\sqrt{30}$

9 a 6 **b** $3\sqrt{5}$ **c** $6\sqrt{6}$ **d** $2\sqrt{3}$ **e** $4\sqrt{5}$ **f** 5
g $7\sqrt{3}$ **h** $2\sqrt{7}$ **i** 6 **j** $2\sqrt{7}$ **k** 5 **l** 24

10 a $2\sqrt{3}$ **b** 4 **c** $6\sqrt{2}$ **d** $4\sqrt{2}$ **e** $6\sqrt{5}$ **f** 24
g $3\sqrt{2}$ **h** $\sqrt{7}$ **i** $10\sqrt{7}$ **j** $8\sqrt{3}$ **k** $10\sqrt{3}$ **l** 6

11 a abc **b** $\frac{a}{c}$ **c** $c\sqrt{b}$

12 a 20 **b** 24 **c** 10 **d** 24 **e** 3 **f** 6

13 a $\frac{3}{4}$ **b** $8\frac{1}{3}$ **c** $\frac{5}{16}$ **d** 12 **e** 2

14 a False **b** False

15 Possible answer: $\sqrt{3} \times 2\sqrt{3}$ (= 6)

Exercise 5D

1 Expand the brackets each time.

2 a $2\sqrt{3} - 3$ **b** $3\sqrt{2} - 8$ **c** $10 + 4\sqrt{5}$
d $12\sqrt{7} - 42$ **e** $15\sqrt{2} - 24$ **f** $9 - \sqrt{3}$

3 a $2\sqrt{3}$ **b** $1 + \sqrt{5}$ **c** $-1 - \sqrt{2}$
d $\sqrt{7} - 30$ **e** $3 - 2\sqrt{2}$ **f** $11 + 6\sqrt{2}$

4 a $\sqrt{3} - 1$ cm^2 **b** $2\sqrt{5} + 5\sqrt{2}$ cm^2 **c** $2\sqrt{3} + 18$ cm^2

5 a $\frac{\sqrt{3}}{3}$ **b** $\frac{\sqrt{2}}{2}$ **c** $\frac{\sqrt{5}}{5}$ **d** $\frac{\sqrt{3}}{6}$ **e** $\sqrt{3}$ **f** $\frac{5\sqrt{2}}{2}$
g $\frac{3}{2}$ **h** $\frac{5\sqrt{2}}{2}$ **i** $\frac{\sqrt{21}}{3}$ **j** $\frac{\sqrt{2}+2}{2}$ **k** $\frac{2\sqrt{3}-3}{3}$ **l** $\frac{5\sqrt{3}}{3}$

6 a i 1 **ii** –4 **iii** 2 **iv** 17 **v** –44
b They become whole numbers. Difference of two squares makes the 'middle terms' (and surds) disappear.

7 a 7 **b** 1 **c** 2 **d** –1 **e** –3 **f** 4

8 a $\sqrt{3}+1$ **b** $4(2-\sqrt{3})$ **c** $\frac{(20 + 4\sqrt{3})}{11}$
d $2-\sqrt{2}$ **e** $5+2\sqrt{5}$ **f** $5 + 3\sqrt{3}$

9 $2(3+\sqrt{3})$

10 $\frac{1}{r} = \frac{2}{\sqrt{5}+1} = \frac{2(\sqrt{5}-1)}{(\sqrt{5}+1)(\sqrt{5}-1)} = \frac{2(\sqrt{5}-1)}{5-1} = \frac{2(\sqrt{5}-1)}{4} = \frac{\sqrt{5}-1}{2}$
and $r - 1 = \frac{\sqrt{5}+1}{2} - 1 = \frac{\sqrt{5}+1-2}{2} = \frac{\sqrt{5}-1}{2}$ so they are equal

---

Answers to Chapter 6

6.1 Inequalities

Exercise 6A

1 a > **b** < **c** < **d** =
e = **f** > **g** > **h** <

2 $\frac{1}{3} < \frac{1}{2} < \frac{3}{5}$

3 a 4,5,6 **b** 1,2 **c** 6 **d** 1,2,3,4,5
e 2,3,4 **f** 4,5 **g** 1,2,3 **h** 6

4 a underweight **b** overweight
c normal **d** normal

5 20, 22, 26, 28

6 a 49 **b** 45
c 3,6,9 **d** 16,17,18,19,20

7 a true **b** false **c** true
d true **e** false **f** true

8 a 6,7,8 **b** 26, 27, 28 **c** –7, –6, –5, –4
d –2, –1, 0, 1 **e** there are none **f** 33

6.2 Sets

Exercise 6B

1 a days of the week **b** odd numbers **c** planets
d compass directions **e** numbers on a dice
(Other descriptions are possible in this question)

2 a {6,7,8,9} **b** {1,2,3,4,6,12} **c** {23, 29} **d** {3,
3 a false **b** true **c** true **d** true
e true **f** false **g** true **h** false

4 a {6, 12} **b** {2} **c** { } **d** {3, 9}

5 a {2, 3, 5, 6, 7, 9, 11, 12}
b {2, 3, 4, 6, 8, 9, 10, 12}
c $\mathscr{E}$

6 {letters of the alphabet} is one possibility

7 a { } **b** {a, i} **c** {e, r}

8 a 9 **b** 9 **c** 8

9 {2, 4, 5, 7, 8}

3 Venn diagrams

Exercise 6C

a i {s,a,n,g} **ii** {s,p,a,r,e} **iii** {a,s}
iv {n,g,s,a,p,r,e} **v** {p,r,e,i,o} **vi** {i,o}
vii {p,r,e} **viii** {g,n}
b Singapore

a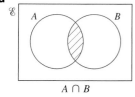

b i {c,n} **ii** {f,s,c,n} **iii** {f,s,c,n,i,t,r}

a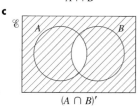

b {multiples of 6} is a possible description
a i {1,2,3,4,6,12} **ii** {1,4,7,10}
iii {1,4} **iv** {1,3}
v {1,3,4,5,7,9,10,11} **vi** {2,4,6,8,10,12}
vii {1} **viii** {1,3,7}
ix {2,6,8,12} **x** {2,6,12}
b i the missing numbers are 1 and 12
ii 12 **iii** odd

a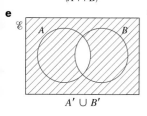
$A \cap B$

b
$A' \cap B$

c
$(A \cap B)'$

d
$(A \cup B)'$

e
$A' \cup B'$

f
$A' \cap B'$

$(A \cap B)'$ is identical to $A' \cup B'$; $(A \cup B)'$ is identical to $A' \cap B'$
a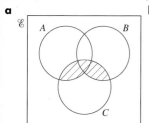

b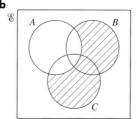

c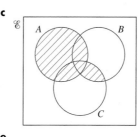

d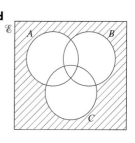

e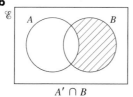

6.4 More notation

Exercise 6D

1 a 6 **b** 4 **c** 14 **d** 2
e 8 **f** 8 **g** 6 **h** 12
2 a There is only one even prime number. It is 2.
b There are no prime square numbers
c {odd square numbers}
d There are many possible answers. 15 and 33 are two of them
3 a false **b** true **c** true **d** false **e** true **f** false
4 63
5 a ℰ **b** ∅ **c** A ∪ B **d** A ∩ B **e** A
6 a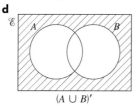

b six **c** multiples of 15
d 30 and 60 are two possible values
e {x : x is a multiple of 30} is one possible description

6.5 Practical problems

Exercise 6E

1 a 40 **b** 12 **c** 20 **d** 72
2 a 33 **b** 19 **c** 16 **d** 23
3 a 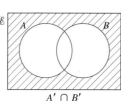 **b** 5

4 24
5 7
6 11

Answers to Chapter 7

7.1 Ratio

Exercise 7A

1 **a** 1 : 3   **b** 1 : 4   **c** 2 : 3   **d** 2 : 1
   e 2 : 5   **f** 2 : 5   **g** 5 : 8   **h** 5 : 1

2 **a** 8 : 1   **b** 12 : 1   **c** 5 : 6   **d** 1 : 24
   e 48 : 1   **f** 5 : 2   **g** 3 : 8   **h** 1 : 5

3 $\frac{7}{10}$

4 $\frac{10}{25} = \frac{2}{5}$

5 **a** $\frac{2}{5}$ **b** $\frac{3}{5}$

6 **a** $\frac{7}{10}$ **b** $\frac{3}{10}$

7 **a** $\frac{1}{2}$ **b** $\frac{7}{20}$ **c** $\frac{3}{20}$

8 3 : 1

9 1 : 4

Exercise 7B

1 **a** 160 g, 240 g   **b** 80 kg, 200 kg
   c 150, 350   **d** 950 m, 50 m
   e 175 min, 125 min   **f** $20, $30, $50
   g $36, $60, $144   **h** 50 g, 250 g, 300 g
   i $1.40, $2, $1.60   **j** 120 kg, 72 kg, 8 kg

2 **a** 175   **b** 30%

3 **a** 28   **b** 42

4 21

5 Joshua $2500, Aicha $3500, Mariam $4000

6 **a** 1 : 400 000   **b** 1 : 125 000   **c** 1 : 250 000
   d 1 : 25 000   **e** 1 : 20 000   **f** 1 : 40 000
   g 1 : 62 500   **h** 1 : 10 000   **i** 1 : 60 000

7 **a** 1 : 1 000 000   **b** 47 km   **c** 8 mm

8 **a** 1 : 250 000   **b** 2 km   **c** 4.8 cm

9 **a** 1 : 20 000   **b** 0.54 km   **c** 40 cm

10 **a** 1 : 1.6   **b** 1 : 3.25   **c** 1 : 1.125
   d 1 : 1.44   **e** 1 : 5.4   **f** 1 : 1.5
   g 1 : 4.8   **h** 1 : 42   **i** 1 : 1.25

Exercise 7C

1 **a** 3 : 2   **b** 32   **c** 80

2 1000 g

3 10 125

4 **a** 14 min   **b** 75 min

5 **a** 11 pages   **b** 32%

6 Ren $2040, Shota $2720

7 **a** lemonade 20 litres, ginger 0.5 litres
   b This one, one-thirteenth is greater than one-fiftieth.

8 100

9 40 cm³

7.2 Speed

Exercise 7D

1 18 km/hour

2 440 kilometres

3 52.5 km/hour

4 11.50 am

5 500 s

6 **a** 75 km/hour **b** 6.5 hours   **c** 175 km   **d** 240 km
   e 64 km/h   **f** 325 km   **g** 4.3 hours (4 h 18 min)

7 **a** 7.75 h   **b** 85.2 km/hour

8 **a** 2.25 h   **b** 157.5 km

9 **a** 1.25 h   **b** 1 h 15 min

10 **a** 48 km/hour **b** 6 h 40 min

11 **a** 120 km   **b** 48 km/h

12 **a** 30 min   **b** 12 km/h

13 **a** 10 m/s   **b** 3.3 m/s   **c** 16.7 m/s   **d** 41.7 m/s
   e 20.8 m/s

14 **a** 90 km/h   **b** 43.2 km/h   **c** 14.4 km/h   **d** 108 km/h
   e 1.8 km/h

15 **a** 64.8 km/h **b** 28 s   **c** 8.07 (37 min journey)

16 **a** 6.7 m/s   **b** 66 km   **c** 5 minutes   **d** 133.3 m

17 6.6 minutes

7.3 Density and pressure

Exercise 7E

1 **a** 0.75 g/cm³

2 4 pa

3 8.3 g/cm³

4 $2\frac{1}{2}$ N

5 32 g

6 5 m²

7 120 cm³

8 156.8 g

9 First statue is the fake as density is approximately 26 g/cm

10 Second piece by 1 cm³

11 0.339 m³

7.4 Direct proportion

Exercise 7F

1 60 g

2 $5.22

3 45

4 $6.72

5 **a** $312.50   **b** 8

6 **a** 56 litres   **b** 350 km

7 **a** 300 kg   **b** 9 weeks

8 40 s

a i 100 g margarine, 200 g sugar, 250 g flour, 150 g ground rice

ii 150 g margarine, 300 g sugar, 375 g flour, 225 g ground rice

iii 250 g margarine, 500 g sugar, 625 g flour, 375 g ground rice

b 24

Peter's shop as I can buy 24. At Paul's shop I can only buy 20.

7.5 Proportional variables

Exercise 7G

1 50

2 30 and 54

3 21.7 and 30.8

4 x is 16
y is 15 and 18

5 360

6 No. The multipliers are not all the same.

7 15.0 and 13.1

8 720 and 2000

Answers to Chapter 8

Rounding whole numbers

Exercise 8A

a 20	**b** 60	**c** 80	**d** 50	**e** 100
f 20	**g** 90	**h** 70	**i** 10	**j** 30

a 200	**b** 600	**c** 800	**d** 500	**e** 1000
f 100	**g** 600	**h** 400	**i** 1000	**j** 1100

a 2000	**b** 6000	**c** 8000	**d** 5000
e 10 000	**f** 1000	**g** 6000	**h** 3000
i 9000	**j** 2000		

a True	**b** False	**c** True	**d** True	**e** True	**f** False

a Highest Germany, lowest Italy

b 36 000, 43 000 , 25 000, 29 000

c 25 499 and 24 500

a 375

b 98 (350 to 449 inclusive, but not Matthew's number which is 375)

A number between 75 and 84 inclusive added to a number between 45 and 54 inclusive with a total not equal to 130, for example 79 + 49 = 128

Rounding decimals

Exercise 8B

a 4.8	**b** 3.8	**c** 2.2	**d** 8.3	**e** 3.7
f 46.9	**g** 23.9	**h** 9.5	**i** 11.1	**j** 33.5

a 5.78	**b** 2.36	**c** 0.98	**d** 33.09	**e** 6.01
f 23.57	**g** 91.79	**h** 8.00	**i** 2.31	**j** 23.92

a 4.6	**b** 0.08	**c** 45.716	**d** 94.85	**e** 602.1
f 671.76	**g** 7.1	**h** 6.904	**i** 13.78	**j** 0.1

a 8	**b** 3	**c** 8	**d** 6	**e** 4
f 7	**g** 2	**h** 47	**i** 23	**j** 96

3 + 9 + 6 + 4 = 22 dollars

3, 3.46, 3.5

4.7275 or 4.7282

8.3 Rounding to significant figures

Exercise 8C

1 a 50 000	**b** 60 000	**c** 30 000	**d** 90 000
e 90 000	**f** 0.5	**g** 0.3	**h** 0.006
i 0.05	**j** 0.0009	**k** 10	**l** 90
m 90	**n** 200	**o** 1000	

2 a 56 000	**b** 27 000	**c** 80 000	**d** 31 000
e 14 000	**f** 1.7	**g** 4.1	**h** 2.7
i 8.0	**j** 42	**k** 0.80	**l** 0.46
m 0.066	**n** 1.0	**o** 0.0098	

3 a 60 000	**b** 5300	**c** 89.7	**d** 110
e 9	**f** 1.1	**g** 0.3	**h** 0.7
i 0.4	**j** 0.8	**k** 0.2	**l** 0.7

4 a 65, 74   **b** 95, 149   **c** 950, 1499

5 Satora 750, 849, Nimral 1150, 1249, Korput 164 500, 165 499

6 One, because there could be 450 then 449.

7 Vashti has rounded to 2 significant figures or nearest 10 000.

8.4 Approximation of calculations

Exercise 8D

1 a 35 000	**b** 15 000	**c** 960	**d** 12 000	**e** 1050
f 4000	**g** 4	**h** 20	**i** 1200	

2 a $3000   **b** $2000   **c** $1500   **d** $700

3 a $15 000   **b** $18 000   **c** $18 000

4 $21 000

5 a 14	**b** 10	**c** 1.1	**d** 1	**e** 5	**f** $\frac{2}{3}$
g 3 or 4	**h** $\frac{1}{2}$	**i** 6	**j** 400	**k** 2	**l** 20

6 a 500   **b** 200   **c** 90   **d** 50   **e** 50   **f** 500

7 8

8 a 200   **b** 2800   **c** 10   **d** 1000

9 1000 or 1200

10 a 28 km   **b** 120 km   **c** 1440 km

11 400 or 500

12 a 3 kg   **b** 200

8.5 Upper and lower bounds

Exercise 8E

1 a 6.5 and 7.5   **b** 115 and 125
  c 3350 and 3450   **d** 49.5 and 50.5
  e 5.50 and 6.50   **f** 16.75 and 16.85
  g 75 550 and 76 499
  h 14 450 and 14 549
  i 28 500 and 29 499
  j 23 500 000 and 24 499 999

2 a $5.5 \leqslant$ length in cm < 6.5
  b $16.5 \leqslant$ mass in kg < 17.5
  c $31.5 \leqslant$ time in minutes < 32.5
  d $237.5 \leqslant$ distance in km < 238.5
  e $7.25 \leqslant$ distance in m < 7.35
  f $25.75 \leqslant$ mass in kg < 25.85
  g $3.35 \leqslant$ time in hours < 3.45
  h $86.5 \leqslant$ mass in g < 87.5
  i $4.225 \leqslant$ distance in mm < 4.235
  j $2.185 \leqslant$ mass in kg < 2.195
  k $12.665 \leqslant$ time in minutes < 12.675
  l $24.5 \leqslant$ distance in metres < 25.5
  m $35 \leqslant$ length in cm < 45
  n $595 \leqslant$ mass in g < 605
  o $25 \leqslant$ time in minutes < 35
  p $995 \leqslant$ distance in metres < 1050
  q $3.95 \leqslant$ distance in metres < 4.05
  r $7.035 \leqslant$ mass in kg < 7.045
  s $11.95 \leqslant$ time in seconds < 12.05
  t $6.995 \leqslant$ distance in metres < 7.005

3 a 7.5, 8.5   **b** 25.5, 26.5
  c 24.5, 25.5   **d** 84.5, 85.5
  e 2.395, 2.405   **f** 0.15, 0.25
  g 0.055, 0.065   **h** 250 g, 350 g
  i 0.65, 0.75   **j** 365.5, 366.5
  k 165, 175   **l** 205, 215

4 C: The chain and distance are both any value between 29.5 and 30.5 metres, so there is no way of knowing if the chain longer or shorter than the distance.

5 2 kg 450 grams

6 a <65.5 g   **b** 64.5 g
  c <2620 g   **d** 2580 g

8.6 Upper and lower bounds for calculations

Exercise 8F

1 65 kg and 75 kg

2 a 12.5 kg   **b** 20

3 9kg   53.5 – 44.5

4 a 26 cm $\leqslant$ perimeter < 30 cm
  b 25.6 cm $\leqslant$ perimeter < 26.0 cm
  c 50.5 cm $\leqslant$ perimeter < 52.7 cm

5 a 38.25 cm^2 $\leqslant$ area < 52.25 cm^2
  b 37.1575 cm^2 $\leqslant$ area < 38.4475 cm^2
  c 135.625 cm^2 $\leqslant$ area < 145.225 cm^2

6 79.75 m^2 $\leqslant$ area < 100.75 m^2

7 216.125 cm^3 $\leqslant$ volume < 354.375 cm^3

8 12.5 metres

9 Yes, because they could be walking at 4.5 km/h and 2.5 km/h meaning that they would cover 4.5 km + 2.5 km = 7 in 1 hour

10 20.9 m $\leqslant$ length < 22.9 m (3 sf)

11 a 14.65 s $\leqslant$ time < 14.75 s
  b 99.5 m $\leqslant$ length < 100.5 m
  c 6.86 m/s (3 sf)

12 14 s $\leqslant$ time < 30 s

13 337.75 and 334.21

14 177.3 and 169.4

Answers to Chapter 9

9.1 Standard form

Exercise 9A

1 a 250   **b** 34.5
  c 0.00467   **d** 34.6
  e 0.020789   **f** 5678
  g 246   **h** 7600
  i 897 000   **j** 0.008 65
  k 60 000 000   **l** 0.000567

2 a 2.5×10^2 **b** 3.45×10^{-1} **c** 4.67×10^4
  d 3.4×10^9 **e** 2.078×10^{10} **f** 5.678×10^{-4}
  g 2.46×10^3 **h** 7.6×10^{-2} **i** 7.6×10^{-4}
  j 6×10^{-4} **k** 5.67×10^{-3} **l** 5.60045×10^1

3 1.065×10^9

4 4.504×10^7

5 1.298×10^7, 2.997×10^9, 9.3×10^4

6 100

7 7.78×10^8; 5.8×10^7; 5.92×10^9

8 Width 1.2×10^{-6} Mass 9.5×10^{-13}

2 Calculating with standard form

xercise 9B

a 5.67×10^3 **b** 6×10^2 **c** 3.46×10^{-1}
d 7×10^{-4} **e** 5.6×10^2 **f** 6×10^5
g 7×10^3 **h** 2.3×10^7

a 1.08×10^8 **b** 4.8×10^6 **c** 1.2×10^9
d 1.08 **e** 6.4×10^2 **f** 1.2×10^1
g 2.5×10^7 **h** 8×10^{-6}

a 2.7×10 **b** 1.6×10^{-2} **c** 2×10^{-1}
d 4×10^{-8} **e** 2×10^5 **f** 6×10^{-2}

2×10^{13}, 1×10^{-10}, mass $= 2 \times 10^3$ g (2 kg)

3.80×10^7 sq km

5×10^4

2.3×10^5

Any value from 1.00000001×10^8 to 1×10^9 (excluding 1×10^9), i.e. any value of the form $a \times 10^8$ where $1 < a < 10$

9.3 Solving problems

Exercise 9C

1 **a** $(2^{63}) = 9.2 \times 10^{18}$ grains **b** $2^{64} - 1 = 1.8 \times 10^{19}$
2 **a** 1.0×10^8 sq km **b** 31%
3 455 070 000 kg or 455 070 tonnes or 4.55×10^8
4 **a** 80 000 000 (80 million) **b** 1.2%
5 **a** 2.048×10^6 **b** 4.816×10^6
6 9.41×10^4
7 **a** India **b** Tunisia and Senegal
 c 1.8×10^7 **d** 19 or 20 **e** 400
8 **a** Togo **b** Sri Lanka
 c Sri Lanka **d** Russian Federation **e** $\frac{1}{261}$

Answers to Chapter 10

.1 Units of measurement

xercise 10A

a metres **b** kilometres
c millimetres **d** kilograms or grams
e litres **f** tonnes
g millilitres **h** metres
i kilograms **j** millimetres

Check individual answers.

The 5 metre since his height is about 175 cm, the lamp post will be about 525 cm

.2 Converting between metric units

xercise 10B

a 1.25 m **b** 8.2 cm **c** 0.55 m **d** 4.2 kg
e 5.75 t **f** 8.5 cl **g** 0.755 kg **h** 0.8 l
i 2 l **j** 1.035 m³ **k** 0.53 m³ **l** 34 000 m
a 3400 mm **b** 135 mm **c** 67 cm **d** 640 m
e 2400 ml **f** 590 cl **g** 3750 kg **h** 0.00094 l
i 2160 cl **j** 15 200 g **k** 14 000 l **l** 0.19 ml

He should choose the 2000 mm × 15 mm × 20 mm

1 000 000

.3 Reading scales

xercise 10C

i 8 kg **ii** 71.6 **iii** 64

a

b
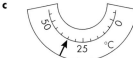

c

3 **a** 360 g
 b weigh out 400 g, then weigh out 300 g
4 **a** 1.2 kg
 b 125 g

10.4 Time

Exercise 10D

1 **a** 1 hour 10 minutes; 2 hours 3 minutes; 2 hours 9 minutes; 1 hour 45 minutes
 b the 0900
2 **a** 9:45am, 10:36am, 1:33pm, 4:49pm
 b 3 hours and 48 minutes, 6 hours and 13 minutes
3 **a** 1605 **b** 0815 **c** 6 hours 45minutes
4 **a** 1050 **b** 1635 **c** 5 hours 45 minutes
5 **a** 1210 **b** 2hours 50 minutes
6 **a** 12 minutes **b** 40 minutes **c** 54 minutes
7 1 hour 13 minutes
8 1415
9 0715 the next day

10.5 Currency conversions

Exercise 10E

1 3197.41

2 164

3 The missing values are 3.88, 7.76, 38.78, 193.88, 387.75, 775.50

4 43.01

5 a 224.91   **b** 172.74

6 a i 349.83   **ii** 24692   **iii** 432.90
   b 54000 yen, 500 euros, 650 dollars

7 a 2391.38   **b** 3489.75
   c Taiwan dollar   **d** 1.4593

8 a 74.7755   **b** 0.14747

10.6 Using a calculator efficiently

Exercise 10F

1 a 144   **b** 108

2 a 12.54   **b** 27.45

3 a 196.48   **b** 1.023
   c 0.236   **d** 4.219

4 a 3.58   **b** 6

5 a 497.952   **b** 110.98

6 a 3.12   **b** 0.749
   c 90.47   **d** 184.96
   e 6.45   **f** 27.52

Answers to Chapter 11

11.1 The language of algebra

Exercise 11A

1 a $x + 2$ **b** $x - 6$ **c** $k + x$ **d** $x - t$
  e $x + 3$ **f** $d + m$ **g** $b - y$ **h** $p + t + w$
  i $8x$ **j** hj **k** $x \div 4$ or $\frac{x}{4}$ **l** $2 \div x$ or $\frac{2}{x}$
  m $y \div t$ or $\frac{y}{t}$ **n** wt **o** a^2 **p** g^2

2 a $x + 3$ yr   **b** $x - 4$ yr

3 $F = 2C + 30$

4 Rule **c**

5 a $3n$ **b** $3n + 3$ **c** $n + 1$ **d** $n - 1$

6 Anil: $2n$, Reza: $n + 2$, Dale: $n - 3$, Chen: $2n + 3$

7 a \$4   **b** \$(10 − x)   **c** \$(y − x)   **d** \$2x

8 a \$75   **b** \$15x   **c** \$4A   **d** \$Ay

9 $(A − B)$ dollars

10 $A \div 5$ or $\frac{A}{5}$.

11 a Dad: $(72 + x)$ yr, me: $(T + x)$ yr   **b** 31

12 a $T \div 2$ or $\frac{T}{2}$ **b** $T \div 2 + 4$ or $\frac{T}{2} + 4$ **c** $T − x$

13 a $8x$ **b** $12m$ **c** $18t$

14 Andrea: $3n − 3$, Barak: $3n − 1$, Ahmed: $3n − 6$ or $3(n − 2)$, Dina: 0, Emma: $3n − n = 2n$. Hana: $3n − 3m$

15 For example, $2 \times 6m$, $1 \times 12m$, $6m + 6m$, etc.

11.2 Substitution into formulae

Exercise 11B

1 a 8   **b** 17   **c** −28

2 a 13   **b** 11   **c** 43

3 a 21   **b** −7   **c** 11.8

4 a 9   **b** 3.8   **c** 23

5 a 13   **b** $5\frac{1}{2}$ **c** $6\frac{3}{4}$

6 a −20   **b** 13   **c** 10.9

7 a \$4   **b** 13 km   **c** Yes, the fare is \$5.00

8 a $2 \times 8 + 6 \times 11 − 3 \times 2 = 76$
  b $5 \times 2 − 2 \times 11 + 3 \times 8 = 12$

9 Any values such that $lw = \frac{1}{2}bh$ or $bh = 2lw$

10 a 32   **b** 64   **c** 16

11 a 6.5   **b** 18.5   **c** −2.5

12 a 2   **b** 8   **c** −10

13 a 3   **b** 2.5   **c** −5

14 a 6   **b** 24   **c** −2

15 a 12   **b** 8   **c** $1\frac{1}{2}$

16 a $\frac{1050}{n}$ **b** \$925

17 a i odd   **ii** odd   **iii** even   **iv** odd
  b Any valid expression such as $xy + z$

18 a \$20
  b i −\$40   **ii** Delivery cost will be zero.
  c 40 kilometres

11.3 Rearranging formulae

Exercise 11C

1 $k = \frac{T}{3}q$ **2** $y = X + 1$

3 $p = 3Q$ **4** $r = \frac{A - 9}{4}$

5 $n = \frac{W + 1}{3}$ **6 a** $m = p − t$

b $t = p − m$ **7** $m = gv$

8 $m = \sqrt{t}$ **9** $r = \frac{C}{2\pi}$

10 $b = \frac{A}{h}$ **11** $l = \frac{P - 2w}{2}$

12 $p = \sqrt{m - 2}$

13 a $−40 − 32 = −72$, $−72 \div 9 = −8$, $5 \times −8 = −40$
  b $68 − 32 = 36$, $36 \div 9 = 4$, $4 \times 5 = 20$
  c student's own demonstration

14 a $a = \frac{v - u}{t}$ **b** $t = \frac{v - u}{a}$

15 $d = \sqrt{\frac{4A}{\pi}}$

16 a $n = \frac{W - t}{3}$ **b** $t = W − 3n$

a $y = \frac{x + w}{5}$ **b** $w = 5y - x$

p $p = \sqrt{\frac{k}{2}}$

a $t = u^2 - v$ **b** $u = \sqrt{v + t}$

a $m = k - n^2$ **b** $n = \sqrt{k - m}$

$r = \sqrt{\frac{T}{5}}$

a $w = K - 5n^2$ **b** $n = \sqrt{\frac{K - w}{5}}$

.4   More complicated formulae

xercise 11D

a 2.5   **b** $a = \sqrt{c^2 - b^2}$

a 60   **b** $a = \frac{2(s - ut)}{t^2}$

3 a $b = ac - 2$ **b** $c = \frac{b + 2}{a}$

4 $t = \frac{r}{p} + 3$

5 $e = \left(\frac{12}{a} - 1\right)^2$

6 a 5   **b** $u = \sqrt{v^2 - 2as}$ **c** $s = \frac{v^2 - u^2}{2a}$

7 a $L = \left(\frac{T}{2\pi}\right)^2 G$ **b** Student's proof

8 a $R = \sqrt{\frac{D + \pi r^2}{\pi}}$ **b** $r = \sqrt{\frac{\pi R^2 - D}{\pi}}$ **c** $\pi = \frac{D}{R^2 - r^2}$

9 a $x = 5$ or -5 **b** $x = \sqrt{\frac{11 + 4y^2}{3}}$ **c** $y = \sqrt{\frac{3x^2 - 11}{4}}$

10 a $a = \left(\frac{T}{2}\right)^2 (c + 3)$ **b** $c = a\left(\frac{2}{T}\right)^2 - 3$

11 $T = \frac{b^2 + c^2 - a^2}{2bc}$

12 a 12   **b** $f = \frac{uv}{u + v}$ **c** $u = \frac{fv}{v - f}$ **d** $v = \frac{fu}{u - f}$

Answers to Chapter 12

.1   Simplifying expressions

xercise 12A

a $6t$ **b** $15y$ **c** $8w$ **d** $5b^2$ **e** $2w^2$
f $8p^2$ **g** $6t^2$ **h** $15t^2$ **i** $2mt$ **j** $5qt$
k $6mn$ **l** $6qt$ **m** $10hk$ **n** $21pr$

a All except $2m \times 6m$
b 2 and 0

$4x$ cm

a y^3 **b** $3m^3$ **c** $4t^3$ **d** $6n^3$ **e** t^4
f h^5 **g** $12n^5$ **h** $6a^7$ **i** $4k^7$ **j** t^3
k $12d^3$ **l** $15p^6$ **m** $3mp^2$ **n** $6m^2n$ **o** $8m^2p^2$

xercise 12B

a $\$t$ **b** $\$(4t + 3)$

a $10x + 2y$ **b** $7x + y$ **c** $6x + y$

a $5a$ **b** $6c$ **c** $9e$
d $6f$ **e** $4j$ **f** $3q$
g 0   **h** $-w$ **i** $6x^2$
j $5y^2$ **k** 0

a $7x$ **b** $3t$ **c** $-5x$
d $-5k$ **e** $2m^2$ **f** 0

a $7x + 5$ **b** $5x + 6$ **c** $5p$
d $5x + 6$ **e** $5p + t + 5$ **f** $8w - 5k$
g c **h** $8k - 6y + 10$

a $2c + 3d$ **b** $5d + 2e$ **c** $f + 3g + 4h$
d $6u - 3v$ **e** $7m - 7n$ **f** $3k + 2m + 5p$
g $2v$ **h** $2w - 3y$ **i** $11x^2 - 5y$
j $-y^2 - 2z$ **k** $x^2 - z^2$

a $8x + 6$ **b** $3x + 16$ **c** $2x + 2y + 8$

Any acceptable answers, e.g. $x + 4x + 2y + 2y$
or $6x - x + 6y - 2y$

a $2x$ and $2y$ **b** a and $7b$

10 a $3x - 1 - x$ **b** $10x$ **c** 25 cm

11 Maria is correct, as the two short horizontal lengths are equal
to the bottom length and the two short vertical lengths are
equal to the side length.

12.2   Expanding brackets

Exercise 12C

1 a $6 + 2m$ **b** $10 + 5l$
   c $12 - 3y$ **d** $20 + 8k$ **e** $6 - 12f$
   f $10 - 6w$ **g** $10k + 15m$ **h** $12d - 8n$
   i $t^2 + 3t$ **j** $k^2 - 3k$ **k** $4t^2 - 4t$
   l $8k - 2k^2$ **m** $8g^2 + 20g$ **n** $15h^2 - 10h$
   o $y^3 + 5y$ **p** $h^4 + 7h$ **q** $k^3 - 5k$
   r $3t^3 + 12t$ **s** $15d^3 - 3d^4$ **t** $6w^3 + 3tw$
   u $15a^3 - 10ab$ **v** $12p^4 - 15mp$
   w $12h^3 + 8h^2g$ **x** $8m^3 + 2m^4$

2 a $5(t - 1)$ and $5t - 5$
   b Yes, as $5(t - 1)$ when $t = 4.50$ is $5 \times 3.50 = \$17.50$

3 He has worked out 3×5 as 8 instead of 15 and he has not
multiply the second term by 3. Answer should be $15x - 12$.

4 a $3(2y + 3)$ **b** $2(6z + 4)$ or $4(3z + 2)$

Exercise 12D

1 a $7t$ **b** $9d$ **c** $3e$ **d** $2t$
   e $5t^2$ **f** $4y^2$ **g** $5ab$ **h** $3a^2d$

2 a $2x$ and $11y$ **b** a and $8b$

3 a $3x - 1 - x$ **b** $10x$ **c** 25 cm

4 a $22 + 5t$ **b** $21 + 19k$ **c** $22 + 2f$ **d** $14 + 3g$

5 a $2 + 2h$ **b** $9g + 5$ **c** $17k + 16$ **d** $6e + 20$

6 a $4m + 3p + 2mp$ **b** $3k + 4h + 5hk$
   c $12r + 24p + 13pr$ **d** $19km + 20k - 6m$

7 a $9t^2 + 13t$ **b** $13y^2 + 5y$ **c** $10e^2 - 6e$ **d** $14k^2 - 3kp$

8 a $17ab + 12ac + 6bc$ **b** $18wy + 6ty - 8tw$
 c $14mn - 15mp - 6np$ **d** $8r^3 - 6r^2$

9 For x-coefficients, 3 and 1 or 1 and 4; for y-coefficients, 5 and 1 or 3 and 4 or 1 and 7.

10 $5(3x + 2) - 3(2x - 1) = 9x + 13$

12.3 Factorisation

Exercise 12E

1 a $6(m + 2t)$ **b** $3(3t + p)$
 c $4(2m + 3k)$ **d** $4(r + 2t)$
 e $m(n + 3)$ **f** $g(5g + 3)$
 g $2(2w - 3t)$ **h** $y(3y + 2)$
 i $t(4t - 3)$ **j** $3m(m - p)$
 k $3p(2p + 3t)$ **l** $2p(4t + 3m)$
 m $4b(2a - c)$ **n** $5bc(b - 2)$
 o $2b(4ac + 3de)$ **p** $2(2a^2 + 3a + 4)$
 q $3b(2a + 3c + d)$ **r** $t(5t + 4 + a)$
 s $3mt(2t - 1 + 3m)$ **t** $2ab(4b + 1 - 2a)$
 u $5pt(2t + 3 + p)$

2 a Suni has taken out a common factor.
 b Because the bracket adds up to $10.
 c $30

3 a, d, f and **h** do not factorise.
 b $m(5 + 2p)$
 c $t(t - 7)$
 e $2m(2m - 3p)$
 g $a(4a - 5b)$
 i $b(5a - 3bc)$

4 a Bernice
 b Aidan has not taken out the largest possible common factor. Craig has taken m out of both terms but there isn't an m in the second term.

5 There are no common factors.

12.4 Expanding two brackets

Exercise 12F

1 $x^2 + 5x + 6$ **2** $t^2 + 7t + 12$
3 $w^2 + 4w + 3$ **4** $m^2 + 6m + 5$
5 $k^2 + 8k + 15$ **6** $a^2 + 5a + 4$
7 $x^2 + 2x - 8$ **8** $t^2 + 2t - 15$
9 $w^2 + 2w - 3$ **10** $f^2 - f - 6$
11 $g^2 - 3g - 4$ **12** $y^2 + y - 12$
13 $x^2 + x - 12$ **14** $p^2 - p - 2$
15 $k^2 - 2k - 8$ **16** $y^2 + 3y - 10$
17 $a^2 + 2a - 3$ **18** $x^2 - 9$
19 $t^2 - 25$ **20** $m^2 - 16$
21 $t^2 - 4$ **22** $y^2 - 64$
23 $p^2 - 1$ **24** $25 - x^2$
25 $49 - g^2$ **26** $x^2 - 36$
27 $(x + 2)$ and $(x + 3)$

28 a B: $1 \times (x - 2)$ C: 1×2 D: $2 \times (x - 1)$
 b $(x - 2) + 2 + 2(x - 1) = 3x - 2$
 c Area A $= (x - 1)(x - 2) =$ area of square minus areas
      $(B + C + D)$

$$= x^2 - (3x - 2)$$
$$= x^2 - 3x + 2$$

29 a $x^2 - 9$
 b i 9991 **ii** 39991

12.5 Multiplying more complex expression

Exercise 12G

1 $6x^2 + 11x + 3$ **2** $12y^2 + 17y + 6$
3 $6t^2 + 17t + 5$ **4** $8t^2 + 2t - 3$
5 $10m^2 - 11m - 6$ **6** $12k^2 - 11k - 15$
7 $6p^2 + 11p - 10$ **8** $10w^2 + 19w + 6$
9 $6a^2 - 7a - 3$ **10** $8r^2 - 10r + 3$
11 $15g^2 - 16g + 4$ **12** $12d^2 + 5d - 2$
13 $8p^2 + 26p + 15$ **14** $6t^2 + 7t + 2$
15 $6p^2 + 11p + 4$ **16** $6 - 7t - 10t^2$
17 $12 + n - 6n^2$ **18** $6f^2 - 5f - 6$
19 $12 + 7q - 10q^2$ **20** $3 - 7p - 6p^2$
21 $4 + 10t - 6t^2$
22 a $x^2 - 1$ **b** $4x^2 - 1$ **c** $4x^2 - 9$ **d** $9x^2 - 25$
23 a $(3x - 2)(2x + 1) = 6x^2 - x - 2$
      $(2x - 1)(2x - 1) = 4x^2 - 4x + 1$
      $(6x - 3)(x + 1) = 6x^2 + 3x - 3$
      $(3x + 2)(2x + 1) = 6x^2 + 7x + 2$
 b Multiply the x terms to match the x^2 term and/or multiply the constant terms to get the constant term in the answer

Exercise 12H

1 $4x^2 - 1$ **2** $9t^2 - 4$
3 $25y^2 - 9$ **4** $16m^2 - 9$
5 $4k^2 - 9$ **6** $16h^2 - 1$
7 $4 - 9x^2$ **8** $25 - 4t^2$
9 $36 - 25y^2$ **10** $a^2 - b^2$
11 $9t^2 - k^2$ **12** $4m^2 - 9p^2$
13 $25k^2 - g^2$ **14** $a^2b^2 - c^2d^2$
15 $a^4 - b^4$
16 a $a^2 - b^2$
 b Dimensions: $a + b$ by $a - b$; Area: $a^2 - b^2$
 c Areas are the same, so $a^2 - b^2 = (a + b) \times (a - b)$
17 First shaded area is $(2k)^2 - 1^2 = 4k^2 - 1$
      Second shaded area is $(2k + 1)(2k - 1) = 4k^2 - 1$

Exercise 12I

1 $x^2 + 10x + 25$ **2** $m^2 + 8m + 16$
3 $t^2 + 12t + 36$ **4** $p^2 + 6p + 9$
5 $m^2 - 6m + 9$ **6** $t^2 - 10t + 25$
7 $m^2 - 8m + 16$ **8** $k^2 - 14k + 49$
9 $9x^2 + 6x + 1$ **10** $16t^2 + 24t + 9$
11 $25y^2 + 20y + 4$ **12** $4m^2 + 12m + 9$
13 $16t^2 - 24t + 9$ **14** $9x^2 - 12x + 4$
15 $25t^2 - 20t + 4$ **16** $25r^2 - 60r + 36$
17 $x^2 + 2xy + y^2$ **18** $m^2 - 2mn + n^2$

$4t^2 + 4ty + y^2$ **20** $m^2 - 6mn + 9n^2$

$x^2 + 4x$ **22** $x^2 - 10x$

$x^2 + 12x$ **24** $x^2 - 4x$

a Marcela has just squared the first term and the second term. She hasn't written down the brackets twice.

b Paulo has written down the brackets twice but has worked out $(3x)^2$ as $3x^2$ and not $9x^2$.

c $9x^2 + 6x + 1$

Whole square is $(2x)^2 = 4x^2$.

Three areas are $2x - 1$, $2x - 1$ and 1.

$4x^2 - (2x - 1 + 2x - 1 + 1) = 4x^2 - (4x - 1) = 4x^2 - 4x + 1$

2.6   Quadratic factorisation

Exercise 12J

$(x + 2)(x + 3)$ **2** $(t + 1)(t + 4)$

$(m + 2)(m + 5)$ **4** $(k + 4)(k + 6)$

$(p + 2)(p + 12)$ **6** $(r + 3)(r + 6)$

$(w + 2)(w + 9)$ **8** $(x + 3)(x + 4)$

$(a + 2)(a + 6)$ **10** $(k + 3)(k + 7)$

$(f + 1)(f + 21)$ **12** $(b + 8)(b + 12)$

$(t - 2)(t - 3)$ **14** $(d - 4)(d - 1)$

$(g - 2)(g - 5)$ **16** $(x - 3)(x - 12)$

$(c - 2)(c - 16)$ **18** $(t - 4)(t - 9)$

$(y - 4)(y - 12)$ **20** $(j - 6)(j - 8)$

$(p - 3)(p - 5)$ **22** $(y + 6)(y - 1)$

$(t + 4)(t - 2)$ **24** $(x + 5)(x - 2)$

$(m + 2)(m - 6)$ **26** $(r + 1)(r - 7)$

$(n + 3)(n - 6)$ **28** $(m + 4)(m - 11)$

$(w + 4)(w - 6)$ **30** $(t + 9)(t - 10)$

$(h + 8)(h - 9)$ **32** $(t + 7)(t - 9)$

$(d + 1)^2$ **34** $(y + 10)^2$

$(t - 4)^2$ **36** $(m - 9)^2$

$(x - 12)^2$ **38** $(d + 3)(d - 4)$

$(t + 4)(t - 5)$ **40** $(q + 7)(q - 8)$

$(x + 2)(x + 3)$, giving areas of $2x$ and $3x$, or $(x + 1)(x + 6)$, giving areas of x and $6x$.

2.7   Factorising $ax^2 + bx + c$

Exercise 12K

$(2x + 1)(x + 2)$ **2** $(7x + 1)(x + 1)$

$(4x + 7)(x - 1)$ **4** $(3t + 2)(8t + 1)$

$(3t + 1)(5t - 1)$ **6** $(4x - 1)^2$

$3(y + 7)(y - 3)$ **8** $4(y + 6)(y - 4)$

$(2x + 3)(4x - 1)$ **10** $(2t + 1)(3t + 5)$

$(x - 6)(3x + 2)$ **12** $(x - 5)(7x - 2)$

$4x + 1$ and $3x + 2$

a All the terms in the quadratic have a common factor of 6.

b $6(x + 2)(x + 3)$. This has the highest common factor taken out.

c For example, 'A rectangle could be split in many different ways.'

12.8   More than two brackets

Exercise 12L

1 **a** $x^2 + 3x + 2$ **b** $x^3 + 6x^2 + 11x + 6$

2 **a** $x^3 + 3x^2 - 4$ **b** $x^3 + 3x^2 - 13x - 15$

3 **a** $x^2 + 6x + 9$ **b** $x^3 + 6x^2 + 9x$

   c $x^3 + 5x^2 + 3x - 9$

4 **a** $x^3 + 10x^2 - 4x - 40$ **b** $x^3 + 2x^2 - 7x + 4$

5 **a** $x^3 + 5x^2 + 7x + 3$ **b** $x^3 + x^2 - 5x + 3$

6 **a** $x^3 + 3x^2 + 3x + 1$ **b** $x^3 - 3x^2 + 3x - 1$

   c $x^3 + 6x^2 + 12x + 8$ **d** $x^3 - 6x^2 + 12x - 8$

7 **a** The square is divided into 4 parts. One has area x^2, two have area x and one has area 1.

   b The cube is divided into 8 parts. One has volume x^3, three have volume x^2, three have volume x and one has volume 1.

8   c = 3

9 **a** $2x^2 - 3x - 2$ **b** $2x^3 + 3x^2 - 11x - 6$

10 a $12x^2 - x - 1$ **b** $12x^3 - 25x^2 + x + 2$

11 a $6x^2 - 19x + 10$ **b** $12x^3 - 44x^2 + 39x - 10$

12.9   Algebraic fractions

Exercise 12M

1 **a** $\frac{5x}{6}$ **b** $\frac{19x}{20}$ **c** $\frac{23x}{20}$ **d** $\frac{3x + 2y}{6}$

   e $\frac{x^2y + 8}{4x}$ **f** $\frac{5x + 7}{6}$ **g** $\frac{7x + 3}{4}$ **h** $\frac{13x + 5}{15}$

   i $\frac{3x - 7}{4}$ **j** $\frac{5x - 10}{4}$

2 **a** $\frac{x}{6}$ **b** $\frac{11x}{20}$ **c** $\frac{7x}{20}$ **d** $\frac{3x - 2y}{6}$

   e $\frac{xy^2 - 8}{4y}$ **f** $\frac{x - 1}{6}$ **g** $\frac{x + 1}{4}$ **h** $\frac{-7x - 5}{15}$

   i $\frac{x - 1}{4}$ **j** $\frac{2 - 3x}{4}$

3 **a** $\frac{x^2}{6}$ **b** $\frac{3xy}{14}$ **c** $\frac{8}{3}$ **d** $\frac{2xy}{3}$

   e $\frac{x^2 - 2x}{10}$ **f** $\frac{1}{6}$ **g** $\frac{6x^2 + 5x + 1}{8}$ **h** $\frac{2x^2 + x}{15}$

   i $\frac{2x - 4}{x - 3}$ **j** $\frac{1}{2x}$

4 **a** x **b** $\frac{x}{2}$ **c** $\frac{3x^2}{16}$ **d** 3

   e $\frac{17x + 1}{10}$ **f** $\frac{13x + 9}{10}$ **g** $\frac{3x^2 - 5x - 2}{10}$ **h** $\frac{x + 3}{2}$

   i $\frac{2x^2 - 6y^2}{9}$

5 **a** $\frac{7x + 9}{(x + 1)(x + 2)}$ **b** $\frac{(11x - 10)}{(x - 2)(x + 1)}$ **c** $\frac{10 - 13x}{(4x + 1)(x + 2)}$

   d $\frac{8 - 10x}{(2x - 1)(x + 1)}$ **e** $\frac{x + 1}{(2x - 1)(3x - 1)}$

6 First, he did not factorise and just cancelled the x^2s. Then he cancelled 2 and 6 with the wrong signs. Then he said two minuses make a plus when adding, which is not true.

7 $\frac{2x^2 + x - 3}{4x^2 - 9}$

8 **a** $\frac{9x + 13}{(x + 1)(x + 2)}$ **b** $\frac{14x + 19}{(4x - 1)(x + 1)}$

   c $\frac{2x^2 + x - 13}{2(x + 1)}$ **d** $\frac{x + 1}{(2x - 1)(3x - 1)}$

9 **a** $\frac{x - 1}{2x + 1}$ **b** $\frac{2x + 1}{x + 3}$ **c** $\frac{2x - 1}{3x - 2}$

   d $\frac{x + 1}{(x - 1)}$ **e** $\frac{2x + 5}{4x + 1}$

13.1 Solving simple linear equations

Exercise 13A

1 **a** 56   **b** 2   **c** 6   **d** 3   **e** 4
   f $2\frac{1}{2}$ **g** $3\frac{1}{2}$ **h** $2\frac{1}{2}$ **i** 4   **j** 21
   k 18   **l** 56   **m** 0   **n** –7   **o** –18
   p 36   **q** 36   **r** 60   **s** 7   **t** 11
   u 2   **v** 7   **w** 2.8   **x** 1   **y** 11.5
   z 1.4

2 **a** –4   **b** 15

3 **a** Elif
   b Second line: Mustafa subtracts 1 instead of adding 1; fourth line: Mustafa subtracts 2 instead of dividing by 2.

Exercise 13B

1 **a** 3   **b** 7   **c** 5   **d** 3   **e** 4   **f** 6
   g 8   **h** 1   **i** $1\frac{1}{2}$ **j** $2\frac{1}{2}$ **k** $\frac{1}{2}$ **l** $1\frac{1}{5}$
   m 2   **n** –2   **o** –1   **p** –2   **q** –2   **r** –1

2 Any values that work, e.g. $a = 2$, $b = 3$ and $c = 30$.

Exercise 13C

1 **a** $x = 2$ **b** $y = 1$ **c** $a = 7$ **d** $t = 4$
   e $p = 2$ **f** $k = -1$ **g** $m = 3$ **h** $s = -2$

2 $3x - 2 = 2x + 5$, $x = 7$

3 **a** $d = 6$ **b** $x = 11$ **c** $y = 1$ **d** $h = 4$
   e $b = 9$ **f** $c = 6$

4 $6x + 3 = 6x + 10$; $6x - 6x = 10 - 3$; $0 = 7$, which is obviously false. Both sides have $6x$, which cancels out.

5 Check student's example.

13.2 Setting up equations

Exercise 13D

1 $0.90 or 90 cents

2 **a** 1.5   **b** 2

3 **a** 1.5 cm   **b** 6.75 cm^2

4 17

5 8

6 **a** $8c - 10 = 56$ **b** $8.25

7 **a** B: 450 cars, C: 450 cars, D: 300 cars   **b** 800   **c** 750

8 Length is 5.5 m, width is 2.5 m and area is 13.75 m^2. Tiles cost 123.75 dollars

9 3 years

10 9 years

11 3 cm

12 5

13 **a** $4x + 40 = 180$ **b** $x = 35°$

14 **a** $\frac{x + 10}{5} = 9.50$ **b** $37.50

15 No, as $x + x + 2 + x + 4 + x + 6 = 360$ gives $x = 87°$ so the consecutive numbers (87, 89, 91, 93) are not even but odd

16 $4x + 18 = 3x + 1 + 50$, $x = 33$. Large bottle 1.5 litres, small bottle 1 litre

13.3 More complex equations

Exercise 13E

1 **a** 7   **b** 9   **c** 14
   d 5   **e** 2.5   **f** –2

2 **a** $2\frac{2}{3}$ **b** 6   **c** 2
   d 5   **e** $\frac{5}{16}$ **f** 3

3 **a** 6   **b** 14   **c** 7

4 5, 6 and 7

5 50, 55 and 75 degrees

13.4 Solving quadratic equations by factorisation

Exercise 13F

1 –2, –5   **2** –3, –1
3 –6, –4   **4** –3, 2
5 –1, 3   **6** –4, 5
7 1, –2   **8** 2, –5
9 7, –4   **10** 3, 2
11 1, 5   **12** 4, 3
13 –4, –1   **14** –9, –2
15 2, 4   **16** 3, 5
17 –2, 5   **18** –3, 5
19 –6, 2   **20** –6, 3
21 –1, 2   **22** –2
23 –5   **24** 4
25 –2, –6   **26** 7
27 **a** $x(x - 3) = 550$, $x^2 - 3x - 550 = 0$
   b $(x - 25)(x + 22) = 0$, $x = 25$
28 $x(x + 40) = 48000$, $x^2 + 40x - 48000 = 0$, $(x + 240)(x - 200) = 0$. Fence is $2 \times 200 + 2 \times 240 = 880$
29 –6, –4   **30** 2, 16
31 –6, 4   **32** –9, 6
33 –10, 3   **34** –4, 11
35 –8, 9   **36** 8, 9
37 1
38 Mario was correct. Sylvan did not make it into a standard quadratic and only factorised the x terms. She also incorrec[t] solved the equation $x - 3 = 4$.

13.5 More factorisation in quadratic equations

Exercise 13G

1 **a** $\frac{1}{3}$, –3   **b** $1\frac{1}{3}$, $-\frac{1}{2}$ **c** $-\frac{1}{5}$, 2
   d $-2\frac{1}{2}$, $3\frac{1}{2}$ **e** $-\frac{1}{6}$, $-\frac{1}{3}$ **f** $\frac{2}{3}$, 4
   g $\frac{1}{2}$, –3   **h** $\frac{5}{2}$, $-\frac{7}{6}$ **i** $-1\frac{2}{3}$, $1\frac{2}{5}$
   j $1\frac{3}{4}$, $1\frac{2}{7}$ **k** $\frac{2}{3}$, $\frac{1}{8}$ **l** $\pm\frac{1}{4}$

m $-2\frac{1}{4}$, 0   **n** $\pm 1\frac{2}{5}$ **o** $-\frac{1}{3}$, 3

a $-6, 7$ **b** $-\frac{5}{2}, \frac{3}{2}$ **c** $-6, 7$

d $-1, \frac{11}{13}$ **e** $-2, 3$ **f** $-\frac{2}{5}, \frac{1}{2}$

g $-\frac{1}{2}, \frac{1}{3}$ **h** $-2, \frac{1}{5}$ **i** 4

j $-2, \frac{1}{8}$ **k** $-\frac{1}{3}$, 0   **l** $-5, 5$

m $-\frac{5}{3}$ **n** $-\frac{7}{2}, \frac{7}{2}$ **o** $-\frac{5}{2}$, 3

a Both have only one solution: $x = 1$.

b B is a linear equation, but A and C are quadratic equations.

a $(5x - 1)^2 = (2x + 3)^2 + (x + 1)^2$, when expanded and collected into the general quadratics, gives the required equation.

b $(10x + 3)(2x - 3)$, $x = 1.5$; area $= 7.5$ cm^2.

13.6 Solving quadratic equations by completing the square

Exercise 13H

a $(x + 2)^2 - 4$ **b** $(x + 7)^2 - 49$

c $(x - 3)^2 - 9$ **d** $(x + 3)^2 - 9$

e $(x - 1.5)^2 - 2.25$ **f** $(x - 4.5)^2 - 20.25$

g $(x + 6.5)^2 - 42.25$ **h** $(x + 5)^2 - 25$

i $(x + 4)^2 - 16$ **j** $(x - 1)^2 - 1$

k $(x + 1)^2 - 1$

a $(x + 2)^2 - 5$ **b** $(x + 7)^2 - 54$

c $(x - 3)^2 - 6$ **d** $(x + 3)^2 - 2$

e $(x - 1.5)^2 - 3.25$ **f** $(x + 3)^2 - 6$

g $(x - 4.5)^2 - 10.25$ **h** $(x + 6.5)^2 - 7.25$

i $(x + 4)^2 - 22$ **j** $(x + 1)^2 - 2$

k $(x - 1)^2 - 8$ **l** $(x + 1)^2 - 10$

a $-2 \pm \sqrt{5}$ **b** $-7 \pm 3\sqrt{6}$ **c** $3 \pm \sqrt{6}$

d $-3 \pm \sqrt{2}$ **e** $1.5 \pm \sqrt{3.25}$ **f** $3 \pm \sqrt{6}$

g $4.5 \pm \sqrt{10.25}$ **h** $-6.5 \pm \sqrt{7.25}$ **i** $-4 \pm \sqrt{22}$

j $-1 \pm \sqrt{2}$ **k** $1 \pm 2\sqrt{2}$ **l** $-1 \pm \sqrt{10}$

a $1.45, -3.45$ **b** $5.32, -1.32$ **c** $-4.16, 2.16$

a $x = 1.5 \pm \sqrt{3.75}$ **b** $x = 1 \pm \sqrt{0.75}$

c $x = -1.25 \pm \sqrt{6.5625}$ **d** $x = 7.5 \pm \sqrt{40.25}$

$p = -14$, $q = -3$

a 3rd, 1st, 4th and 2nd – in that order

13.7 Solving quadratic equations by the quadratic formula

Exercise 13I

1 $1.77, -2.27$ **2** $-0.23, -1.43$

3 $3.70, -2.70$ **4** $0.29, -0.69$

5 $-0.19, -1.53$ **6** $-1.23, -2.43$

7 $-0.41, -1.84$ **8** $-1.39, -2.27$

9 $1.37, -4.37$ **10** $2.18, 0.15$

11 $-0.39, -5.11$ **12** $0.44, -1.69$

13 $1.64, 0.61$ **14** $0.36, -0.79$

15 $1.89, 0.11$

19 (the lawn perimeter is 18.8 m)

17 $x^2 - 3x - 7 = 0$

18 Hasan gets $x = \frac{4 \pm \sqrt{0}}{8}$ and Mirian gets $(2x - 1)^2 = 0$; each method only gives one solution, $x = \frac{1}{2}$

13.8 Simple simultaneous equations

Exercise 13J

1 **a** $x = 5, y = 10$ **b** $x = 18, y = 6$ **c** $x = 12, y = 48$

2 **a** $x = 6, y = 18$ **b** $x = 12.5, y = 2.5$ **c** $x = 0.5, y = 4.5$

3 **a** $x = 13, y = 7$ **b** $x = 9, y = 14$ **c** $x = 10, y = -4$

4 **a** $x = 0.5, y = 4$ **b** $x = 5.5, y = 14.5$ **c** $x = 2, y = 8$

5 Carmen 32, Anish 8

6 11.5 and 25.5

7 8 and -3

Exercise 13K

1 **a** $x = 4, y = 1$ **b** $x = 1, y = 4$

   c $x = 3, y = 1$ **d** $x = 5, y = -2$

   e $x = 7, y = 1$ **f** $x = 5, y = \frac{1}{2}$

   g $x = 4\frac{1}{2}, y = 1\frac{1}{2}$ **h** $x = -2, y = 4$

   i $x = 2\frac{1}{2}, y = -1\frac{1}{2}$ **j** $x = 2\frac{1}{4}, y = 6\frac{1}{2}$

   k $x = 4, y = 3$ **l** $x = 5, y = 3$

2 **a** 3 is the first term. The next term is $3 \times a + b$, which equals 14.

   b $14a + b = 47$

   c $a = 3, b = 5$

   d 146, 443

Exercise 13L

1 **a** $x = 2, y = -3$ **b** $x = 7, y = 3$

   c $x = 4, y = 1$ **d** $x = 2, y = 5$

   e $x = 4, y = -3$ **f** $x = 1, y = 7$

   g $x = 2\frac{1}{2}, y = 1\frac{1}{2}$ **h** $x = -1, y = 2\frac{1}{2}$

   i $x = 6, y = 3$ **j** $x = \frac{1}{2}, y = -\frac{3}{4}$

   k $x = -1, y = 5$ **l** $x = 1\frac{1}{2}, y = \frac{3}{4}$

2 **a** They are the same equation. Divide the first by 2 and it is the second, so they have an infinite number of solutions.

   b Double the second equation to get $6x + 2y = 14$ and subtract to get $9 = 14$. The left-hand sides are the same if the second is doubled so they cannot have different values.

13.9 More complex simultaneous equations

Exercise 13M

1 **a** $x = 5, y = 1$ **b** $x = 3, y = 8$

   c $x = 9, y = 1$ **d** $x = 7, y = 3$

   e $x = 4, y = 2$ **f** $x = 6, y = 5$

   g $x = 3, y = -2$ **h** $x = 2, y = \frac{1}{2}$

   i $x = -2, y = -3$ **j** $x = -1, y = 2\frac{1}{2}$

   k $x = 2\frac{1}{2}, y = -\frac{1}{2}$ **l** $x = -1\frac{1}{2}, y = 4\frac{1}{2}$

   m $x = -\frac{1}{2}, y = -6\frac{1}{2}$ **n** $x = 3\frac{1}{2}, y = 1\frac{1}{2}$

   o $x = -2\frac{1}{2}, y = -3\frac{1}{2}$

2 (1, −2) is the solution to equations A and C; (−1, 3) is the solution to equations A and D; (2, 1) is the solution to B and C; (3, −3) is the solution to B and D.

3 Intersection points are (0, 6), (1, 3) and (2, 4). Area is 2 cm²

4 Intersection points are (0, 3), (6, 0) and (4, −1). Area is 6 cm²

13.10 Linear and non-linear simultaneous equations

Exercise 13N

1 a (5, −1)　　　**b** (4, 1)　　　**c** (8, −1)

2 a (1, 2) and (−2, −1)　　　**b** (−4, 1) and (−2, 2)

3 a (3, 4) and (4, 3)　　　**b** (0, 3) and (−3, 0)
c (3, 2) and (−2, 3)

4 a (2, 5) and (−2, −3)　　　**b** (−1, −2) and (4, 3)
c (3, 3) and (1, −1)

5 a (−3, −3), (1, 1)　　　**b** (3, −2), (−2, 3)
c (−2, −1), (1, 2)　　　**d** (2, −1), (3, 1)
e (−2, 1), (3, 6)　　　**f** (1, −4), (4, 2)
g (4, 5), (−5, −4)

Answers to Chapter 14

14.1　Conversion graphs

Exercise 14A

1 a i $8\frac{1}{4}$ kg　**ii** $2\frac{1}{4}$ kg　**iii** 9 lb　**iv** 22 lb
b 2.2 lb
c Read off the value for 12 lb (5.4 kg) and multiply this by 4 (21.6 kg)

2 a i 10 cm　**ii** 23 cm　**iii** 2 in　**iv** $8\frac{3}{4}$ in
b $2\frac{1}{2}$ cm
c Read off the value for 9 in (23 cm) and multiply this by 2 (46 cm)

3 a i $320　**ii** $100　**iii** £45　**iv** £78
b $3.20
c It would become more steep.

4 a i $120　**ii** $82　**b i** 32　**ii** 48

5 a i $100　**ii** $325　**b i** 500　**ii** 250

6 a i $70　**ii** $29　**b i** $85　**ii** $38

7 a i 95 °F　**ii** 68 °F　**iii** 10 °C　**iv** 32 °C
b 32°F

8 a Check student's graph　**b** $50

9 a Student's own graph　**b** about 48 kilometres
c about 16 miles

10 a Student's own graph　**b** about 9 centimetres
c about 4 hours

11 a Student's own graph　**b** about 23 minutes

14.2　Travel graphs

Exercise 14B

1 a i 2 h　**ii** 3 h　**iii** 5 h
b i 40 km/h　**ii** 120 km/h　**iii** 40 km/h
c 5.30 am

2 a i 125 km　**ii** 125 km/h
b i Between 2 pm and 3 pm　**ii** About 12 km/h

3 a 30 km　**b** 40 km　**c** 100 km/h

4 a i 263 m/min (3 sf)　**ii** 15.8 km/h (3 sf)

b 500 m/min　　　**c** Yuto by 1 minute

5 a Patrick ran quickly at first, then had a slow middle section but he won the race with a final sprint. Araf ran steadily all the way and came second. Sean set off the slowest, speeded up towards the end but still came third.
b i 1.67 m/s　**ii** 6 km/h

6

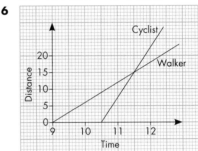

b At 1130

7 a i Because it stopped several times　**ii** Ravinder
b Ravinder at 1558, Sue at 1620, Michael at 1635
c i 24 km/h　**ii** 20.5 km/h　**iii** 5

8 a 50 metres　**b** Student's graph　**c** 1 metre/second

9 a student's graph　**b** 80 km/hour

10 a 1300　　　**b** 15 km
c student's graph
d For the three stages, 5 km/hour, 4 km/hour and 2 km/hour. For the whole trip 3.75 km/hour

14.3　Speed–time graphs

Exercise 14C

1 a 20 m/s　　　**b** 5 m/s
c Between 30 and 45 seconds

2 a 50 seconds　　　**b** About 33 seconds

3 a For the first 10 seconds　**b** 10 seconds
c 4 m/s

4 a 40 m/s　　　**b** 150 seconds

5 a 15 m/s　　　**b** After 12 seconds
c 450 metres　　　**d** 300 metres

Answers to Chapter 15

15.1 Using coordinates

Exercise 15A

a A(4, 2); B(–2, 5); C(–3, 2); D(–2, –3); E(3, –3)
b i $y = 2$ **ii** $x = -2$ **iii** $y = -3$
c B
d C

a P(4, 3); Q(–2, 1)
b (1, 2)
c i (–2, –1)   **ii** (0, –2)   **iii** (3, 1)
d R

a $(3\frac{1}{2}, -\frac{1}{2})$ and $(-2\frac{1}{2}, -\frac{1}{2})$
b $y = -\frac{1}{2}$

15.2 Drawing straight line graphs

Exercise 15B

Extreme points are (0, 4), (5, 19)
Extreme points are (0, –5), (5, 5)
Extreme points are (0, –3), (10, 2)
Extreme points are (–3, –4), (3, 14)
Extreme points are (–6, 2), (6, 6)
a Extreme points are (0, –2), (5, 13) and (0, 1), (5, 11)
b (3, 7)
a Extreme points are (0, –5), (5, 15) and (0, 3), (5, 13)
b (4, 11)
a Extreme points are (0, –1), (12, 3) and (0, –2), (12, 4)
b (6, 1)
a Extreme points are (0, 1), (4, 13) and (0, –2), (4, 10)
b Do not cross because they are parallel
a Values of y: 5, 4, 3, 2, 1, 0. Extreme points are (0, 5), (5, 0)
b Extreme points are (0, 7), (7, 0)
a yes   **b** no   **c** yes   **d** no   **e** yes   **f** no

15.3 More straight line graphs

Exercise 15C
a All of them.
b

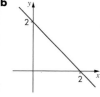

2 a 5 – 5 = 0
b Student's points
c

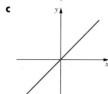

3 a 6 – 3 = 3; 0 – (–3) = 3
b Student's points
c

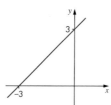

4

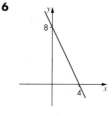

5 a At $x = 6$
b At $y = 3$
c

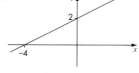

6

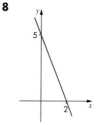

7

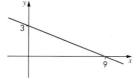

8

9

Exercise 15D

1 a A $\frac{1}{20}$ B $\frac{4}{25}$ C $\frac{1}{8}$ **b** B   **c** A

2 $\frac{1}{5}$

3 A 3   B $\frac{1}{3}$ C –1

4 a 4   **b** $\frac{1}{4}$ **c** 2.5   **d** 10
   e –2   **f** $-\frac{1}{5}$ **g** 0   **h** –1.5

15.4 The equation $y = mx + c$

Exercise 15E

1 Student's own check;
   a 3, (0, 4)   **b** 2, (0, –5)   **c** $\frac{1}{2}$, (0, –3)
   d 3, (0, 5)   **e** $\frac{1}{3}$, (0,4)

2 a $y = 2x - 2$ **b** $y = x + 1$
   c $y = 2x - 3$ **d** $2y = x + 6$
   e $y = x$ **f** $y = 2x$

3 a $y = 2x + 1, y = -2x + 1$
   b $5y = 2x - 5, 5y = -2x - 5$
   c $y = x + 1, y = -x + 1$

4 **a** $y = -2x + 1$ **b** $2y = -x$
c $y = -x + 1$ **d** $5y = -2x - 5$
e $y = -\frac{3}{2}x - 3$ or $2y = -3x - 6$

5 **a** 3   **b** $(0, 3)$

6 **a** 4   **b** 4

15.5 Finding equations

Exercise 15F

1 **a** 3   **b** $\frac{1}{2}$ **c** 4   **d** -1 **e** $-\frac{1}{2}$ **f** $\frac{2}{3}$

2 **a** $y = 2x - 3$ **b** $y = \frac{1}{2}x + 4$
c $y = 4x - 2$ **d** $y = -3x + 8$

3 **a** $(5, 3)$ **b** $(4, 5)$
c $(3, 2)$ **d** $(3, 3)$
e $(1, 3.5)$ **f** $(-0.5, 0)$

4 **a** student's graph   **b** $y = 0.5x + 6.5$
c $(-1, 3)$ **d** $y = -x + 8$

5 **a** $y = \frac{3}{4}x + \frac{1}{2}$ **b** $y = \frac{1}{2}x + 3\frac{1}{2}$
c $y = -2x + 7$ **d** $y = x + 2$

15.6 Parallel and perpendicular lines

Exercise 15G

1 **a** Line A does not pass through $(0, 1)$.
b Line C is perpendicular to the other two.
c (i)

2 **a** $-\frac{1}{2}$ **b** $\frac{1}{3}$ **c** -2 **d** $\frac{3}{2}$ **e** $-\frac{2}{3}$ **f** $-\frac{3}{4}$

3 $y = 3x + 5$, $x + 3y = 10$, $y = 8 - \frac{1}{3}x$, $y = 3(x + 2)$

4 $x = 6$ and $y = -2$
$x + y = 5$ and $y = x + 4$
$y = 8x - 9$ and $y = -\frac{1}{8}x + 6$
$2y = x + 4$ and $2x + y = 9$
$5y = 2x + 15$ and $2y + 5x = 2$
$y = 0.1x + 2$ and $y = 33 - 10x$

5 **a** $y = \frac{1}{2}x - 2$ **b** $y = -x + 3$
c $y = -\frac{1}{3}x - 1$ **d** $y = 3x + 5$

6 **a** -4 **b** $\frac{1}{4}$ **c** $(11, 7)$ **d** $y = \frac{1}{4}x +$
Substitute in $(11, 7)$ and solve to get $c = \frac{17}{4}$,
so $4y - x = 17$

7 $y = -\frac{1}{4}x + 2$

8 **i a** AB: $-\frac{1}{5}$, BC: 1, CD: $-\frac{1}{5}$, DA: 1
b Parallelogram (two pairs of parallel sides)
ii a AB: $\frac{2}{3}$, BC: $-\frac{3}{2}$, CD: $\frac{2}{3}$, DA: $-\frac{3}{2}$
b Rectangle (two pairs of perpendicular sides)
iii a AB: $\frac{2}{5}$, BC: $\frac{1}{4}$, CD: $\frac{2}{5}$, DA: 1
b Trapezium (one pair of parallel sides)

9 $y = -\frac{1}{2}x + 5$

10 **a** $y = 3x - 6$
b Bisector of AB is $y = -2x + 9$, bisector of AC is $y = \frac{1}{2}x +$
solving these equations shows the lines intersect at $(3, 3)$
c $(3, 3)$ lies on $y = 3x - 6$ because $(3 \times 3) - 6 = 3$

11 $(3, 10)$

15.7 Graphs and simultaneous equation

Exercise 15H

1 **a** $x = 2$, $y = 6$ **b** $x = 6$, $y = 2$
c $x = 4$, $y = 4$ **d** $x = 0$, $y = 4$

2 **a** $x = 4$, $y = 8$ **b** $x = 6$, $y = 4$ **c** $x = 6$, $y = 0$

Answers to Chapter 16

16.1 Quadratic graphs

Exercise 16A

1 x: $-3, -2, -1, 0, 1, 2, 3$
y: $11, 6, 3, 2, 3, 6, 11$

2 **a** x: $-3, -2, -1, 0, 1, 2, 3, 4, 5$
x^2: $9, 4, 1, 0, 1, 4, 9, 16, 25$
$-3x$: $9, 6, 3, 0, -3, -6, -9, -12, -15$
y: $18, 10, 4, 0, -2, -2, 0, 4, 10$
b 1.8   **c** $(1.5, -2.25)$
d $x = 1.5$ **e** $x = 4.2$ or -1.2

3 **a** y: $7, 0, -5, -8, -9, -8, -5, 0, 7$
b The graph should give a value of about -8.75
c The graph should give values of about 4.5 and -2.5

4 **a** h: $10, 4, 0, -2, -2, 0, 4, 10$ **b** $(2.5, -2.25)$
c $t = 2.5$

d The graph should give a value of about 6.75
e The graph should give values of about 0.2 and 4.8

5 **a**

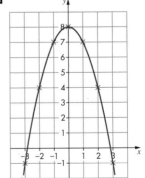

b $x = 1.4$ or -1.4

a

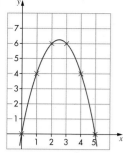

b $x = 2.5$

2 Solving equations with quadratic graphs

ercise 16B

a x: –4, –3, –2, –1, 0, 1, 2
  y: 7, 2, –1, –2, –1, 2, 7
b 0.25
c [The line on the graph goes through (–2, –1) and (1, 2)]
d $x = -2$ or 1
e Student's own explanation and check

a x: –4, –3, –2, –1, 0, 1, 2, 3, 4
  y: –4, 3, 8, 11, 12, 11, 8, 3, –4
b 9.75
c ± 3.5
d The line passes through 6 on each axis
e $x = -2$ or 3

a

x	–5	–4	–3	–2	–1	0	1	2
x^2	25	16	9	4	1	0	1	4
$+4x$	–20	–16	–12	–8	–4	0	4	8
y	5	0	–3	–4	–3	0	5	12

b $x = -4$ and 0   **c** –3.8
d 0.6, –4.6   **e** $x = -1$ or –3
f $x = 0$ or –3   **g** $x = -4$ or 1
a t: –1, 0, 1, 2, 3, 4, 5, 6, 7
  s: 10, 3, –2, –5, –6, –5, –2, 3, 10
b $t = 0.6$ or 5.4
c –5.8
d –0.3 and 6.3
a y values: –6, 0, 4, 6, 6, 4, 0, –6
b Student's graph
c (2.5, 6.25)
d $x = 2.5$
e $x = 4.6$ and 0.4

3 Other graphs

ercise 16C

Student's own graph.
a y values: –7.81, –4, –1.69, –0.5, –0.06, 0, 0.06, 0.5, 1.69, 4, 7.81
b 2.3

3 a y values: –12.63, –5, –0.38, 2, 2.89, 3, 3.13, 4, 6.38, 11, 18.63
   b –1.4
4 a

x	1	2	3	4	5	6
y	12	8	6.67	6	5.6	5.33

   b

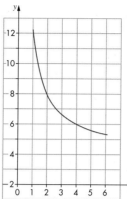

5 a y values: 1, 4.63, 6, 5.88, 5, 4.13, 4, 5.38, 9
   b $x = -1.8$
6 a y values: 20, 5, 2.22, 1.25, 0.8
   b student's graph
   c student's graph, a reflection of the previous one in the y-axis.
7 a y values: 4.25, 1.5, –0.22, –1.46, –2.44
   b student's graph
   c About 5.85

16.4   Estimating gradients

Exercise 16D

Gradients found in this exercise may vary from the answers given due to variations in drawings of the tangents
1   0.67
2   A: 0.5   B: –2
3 a student's drawing   **b** student's drawing
   c about 2   **d** (1, 1.5)
4 a y values: 0, 0.01, 0.1, 0.34, 0.8, 1.56, 2.7
   b student's drawing
   c student's drawing
   d about 1.2
5 a y values: 2.5, 1.67, 1.25, 1, 0.83
   b student's drawing
   c about 0.3

16.5   Graphs of sin x, cos x and tan x

Exercise 16E

1   360°, 540°, –180°, etc.   **2**   90°, 270°, 450°, etc.
3 a 1   **b** –1   **4 a** 1   **b** –1
5   90°, 450°, –270°, etc.   **6**   270°, -90°, –450°, etc.
7   0°, 360°, -360°, etc.   **8**   180°, –180°, 540°, etc.
9   180°   **10** 0°, 180°, 360°, –180°, etc.
11 $x = 270°$, $x = -270°$, etc

12 $x = 0°$, $x = 180°$, $x = -180°$, etc

13 (180, 0), (–180, 0), etc

14 The graph of $y = \cos x$ has rotational symmetry about (90, 0), (270, 0), etc

15 150°, 390°, etc

16 –60°, 300°, etc

16.6   Transformations of graphs

Exercise 16F

1 b Translation by $\begin{pmatrix} 0 \\ 3 \end{pmatrix}$ **c** Translation by $\begin{pmatrix} 0 \\ -1 \end{pmatrix}$

   d Translation by $\begin{pmatrix} -3 \\ 0 \end{pmatrix}$

2 b Translation by $\begin{pmatrix} -90 \\ 0 \end{pmatrix}$ **c** Translation by $\begin{pmatrix} 45 \\ 0 \end{pmatrix}$

   d Translation by $\begin{pmatrix} 0 \\ 2 \end{pmatrix}$

3 a

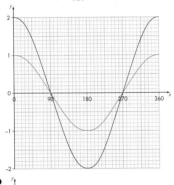

   b

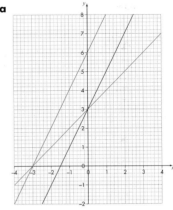

   c

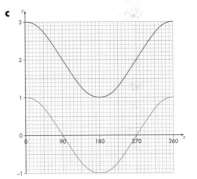

4 a

   b A stretch, factor $\frac{1}{2}$, from the y-axis

   c A stretch, factor 2, from the x-axis

5 a $y = 3 + \cos x$ or $y = \cos x + 3$

   b $y = \cos (x + 30)$

6 $y = x^2$ is B; $y = (2x)^2$ is A; $y = (\frac{1}{2}x)^2$ is C

7 The graph $y = \cos(x - 90)$ is a translation of $y = \cos x$ by $\begin{pmatrix} 9 \\ \end{pmatrix}$ and this is the same as $y = \sin x$

8 a $y = (x - 2)^2$ **b** $y = x^2 + 2$

Answers to Chapter 17

17.1   Number sequences

Exercise 17A

1 a 9, 11, 13: add 2
   b 10, 12, 14: add 2
   c 80, 160, 320: double
   d 81, 243, 729: multiply by 3
   e 28, 34, 40: add 6
   f 23, 28, 33: add 5
   g 20 000, 200 000, 2 000 000: multiply by 10
   h 19, 22, 25: add 3
   i 46, 55, 64: add 9
   j 405, 1215, 3645: multiply by 3
   k 18, 22, 26: add 4
   l 625, 3125, 15 625: multiply by 5

2 a 16, 22   **b** 26, 37
   c 31, 43   **d** 46, 64
   e 121, 169   **f** 782, 3907
   g 22 223, 222 223   **h** 11, 13
   i 33, 65   **j** 78, 108

3 a 48, 96, 192   **b** 33, 39, 45
   c 4, 2, 1   **d** 38, 35, 32
   e 37, 50, 65   **f** 26, 33, 41
   g 14, 16, 17   **h** 19, 22, 25
   i 28, 36, 45   **j** 5, 6, 7
   k 0.16, 0.032, 0.0064
   l 0.0625, 0.031 25, 0.015 625

4 a 21, 34: add previous 2 terms
   b 49, 64: next square number
   c 47, 76: add previous 2 terms
   d 216, 343: cube numbers

5 15, 21, 28, 36

61, 91, 127

29 and 41

No, they both increase by the same number (3).

7.2 The nth term of a sequence

Exercise 17B

a 3, 5, 7, 9, 11   **b** 1, 4, 7, 10, 13
c 7, 12, 17, 22, 27

a 4, 5, 6, 7, 8   **b** 2, 5, 8, 11, 14
c 3, 8, 13, 18, 23   **d** 9, 13, 17, 21, 25

a $305   **b** $600   **c** 3   **d** 5

$4n - 2 = 3n + 7$ rearranges as $4n - 3n = 7 + 2$, $n = 9$

a $n + 1$ **b** $n + 3$ **c** $n + 10$ **d** $n - 1$ **e** $2n$ **f** $3n$

a $2n$ **b** $2n - 1$ **c** $2n + 2$ **d** $2n + 1$ **e** $2n + 3$

a $4n$ **b** $4n + 1$ **c** $4n + 3$ **d** $4n - 1$ **e** $4n + 2$

a $5n$ **b** $6n$ **c** $8n$

7.3 Finding the nth term of an arithmetic sequence

Exercise 17C

a 13, 15, $2n + 1$ **b** 25, 29, $4n + 1$
c 33, 38, $5n + 3$ **d** 32, 38, $6n - 4$

e 20, 23, $3n + 2$ **f** 37, 44, $7n - 5$
g 21, 25, $4n - 3$ **h** 23, 27, $4n - 1$
i 17, 20, $3n - 1$ **j** 42, 52, $10n - 8$
k 24, 28, $4n + 4$ **l** 29, 34, $5n - 1$

2 a $3n + 1$, 151   **b** $2n + 5$, 105
c $5n - 2$, 248   **d** $4n - 3$, 197
e $8n - 6$, 394   **f** $n + 4$, 54
g $5n + 1$, 251   **h** $8n - 5$, 395
i $3n - 2$, 148   **j** $3n + 18$, 168
k $7n + 5$, 355   **l** $8n - 7$, 393

3 a i $4n + 1$ **ii** 401   **b i** $2n + 1$ **ii** 201
c i $3n + 1$ **ii** 301   **d i** $2n + 6$ **ii** 206
e i $4n + 5$ **ii** 405   **f i** $5n + 1$ **ii** 501
g i $3n - 3$ **ii** 297   **h i** $6n - 4$ **ii** 596
i i $8n - 1$ **ii** 799   **j i** $2n + 23$ **ii** 223

17.4 The sum of an arithmetic sequence

Exercise 17D

1 a $a = 10$ $d = 6$ **b** 43   **c** 318
2 a 34   **b** 120   **c** 1100
3 a 57   **b** 300   **c** 675
4 a 420   **b** 2550
5 a i 64   **ii** 144   **iii** 225
  b the sum of the first n odd numbers is n²
6 $1710

Answers to Chapter 18

3.1 Using indices

Exercise 18A

a 2^4 **b** 3^5 **c** 7^2 **d** 5^3 **e** 10^7
f 6^4 **g** 4^1 **h** 1^7 **i** 0.5^4 **j** 100^3

a $3 \times 3 \times 3 \times 3$
b $9 \times 9 \times 9$
c 6×6
d $10 \times 10 \times 10 \times 10 \times 10$
e $2 \times 2 \times 2 \times 2 \times 2 \times 2 \times 2 \times 2 \times 2 \times 2$
f 8
g $0.1 \times 0.1 \times 0.1$
h 2.5×2.5
i $0.7 \times 0.7 \times 0.7$
j 1000×1000

a 16   **b** 243
c 49   **d** 125
e 10 000 000   **f** 1296
g 4   **h** 1
i 0.0625   **j** 1 000 000

a 81   **b** 729
c 36   **d** 10 000
e 1024   **f** 8
g 0.001   **h** 6.25
i 0.343   **j** 1 000 000

5 a 125 m³   **b** 10^2 **c** 2^3 **d** 5^2
6 a 1   **b** 4   **c** 1   **d** 1   **e** 1
7 Any power of 1 is equal to 1.
8 10^6

18.2 Negative indices

Exercise 18B

1 a $\frac{1}{5^3}$ **b** $\frac{1}{6}$ **c** $\frac{1}{10^5}$ **d** $\frac{1}{3^2}$

e $\frac{1}{8^2}$ **f** $\frac{1}{9}$ **g** $\frac{1}{w^2}$ **h** $\frac{1}{t}$

i $\frac{1}{x^m}$ **j** $\frac{4}{m^3}$

2 a 3^{-2} **b** 5^{-1} **c** 10^{-3} **d** m^{-1} **e** t^{-n}

3 a i 2^4 **ii** 2^{-1} **iii** 2^{-4} **iv** -2^3
  b i 10^3 **ii** 10^{-1} **iii** 10^{-2} **iv** 10^6
  c i 5^3 **ii** 5^{-1} **iii** 5^{-2} **iv** 5^0
  d i 3^2 **ii** 3^{-3} **iii** 3^0 **iv** -3^5

4 a $\frac{5}{x^3}$ **b** $\frac{6}{t}$ **c** $\frac{7}{m^2}$ **d** $\frac{4}{q^4}$

e $\frac{10}{y^5}$ **f** $\frac{1}{2x^3}$ **g** $\frac{1}{2m}$ **h** $\frac{3}{4r^4}$

i $\frac{4}{5y^3}$ **j** $\frac{7}{8x^5}$

5 a $7x^{-3}$ **b** $10p^{-1}$ **c** $5t^{-2}$
  d $8m^{-5}$ **e** 2^{-6} or 4^{-3} or 8^{-2}

6 a i 25   **ii** $\frac{1}{125}$ **iii** $\frac{4}{5}$

  b i 64   **ii** $\frac{1}{16}$ **iii** $\frac{5}{256}$

  c i 8   **ii** $\frac{1}{32}$ **iii** $\frac{9}{2}$ or $4\frac{1}{2}$

  d i 1 000 000 **ii** $\frac{1}{1000}$ **iii** $\frac{1}{4}$

7 a a^{-7} **b** a^2 **c** a^4

  d a^{-5} **e** a^{-6} **f** a^6

18.3 Multiplying and dividing with indices

Exercise 18C

1 a 5^4 **b** 5^3 **c** 5^{-2} **d** 5^{-4} **e** 5

2 a 6^3 **b** 6^3 **c** 6^{-2} **d** 6^5 **e** 6^2

3 a a^3 **b** a^5 **c** a^{-4} **d** a^4 **e** b^{-3} **f** b^3

4 Many possible answers

5 a 4^6 **b** 4^{15} **c** 4^{-2}
  d 4^4 **e** 4^3 **f** 4^{-9}

6 a $\frac{1}{a}$ **b** $\frac{1}{a^2}$ **c** a^3 **d** $\frac{1}{a^3}$ **e** a

7 −3

8 −2

9 a a^6 **b** t^{-3} **c** k^2 **d** x^3

10 a $3a$ **b** $4a^3$ **c** $3a^4$
  d $\frac{6}{a}$ **e** $4a^3$ **f** $\frac{5}{a^4}$

11 a Possible answer: $6x^2 \times 2y^5$ and $3xy \times 4xy^4$
  b Possible answer: $24x^2y^7 \div 2y^2$ and $12x^6y^8 \div x^4y^3$

12 12 $a = 2$, $b = 1$, $c = 3$

18.4 Fractional indices

Exercise 18D

1 a 5   **b** 10   **c** 8   **d** 9   **e** 25
  f 3   **g** 4   **h** 10   **i** 5   **j** 8
  k 12   **l** 20   **m** 5   **n** 3   **o** 10
  p 3   **q** 2   **r** 2   **s** 6   **t** 6
  u $\frac{1}{4}$ **v** $\frac{1}{2}$ **w** $\frac{1}{3}$ **x** $\frac{1}{5}$ **y** $\frac{1}{10}$

2 a $\frac{5}{6}$ **b** $1\frac{2}{3}$ **c** $\frac{8}{9}$ **d** $1\frac{4}{5}$ **e** $\frac{5}{8}$
  f $\frac{3}{5}$ **g** $\frac{1}{4}$ **h** $2\frac{1}{2}$ **i** $\frac{4}{5}$ **j** $1\frac{1}{7}$

3 $(x^{\frac{1}{n}})^n = x^{\frac{1}{n} \times n} = x^1 = x$, but $(\sqrt[n]{x})^n = \sqrt[n]{x} \times \sqrt[n]{x} \ldots n$ times $= x$,
  so $x^{\frac{1}{n}} = \sqrt[n]{x}$

4 $64^{-\frac{1}{2}} = \frac{1}{8}$, others are both $\frac{1}{2}$

5 Possible answer: The negative power gives the reciprocal, s•
$27^{-\frac{1}{3}} = \frac{1}{27^{\frac{1}{3}}}$. The power one-third means cube root, so you
need the cube root of 27 which is 3, so $27^{\frac{1}{3}} = 3$ and $\frac{1}{27^{\frac{1}{3}}} =$

6 Possible answer: $x = 1$ and $y = 1$, $x = 8$ and $y = \frac{1}{64}$.

7 a 3   **b** $\frac{1}{3}$ **c** 0   **d** $\frac{1}{2}$
  e $\frac{1}{2}$ **f** $\frac{1}{4}$ **g** $\frac{1}{4}$ **h** $\frac{1}{3}$
  i $\frac{1}{3}$ **j** $\frac{1}{2}$ **k** $\frac{1}{3}$ **l** $\frac{1}{7}$

Exercise 18E

1 a 16   **b** 25   **c** 216   **d** 81

2 a $t^{\frac{2}{3}}$ **b** $m^{\frac{3}{4}}$ **c** $k^{\frac{2}{5}}$ **d** $x^{\frac{3}{2}}$

3 a 4   **b** 9   **c** 64   **d** 3125

4 a $\frac{1}{5}$ **b** $\frac{1}{6}$ **c** $\frac{1}{2}$ **d** $\frac{1}{3}$
  e $\frac{1}{4}$ **f** $\frac{1}{2}$ **g** $\frac{1}{2}$ **h** $\frac{1}{3}$

5 a $\frac{1}{125}$ **b** $\frac{1}{216}$ **c** $\frac{1}{8}$ **d** $\frac{1}{27}$
  e $\frac{1}{256}$ **f** $\frac{1}{4}$ **g** $\frac{1}{4}$ **h** $\frac{1}{9}$

6 a $\frac{1}{100\,000}$ **b** $\frac{1}{12}$ **c** $\frac{1}{25}$ **d** $\frac{1}{27}$
  e $\frac{1}{32}$ **f** $\frac{1}{32}$ **g** $\frac{1}{81}$ **h** $\frac{1}{13}$

7 $8^{-\frac{2}{3}} = \frac{1}{4}$, others are both $\frac{1}{8}$

8 Possible answer: The negative power gives the reciprocal, s•
the power one-third means cube root, so we need the cube
root of 27 which is 3 and the power 2 means square, so
$3^2 = 9$, so $27^{\frac{2}{3}} = 9$ and $\frac{1}{27^{\frac{2}{3}}} = \frac{1}{9}$

9 a $\frac{27}{8}$ **b** $\frac{9}{25}$ **c** $\frac{1024}{243}$ **d** $\frac{8}{343}$
  e $\frac{16}{9}$ **f** $\frac{8}{27}$ **g** $\frac{625}{256}$ **h** $\frac{32}{243}$

10 a $\frac{25}{9}$ **b** $\frac{27}{64}$ **c** $\frac{125}{729}$ **d** $\frac{243}{32}$
  e $\frac{8}{27}$ **f** $\frac{243}{32}$ **g** $\frac{9}{4}$ **h** $\frac{125}{343}$
  i $\frac{16}{25}$ **j** $\frac{512}{125}$ **k** $\frac{243}{32}$ **l** $\frac{32}{243}$

11 a x^4 **b** x^{-1} **c** $4y^2$
  d $10x^2$ **e** $20x^{-1}$ **f** $\frac{1}{3}y$

12 a x **b** d^{-1} **c** $t^{\frac{3}{2}}$
  d x^2 **e** $y^{\frac{1}{2}}$ **f** a^4

13 a $x^{\frac{1}{2}}$ **b** y^{-1} **c** $a^{\frac{5}{3}}$
  d t^{-2} **e** d^2 **f** 1

Answers to Chapter 19

19.1 Direct proportion

Exercise 19A

a 15   **b** 2

a 75   **b** 6

a 150   **b** 6

a 22.5   **b** 12

a 175 kilometres   **b** 8 hours

a 66.50 dollars   **b** 175 kg

a 44   **b** 84 m^2

a 33 spaces

b 66 spaces since new car park has 366 spaces

17 minutes 30 seconds

Exercise 19B

a 100   **b** 10

a 27   **b** 5

a 56   **b** 1.69

a 192   **b** 2.25

a 25.6   **b** 5

a $50   **b** 225

a 3.2°C   **b** 10 atm

8 a 388.8 g   **b** 3 mm

9 a 2 J   **b** 40 m/s

10 a 78 dollars   **b** 400 miles

11 4000 cm^3

12 $250

13 a B   **b** A   **c** C

14 a B   **b** A

19.2 Inverse proportion

Exercise 19C

1 $Tm = 12$ **a** 3   **b** 2.5

2 $Wx = 60$ **a** 20   **b** 6

3 $Q(5 - t) = 16$ **a** −3.2   **b** 4

4 $Mt^2 = 36$ **a** 4   **b** 5

5 $x^2y = 16$ **a** 16   **b** 8

6 $VR^3 = 2700$, $V = 2.7$ and $R = 15$

7 $gp = 1800$ **a** $15   **b** 36

8 $td = 24$ **a** 3°C   **b** 12 km

9 $ds^2 = 432$ **a** 1.92 km   **b** 8 m/s

10 B – This is inverse proportion, as x increases y decreases.

Answers to Chapter 20

20.1 Linear inequalities

Exercise 20A

a $x < 3$ **b** $t > 8$ **c** $p \geqslant 10$ **d** $x < 5$

e $y \leqslant 3$ **f** $t > 5$ **g** $x < 6$ **h** $y \leqslant 15$

i $t \geqslant 18$ **j** $x < 7$ **k** $x \leqslant 3$ **l** $t \geqslant 5$

a 8   **b** 6   **c** 16   **d** 3   **e** 7

a 11   **b** 16   **c** 16

$2x + 3 < 20$, $x < 8.50$, so the most each could cost is $8.49

a Because $3 + 4 = 7$, which is less than the third side of length 8

b $x + x + 2 > 10$, $2x + 2 > 10$, $2x > 8$, $x > 4$, so smallest value of x is 5

a $x = 6$ and $x < 3$ scores −1 (nothing in common), $x < 3$ and $x > 0$ scores 1 (1 in common for example), $x > 0$ and $x = 2$ scores 1 (2 in common), $x = 2$ and $x \geqslant 4$ scores −1 (nothing in common), so we get $-1 + 1 + 1 - 1 = 0$

b $x > 0$ and $x = 6$ scores +1 (6 in common), $x = 6$ and $x \geqslant 4$ scores +1 (6 in common), $x \geqslant 4$ and $x = 2$ scores −1 (nothing in common), $x = 2$ and $x < 3$ scores +1 (2 in common), $+1 + 1 - 1 + 1 = 2$

c Any acceptable combination, e.g. $x = 2$, $x < 3$, $x > 0$, $x \geqslant 4$, $x = 6$

a $x \geqslant -6$ **b** $t \leqslant \frac{8}{3}$

c $y \leqslant 4$ **d** $x \geqslant -2$

e $w \leqslant 5.5$ **f** $x \leqslant \frac{14}{5}$

8 a $x \leqslant 2$ **b** $x > 38$

   c $x < 6\frac{1}{2}$ **d** $x \geqslant 7$

   e $t > 15$ **f** $y \leqslant \frac{7}{5}$

9 a 4   **b** 99   **c** 11   **d** 11   **e** 6

10 a 0, 10, −10   **b** $x < 16$

11 a $x < 9$ **b** $x \geqslant 11$ **c** $x \geqslant 3$

12 a $x \geqslant 7.5$ **b** $x \leqslant -2$

   c $x < 6$ **d** $x > 1.5$

   e $x \geqslant -5$ **f** $x < 0.5$

13 a $4 < x \leqslant 8$ **b** $15 < x < 23$ **c** $2 \leqslant x \leqslant 21$

   d $-6 < x < -2$ **e** $0.5 \leqslant x \leqslant 1.75$ **f** $3 < x < 13$

14 11, 12, 13, 14

15 a $3n + 3$ **b** $90 < 3n + 3 < 100$

   c 30, 31, 32

Exercise 20B

1 a $x > 1$ **b** $x \leqslant 3$ **c** $x < 2$ **d** $x \geqslant -1$ **e** $x \leqslant -1$ **f** $x \geqslant 1$

2 a

b

c

d

e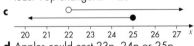

f number line

g number line

h number line

3 a $x \geqslant 4$

number line

b $x < -2$

number line

c $x \geqslant 3.5$

number line

d $x < -1$

number line

e $x < 1.5$

number line

f $x \leqslant -2$

number line

g $x > 50$

number line

h $x \geqslant -6$

number line

4 a Because 3 apples plus the chocolate bar cost more that £1.20: $x > 22$

b Because 2 apples plus the chocolate bar left Max with at least 16p change: $x \leqslant 25$

c number line

d Apples could cost 23p, 24p or 25p.

5 Any two inequalities that overlap only on the integers –1, 0, 1 and 2 – for example, $x \geqslant -1$ and $x < 3$

6 4

7 a $x > 2$

number line

b $x \geqslant 6$

number line

c $x \leqslant -1$

number line

d $x \geqslant -4$

number line

8 a $2 < x < 8$

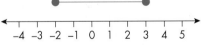

number line

b $-2 \leqslant x \leqslant 3$

number line

20.2 Quadratic inequalities

Exercise 20C

1 a $-4 \leqslant x \leqslant 4$

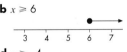

b $-2 < x < 2$

number line

c $x < -2.5$ or $x > 2.5$

number line

d $x \leqslant -1$ or $x \geqslant 1$

number line

2 a $-3 < x < 3$

b $x < -5$ or $x > 5$

c $x \leqslant -1.5$ or $x \geqslant 1.5$

d $-0.5 < x < 0.5$

3 a $-2 < x < 2$

b $x < -3.5$ or $x > 3.5$

c $-2.5 < x < 2.5$

d $-3 \leqslant x \leqslant 3$

4 $x^2 > 100$

5

20.3 Graphical inequalities

Exercise 20D

1 a & b

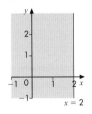

2 a & b

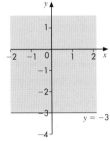

3 a–c

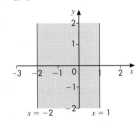

4 a–c

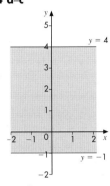

5

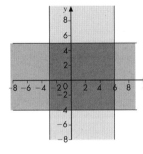

6 a & b

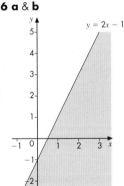

b i Yes **ii** Yes **iii** No

7 a & b

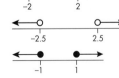

8 a & b

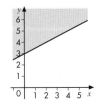

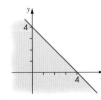

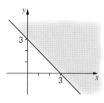

0.4 More than one inequality

xercise 20E

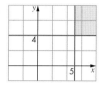

2

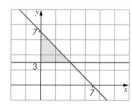

4

6

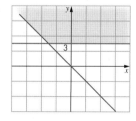

$x \geqslant -2$ and $y \geqslant 3$. The $>$ sign can be used instead.

$x + y \geqslant 5$ and $x \geqslant 2$ and $y \geqslant 1$. The $>$ sign can be used instead.

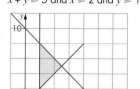

10

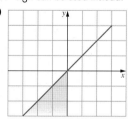

20.5 More complex inequalities

Exercise 20F

1 a–e

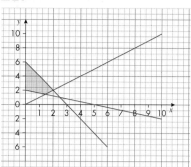

f i No   **ii** Yes   **iii** Yes

2 a

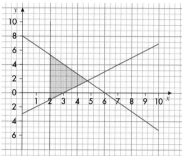

b i No   **ii** Yes   **iii** Yes   **iv** No

3 a & b

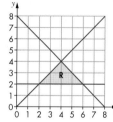

4 a & b

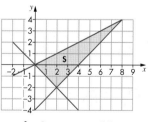

c 4   **d** –2   **e** 12

5 Test a point such as the origin (0, 0), so $0 < 0 + 2$, which is
true. So the side that includes the origin is the required side.

6   a $x + y \geqslant 3$, $y \leqslant \frac{1}{2}x + 3$ and $y \geqslant 5x - 15$ ($>$ and $<$ could be
used instead)

b 9   **c** 3 at (3,0)

Answers to Chapter 21

21.1 Function notation

Exercise 21A

1 **a** 12   **b** 26   **c** 7
   d –2   **e** 3

2 **a** 0.5   **b** 5   **c** 50.5
   d 2.5   **e** 0.625 or $\frac{5}{8}$

3 **a** 5   **b** –3   **c** 999801
   d 1   **e** $\frac{1}{8}$

4 **a** 4   **b** 32   **c** 1
   d $\frac{1}{2}$ **e** $\frac{1}{8}$

5 **a** 3   **b** 2   **c** 0
   d –1   **e** 5

6 **a** 7.5   **b** –2.5   **c** –5

7 **a** 6   **b** 97   **c** 3.25

8 **a** 6   **b** at (6, 4)

9 **a** 3   **b** –3   **c** $x = 3$
   d $x = \frac{1}{2}$

21.2 Domain and range

Exercise 21B

1 **a** $x < 0$ **b** –1   **c** $x < -1$ **d** $-\frac{1}{2}$

2 **a** {10, 17, 26}   **b** {1, 2, 5}   **c** $\{y : 2 \leqslant y \leqslant 5\}$
   d $\{y : y \geqslant 101\}$ **e** Same as d

3 **a** {0, 1, 4}   **b** $\{1, \frac{1}{2}, \frac{1}{3}, \frac{1}{4}\}$ **c** {5, 7, 9, 11}
   d {5, 4, 3, 2}   **e** {0, –2}

4 –2 can be squared so it could be in the domain.
$x^2 = -2$ has no solution so –2 cannot be in the range.

5 **a** Yes   **b** No   **c** Yes

21.3 Inverse functions

Exercise 21C

1 **a** $x - 7$ **b** $\frac{x}{8}$ **c** $5x$ **d** $x + 3$

2 **a** 8   **b** 4   **c** 5   **d** –2

3 **a** $3(x + 2)$ **b** $\frac{x}{4} + 5$ **c** $5x - 4$ **d** $\frac{2x + 6}{3}$

   e $2(\frac{x}{3} - 4)$ **f** $\sqrt[3]{\frac{x}{4}}$

4 **a** 2   **b** $\frac{1}{2}$ **c** –2.5

5 **a** $10 - x$ **b** They are identical

6 **a** $\frac{8}{x}$ **b** $\frac{20}{x + 1}$ **c** $\frac{2}{x} - 1$

7 **a** $\frac{x + 4}{2}$ **b** Student's graph   **c** (4,4)

8 5

9 $\frac{x + 2}{3}$

21.4 Composite functions

Exercise 21D

1 **a** 6 and 3   **b** 7 and 3.5
   c 10 and 5   **d** $\frac{x + 4}{2}$
   e 1 and 5   **f** 1.5 and 5.5
   g –5 and –1   **h** $\frac{x}{2} + 4$

2 **a** 6 and 216   **b** 10 and 1000   **c** $(2x)^3$ or $8x^3$
   d 64 and 128   **e** $2x^3$

3 **a** 1, 9, 25   **b** 1, 3, 5   **c** $\sqrt{2x + 1}$

4 **a** 6 and 18   **b** 12 and 36   **c** $9x$

5 **a** $3(x - 6)$ **b** $3x - 6$

6 Both are $x - 3$

21.5 More about composite functions

Exercise 21E

1 **a** 3.5   **b** 1   **c** 8   **d** 5.5

2 **a** 20   **b** 9   **c** 8.75   **d** 3

3 **a** 7   **b** 8   **c** 256   **d** 21

4 **a** $6x$ **b** $6x - 5$

5 **a** $9x^2 + 24x + 16$ or $(3x + 4)^2$ **b** $6x - 5$
   c $2x + 3$ **d** $4 - 2x$

6 **a** $x - 10$ **b** $x + 10$ **c** x

7 **a** x^4 **b** $(\frac{12}{x})^2$ or $\frac{144}{x^2}$
   c $\frac{12}{x^2}$ **d** x

8 **a** 80
   b $2(2x - 1) - 1$ simplified
   c $(2x - 1)^2 + 2(2x - 1)$ simplified

9 **a** $\frac{1}{3x - 3}$ **b** $\frac{x - 1}{x - 4}$

10 **a** $6x - 14$ **b** $\frac{x + 12}{4}$

11 **a** 0.5 (1 + 9) = 5   **b** 7   **c** 8
   d 8.5, 8.75, 8.875, 8.9375, 8.96875, 8.984375
   e Getting closer and closer to 9, halving the difference fro
     9 each time.

12 Student's own description of the convergence towards 9.

13 **a** $f^{-1}(x) = \frac{x - 3}{2}$
   b 100
   c $ff^{-1}(x) = x$
   d $ff^{-1}(x) = x$
   e 100

Answers to Chapter 22

2.1 The gradient of a curve

xercise 22A

a The missing numbers are 0, –1, 0

b

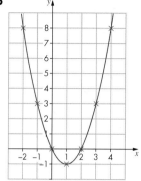

c $2x - 2$ **d** 4

e 6   **f** Student's choice

g $(1, -1)$ **h** Student's check

a $2x - 6$ **b** –6   **c** 4   **d** $(4, 7)$

a $4x$ **b** 8   **c** –4   **d** $(3, 8)$

a $4 - 2x$ **b** 4 and –4   **c** $(1, 3)$ **d** $(1.5, 3.75)$

a $2x + 1$ **b** $2x - 7$ **c** $8x - 1$ **d** $0.6x - 1.5$

e $-2 + 2x$ **f** $3 - 2x$ **g** 2   **h** 0

$2x + 2$

a $4x + 2$ **b** $2x + 7$ **c** $2x$

a $(0, -5)$ **b** 2

.2 More complex curves

ercise 22B

a $6x^2$ **b** 6 and 24

a $3x^2 - 12x + 8$

b If $x = 0$ or 2 or 4, $y = 0$

c 8; –4; 8

a $\frac{-12}{x^2}$ or $12x^{-2}$

b –3 and –0.75

c When $x = \sqrt{12}$. At $(3.46, 3.46)$

a $8x^3$ **b** $1 - \frac{3}{x^2}$

c $15x^2 - 2$ **d** $-\frac{8}{x^3}$

e $9x^2 + 5$ **f** $-3x^2$

g $4x^3 - 1$ **h** $1 - \frac{1}{x^2}$

16 at $(2, 0)$; –16 at $(-2, 0)$; 0 at $(0, 0)$

a $(2, 5)$ and $(-2, 5)$

b –5 at $(2, 5)$; 5 at $(-2, 5)$

$x^2 - 5 = 4$ has two solutions, $x = 3$ or –3.
Points are $(3, -2)$ and $(-3, 10)$

22.3 Turning points

Exercise 22C

1   **a** $2x - 4$ **b** $2x - 4 = 0 \Rightarrow x = 2$; $(2, -1)$
    c Minimum

2   **a** $(-3, -12)$ **b** Minimum

3   **a** $5 - 2x$ **b** $(2.5, 7.25)$ **c** Maximum

4   **a** $1 - \frac{1}{x^2}$ **b** $(1, 2)$

5   **a** $3x^2 - 6x$ **b** $x = 0$ or 2   **c** $(0, 0)$ and $(2, -4)$

6   **a** If $x = -2$ or 5, $y = 0$ **b** $2x - 3$
    c $(1.5, -12.25)$; Minimum   **d** $x = 1.5$

7   **a** $3 - \frac{24}{x^3}$ **b** $(2, 9)$ **c** Minimum

8   **a** $6x^2 - 6$ **b** $(1, 0)$ minimum, $(-1, 8)$ maximum

9   **a** The two sides add up to half the perimeter
    b $15 - 2x$
    c $(7.5, 56.25)$
    d Maximum
    e The largest possible area is 56.25 cm^2, when the rectangle is a square of side 7.5 cm

22.4 Motion of a particle

Exercise 22D

1   **a** $2t - 3$ **b** $s = 3, v = -1$
    c $s = 5, v = 3$ **d** $s = 9, v = 5$
    e 2 m/s^2 **f** When $t = 1.5$

2   **a** $3t^2 - 2t$ **b** $6t - 2$
    c $s = 0, v = 0, a = -2$ **d** $s = 0, v = 1, a = 4$
    e $s = 4, v = 8, a = 10$

3   **a** $v = 2t - 4$; $a = 2$
    b 18 m and 8 m/s
    c When $t = 2$

4   **a** $v = 24t^2 - 24t$; $a = 48t - 24$
    b 72 m/s
    c When $t = 0.5$

5   **a** $v = 5 - \frac{8}{t^2}$ **b** $\frac{16}{t^3}$
    c 14 m; 3 m/s; 2m/s^2

6   **a** $v = 48 - 12t^2$; $a = -24t$ **b** When $t = 2.5$
    c When $t = 2$ **d** 64 metres

7   **a** $v = 60 - 24t$
    b $t = 2.5$
    c 75 metres

8   **a** $v = 12t - 3t^2$; $a = 12 - 6t$ **b** When $t = 2$
    c When $t = 0$ or 4   **d** 32 metres

9   **a** $t = 10$ **b** $v = 4t^3 - 30t^2$; $a = 12t^2 - 60t$
    c When $t = 0$ or 7.5   **d** When $t = 0$ or 5

23.1 Angle facts

Exercise 23A

1 **a** 48° **b** 307° **c** 108° **d** 52°
   e 59° **f** 81° **g** 139° **h** 58°

2 **a** 82° **b** 105° **c** 75°

3 45° + 125° = 170° and for a straight line it should be 180°.

4 **a** $x = 100°$ **b** $x = 110°$ **c** $x = 30°$

5 **a** $x = 55°$ **b** $x = 45°$ **c** $x = 12.5°$

6 **a** $x = 34°, y = 98°$ **b** $x = 70°, y = 120°$ **c** $x = 20°, y = 80°$

7 6 × 60° = 360°; imagine six of the triangles meeting at a point.

8 $x = 35°, y = 75°$; $2x = 70°$ (opposite angles), so $x = 35°$ and $x + y = 110°$ (angles on a line), so $y = 75°$

23.2 Parallel lines

Exercise 23B

1 **a** 40°
   c $d = 75, e = f = 105°$
   e $j = k = l = 70°$
   b $b = c = 70°$
   d $g = 50°, h = i = 130°$
   f $n = m = 80°$

2 **a** $a = 50°, b = 130°$
   c $g = i = 65°, h = 115°$
   e $m = n = o = p = 105°$
   b $c = d = 65°, e = f = 115°$
   d $j = k = 72°, l = 108°$
   f $q = r = s = 125°$

3 **a** $a = 95°$ **b** $b = 66°, c = 114°$

4 **a** $x = 30°, y = 120°$ **b** $x = 25°, y = 105°$
   c $x = 30°, y = 100°$

5 **a** $x = 50°, y = 110°$ **b** $x = 25°, y = 55°$
   c $x = 20, y = 140°$

6 290°; x is double the angle allied to 35°, so is 2 × 145°

7 $a = 66$

8 Angle PQD = 64° (alternate angles), so angle DQY = 116° (angles on a line = 180°)

9 Use alternate angles to see b, a and c are all angles on a straight line, and so total 180°.

10 Third angle in triangle equals q (alternative angle), angle sum of triangle is 180°.

23.3 Angles in a triangle

Exercise 23C

1 **a** 70° **b** 50° **c** 80° **d** 60°
   e 75° **f** 109° **g** 38° **h** 63°

2 **a** No, total is 190° **b** Yes, total is 180° **c** No, total is 170°
   d Yes, total is 180° **e** Yes, total is 180° **f** No, total is 170°

3 **a** 60° **b** Equilateral triangle **c** Same length

4 **a** 70° each **b** Isosceles triangle **c** Same length

5 **a** 109° **b** 130° **c** 135°

6 Isosceles triangle; angle DFE ∠30° (opposite angles), angle DEF ∠75° (angles on a line), angle FDE ∠75° (angles in a triangle), so there are two equal angles in the triangle and hence it is an isosceles triangle

7 a is ∠80° (opposite angles), b is 65° (angles on a line), c is 35° (angles in a triangle)

8 Missing angle = y, $x + y = 180°$ and $a + b + y = 180°$ so x = a + b

23.4 Angles in a quadrilateral

Exercise 23D

1 **a** $a = 110°, b = 55°$ **b** $c = e = 105°, d = 75°$
   c $f = 135°, g = 25°$ **d** $h = i = 94°$
   e $j = l = 105°, k = 75°$ **f** $m = o = 49°, n = 131°$

2 **a** $x = 25°, y = 15°$ **b** $x = 7°, y = 31°$ **c** $x = 60°, y = 30$

3 **a** $x = 50°$: 60°, 70°, 120°, 110° – possibly trapezium
   b $x = 60°$: 50°, 130°, 50°, 130° – parallelogram or isosceles trapezium
   c $x = 30°$: 20°, 60°, 140°, 140° – possibly kite
   d $x = 20°$: 90°, 90°, 90°, 90° – square or rectangle

4 52°

5 Both 129°

6 $y = 360° – 4x$

23.5 Regular polygons

Exercise 23E

1 **a** i 45° ii 8 iii 1080°
   b i 20° ii 18 iii 2880°
   c i 15° ii 24 iii 3960°
   d i 36° ii 10 iii 1440°

2 **a** i 172° ii 45 iii 7740°
   b i 174° ii 60 iii 10 440°
   c i 156° ii 15 iii 2340°
   d i 177° ii 120 iii 21 240°

3 **a** Exterior angle is 7°, which does not divide exactly into 360°
   b Exterior angle is 19°, which does not divide exactly into 360°
   c Exterior angle is 11°, which does not divide exactly into 360°
   d Exterior angle is 70°, which does not divide exactly into 360°

4 **a** 7° does not divide exactly into 360°
   b 26° does not divide exactly into 360°
   c 44° does not divide exactly into 360°
   d 13° does not divide exactly into 360°

5 $x = 45°$, they are the same, true for all regular polygons

6 **a** 36° **b** 10

23.6 Irregular polygons

Exercise 23F

1 **a** 1440° **b** 2340° **c** 17 640° **d** 7740°

2 **a** 9 **b** 15 **c** 102 **d** 50

3 **a** 130° **b** 95° **c** 130°

4 **a** 50° **b** 40° **c** 59°

5 Hexagon

a Octagon   **b** 89°

a i 71°   **ii** 109°   **iii** Equal

b If S = sum of the two opposite interior angles, then $S + I$ = 180° (angles in a triangle), and we know $E + I = 180°$ (angles on a straight line), so $S + I = E + I$, therefore $S = E$

144°; 360 – (2 × 108)°

Three angles are 135° and two angles are 67.5°

3.7 Tangents and chords

Exercise 23G

a 38°   **b** 110°   **c** 15°   **d** 45°

a $x = 12°$, $y = 156°$ **b** $x = 100°$, $y = 50°$

c $x = 62°$, $y = 28°$ **d** $x = 30°$, $y = 60°$

a $\angle OCD = 58°$ (triangle OCD is isosceles), $\angle OCB = 90°$ (tangent/radius theorem), so $\angle DCB = 32°$, hence triangle BCD is isosceles (2 equal angles)

b CD is a chord; the part of the circle between chord AD and arc AD is a segment; the part of the circle between chord AD and the radii OC and OD is a segment.

a 62°   **b** 66°   **c** 19°   **d** 20°

3.8 Angles in a circle

Exercise 23H

a 56°   **b** 62°   **c** 105°   **d** 55°

e 45°   **f** 30°   **g** 60°   **h** 145°

a 55°   **b** 52°   **c** 50°   **d** 24°

e 39°   **f** 80°   **g** 34°   **h** 30°

a 41°   **b** 49°   **c** 41°

a 72°   **b** 37°   **c** 72°

$\angle AZY = 35°$ (angles in a triangle), a = 55° (angle in a semicircle = 90°)

a $x = y = 40°$ **b** $x = 131°$, $y = 111°$

c $x = 134°$, $y = 23°$ **d** $x = 32°$, $y = 19°$

e $x = 59°$, $y = 121°$ **f** $x = 155°$, $y = 12.5°$

68°

$\angle ABC = 180° - x$ (angles on a line), $\angle AOC = 360° - 2x$ (angle at centre is twice angle at circumference), reflex $\angle AOC = 360° - (360° - 2x) = 2x$ (angles at a point)

a x **b** $2x$

c From part b, angle AOD = $2x$

Similarly, angle COD = $2y$

So angle AOC = AOD + COD = $2x + 2y = 2(x + y)$ = 2 × angle ABC

9 Cyclic quadrilaterals

Exercise 23I

a $a = 50°$, $b = 95°$ **b** $c = 92°$, $x = 90°$

c $d = 110°$, $e = 110°$, $f = 70°$ **d** $g = 105°$, $h = 99°$

e $j = 89°$, $k = 89°$, $l = 91°$ **f** $m = 120°$, $n = 40°$

g $p = 44°$, $q = 68°$ **h** $x = 40°$, $y = 34°$

a $x = 26°$, $y = 128°$ **b** $x = 48°$, $y = 78°$

c $x = 133°$, $y = 47°$ **d** $x = 36°$, $y = 72°$

e $x = 55°$, $y = 125°$ **f** $x = 35°$

g $x = 48°$, $y = 45°$ **h** $x = 66°$, $y = 52°$

3 a $x = 49°$, $y = 49°$ **b** $x = 70°$, $y = 20°$

    c $x = 80°$, $y = 100°$ **d** $x = 100°$, $y = 75°$

4 a $x = 50°$, $y = 62°$ **b** $x = 92°$, $y = 88°$

    c $x = 93°$, $y = 42°$ **d** $x = 55°$, $y = 75°$

5 a $x = 95°$, $y = 138°$ **b** $x = 14°$, $y = 62°$

    c $x = 32°$, $y = 48°$ **d** 52°

6 a 71°   **b** 125.5°   **c** 54.5°

7 a $x + 2x - 30° = 180°$ (opposite angles in a cyclic quadrilateral), so $3x - 30° = 180°$

    b $x = 70°$, so $2x - 30° = 110°$ $\angle DOB = 140°$ (angle at centre equals twice angle at circumference), $y = 80°$ (angles in a quadrilateral)

8 a x **b** $360° - 2x$

    c $\angle ADC = \frac{1}{2}$reflex $\angle AOC = 180° - x$, so $\angle ADC + \angle ABC = 180°$

9 Let $\angle AED = x$, then $\angle ABC = x$ (opposite angles are equal in a parallelogram), $\angle ADC = 180° - x$ (opposite angles in a cyclic quadrilateral), so $\angle ADE = x$ (angles on a line)

10 Let $\angle ABC = x$ and $\angle EFG = y$.

Then $\angle ADC = 180 - x°$ (opposite angles in a cyclic quadrilateral) and $\angle EDG = 180 - y°$.

But $\angle ADC = \angle EDG$ (opposite angles).

$180 - x° = 180 - y°$ and therefore $x = y$.

23.10 Alternate segment theorem

Exercise 23J

1 a $a = 65°$, b = 75°, c = 40°

    b $d = 79°$, e = 58°, f = 43°

    c $g = 41°$, h = 76°, i = 76°

    d $k = 80°$, m = 52°, n = 80°

2 a $a = 75°$, b = 75°, c = 75°, d = 30°

    b $a = 47°$, b = 86°, c = 86°, d = 47°

    c $a = 53°$, b = 53°

    d $a = 55°$

3 a 36°   **b** 70°

4 a $x = 25°$ **b** $x = 46°$, $y = 69°$, $z = 65°$

    c $x = 38°$, $y = 70°$, $z = 20°$ **d** $x = 48°$, $y = 42°$

5 $\angle ACB = 64°$ (angle in alternate segment), $\angle ACX = 116°$ (angles on a line), $\angle CAX = 32°$ (angles in a triangle), so triangle ACX is isosceles (two equal angles)

6 $\angle AXY = 69°$ (tangents equal and so triangle AXY is isosceles), $\angle XZY = 69°$ (alternate segment), $\angle XYZ = 55°$ (angles in a triangle)

7 a $2x$ **b** $90° - x$

    c $\angle OPT = 90°$, so $\angle APT = x$

23.11 Intersecting chords

Exercise 23K

1 a 12   **b** 4.8   **c** 11.25   **d** $10\frac{2}{3}$

2 a 22.5 cm   **b** 10.5 cm

3 14.4 cm

4 47.5 cm

5 5.3

6 20.25 cm

7 12

8 7.1

Answers to Chapter 24

24.1 Measuring and drawing angles

Exercise 24A

1 a 40° **b** 125° **c** 340° **d** 225°

2 Student's drawings of angles.

3 AC and BE; AD and CE; AE and CF.

4 Yes, the angle is 75°.

5 Any angle between 90° and 100°.

6 a 80° **b** 50° **c** 25°

24.2 Bearings

Exercise 24B

1 a 110° **b** 250° **c** 091° **d** 270° **e** 130° **f** 180°

2 Student's sketches.

3
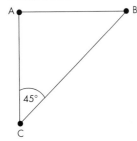

4 a 090°, 180°, 270° **b** 000°, 270°, 180°

5

Leg	Actual distance	Bearing
1	50 km	060°
2	70 km	355°
3	65 km	260°
4	46 km	204°
5	60 km	130°

6 a 045° **b** 286°

7 a 250° **b** 325° **c** 144°

8 a 900 m **b** 280°
  c ∠NHS = 150° and HS = 3 cm

9 108°

10 255°

24.3 Congruent shapes

Exercise 24C

1 a yes **b** yes **c** no **d** yes **e** no **f** yes

2 a triangle ii **b** triangle iii **c** sector i

3

PQR to QRS to RSP to SPQ;
SXP to PXQ to QXR to RXS

4

EGF to FHE to GEH to HFG;
EFX to HGX; EXH to FXG

5
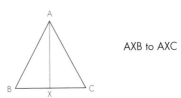

ABC to CDA; BDC to DB
BXA to DXC; BXC to DX

6

AXB to AXC

24.4 Similar shapes

Exercise 24D

1 a 2 **b** 3

2 a Yes, 4
  b No, corresponding sides have different ratios.

3 a PQR is an enlargement of ABC
  b 1 : 3 **c** Angle R **d** BA

4 a Sides in same ratio **b** Angle P **c** PR

5 a Same angles **b** Angle Q **c** AR

6 a 8 cm **b** $x = 45$cm, $y = 9$cm
  c $x = 19.5$ cm, $y = 10.8$ cm
  d 4.2 cm

7 a The angles are all 90 degrees. The sides of a square are all equal so the ratio between sides of two different squares will be the same, whatever two sides are chose
  b No. They will only be similar if they have the same ratio length to width.

8 5.2 m

24.5 Areas of similar triangles

Exercise 24E

1 a 2.5 **b** 125 cm²

2 a All equilateral triangles are similar **b** 7.81 cm

3 40.32 cm²

4 75 cm²

5 a 144 cm2 **b** 69.4 cm²

6 a All angles are the same and corresponding sides are in same ratio
  b 247.68 cm²

7 a 2 **b** 10 cm **c** 7.1 cm

8 354.9 cm²

9 It will double the area

10 28.3 cm²

4.6 Areas and volumes of similar shapes

a 4 : 25   **b** 8 : 125
a 16 : 49   **b** 64 : 343

Linear scale factor	Linear ratio	Linear fraction	Area scale factor	Volume scale factor
2	1 : 2	$\frac{2}{1}$	4	8
3	1 : 3	$\frac{3}{1}$	9	27
$\frac{1}{4}$	4 : 1	$\frac{1}{4}$	$\frac{1}{16}$	$\frac{1}{64}$
5	1 : 5	$\frac{5}{1}$	25	125
$\frac{1}{10}$	10 : 1	$\frac{1}{10}$	$\frac{1}{100}$	$\frac{1}{1000}$

135 cm^2
a 56 cm^2 **b** 126 cm^2
a 48 m^2 **b** 3 m^2
a 2400 cm^3 **b** 8100 cm^3
4 litres
1.38 m^3

10 $6
11 4 cm
12 8 × 0.60 = 4.80 which is greater than 4.00 so the large tub is better value
13 a 3 : 4   **b** 9 : 16   **c** 27 : 64
14 720 ÷ 8 = 90 cm^3

Exercise 24G

1 6.2 cm, 10.1 cm
2 4.26 cm, 6.74 cm
3 9.56 cm
4 3.38 m
5 8.39 cm
6 26.5 cm
7 16.9 cm
8 a 4.33 cm, 7.81 cm   **b** 143 g, 839 g
9 53.8 kg
10 1.73 kg
11 8.8 cm
12 7.9 cm and 12.6 cm
13 b

Answers to Chapter 25

5.1 Constructing shapes

a BC = 2.9 cm, ∠B = 53°, ∠C = 92°
b EF = 7.4 cm, ED = 6.8 cm, ∠E = 50°
c ∠G = 105°, ∠H = 29°, ∠I = 46°
d ∠J = 48°, ∠L = 32°, JK = 4.3 cm
e ∠N = 55°, ON = OM = 7 cm
f ∠P = 51°, ∠R = 39°, QP = 5.7 cm
a Students can check one another's triangles.
b ∠ABC = 44°, ∠BCA = 79°, ∠CAB = 57°
5.9 cm
Student drawing.
Student drawing.
Student drawing.
BC can be 2.6 cm or 7.8 cm.
4.3 cm
4.3 cm
a Right-angled triangle constructed with sides 3, 4, 5 and 4.5, 6, 7.5, and scale marked 1 cm : 1 m
b Right-angled triangle constructed with 12 equally spaced dots.
An equilateral triangle of side 4 cm.
Even with all three angles, you need to know at least one length.

25.2 Bisectors

Exercise 25B

1–9 Practical work; check students' constructions
10 The centre of the circle

25.3 Scale drawings

Exercise 25C

1 a pond: 40 m × 10 m, fruit: 50 m × 10 m, trees: 20 m × 20 m, lawn: 30 m × 20 m, vegetables: 50 m × 20 m
   b pond: 400 m^2, fruit: 500 m^2, trees: 400 m^2, grass: 600 m^2, vegetables: 1000 m^2
2 a 33 cm   **b** 9 cm
3 a 30 cm × 30 cm   **b** 40 cm × 10 cm   **c** 20 cm × 15 cm
   d 30 cm × 20 cm   **e** 30 cm × 20 cm   **f** 10 cm × 5 cm
4 a Student's scale drawing.   **b** 38 plants
5 a 8.4 km   **b** 4.6 km   **c** 6.2 km
   d 6.4 km   **e** 7.6 km   **f** 2.4 km
6 a student's drawing   **b** 12.9 metres
7 a 900 km   **b** 1100 km   **c** 860 km
8 c – 7 cm represents 210 m, so 1 cm represents 30 m

26.1 Pythagoras' theorem

Exercise 26A

1 10.3 cm

2 5.9 cm

3 8.5 cm

4 20.6 cm

5 18.6 cm

6 17.5 cm

7 5 cm

8 13 cm

9 10 cm

10 The square in the first diagram and the two squares in the second have the same area.

Exercise 26B

1 **a** 15 cm **b** 14.7 cm **c** 6.3 cm **d** 18.3 cm

2 **a** 20.8 m **b** 15.5 cm **c** 15.5 m **d** 12.4 cm

3 **a** 5 m **b** 6 m **c** 3 m **d** 50 cm

4 There are infinite possibilities, e.g. any multiple of 3, 4, 5 such as 6, 8, 10; 9, 12, 15; 12, 16, 20; multiples of 5, 12, 13 and of 8, 15, 17.

5 42.6 cm

26.2 Trigonometric ratios

Exercise 26C

1 **a** 0.682 **b** 0.829 **c** 0.922 **d** 1

2 **a** 0.731 **b** 0.559 **c** 0.388 **d** 0

3 **a i** 0.574 **ii** 0.574

  b i 0.208 **ii** 0.208

  c i 0.391 **ii** 0.391

  d Same

  e i sin 15° is the same as cos 75°

  ii cos 82° is the same as sin 8°

  iii sin x is the same as cos (90° − x)

4 **a** 0.933 **b** 1.48 **c** 2.38 **d** Infinite

  e 1 **f** 0.364 **g** 0.404 **h** 0

5 Has values > 1

6 **a** 3.56 **b** 8.96 **c** 28.4 **d** 8.91

7 **a** 5.61 **b** 7.08 **c** 1.46 **d** 7.77

8 **a** $\frac{4}{5}, \frac{3}{5}, \frac{4}{3}$

  b $\frac{5}{13}, \frac{12}{13}, \frac{5}{12}$

  c $\frac{7}{25}, \frac{24}{25}, \frac{7}{24}$

26.3 Calculating angles

Exercise 26D

1 **a** 30° **b** 51.7° **c** 39.8°

  d 61.3° **e** 87.4° **f** 45.0°

2 **a** 60° **b** 50.2° **c** 2.6°

  d 45.0 **e** 78.5° **f** 45.6°

3 **a** 31.0° **b** 20.8° **c** 41.8°

  d 46.4° **e** 69.5° **f** 77.1°

4 Error message, largest value 1, smallest value −1

5 **a i** 17.5° **ii** 72.5° **iii** 90°

  b Yes

26.4 Using sine, cosine and tangent functions

Exercise 26E

1 **a** 17.5° **b** 22.0° **c** 32.2°

2 **a** 5.29 cm **b** 5.75 cm **c** 13.2 cm

3 **a** 4.57 cm **b** 6.86 cm **c** 100 cm

4 **a** 5.12 cm **b** 9.77 cm **c** 11.7 cm **d** 15.5 cm

Exercise 26F

1 **a** 51.3° **b** 75.5° **c** 51.3°

2 **a** 5.35 cm **b** 14.8 cm **c** 12.0 cm **d** 8.62 cm

3 **a** 5.59 cm **b** 46.6° **c** 9.91 cm **d** 40.1°

Exercise 26G

1 **a** 33.7° **b** 36.9° **c** 52.1°

2 **a** 9.02 cm **b** 7.51 cm **c** 7.14 cm **d** 8.90 cm

3 **a** 13.7 cm **b** 48.4° **c** 7.03 cm **d** 41.2°

26.5 Which ratio to use

Exercise 26H

1 **a** 12.6 **b** 59.6 **c** 74.7 **d** 16.0

  e 67.9 **f** 20.1

2 **a** 44.4° **b** 39.8° **c** 44.4° **d** 49.5°

  e 58.7° **f** 38.7°

3 **a** 67.4° **b** 11.3 **c** 134 **d** 28.1°

  e 39.7 **f** 263 **g** 50.2° **h** 51.3°

  i 138 **j** 22.8

4 **a** Sides of right-hand triangle are sine θ and cosine θ

  b Pythagoras' theorem

  c Students should check the formulae

26.6 Solving problems using trigonometry

Exercise 26I

1 65°

2 The safe limits are between 1.04 m and 2.05 m. The ladder will reach between 5.64 m and 5.91 m up the wall.

3 44°

4 6.82 m

5 31°

a 25°

b 2.10 m

c Thickness of wood has been ignored

a 20°

b 4.78 m

She would calculate 100 tan 23°. The answer is about 42.4 m

21.1 m

a 338 km

b 725 km

43 km

170 km

426 km

One way is stand opposite a feature, such as a tree, on the opposite bank, move a measured distance, x, along your bank and measure the angle, θ, between your bank and the feature. Width of river is x tan θ. This of course requires measuring equipment! An alternative is to walk along the bank until the angle is 45° (if that is possible). This angle is easily found by folding a sheet of paper. This way an angle measurer is not required.

6.7 Angles of elevation and depression

Exercise 26J

10.1 km

22°

429 m

a 156 m

b No. the new angle of depression is $\tan^{-1}\left(\frac{200}{312}\right) = 33°$ and half of 52° is 26°

a 222 m

b 42°

a 21.5 m

b 17.8 m

13.4 m

19°

The angle is 16° so Cara is not quite correct.

6.8 Problems in three dimensions

Exercise 26K

25.1°

a 25cm

b 58.6°

c 20.5 cm

a 3.46 m

b 75.5°

c 73.2°

a 24.0°

b 48.0°

c 13.5 cm

d 16.6°

It is 44.6°; use triangle XDM where M is the midpoint of BD; triangle DXB is isosceles, as X is over the point where the diagonals of the base cross; the length of DB is $\sqrt{656}$. and the cosine of the required angle is $0.5\sqrt{656} \div 18$

26.9 Sine, cosine and tangent of obtuse angles

Exercise 26L

1 a The bottom row of the table is 0.174, 0.5, 0.766, 0.996, 1, 0.996, 0.766, 0.5, 0.174.

b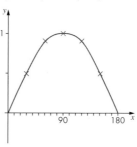

c It has reflection symmetry. The line of symmetry is x = 90.

d You could choose 10° and 170°, 30° ad 150°, 50° and 130° or 85° and 95°

2 30° and 150°.

3 46° and 134°.

4 122.9°

5 a The bottom row of the table is 0.966, 0.819, 0.5, 0.174, 0, –0.174, –0.5, –0.819, –0.966.

b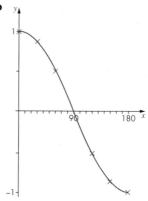

c It has rotational symmetry of order 2 about the point (90, 0)

6 a 31.8°   **b** 148.2°   **c** 120°

d 90°   **e** 82.8°   **f** 97.2°

7 a 53°   **b** 104°   **c** 49°, 131°   **d** 90°

e 90°   **f** 72°, 108°   **g** no solution   **h** 45°

8 The bottom row of the table is 0.577, 1.428, 5.671, –5.671, –1.428, – 0.577

9 –0.176

10 a Student's check

b Cannot find $\frac{\sin 90°}{\cos 90°}$ because cos 90° = 0

11 135

26.10 The sine rule and the cosine rule

Exercise 26M

1 a 3.64 m   **b** 8.05 cm   **c** 19.4 cm

2 a 46.6°   **b** 112.0°   **c** 36.2°

3 a i 30°   **ii** 40°

b 19.4 m

4 36.5 m

5 22.2 m

6 3.47 m

7 64.6 km

8 134°

1 **a** 7.71 m   **b** 29.1 cm   **c** 27.4 cm

2 **a** 76.2°   **b** 125.1°   **c** 90°
    d Right-angled triangle

3 5.16 cm

4 65.5 cm

5 **a** 10.7 cm   **b** 41.7°   **c** 38.3°   **d** 6.69 cm

6 58.4 km at 092.5°

7 21.8°

8 42.5 km

9 111°; the largest angle is opposite the longest side

1 **a** 8.60 m   **b** 90°   **c** 27.2 cm
    d 26.9°   **e** 27.5°   **f** 62.4 cm
    g 90.0°   **h** 866 cm   **i** 86.6 cm

2 7 cm

3 11.1 km

4 **a** $A = 90°$; this is Pythagoras' theorem
    b A is acute
    c A is obtuse

5 142 m

26.11 Using sine to find the area of a triangle

1 **a** 24.0 cm^2 **b** 26.7 cm^2 **c** 243 cm^2
    d 21 097 cm^2 **e** 1224 cm^2

2 4.26 cm

3 **a** 42.3°   **b** 49.6°

4 2033 cm^2

5 21.0 cm^2

6 726 cm^2

7 149 km^2

8 **a** 66.4 m   **b** 118.9°   **c** 1470 m^2

Answers to Chapter 27

27.1 Perimeter and area of a rectangle

1 **a** 35 cm^2, 24 cm   **b** 33 cm^2, 28 cm
    c 45 cm^2, 36 cm   **d** 70 cm^2, 34 cm
    e 56 cm^2, 30 cm   **f** 10 cm^2, 14 cm

2 **a** 53.3 cm^2, 29.4 cm   **b** 84.96 cm^2, 38 cm

3 39

4 **a** 4   **b** 1 h 52 min

5 40 cm

6 Area B, 44 cm^2; perimeter B, 30 cm

7 Never (the area becomes four times greater).

8 **a** 28 cm, 30 cm^2 **b** 28 cm, 40 cm^2
    c 40 cm, 51 cm^2 **d** 30 cm, 35 cm^2
    e 32 cm, 43 cm^2 **f** 34 cm, 51 cm^2
    g cannot tell what the perimeter is; 48 cm^2
    h 34 cm, 33 cm^2

9 72 cm^2

10 48 cm

27.2 Area of a triangle

1 **a** 21 cm^2 **b** 12 cm^2 **c** 14 cm^2
    d 55 cm^2 **e** 90 cm^2 **f** 140 cm^2

2 **a** 28 cm^2 **b** 8 cm   **c** 4 cm
    d 3 cm   **e** 7 cm   **f** 44 cm^2

3 64 cm^2

4 **a** 40 cm^2 **b** 65 m^2 **c** 80 cm^2

5 **a** 65 cm^2 **b** 50 m^2

6 For example: height 10 cm, base 10 cm; height 5 cm, base 20 cm; height 25 cm, base 4 cm; height 50 cm, base 2 cm

7 Triangle c; a and b each have an area of 15 cm^2 but c has area of 16 cm^2

27.3 Area of a parallelogram

1 **a** 96 cm^2 **b** 70 cm^2 **c** 20 m^2
    d 125 cm^2 **e** 10 cm^2 **f** 112 m^2

2 No, it is 24 cm^2, she used the slanting side instead of the perpendicular height.

3 16 cm

4 **a** 500 cm^2 **b** 15

27.4 Area of a trapezium

1 **a** 30 cm^2 **b** 77 cm^2 **c** 24 cm^2 **d** 42 cm^2
    e 40 cm^2 **f** 6 cm   **g** 3 cm

2 **a** 27.5 cm, 36.25 cm^2
    b 33.4 cm, 61.2 cm^2
    c 38.6 m, 88.2 m^2

3 Any pair of lengths that add up to 10 cm. For example: 1 cm, 9 cm; 2 cm, 8 cm; 3 cm, 7 cm; 4 cm, 6 cm; 4.5 cm, 5.5 cm

Shape c. Its area is 25.5 cm^2

Shape a. Its area is 28 cm^2

a

2 cm

1.4 m^2

.5   Circumference and area of a circle

a 31.4 cm, 78.5 cm^2 **b** 18.8 cm, 28.3 cm^2

c 9.4 cm, 7.1 cm^2 **d** 25.1 cm, 50.3 cm^2

a 25.1 cm and 50.3 cm^2

b 15.7 cm and 19.6 cm^2

c 28.9 cm and 66.5 cm^2

d 14.8 cm and 17.3 cm^2

a i 353 cm^2 **ii** 77.1 cm

b i 100.5 cm^2 **ii** 41.1 cm

c i 2.26 m^2 **ii** 6.17 m

d i 0.39 m^2 **ii** 2.57 m

a 19.1 cm   **b** 9.5 cm

c 286.5 cm^2 (or 283.5 cm^2)

962.9 cm^2 (or 962.1 cm^2)

a 30.8 cm and 56.5 cm^2 **b** 17.9 cm and 19.6 cm^2

a 50.3 m^2 **b** 44.0 cm^2 **c** 28.3 cm^2

141.4 cm^2; A = π × 9^2 − π × 6^2 = 141.4 cm^2

$a^2 = \pi r^2$, so $r^2 = \frac{a^2}{\pi}$ therefore $r = \frac{a}{\sqrt{\pi}}$

21.5 cm^2

.6   Surface area and volume of a cuboid

a i 198 cm^3 **ii** 234 cm^2

b i 90 cm^3 **ii** 146 cm^2

c i 1440 cm^3 **ii** 792 cm^2

d i 525 cm^3 **ii** 470 cm^2

24 litres

a 160 cm^3 **b** 480 cm^3 **c** 150 cm^3

a i 64 cm^3 **ii** 96 cm^2

b i 343 cm^3 **ii** 294 cm^2

c i 1000 mm^3 **ii** 600 mm^2

d i 125 m^3 **ii** 150 m^2

e i 1728 m^3 **ii** 864 m^2

5 86

1.6 m

48 m^2

a 3 cm   **b** 5 m   **c** 2 mm   **d** 1.2 m

a 148 cm^3 **b** 468 cm^3

If this was a cube, the side length would be 5 cm, so total surface area would be 5 × 5 × 6 = 150 cm^2; no this particular cuboid is not a cube.

600 cm^2

27.7   Volume of a prism

1 **a i** 21 cm^2 **ii** 63 cm^3

　　b i 48 cm^2 **ii** 432 cm^3

　　c i 36 m^2 **ii** 324 m^3

2 **a** 432 m^3 **b** 225 m^3 **c** 1332 m^3

3 **a** A cross-section parallel to the side of the pool always has the same shape.

　　b About 3$\frac{1}{2}$ hrs

4 7.65 m^3

5 **a** 21 cm^3, 210 cm^3 **b** 54 cm^2, 270 cm^2

6 146 cm^3

7 327 litres

8 **a** 672 cm^3 **b i** 5   **ii** 6   **iii** 9

27.8   Volume and surface area of a cylinder

1 **a i** 226 cm^3 **ii** 207 cm^2

　　b i 14.9 cm^3 **ii** 61.3 cm^2

　　c i 346 cm^3 **ii** 275 cm^2

　　d i 1060 cm^3 **ii** 636 cm^2

2 **a i** 72π cm^3 **ii** 48π cm^2

　　b i 112π cm^3 **ii** 56π cm^2

　　c i 180π cm^3 **ii** 60π cm^2

　　d i 600π m^3 **ii** 120π m^2

3 665 cm^3

4 Label should be less than 10.5 cm wide so that it fits the can and does not overlap the rim and more than 23.3 cm long to allow an overlap.

5 332 litres

6 There is no right answer. Students could start with the dimensions of a real can. Often drinks cans are not exactly cylindrical. One possible answer is height of 6.6 cm and diameter of 8 cm.

7 About 127 cm

8 A diameter of 10 cm and a length of 5 cm give a volume close to 400 cm^3 (0.4 litres).

27.9   Arcs and sectors

1 **a i** 5.59 cm   **ii** 22.3 cm^2

　　b i 8.29 cm   **ii** 20.7 cm^2

　　c i 16.3 cm   **ii** 98.0 cm^2

　　d i 15.9 cm   **ii** 55.6 cm^2

2 2π cm, 6π cm^2

3 **a** 73.8 cm   **b** 20.3 cm

4 **a** 107 cm^2 **b** 173 cm^2

5 43.6 cm

6 (36π − 72) cm^2

7 36.5 cm^2

8 **a** 13.9 cm   **b** 7.07 cm^2

27.10 Volume and surface area of a cone

Exercise 27J

1 a i 3560 cm³   **ii** 1430 cm²
  b i 314 cm³   **ii** 283 cm²
  c i 1020 cm³   **ii** 679 cm²

2 24π cm²

3 a 816π cm³   **b** 720π mm³

4 a 4 cm   **b** 6 cm
  c Various answers, e.g. 60° gives 2 cm, 240° gives 8 cm

5 24π cm²

6 If radius of base is r, slant height is 2r.
  Area of curved surface = πr × 2r = 2πr², area of base = πr²

7 140 g

8 2.81 cm

27.11 Volume and surface of a sphere

Exercise 27K

1 a 36π cm³ and 36 π cm²
  b 288π cm³ and 144 π cm²
  c 1330π cm³ and 400 π cm²

2 65 400 cm³, 7850 cm²

3 a 1960 cm²
  b 8180 cm³

4 125 cm

5 6231

6 7.8 cm

7 48%

Answers to Chapter 28

28.1 Lines of symmetry

Exercise 28A

1 a   **b**   **c**

d   **e**   **f**

g

2 a i 5   **ii** 6   **iii** 8
  b 10

3 2, 1, 1, 2, 0

4

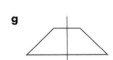

5 a 1   **b** 5   **c** 1   **d** 6

6

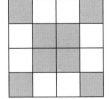

28.2 Rotational symmetry

Exercise 28B

1 a 4   **b** 2   **c** 2   **d** 3   **e** 6
2 a 4   **b** 5   **c** 6   **d** 4   **e** 6
3 a 2   **b** 2   **c** 2   **d** 2   **e** 2
4 a 6
  b 9 (the small red circle surrounded by nine 'petals') and
  (the centre pattern)

5 For example:

28.3 Symmetry of special two-dimensional shapes

Exercise 28C

1

2 a kite
  b rectangle 2, square 4, equilateral triangle 3, rhombus
3 a isosceles   **b** no
4 a parallelogram   **b** square
5 a rectangle and rhombus   **b** no
6 a B and D   **b** AB and AD; CB and CD   **c** kite
7 a diameter   **b** infinite   **c** infinite
8 a A and C; B and D   **b** AD and BC; AB and DC
  c Parallelogram
9 It will have two pairs of equal angles

Answers to Chapter 29

.1 Introduction to vectors

a i $\begin{pmatrix} 4 \\ -1 \end{pmatrix}$ **ii** $\begin{pmatrix} 3 \\ 1 \end{pmatrix}$

iii $\begin{pmatrix} 2 \\ 3 \end{pmatrix}$ **iv** $\begin{pmatrix} -2 \\ 4 \end{pmatrix}$

v $\begin{pmatrix} 2 \\ -4 \end{pmatrix}$ **vi** $\begin{pmatrix} -1 \\ 2 \end{pmatrix}$

b Both are $\begin{pmatrix} 1 \\ -2 \end{pmatrix}$. D is the midpoint of AC.

b i $\begin{pmatrix} -2 \\ 4 \end{pmatrix}$ **ii** $\begin{pmatrix} 2 \\ -4 \end{pmatrix}$ **iii** $\begin{pmatrix} 3 \\ 4 \end{pmatrix}$

iv $\begin{pmatrix} -1 \\ 2 \end{pmatrix}$ **v** $\begin{pmatrix} -2 \\ -6 \end{pmatrix}$

he diagrams should show the following vectors:

a $\begin{pmatrix} 3 \\ -1 \end{pmatrix}$ **b** $\begin{pmatrix} -1 \\ 4 \end{pmatrix}$

c $\begin{pmatrix} 1 \\ 7 \end{pmatrix}$ **d** $\begin{pmatrix} 3 \\ -1 \end{pmatrix}$

e $\begin{pmatrix} -1 \\ -7 \end{pmatrix}$ **f** $\begin{pmatrix} 2 \\ -8 \end{pmatrix}$

a $\begin{pmatrix} -1 \\ 2 \end{pmatrix}$ **b** $\begin{pmatrix} 6 \\ 12 \end{pmatrix}$

c $\begin{pmatrix} -5 \\ -6 \end{pmatrix}$ **d** $\begin{pmatrix} 5 \\ 6 \end{pmatrix}$

e $\begin{pmatrix} -12 \\ -8 \end{pmatrix}$ **f** $\begin{pmatrix} 1 \\ 6 \end{pmatrix}$

–e

k is 4.

29.2 Using vectors

1 a Any three, of: $\overrightarrow{AC}, \overrightarrow{CF}, \overrightarrow{BD}, \overrightarrow{DG}, \overrightarrow{GI}, \overrightarrow{EH}, \overrightarrow{HJ}, \overrightarrow{JK}$

   b Any three of: $\overrightarrow{BE}, \overrightarrow{AD}, \overrightarrow{DH}, \overrightarrow{CG}, \overrightarrow{GJ}, \overrightarrow{FI}, \overrightarrow{IK}$

   c Any three of: $\overrightarrow{AO}, \overrightarrow{CA}, \overrightarrow{FC}, \overrightarrow{IG}, \overrightarrow{GD}, \overrightarrow{DB}, \overrightarrow{KJ}, \overrightarrow{JH}, \overrightarrow{HE}$

   d Any three of: $\overrightarrow{BO}, \overrightarrow{EB}, \overrightarrow{HD}, \overrightarrow{DA}, \overrightarrow{JG}, \overrightarrow{GC}, \overrightarrow{KI}, \overrightarrow{IF}$

2 a 2a   **b** 2b   **c** 3a + 2b   **d** a + 2b

   e a + b   **f** 2a + 2b   **g** 3a + b

3 $\overrightarrow{AI}, \overrightarrow{BJ}, \overrightarrow{DK}$

4 $\overrightarrow{OF}, \overrightarrow{BI}, \overrightarrow{EK}$

5

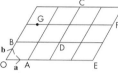

6 a – b   **b** 3a – b   **c** 2a – b   **d** a – b

   e a + b   **f** –a – b   **g** 2a – b   **h** –a – 2b

   i a + 2b   **j** –a + b   **k** 2a – 2b   **l** a – 2b

7 a $\overrightarrow{BJ}, \overrightarrow{CK}$

   b $\overrightarrow{EB}, \overrightarrow{GO}, \overrightarrow{KH}$

8

9 a i 3a + 2b   **ii** 3a + b

    iii 2a –b   **iv** 2b – 2a

   b $\overrightarrow{DG}$ and $\overrightarrow{BC}$

10 a 2a + b   **b** 2b – a   **c** a + 1.5b

    d 0.5a + 0.5b   **e** 1.5a + 1.5b   **f** 1.5a – 0.5b

11 a i –a + b   **ii** $\frac{1}{2}$(b – a)

    iii   **iv** $\frac{1}{2}$a + $\frac{1}{2}$b

   b

   c M is midpoint of parallelogram of which OA and OB are two sides.

12 a i –a + b   **ii** $\frac{1}{3}$(–a + b)   **iii** $\frac{2}{3}$a + $\frac{1}{3}$b

   b $\frac{3}{4}$a + $\frac{1}{4}$b

13 a i $\frac{2}{3}$b   **ii** $\frac{1}{2}$a + $\frac{1}{2}$b   **iii** $\frac{2}{3}$b

   b $\frac{1}{2}$a – $\frac{1}{6}$b

   c $\overrightarrow{DE} = \overrightarrow{DO} + \overrightarrow{OE} = \frac{3}{2}$a – $\frac{1}{2}$b

   d $\overrightarrow{DE}$ parallel to $\overrightarrow{CD}$ = (multiple of $\overrightarrow{CD}$) and D is a common point

29.3   The magnitude of a vector

Exercise 29C

1 **a**

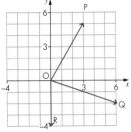

b $\sqrt{34}$; $\sqrt{40}$; 4

c $\sqrt{58}$

d $\sqrt{40}$

2 **a** 10 and 13   **b** $\binom{11}{-4}$ **c** $\sqrt{137}$

d No. 10 + 13 does not equal $\sqrt{137}$

e $\sqrt{401}$ **f** $\sqrt{401}$

g Yes. They are vectors in opposite directions but the sam

length.

3 **a** 10, 10, 10

b Because they are all the same distance from A. The rad

is 10.

4 **a** $\sqrt{17}$ **b** $\sqrt{261}$ **c** 13   **d** 10

Answers to Chapter 30

30.1   Translations

Exercise 30A

1 **a i** $\binom{1}{3}$ **ii** $\binom{4}{2}$ **iii** $\binom{2}{-1}$

b i $\binom{4}{-2}$ **ii** $\binom{-2}{3}$ **iii** $\binom{3}{3}$

c i $\binom{-4}{-2}$ **ii** $\binom{1}{-1}$ **iii** $\binom{0}{4}$

d i $\binom{-2}{-7}$ **ii** $\binom{5}{0}$ **iii** $\binom{1}{-5}$

2

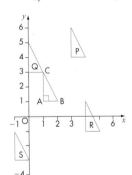

3 **a** $\binom{-3}{-1}$ **b** $\binom{4}{-4}$ **c** $\binom{-5}{-2}$

d $\binom{4}{7}$ **e** $\binom{-1}{5}$ **f** $\binom{1}{6}$

g $\binom{-4}{4}$ **h** $\binom{-4}{-7}$

4 $\binom{-x}{-y}$

5 $\binom{-1}{4}$

30.2   Reflections

Exercise 30B

1

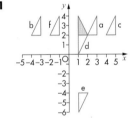

2 **a–e**

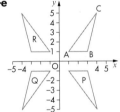

f reflection in the y-axis

3 **a–b**

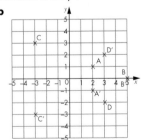

c The y-value changes sign

d (a, –b)

a–b

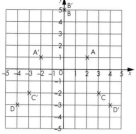

c *x*-value changes sign

d (−*a*, *b*)

2 Possible answer: Take the centre square as ABCD then reflect this square each time in the line, AB, then BC, then CD and finally AD.

3 *x* = −1

.3 Further reflections

ercise 30C

1 Possible answer:

2 a–i

reflection in *y* = *x*

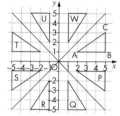

3 a–c

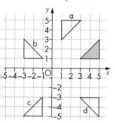

b Coordinates are reversed: *x* becomes *y* and *y* becomes *x*

c (*b*, *a*)

6 a–c

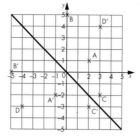

d Coordinates are reversed and change sign: *x* becomes −*y* and *y* becomes −*x*

e (−*b*, −*a*)

30.4 Rotations

1 a

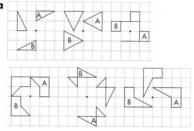

b i Rotation 90° anticlockwise

ii Rotation 180°

2 For example:

3 Possible answer: Label ABCD in an anticlockwise direction around a square. If ABCD is the centre square, rotate about A 90° anticlockwise, rotate about new B 180°, now rotate about new C 180°, and finally rotate about new D 180°.

4

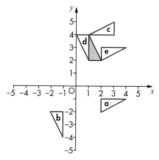

5 a (4,5) 180°

b (5,5) 90° anticlockwise

c (3,3) 180°

d (3,5) 90° clockwise

6 a E

b H

7 i

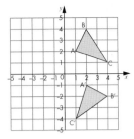

ii A′ (2, –1), B′ (4, –2), C′ (1, –4)
iii Original coordinates (x, y) become (y, –x)
iv Yes

8 i

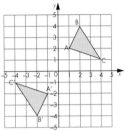

ii A′ (–1, –2), B′ (–2, –4), C′ (–4, –1)
iii Original coordinates (x, y) become (–x, –y)
iv Yes

9 Show by drawing a shape or use the fact that (a, b) becomes
(a, –b) after reflection in the x-axis, and (a, –b) becomes
(–a, –b) after reflection in the y-axis, which is equivalent to a
single rotation of 180°.

30.5 Further rotations

Exercise 30E

1 a–c

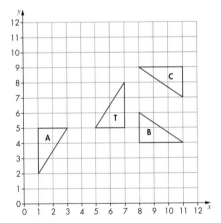

d Rotation of 180° about (9.5, 6.5).

2 a (3,0)   **b** (0,0)   **c** (6,0)
3 a (0, –1.5) 180°   **b** (–0.5, –1.5) 90° clockwise
c (–3,5,2.5) 90° anti clockwise
d (0.5, 2) 180°

4 Show by drawing a shape or use the fact that (a, b) becomes
(b, a) after reflection in the line y = x, and (b, a) becomes
(–a, –b) after reflection in the line y = –x, which is equivalent
to a single rotation of 180°.

5 a

b i Rotation 60° clockwise about O
ii Rotation 120° clockwise about O
iii Rotation 180° about O
iv Rotation 240° clockwise about O, or 120° anticlockw
about O
c i Rotation 60° clockwise about O
ii Rotation 180° about O

6 Rotation 90° anticlockwise about (3, –2).

30.6 Enlargements

Exercise 30F

1 a–d

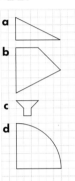

2

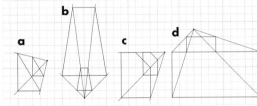

3 a   **b**

c

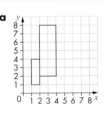

4

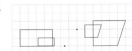

a
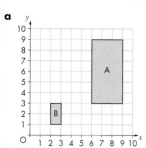

b 3 : 1
c 3 : 1
d 9 : 1

6 a
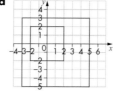
b
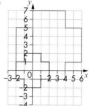

7 a–b
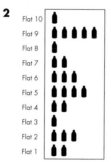

Answers to Chapter 31

31.1 Frequency tables

Exercise 31A

a

Goals	0	1	2	3
Frequency	6	8	4	2

b 1 goal   **c** 22

a

Temperature (°C)	14–16	17–19	20–22	23–25	26–28
Frequency	5	10	8	5	2

b 17–19° C
c Getting warmer in the first half and then getting cooler towards the end.

a

Score	1	2	3	4	5	6
Frequency	5	6	6	6	3	4

b 30   **c** Yes, frequencies are similar.

a

Height (cm)	151–155	156–160	161–165	166–170
Frequency	2	5	5	7
Height (cm)	171–175	176–180	181–185	186–190
Frequency	5	4	3	1

b 166–170 cm
c Student's survey results.

Various answers such as 1–10, 11–20, etc. or 1–20, 21–40, 41–60

The ages 20 and 25 are in two different groups.

31.2 Pictograms

Exercise 31B

1
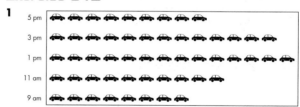

2
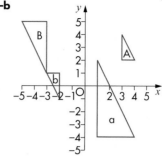

3 a May 10 h, Jun 12 h, Jul 12 h, Aug 12 h, Sep 10 h
b Visual impact, easy to understand.
4 a Simon
b $165
c Difficult to show fractions of a symbol.
5 a i 12   **ii** 6   **iii** 13
b Check students' pictograms.
c 63

31.3   Bar charts

Exercise 31C

1   **a** Swimming   **b** 74

2   **a**

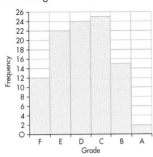

b $\frac{40}{100} = \frac{2}{5}$

c Easier to read the exact frequency.

3   **a**

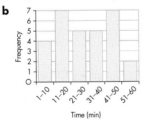

Amir
Hasrul

b Amir got more points overall, but Hasrul was more consistent.

4   **a**

Time (min)	1–10	11–20	21–30	31–40	41–50	51–60
Frequency	4	7	5	5	7	2

b

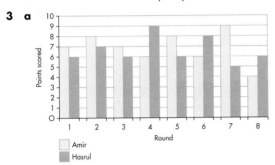

c Some live close to the school. Some live a good distance away and probably travel to school by bus.

5   **a**

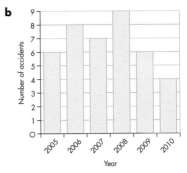

Key 🚚= 1 accident

b

c Use the pictogram because an appropriate symbol make more impact.

6 Yes. If you double the minimum temperature each time, it is very close to the maximum temperature.

31.4   Pie charts

Exercise 31D

1   **a**

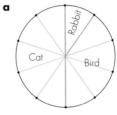

b

c

2 Pie charts with following angles:
   a 36°, 90°, 126°, 81°, 27°
   b 168°, 52°, 100°, 40°

3 Pie charts with these angles: 60°, 165°, 45°, 15°, 75°

4   **a** 36
   b Pie charts with these angles: 50°, 50°, 80°, 60°, 60°, 40 20°
   c Student's bar chart.
   d Bar chart, because easier to make comparisons.

5   **a** Pie charts with these angles: 124°, 132°, 76°, 28°
   b Split of total data seen at a glance.

6   **a** 55°
   b 22

7   **a** Pie charts with these angles: Strings: 36°, 118°, 126°, 8°
   Brass: 82°, 118°, 98°, 39°, 23°
   b Overall, the strings candidates did better, as a smaller proportion failed. A higher proportion of Brass candid scored very good or excellent.

8 $\frac{40}{360} = \frac{1}{9}$

.5 Histograms

The respective frequency densities on which each histogram should be based are:

a 2.5, 6.5, 6, 2, 1, 1.5
b 4, 27, 15, 3
c 17, 18, 12, 6.67
d 0.4, 1.2, 2.8, 1
e 9, 21, 13.5, 9

a

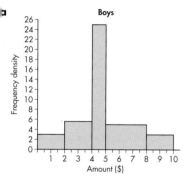

Boys

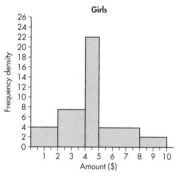

Girls

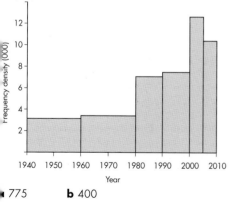

a 775   **b** 400

5 a

Age, y (years)	$9 < y \leq 10$	$10 < y \leq 12$	$12 < y \leq 14$
Frequency	4	12	8
Age, y (years)	$14 < y \leq 17$	$17 < y \leq 19$	$19 < y \leq 20$
Frequency	9	5	1

b

Temperature, t (°C)	$10 < t \leq 11$	$11 < t \leq 12$	$12 < t \leq 14$
Frequency	15	15	50
Temperature, t (°C)	$14 < t \leq 16$	$16 < t \leq 19$	$19 < t \leq 21$
Frequency	40	45	15

c

Weight, w (kg)	$50 < w \leq 70$	$70 < w \leq 90$	$90 < w \leq 100$
Frequency	160	200	120
Weight, w (kg)	$100 < w \leq 120$	$120 < w \leq 170$	
Frequency	120	200	

6 a

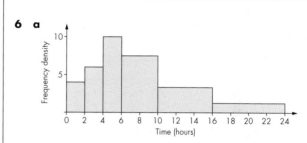

b 45

7 a

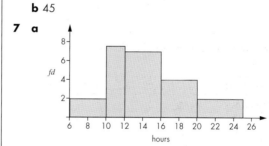

b 33 plants

8 a

Speed, v (mph)	$0 < v \leq 40$	$40 < v \leq 50$	$50 < v \leq 60$
Frequency	80	10	40
Weight, w (kg)	$60 < v \leq 70$	$70 < v \leq 80$	$80 < v \leq 100$
Frequency	110	60	60

b 360

9 a 80
b 31.25%

10 a

Adrienne   Bernice

b Student's own description.

Answers to Chapter 32

32.1 The mode

Exercise 32A

1 a 4 **b** 48 **c** −1
d $\frac{1}{4}$ **e** no mode **f** 3.21

2 a red **b** Sun **c** β **d** ★

3 a 32 **b** 6 **c** no
d no; boys generally take larger shoe sizes

4 a 5
b no; more than half the form got a higher mark

5 The mode will be the most popular item or brand sold in a shop.

6 a 28
b i brown **ii** blue **iii** brown
c Both students had blue eyes.

7 a May lose count.
b Put in a table, or arrange in order
c 4

32.2 The median

Exercise 32B

1 a 5 **b** 33 **c** $7\frac{1}{2}$ **d** 24
e $8\frac{1}{2}$ **f** 0 **g** 5.25

2 a $2.20 **b** $2.25 **c** median, because it is the central value

3 a 5
b i 15 **ii** 215 **iii** 10 **iv** 10

4 a 13, Ella **b** 162 cm, Pat **c** 41 kg, Elisa
d Ella, because she is closest to the 3 medians

5 a 12 **b** 13

6 Answers will vary

7 12, 14, 14, 16, 20, 22, 24

8 a Possible answer: 11, 15, 21, 21 (one below or equal to 12 and three above or equal)
b Any four numbers higher than or equal to 12, and any two lower or equal
c Eight, all 4 or under

9 A median of $8 does not take into account the huge value of the $3000 so is in no way representative.

32.3 The mean

Exercise 32C

1 a 6 **b** 24 **c** 45 **d** 1.57 **e** 2
2 a 55.1 **b** 324.7 **c** 58.5 **d** 44.9 **e** 2.3
3 a 61 **b** 60 **c** 59 **d** Badru **e** 2
4 42 min
5 a $200 **b** $260 **c** $278
d Median, because the extreme value of $480 is not taken into account

6 a 35 **b** 36
7 a 6
b 16; all the numbers and the mean are 10 more than those in part a
c i 56 **ii** 106 **iii** 7

8 Possible answers: Speed – Kath, James, John, Joseph; Rob – Frank, James, Helen, Evie. Other answers are possible.

9 36

10 24

32.4 The range

Exercise 32D

1 a 7 **b** 26 **c** 5 **d** 2.4 **e** 7
2 a 5°, 3°, 2°, 7°, 3°
b Variable weather over England
3 a $31, $28, $33 **b** $8, $14, $4
c Not particularly consistent
4 a 82 and 83 **b** 20 and 12
c Fay, because her scores are more consistent
5 a 5 min and 4 min
b 9 min and 13 min
c Number 50, because times are more consistent
6 a Isaac, Oliver, Evrim, Chloe, Lilla, Badru and Isambard
b 70 cm to 90 cm
7 a Teachers because they have a high mean and students could not have a range of 20.
b Year 11 students as the mean is 15–16 and the range i

32.5 Which average to use

Exercise 32E

1 a i 29 **ii** 28 **iii** 27.1
b 14
2 a i Mode 3, median 4, mean 5
ii 6, 7, $7\frac{1}{2}$
iii 4, 6, 8
b i Mean: balanced data
ii Mode: 6 appears five times
iii Median: 28 is an extreme value
3 a Mode 73, median 76, mean 80
b The mean, because it is the highest average
4 a 150 **b** 20
5 a Mean **b** Median **c** Mode
d Median **e** Mode **f** Mean
6 No. Mode is 31, median is 31, and mean is 31½.
7 a Median **b** Mode **c** Mean
8 Tom mean, David median, Mohamed mode
9 Possible answers: **a** 1, 6, 6, 6, 6 **b** 2, 5, 5, 6, 7
10 Boss chose the mean while worker chose the mode.
11 11.6
12 52.7 kg

2.6   Using frequency tables

Exercise 32F

a i 7   **ii** 6   **iii** 6.4
b i 4   **ii** 4   **iii** 3.7
c i 8   **ii** 8.5   **iii** 8.2
d i 0   **ii** 0   **iii** 0.3

a 668   **b** 1.9   **c** 0   **d** 328

a 2.2, 1.7, 1.3   **b** Better dental care

a 0   **b** 0.96

a 7   **b** 6.5   **c** 6.5

a 1   **b** 1   **c** 0.98

a Roger 5, Brian 4
b Roger 3, Brian 8
c Roger 5, Brian 4
d Roger 5.4, Brian 4.8
e Roger, because he has the smaller range
f Brian, because he has the better mean

3, 4, 15, 3

Add up the weeks to see she travelled in 52 weeks of the year, the median is in the 26th and 27th week. Looking at the weeks in order, the 23rd entry is the end of 2 days in a week so the median must be in the 3 days in a week.

2.7   Grouped data

Exercise 32G

a i $30 < x \leqslant 40$ **ii** 29.5
b i $0 < y \leqslant 100$ **ii** 158.3
c i $5 < z \leqslant 10$ **ii** 9.43
d i 7–9   **ii** 8.4 weeks

a $100 < m \leqslant 120$ g   **b** 10 860 g   **c** 108.6 g
a 207   **b** 19–22 cm   **c** 20.3 cm
a 160   **b** 52.6 min   **c** modal group
d 65%
a $175 < h \leqslant 200$ **b** 31%   **c** 193.25
d No: mode, mean and median are all less than 200 hours

Average price increases: Soundbuy 17.7p, Springfields 18.7p, Setco 18.2p

Yes average distance is 11.7 miles per day.

The first 5 and the 10 are the wrong way round.

$740

As we do not know what numbers are in each group, we cannot say what the median is.

2.8   Measuring spread

Exercise 32H

	Median	Lower quartile	Upper quartile	Inter-quartile range
a	13	5	20	15
b	28	17.5	32.5	15
c	90	76	97	21
d	8	5	12	7

2 a 2
b 1
c Probably not. The distribution is very uneven.

3 a 132
b 28
c 43

4

	Median	Range	Inter-quartile range
a	46	48	11
b	46	19	9

5 a 15.35
b 18.2
c 5.0
d 15.45 and 5.0

32.9   Cumulative frequency diagrams

Exercise 32I

1 a Cumulative frequency 1, 4, 10, 22, 25, 28, 30
b

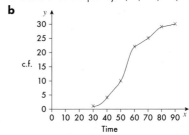

c 54 secs, 16 secs

2 a Cumulative frequency 1, 3, 5, 14, 31, 44, 47, 49, 50
b

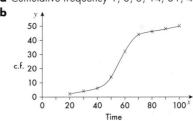

c 56 secs, 17 secs
d Pensioners, median closer to 60 secs

3 a Cumulative frequency 12, 30, 63, 113, 176, 250, 314, 349, 360
b

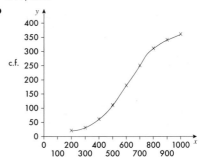

c 605 students, 280 students
d 46–47 schools

4 **a** Cumulative frequency 2, 5, 10, 16, 22, 31, 39, 45, 50

b

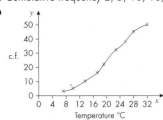
Temperature °C

c 20.5°C, 10°C

d 10.5°C

5 **a**

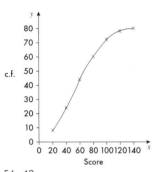
Score

b 56, 43

c about 17.5%

6 **a** Cumulative frequency 6, 16, 36, 64, 82, 93, 98, 100

b

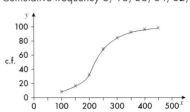
Pocket money (c)

c 225c, 90c

7 **a** Paper A 66, Paper B 57

b Paper A 28, Paper B 18

c Paper B is the harder paper, it has a lower median and lower upper quartile.

d **i** Paper A 43, Paper B 45 **ii** Paper A 78, Paper B 67

8 about 40%

9 Find the top 10% on the cumulative frequency scale, read along to the graph and read down to the marks. The mark seen will be the minimum mark needed for this top grade.

Answers to Chapter 33

33.1 The probability scale

Exercise 33A

1 **a** unlikely **b** unlikely **c** impossible

d very likely **e** evens

2 Answers may vary

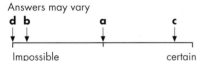
Impossible   certain

3 Answers may vary

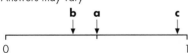
0   1

b For example: 0.5, 0.4, 0.9

4 Student to provide own answers.

5 No. What happens today does not depend on what happened yesterday.

33.2 Calculating probabilities

Exercise 33B

1 **a** $\frac{1}{10}$ **b** $\frac{4}{10}$ or $\frac{2}{5}$ **c** $\frac{7}{10}$

d $\frac{1}{2}$ **e** 0

2 **a** $\frac{1}{8}$ **b** $\frac{5}{8}$ **c** $\frac{1}{2}$

3 **a** 0 **b** 1

4 **a** $\frac{1}{10}$ **b** $\frac{1}{2}$ **c** $\frac{2}{5}$ **d** $\frac{1}{5}$ **e** $\frac{2}{5}$

5 **a** $\frac{6}{11}$ **b** $\frac{5}{11}$ **c** $\frac{6}{11}$

6 **a** $\frac{1}{5}$ **b** $\frac{1}{2}$ **c** $\frac{1}{2}$ **d** $\frac{7}{10}$

7 $\frac{1}{25}$

8 **a** AB, AC, AD, AE, BC, BD, BE, CD, CE, DE

b 1 **c** $\frac{1}{10}$ **d** 6 **e** $\frac{3}{5}$ **f** $\frac{3}{10}$

9 **a** **i** $\frac{12}{25}$ **ii** $\frac{7}{25}$ **iii** $\frac{6}{25}$

b They add up to 1.

c All possible outcomes are mentioned.

10 35%

11 0.5

12 Class U

13 There might not be the same number of boys as girls in the class.

3.3 Probability that an event will not happen

Exercise 33C

a $\frac{3}{4}$ **b** 0.55   **c** 0.2

a $\frac{3}{4}$ **b** $\frac{17}{20}$ **c** $\frac{19}{20}$

a i $\frac{1}{4}$ **ii** $\frac{3}{4}$

b i $\frac{3}{11}$ **ii** $\frac{8}{11}$

Because it might be possible for the game to end in a draw.

3.4 Addition rule for probabilities

Exercise 33D

a $\frac{1}{6}$ **b** $\frac{1}{6}$ **c** $\frac{1}{3}$

a 0.25   **b** 0.45   **c** 0.65

a $\frac{2}{11}$ **b** $\frac{4}{11}$ **c** $\frac{6}{11}$

a $\frac{1}{3}$ **b** $\frac{2}{5}$ **c** $\frac{11}{15}$ **d** $\frac{11}{15}$ **e** $\frac{1}{3}$

a 0.59   **b** 0.36   **c** 0.29

a 0.8   **b** 0.2

a $\frac{17}{20}$ **b** $\frac{2}{5}$ **c** $\frac{3}{4}$

Because these are three separate events. Also, probability cannot exceed 1.

$\frac{3}{4}$

$\frac{8}{45}$

The probability for each day stays the same, at $\frac{1}{4}$.

3.5 Probability from data

Exercise 33E

a 0.3   **b** 0.75

0.6

a 0.32   **b** 0.14   **c** 0.58

a 0.1   **b** $\frac{17}{40}$ or 0.425

a 0.386   **b** 0.462

0.625

a $\frac{27}{39}$ = 0.69   **b** $\frac{35}{47}$ = 0.74

a $\frac{24}{34}$ = 0.71   **b** $\frac{7}{25}$ = 0.28   **c** $\frac{24}{62}$ = 0.39

a 0.08   **b** 0.52   **c** 0.3   **d** 0.1

3.6 Expected frequency

Exercise 33F

a $\frac{1}{6}$ **b** 25

a $\frac{1}{2}$ **b** 1 000

3 a i 6   **ii** 12   **iii** 2
   b i 18   **ii** 9   **iii** 3

4 a 6   **b** 3   **c** 11   **d** 6

5 a 150   **b** 100   **c** 250   **d** 0

6 a 167   **b** 833

7 1050

8 a 10, 10, 10, 10, 10, 10
   b 3.5
   c Find the average of the scores (21 ÷ 6)

9 a 0.111
   b 40

10 281 days

11 Multiply the number of plants by 0.003.

33.7 Combined events

Exercise 33G

1 a 7
   b 2 and 12
   c $\frac{1}{36}, \frac{1}{18}, \frac{1}{12}, \frac{1}{9}, \frac{5}{36}, \frac{1}{6}, \frac{5}{36}, \frac{1}{9}, \frac{1}{12}, \frac{1}{18}, \frac{1}{36}$
   d i $\frac{1}{12}$ **ii** $\frac{1}{3}$
     iii $\frac{1}{2}$ **iv** $\frac{7}{36}$
     v $\frac{5}{12}$ **vi** $\frac{5}{18}$

2 a $\frac{1}{12}$ **b** $\frac{11}{36}$ **c** $\frac{1}{6}$ **d** $\frac{5}{9}$

3 a $\frac{1}{36}$ **b** $\frac{11}{36}$ **c** $\frac{5}{18}$

4

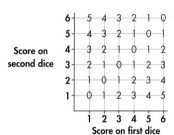

Score on second dice / Score on first dice

a $\frac{5}{18}$ **b** $\frac{1}{6}$ **c** $\frac{1}{9}$

d 0   **e** $\frac{1}{2}$

5 a $\frac{1}{4}$ **b** $\frac{1}{2}$

   c $\frac{3}{4}$ **d** $\frac{1}{4}$

6

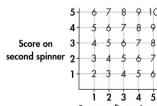

Score on second spinner / Score on first spinner

a 6
b i $\frac{4}{25}$ **ii** $\frac{13}{25}$ **iii** $\frac{1}{5}$ **iv** $\frac{3}{5}$

7 a

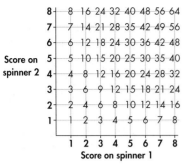

Score on spinner 2								
8	8	16	24	32	40	48	56	64
7	7	14	21	28	35	42	49	56
6	6	12	18	24	30	36	42	48
5	5	10	15	20	25	30	35	40
4	4	8	12	16	20	24	28	32
3	3	6	9	12	15	18	21	24
2	2	4	6	8	10	12	14	16
1	1	2	3	4	5	6	7	8
	1	2	3	4	5	6	7	8

Score on spinner 1

b $\frac{8}{64} = \frac{1}{8}$

8 $\frac{7}{36}$: a diagram will help him to see all possible outcomes

9 a

Type of rose	Colour of rose					
	white	red	orange	yellow	pink	copper
dwarf						
small						
medium						
large climbing						
rambling						

b i $\frac{6}{30} = \frac{1}{5}$ **ii** $\frac{4}{5}$

33.8 Tree diagrams

Exercise 33H

1 The missing probabilities are $\frac{1}{2}, \frac{1}{2}, \frac{1}{2}$.
a $\frac{1}{4}$

b $\frac{1}{2}$

c $\frac{3}{4}$

2 a $\frac{1}{4}$

b $\frac{1}{4}$

c i $\frac{1}{16}$ **ii** $\frac{7}{16}$

3 a $\frac{2}{3}$ **b** $\frac{1}{2}$

c

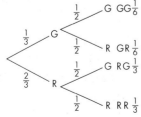

d i $\frac{1}{6}$ **ii** $\frac{1}{2}$ **iii** $\frac{5}{6}$

e 15 days

4 a $\frac{2}{5}$
b i $\frac{4}{25}$ **ii** $\frac{12}{25}$

5 All the missing probabilities are $\frac{1}{2}$.
a $\frac{1}{8}$ **b** $\frac{3}{8}$ **c** $\frac{7}{8}$

6 a

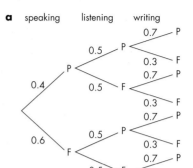

b 0.14
c 0.41
d 0.09

7 a $\frac{3}{5}$

b Student's explanation

c

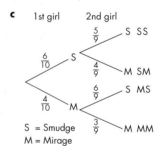

S = Smudge
M = Mirage

d i $\frac{1}{3}$ **ii** $\frac{7}{15}$ **iii** $\frac{8}{15}$

8 a $\frac{1}{6}$ **b** 0

c Student's tree diagram

d i $\frac{2}{3}$ **ii** $\frac{1}{3}$ **iii** 0

9 a i $\frac{3}{8}$ **ii** $\frac{5}{8}$

b i $\frac{5}{12}$ **ii** $\frac{7}{12}$

10 a i $\frac{5}{13}$ **ii** $\frac{8}{13}$

b i $\frac{15}{91}$ **ii** $\frac{4}{13}$

11 a $\frac{1}{120}$ **b** $\frac{7}{40}$ **c** $\frac{21}{40}$ **d** $\frac{7}{24}$

12 a $\frac{1}{9}$ **b** $\frac{2}{9}$ **c** $\frac{2}{3}$ **d** $\frac{7}{9}$

13 a 0.54 **b** 0.38 **c** 0.08 **d** 1

NOTES